Late Quaternary Environmental Change

We work with leading authors to develop the strongest educational materials in Geography, bringing cutting-edge thinking and best learning practice to a global market.

Under a range of well-known imprints, including Prentice Hall, we craft high quality print and electronic publications which help readers to understand and apply their content, whether studying or at work.

To find out more about the complete range of our publishing, please visit us on the World Wide Web at: www.pearsoned.co.uk

Late Quaternary Environmental Change

Physical and Human Perspectives

Second Edition

M. BELL
Professor of Archaeological Science, the University of Reading

M.J.C. WALKER
Professor of Quaternary Science, the University of Wales, Lampeter

Harlow, England • London • New York • Boston • San Francisco • Toronto • Sydney • Singapore • Hong Kong
Tokyo • Seoul • Taipei • New Delhi • Cape Town • Madrid • Mexico City • Amsterdam • Munich • Paris • Milan

Pearson Education Limited
Edinburgh Gate
Harlow
Essex CM20 2JE
England

and Associated Companies throughout the world

Visit us on the World Wide Web at:
www.pearsoned.co.uk

First published 1992
Second edition published 2005

ISBN 0 130 33344 1

British Library Cataloguing-in-Publication Data
A catalogue record for this book is available from the British Library

Library of Congress Cataloging-in-Publication Data
Bell, Martin, 1953–
Late Quaternary environmental change : physical and human perspectives / M. Bell, M. J. C. Walker.—2nd ed.
p. cm.
Includes bibliographical references and index.
ISBN 0-13-033344-1 (alk. paper)
1. Paleoecology—Quaternary. 2. Paleoclimatology—Quaternary. I. Walker, M. J. C. (Mike J. C.), 1947– II. Title.

QE720.B38 2005
560'.45—dc22

2004054493

10 9 8 7 6 5 4 3 2 1
08 07 06 05

Typeset in 10/12.3pt Sabon by 35
Printed and bound by Bell & Bain, Glasgow

The publisher's policy is to use paper manufactured from sustainable forests.

For Jennifer and Gro-Mette

Contents

Preface to Second Edition

The second edition of this book has been almost entirely rewritten to take account of the major advances in Quaternary Science in the 12 years since the first edition was written. These include new ice-core records and improved chronological precision from radiocarbon and dendrochronology. The new book also reflects our evolving understanding of people/environment relationships, the role of disturbance factors, contingency and human coping strategies. The development of geoarchaeology has been another important advance in this period.

For comments and advice on particular sections of the book we are particularly grateful to the following: Professor J.R.L. Allen, Dr S. Andersen, Professor N. Barton, Professor P. Buckland, Dr P. Dark, Professor J.G. Evans, Professor R. Kemp, Professor M.G. Macklin, Professor S. Manning, Drs W. and R. Matthews, Dr A. Olivier, Dr H. Schlichtherle, Professor I. Shennan and Dr N. Whitehouse. We remain of course responsible for any errors. Jennifer Foster kindly prepared the bibliography and index.

We are grateful to Mrs M. Mathews and the University of Reading for preparing some of the new illustrations for this edition. The following have also kindly provided new illustrations and we are most grateful for their help: Dr M. Allen and Wessex Archaeology, Dr P. Allen, Dr N.F. Alley, Dr S. Andersen, Dr J. Arneborg, Professor N. Barton, Professor R. Battarbee, Dr S. Buckley, Professor P. and Mr P. Buckland, Professor B. Cunliffe and the Danebury Trust, Ms J. Davidson, Professor J. Matthews, Ms Nikdic, The National Museum of Denmark, Dr A. Fleckinger and Museo Archaeologico dell Alto Adige, Professor B. Raftery, Dr M. Rasmussen, Historiks-Arkaeologisk Forsøgscenter, Lejre, Ms S. Ripper, The Royal Commission on the Ancient and Historic Monuments of Wales, Dr H. Schlichtherle and the Landesdenkmalamt Baden-Württemberg, Professor I. Shennan, Dr J-P. Steffensen, Professor L. Thompson and Dr. J. Väätäinen.

MB is grateful to the Archaeology Department, University of Reading and the British Academy for a period of research leave during which this second edition was completed. MJCW thanks the University of Wales, Lampeter, also for a period of study leave to spend time on the book. We again dedicate this second edition to our wives (Jennifer Foster and Gro-Mette Gulbrandsen) and families with thanks for their continuing considerable help and forbearance during the time this edition was in preparation.

Martin Bell
Mike Walker
February 2004

Acknowledgements

We are grateful to the following for permission to reproduce copyright material:

Table 1.1, Figure 2.29 and Figure 2.30 after INTCAL98 radiocarbon age calibration, 24-0 cal BP, *Radiocarbon*, 40(3), reprinted by permission of Radiocarbon (Stuiver, M. *et al.*, 1998); Figure 1.2 redrawn from *Human Ecology: basic concepts for sustainable development*, reprinted by permission of James & James/Earthscan (Marten, G.G. 2001); Figure 1.3a redrawn from *Monuments and Landscapes in Atlantic Europe*, reprinted by permission of Routledge (Scarre, C. ed. 2002); Figure 1.4 adapted from Evolution and Environmental Controls: Palaeozoic Black Deaths in *Structure and Contingency: Evolutionary Processes in Life and Human Society* edited by J. Bintliff, pub Leicester University Press, reprinted by permission of T&T Clark International (House, M. 1999); Figure 1.5 photo courtesy of *The Argus*, Brighton; Figure 1.6 modified from *Archaeology as Human Ecology: Method and Theory for a Contextual Approach*, reprinted by permission of Cambridge University Press (Butzer, K.W. 1982); Figure 1.8a redrawn from *The Perception of the Environment: Essays in livelihood, dwelling and skill*, reprinted by permission of Routledge (Ingold, T. 2000); Figure 1.8b redrawn from *Eskimo Essays: Yup'ik lives and how we see them*, reprinted by permission of Rutgers University Press (Feinup-Riordan, A. 1990); Figure 1.8c redrawn from *A Modern Dictionary of Geography, 3rd Edition*, pub Edward Arnold, © 1986, 1989, 1995 John Small and Michael Witherick, reproduced by permission of Hodder Arnold (Small, J. and Witherick, M. 1995); Table 2.1 and Figure 4.6 after Climatic reconstruction of the Weichselian Pleniglacial in north western and central Europe, *Journal of Quaternary Science*, 13, Copyright 1998, © John Wiley & Sons Ltd., reproduced with permission Huijzer, A.S. and Vandenberghe, J. 1998); Plate 2.2 photo courtesy of Peter Allen; Figure 2.3 redrawn from Holocene history of the forest-alpine tundra ecotone in the Scandes mountains (central Sweden), *New Phytologist*, 108, reprinted by permission of Blackwell Publishing (Kullman, L. 1988); Figures 2.4 and 2.7 photos courtesy of Ian Clewes; Figure 2.5 redrawn from *Reconstructing Quaternary Environments, 2nd Edition*, © Longman Group Ltd., 1984. This edition © Addison Wesley Longman Ltd. 1997, reprinted by permission of Pearson Education Ltd. (Lowe, J.J. and Walker, M.J.C. 1997); Plates 2.5a and 2.5b photos courtesy of Department of Geophysics, Niels Bohr Institute, University of Copenhagen; Figure 2.6 photo courtesy of P. Smart; Plate 2.6 photo courtesy of Jari Väätäinen, Geological Survey of Finland (GSF); Plate 2.7 photo courtesy of Lonnie G. Thompson, The Ohio State University; Figure 2.8 photo courtesy of Rick Battarbee/Viv Jones; Plate 2.8 photo courtesy of Neville Alley; Figure 2.9 redrawn from Chironomid-inferred Late Glacial air temperatures at Whitrig Bog, southeast Scotland, *Journal of Quaternary Science*, 15, Copyright 2000, © John Wiley & Sons Ltd., reproduced with permission (Brooks, S.J. and Birks, H.J.B. 2000); Figure 2.10 photo courtesy of Eric Robinson; Figure 2.12 reprinted from *Quaternary Science Reviews*, 19, Barber, K.E. *et al.*, Replicated proxy-climate signals over the last 2000 yr from two distant UK peat bogs: new evidence for regional palaeoclimatic teleconnections, pp. 481–8, Copyright 2000, with permission from Elsevier (Barber, K.E. *et al.*, 2000); Figure 2.13 reprinted from *Quaternary Science Reviews*, 9, Walker, M.J.C. and Lowe, J.J., Reconstructing the environmental history of the last glacial-interglacial transition:

evidence from the Isle of Skye, Inner Hebrides, Scotland, pp. 15–49, Copyright 1990, with permission from Elsevier (Walker, M.J.C. and Lowe, J.J. 1990); Figure 2.14 redrawn from A section of an imaginary bone cave, *Studies in Speleology*, Vol. 2, reproduced by kind permission of William Pengelly Cave Studies Trust, Devon, UK (Sutcliffe, A.J. 1970); Figure 2.17 redrawn from A preliminary history of Holocene colluvial (debris-flow) activity, Leirdalen, Jotunheimen, Norway, *Journal of Quaternary Science*, 12, Copyright 1997, © John Wiley & Sons Ltd., reproduced with permission (Matthews, J.A. 1997); Figure 2.18 photo courtesy J-H. Beck, Museet for Thy og Vester Hanherred; Figure 2.19 photo E. Johansen, courtesy of S.H. Andersen; Figure 2.20 reprinted from *Quaternary Science Reviews*, 19, Raynaud, D. *et al.*, The ice record of greenhouse gases: a view in the context of future changes, pp. 9–18, Copyright 2000, with permission from Elsevier (Raynaud, D. *et al.*, 2000); Figure 2.21 reprinted from *Quaternary Research*, 3, Shackleton, N.J. and Opdyke, N.D., Oxygen isotope and palaeomagnetic stratigraphy of equatorial Pacific core V28–238: oxygen isotope temperatures and ice volumes on a 105 and 106 year scale, pp. 39–55, Copyright 1973, with permission from Elsevier (Shackleton, N.J. and Opdyke, N.D. 1973); Figure 2.22 redrawn from The Greenland Ice-Core Project (GRIP): reducing uncertainties in climatic change, *NERC News*, April 26–30, reprinted by permission of the author (Peel, D.A. 1994); Figure 2.23 redrawn from Calibration of the speleothem delta function: an absolute temperature record for the Holocene in northern Norway, *The Holocene*, 9, reprinted by permission of Hodder Arnold (Lauritzen, S.E. and Lundberg, J. 1999b); Figure 2.24 redrawn from A mid-European decadal isotope-climate record from 15,500 to 5000 years BP, *Science*, 284, reprinted by permission of the American Association for the Advancement of Science (von Grafenstein, U. *et al.* 1999); Figure 2.25 redrawn from Temperature and precipitation reconstruction in southern Portugal during the late Maunder Minimum, *The Holocene*, 10, reprinted by permission of Hodder Arnold (Alcoforado, M-J. *et al.*, 2000); Figure 2.26 redrawn from Grape harvests through the nineteenth century in *Climate and History: Studies in Interdisciplinary History* edited by R.I. Rotberg and T.K. Rabb, reprinted by permission of Princeton University Press (Le Roy Ladurie, E. and Bauland, M. 1981); Figure 2.27 redrawn from The historical temperature series of Bologna (Italy): 1716–1774, *Climatic Change*, Vol. 11, pp. 375–90, Fig. 3, © 1987 Kluwer Academic Publishers, with kind permission of Kluwer Academic Publishers (Comani, S. 1987); Figure 2.28 redrawn from The Minnesota long-term temperature record, *Climatic Change*, Vol. 7, pp. 225–36, Fig. 1, © 1987 Kluwer Academic Publishers, with kind permission of Kluwer Academic Publishers (Baker, D.G. *et al.*, 1985); Figure 2.32 redrawn from Holocene humidity changes in northern Finnish Lapland inferred from lake sediments and submerged Scots pines dated by tree-rings, *The Holocene*, 9, reprinted by permission of Hodder Arnold (Eronen, M. *et al.*, 1999); Figure 2.34 reprinted from *Quaternary Science Reviews*, 5, Lundqvist, J. *et al.*, Late Weichselian glaciation and deglaciation in Scandanavia, pp. 269–92, Copyright 1986, with permission from Elsevier (Lundqvist, J. *et al.*, 1986); Figure 2.36 reprinted from *Quaternary Research*, 27, Martinsson, D.G. *et al.*, Age dating and the orbital theory of ice ages: development of a high-resolution 0-300,000 year chronostratigraphy, pp. 1–29, Copyright 1987, with permission from Elsevier (Martinsson, D.G. *et al.*, 1987); Figure 3.1 reprinted from *Palaeogeography, Palaeoclimatology, Palaeoecology*, 64, Williams, D.F. *et al.*, Chronology of the Pleistocene oxygen isotope record: 0–1.88m.y. BP, pp. 221–40, Copyright 1988, with permission from Elsevier (Williams, D.F. *et al.*, 1988); Table 3.1 reprinted from *Earth and Planetary Science Letters*, 126, Bassinot, F.E. *et al.*, The astronomical theory of climate and the Brunhes-Matuyama magnetic reversal, pp. 91–108, Copyright 1994, with permission from Elsevier (Bassinot, F.E. *et al.*, 1994); Figure 3.2 redrawn from Seasonal reconstructions of the earth's surface at the last glacial maximum, Geological Society of America, Map and Chart Series, MC36, reprinted by permission of The Geological Society of America (CLIMAP 1981); Table 3.2 after Mid- and Late-Holocene climatic change: a test of periodicity and solar forcing in proxy-climatic data from blanket peat bogs, *Journal of*

Quaternary Science, 16, Copyright 2001, © John Wiley & Sons Ltd., reproduced with permission (Chambers, F.M. and Blackford, J.J. 2001); Figure 3.3 reprinted from *Quaternary Science Reviews*, 19, de Vernal, A. and Hillaire-Marcel, C., Sea-ice cover, sea surface salinity and halo-thermocline structure of the northwest North Atlantic: modern versus full glacial conditions, pp. 65–86, Copyright 2000, with permission from Elsevier (de Vernal, A. and Hillaire-Marcel, C. 2000); Table 3.3 after Clausen, H.B. *et al.*, A comparison of the volcanic records over the past 400 years from the Greenland ice core project and Dye 3 Greenland ice cores, *Journal of Geophysical Research*, 102, 26, pp. 707–26, 723–4, Copyright 1998 American Geophysical Union, modified by permission of the American Geophysical Union (Clausen, H.B. *et al.*, 1998); Figure 3.4 reprinted from *Quaternary Science Reviews*, 11, Pons, A. *et al.*, Recent contributions to the climatology of the last glacial-interglacial cycle based on French pollen sequences, pp. 439–48, Copyright 1992, with permission from Elsevier (Pons, A. *et al.*, 1992); Figures 3.5, 3.6 and 3.18 redrawn from Oxygen isotope and palaeotemperature records from six Greenland ice-core stations: Camp Century, Dye-3, GRIP, GISP2, Renland and NorthGRIP, *Journal of Quaternary Science*, 16, Copyright 2001, © John Wiley & Sons Ltd., reproduced with permission (Johnsen, S. *et al.*, 2001); Figure 3.7 redrawn from Vegetation and climate in the Early- and Pleni-Weichselian in northern and central Europe, *Journal of Quaternary Science*, 15, Copyright 2001, © John Wiley & Sons Ltd., reproduced with permission (Caspers, G. and Freund, H. 2001); Figure 3.9 reprinted from *Quaternary Science Reviews*, 19, Jackson, S.T. *et al.*, Vegetation and environment in the eastern North America during the last glacial maximum, pp. 489–508, Copyright 2000, with permission from Elsevier (Jackson, S.T. *et al.*, 2000); Figure 3.12 redrawn from Role of orbital forcing: a two million year perspective in *Global Changes in the Perspective of the Past* edited by J.A. Eddy and H. Oeschger, Copyright 1993, © John Wiley & Sons Ltd., reproduced with permission (Imbrie, J. *et al.*, 1993a); Figure 3.13 reprinted from *Quaternary Science Reviews*, 12, Hooghiemestra, H. *et al.*, Frequency spectra and palaeoclimatic variability of the high-precision 30–1450 ka Funza 1 pollen record (Eastern Cordillera, Colombia), pp. 141–56, Copyright 1993, with permission from Elsevier (Hooghiemstra, H. *et al.*, 1993); Figure 3.14 reprinted from *Quaternary Science Reviews*, 16, Partridge, T.C. *et al.*, Orbital forcing of climate over South Africa: a 200,000-year rainfall record from the Pretoria saltpan, pp. 1125–33, Copyright 1997, with permission from Elsevier (Partridge, T.C. *et al.*, 1997); Figure 3.15 reprinted from *Palaeogeography, Palaeoclimatology, Palaeoecology*, 35, The North Atlantic Ocean during the last deglaciation, pp. 145–214, Copyright 1981, with permission from Elsevier (Ruddiman, W.F. and McIntyre, A. 1981); Figure 3.17 reprinted from *Quaternary Science Reviews*, 20, Lowe, J.J. *et al.*, Inter-regional correlation of palaeoclimatic records for the Last Glacial-Interglacial Transition: a protocol for improved precision recommended by the INTIMATE project group, pp. 1175–88, Copyright 2001, with permission from Elsevier (Lowe, J.J. *et al.*, 2001); Figure 3.19a reprinted from *Quaternary Science Reviews*, 12, Harrison, S.P. and Digerfeldt, G., European lakes as palaeohydrological and palaeoclimatic indicators, pp. 233–48, Copyright 1993, with permission from Elsevier (Harrison, S.P. and Digerfeldt, G. 1993); Figure 3.19b redrawn from Vegetation, lake levels, and climate in eastern Northern America for the past 18,000 years in *Global Climates since the Last Glacial Maximum* edited by H.E. Wright, Jr. *et al.*, reprinted by permission of the University of Minnesota Press (Webb, T. *et al.*, 1993); Figure 3.20 redrawn from Mire-development pathways and palaeoclimatic records from a full Holocene peat archive at Walton Moss, Cumbria, England, *The Holocene*, 10, reprinted by permission of Hodder Arnold (Hughes, P.D.M. *et al.*, 2000); Figure 3.21 reprinted from *Quaternary Science Reviews*, 19, Briffa, K.R., Annual climatic variability in the Holocene: interpreting the message of ancient trees, pp. 87–106, Copyright 2000, with permission from Elsevier (Briffa, K.R. 2000); Figure 3.22 reprinted from *Quaternary Science Reviews*, 19, Bradley, R.S., Past global changes and their significance for the future, pp. 391–402, Copyright 2000, with permission from Elsevier

(Bradley, R.S. 2000); Figure 3.23 reprinted from *Quaternary Science Reviews*, 22, Langdon, P.G. *et al.*, A 7500-year peat-based palaeoclimatic reconstruction and evidence for an 1100-year cyclicity in bog surface wetness from Temple Hill Moss, Pentland Hills, southeast Scotland, pp. 259–74, Copyright 2003, with permission from Elsevier (Langdon, P.G. *et al.*, 2003); Figure 3.24 redrawn from Historical evidence concerning the sun: interpretation of sunspot records during the telescopic and pretelescopic eras, *Philosophical Transactions of the Royal Society*, London, A339, pp. 499–512, Fig. 2, reprinted by permission of The Royal Society (Stephenson, F.R. 1990); Figure 3.25 after Finkel, R.C. and Nishiizumi, N., Beryllium 10 concentrations in the Greenland Ice Sheet Project 2 ice core from 3–40 ka, *Journal of Geophysical Research*, 102, 26, pp. 699–26, 706, Copyright 1997 American Geophysical Union, modified with permission of American Geophysical Union (Finkel, R.C. and Nishiizumi, N. 1997); Figure 3.26a redrawn from Reconstructions of past solar variability in *Climatic Variations and Forcing Mechanisms of the Last 2000 Years* edited by P.D. Jones *et al.*, NATO ASI Series 1, Global Environmental Change, Volume 41, pp. 519–32, Fig. 2, © Springer-Verlag, Berlin, Heidelberg, 1996, reprinted by permission of Springer-Verlag GmbH & Co. KG (Lean, J. 1996); Figure 3.26b redrawn from Changes in atmospheric carbon-14 attributed to a variable sun, *Science,* 207, reprinted by permission of the American Association for the Advancement of Science (Stuiver, M. and Quay, P.D. 1980); Figure 3.27 after Zielinski, G.A. *et al.*, Volcanic aerosol records and tephrochronology of the Summit, Greenland, ice cores, *Journal of Geophysical Research*, 102, 26, pp. 625–6, 640, Copyright 1997 American Geophysical Union, modified by permission of American Geophysical Union (Zielinski, G.A. *et al.*, 1997); Figure 3.28 reprinted from *Quaternary International*, 91, Bucha, V. and Bucha, V., Jr., Geomagnetic forcing and climatic variations in Europe, North America and in the Pacific Ocean, pp. 5–15, Copyright 2002, with permission from Elsevier (Bucha, V. and Bucha, V., Jr. 2002); Figure 3.29a redrawn from Routing of meltwater from the Laurentide Ice Sheet during the Younger Dryas cold episode, *Nature*, 341, reprinted by permission of Macmillan Magazines Ltd. (Broecker, W.S. *et al.*, 1989); Figure 3.29b redrawn from Forcing of the cold event of 8,200 years ago by catastrophic drainage of Laurentide lakes, *Nature*, 400, reprinted by permission of Macmillan Magazines Ltd. (Barber, D.C. *et al.*, 1999); Figure 4.1 redrawn from *Ice Sheets and Late Quaternary Environmental Change*, Copyright 2001, © John Wiley & Sons Ltd., reproduced with permission (Siegert, M.J. 2001); Plate 4.1 from Modelling western North Sea palaeogeographies and tidal changes during the Holocene in *Holocene Land-Ocean Interaction and Environmental Change around the North Sea* edited by I. Shennan and J.E. Andrews, Geological Society of London, Special Publication, No. 166, reprinted by permission of The Geological Society (Shennan, I. *et al.*, 2000b); Figure 4.2 reprinted from *Quaternary Science Reviews*, 5, Bowen, D.Q. *et al.*, Correlation of Quaternary glaciations in England, Ireland, Scotland and Wales, pp. 299–340, Copyright 1986, with permission from Elsevier (Bowen, D.Q. *et al.*, 1986) and *Quaternary Science Reviews*, 21, Bowen, D.Q. *et al.*, New data for the Last Glacial Maximum in Great Britain and Ireland, pp. 89–101, Copyright 2002, with permission from Elsevier (Bowen, D.Q. *et al.*, 1986); Figure 4.3 reprinted from *Quaternary Science Reviews*, 21, Dyke, A.S. *et al.*, The Laurentide and Innuitian ice sheets during the Last Glacial Maximum, pp. 9–31, Copyright 2002, with permission from Elsevier (Dyke, A.S. *et al.*, 2002); Figure 4.5 photo courtesy of John Matthews; Figure 4.7 photo courtesy of Ian Shennan; Figure 4.8 reprinted from *Quaternary Science Reviews*, 19, Shennan, I. *et al.*, Late Devensian and Holocene records of relative sea-level changes in northwest Scotland and their implications for glacio-hydro-isostatic modeling, pp. 1103–35, Copyright 2000, with permission from Elsevier (Shennan, I. *et al.*, 2000a); Figure 4.9 reprinted from *Quaternary Science Reviews*, 21, Shennan, I. *et al.*, Global to local scale parameters determining relative sea-level changes and the post-glacial isostatic adjustment of Great Britain, pp. 397–408, Copyright 2002, with permission from Elsevier (Shennan, I. *et al.*, 2002); Figure 4.10 redrawn from Eustasy and geoid changes as a function of core/mantle changes in *Earth Rheology, Isostasy and Eustasy* edited by N-A. Mörner, Copyright 1980, © John Wiley & Sons Ltd., reproduced with permission (Mörner, N-A. 1980a); Figure 4.11 redrawn from Deglaciation, earth crustal behaviour and sea-level changes in the determination of insularity: a perspective from Ireland in *Ireland Britain: a Quaternary Perspective* edited by R.C. Preece, Geological Society of London, Special Publication, No. 96, reprinted by permission of The Geological Society (Devoy, R.J.N. 1995); Figure 4.12 redrawn from Patterns of isostatic land uplift during the Holocene: evidence from mainland Scotland, *The Holocene*, 10, reprinted by permission of Hodder Arnold (Smith, D.E. *et al.*, 2000); Figure 4.13 reprinted from *Quaternary Science Reviews*, 20, Mix, A. *et al.*, Environmental processes of the ice age: land, oceans, glaciers (EPILOG), pp. 627–59, Copyright 2001, with permission from Elsevier (Mix, A. *et al.*, 2001); Figure 4.14 redrawn from Holocene isostasy and relative sea-level changes on the east coast of England in *Holocene Land-Ocean Interaction and Environmental Change around the North Sea* edited by I. Shennan and J.E. Andrews, Geological Society of London, Special Publication, No.166, reprinted by permission of The Geological Society (Shennan, I. *et al.*, 2000c); Figure 4.15 redrawn from Holocene land- and sea-level changes in Great Britain, *Journal of Quaternary Science*, Copyright 2002, © John Wiley & Sons Ltd., reproduced with permission (Shennan, I. and Horton, B. 2002); Figures 4.17 and 4.18 adapted from *An Atlas of Past and Present Pollen Maps for Europe: 0-13000 Years Ago*, © Cambridge University Press, reprinted with permission of the publisher and authors (Huntley, B. and Birks, H.J.B. 1983); Figure 4.21 after Late Weichselian to early Holocene development of the Baltic Sea – with implications for coastal settlements in the southern Baltic region in *Man and the Sea in the Mesolithic* edited by A. Fischer, pub Oxbow Books, reprinted by permission of the author (Björck, S. 1995a) and *Quaternary International*, 27, Björck, S., A review of the history of the Baltic Sea, 13.0–8.0 ka BP, pp. 19–40, Copyright 1995, with permission from Elsevier (Björck, S. 1995b); Figure 4.22 reprinted from *Quaternary Science Reviews*, 22, Teller, J.T. *et al.*, Freshwater outbursts to the oceans from glacial Lake Agassiz and their role in climate change during the last deglaciation, pp. 879–87, Copyright 2002, with permission from Elsevier (Teller, J.T. *et al.*, 2002); Figure 4.23 reprinted from *Quaternary Science Reviews*, 4, Baker, V.R. and Bunker, R.C., Cataclysmic Late Pleistocene flooding from Glacial lake Missoula: a review, pp. 1–41, Copyright 1985, with permission from Elsevier (Baker, V.R. and Bunker, R.C. 1985); Figure 4.24 reprinted from *Quaternary Science Reviews*, 14, Vendenberghe, J., Timescales, climate and river development, pp. 631–8, Copyright 1995, with permission from Elsevier (Vandenberghe, J. 1995); Figure 4.25 redrawn from Responses of river systems to the Holocene climates in *Late Quaternary Environments of the United States, Volume 2: The Holocene* edited by H.E. Wright, Jr. and Stephen C. Porter, reprinted by permission of the University of Minnesota Press (Knox, J.C. 1983); Figure 5.2 photo Danebury Trust, courtesy of Professor B.W. Cunliffe; Plate 5.3 Photo Archives, South Tyrol Museum of Archaeology – www.iceman.it; Figure 5.5 redrawn from *Thoughtful Foragers: a study of prehistoric decision-making*, reprinted by permission of Cambridge University Press (Mithen, S. 1990); Figure 5.6 modified and redrawn from *Changing the Face of the Earth*, reprinted by permission of Blackwell Publishing Ltd. (Simmons, I.G. 1989); Figure 5.8 redrawn from Domestication of the Southwest Asian Neolithic crop assemblage of cereals, pulses and flax: the evidence from the living plants in *Foraging and Framing* edited by D.R. Harris and G.C. Hillman, reprinted by permission of Routledge (Zohary, D. 1989); Figure 5.9 'Figure 3.18: Maps', from *Village on the Euphrates: The Excavation of Abu Hureya* by Andrew M.T. Moore and A. Legge, copyright © 1999 by Oxford University Press, Inc. Used by permission of Oxford University Press, Inc. (Moore, A.M.T. 2000); Figure 5.10 adapted from Human disturbance of North American forests and grasslands: the fossil pollen record in *Vegetation History* edited by B. Huntley and T. Webb III, Fig. 1, © 1988 Kluwer Academic Publishers, with kind permission of Springer Science and Business Media (McAndrews, J.H. 1988) and *The Emergence of Agriculture*, pub Scientific American Library, by permission of the author (Smith, B.D. 1995a); Figure 5.12 redrawn from Coastal adaptation and marine exploitation in late Mesolithic Denmark – with some special emphasis on the Limfjord Project in *Man and the Sea in the Mesolithic* edited by A. Fischer, pub Oxbow Books, reprinted by permission of the author (Andersen, S.H. 1995); Figure 5.13 photo courtesy of S.H. Andersen; Figure 5.14 redrawn from Neolithic settlement and subsidence in the wetlands of the Rhine-Meuse Delta of the Netherlands in *European Wetlands in Prehistory* edited by J.M. Coles and A.J. Lawson, reprinted by permission of Oxford University Press (Louwe Koojimans, L.P. 1987); Figure 5.16b adapted from Subfossil mammalian tracks (Flandrian) in the Severn Estuary, S.W. Britain: mechanics of formation, preservation and distribution, *Philosophical Transactions of the Royal Society of London B*, 352, pp. 481–518, Fig. 17a, reprinted by permission of The Royal Society (Allen, J.R.L. 1997); Figure 5.17 photo courtesy of Edward Sacre; Figure 5.18a reconstruction drawing from *Prehistoric Intertidal Archaeology in the Welsh Severn Estuary*, Council for British Archaeology Report 120, reprinted by permission of Steven Allen (Bell, M.G. *et al.*, 2000); Figure 5.19a modified and redrawn from The Late-Bronze Age explosive eruption of Thera (Santorini), Greece: regional and local effects in *Volcanic Hazards and Disasters in Human Antiquity* edited by F.W. McCoy and G. Heiken, reprinted by permission of The Geological Society of America (McCoy, F.W. and Heiken, G. 2000a); Figure 5.19b redrawn from The Minoan eruption of Santorini in Greece dated to 1645 BC, *Nature*, 328, reprinted by permission of Macmillan Magazines Ltd. (Hammer, C.U. *et al.*, 1987); Figure 5.19c redrawn from Irish tree rings, Santorini and volcanic dust veils, *Nature*, 332, reprinted by permission of Macmillan Magazines Ltd. (Baillie, M.G.L. and Munro, M.A.R. 1988); Figure 5.21 redrawn from The Storegga tsunami along the Norwegian coast, its age and run up, *BOREAS*, 26, reprinted by permission of Taylor & Francis AS (Bondevik, S. *et al.*, 1997); Figure 5.23b redrawn from Climatic changes, Norseman and modern man, *Nature*, 255, reprinted by permission of Macmillan Magazines Ltd. (Dansgaard, W. *et al.*, 1975); Figures 5.23c and 5.23d redrawn from Interdisciplinary investigations of the end of the Norse Western settlement in Greenland, *The Holocene*, 7, 4, reprinted by permission of Hodder Arnold (Barlow, L.K. *et al.*, 1997); Figure 5.24 excavations by the National Museums of Denmark and Greenland, photo courtesy of P. Buckland; Plate 6.1 from *Pfahlbauten rund um die Alpen*, pub Konrad Theiss Verlag, reprinted by permission of Landesdenkmalamt Baden-Württemberg, T. Leonhardt and H. Schlichtherle (Schlichtherle, H. 1997); Plate 6.2 photo published by permission of the National Museum of Denmark; Plate 6.3 photo courtesy of P. Ashbee; Figure 6.4 modified and redrawn from Late Pleistocene megafaunal extinctions in *Extinctions in Near Time: causes, contexts and consequences* edited by R.D.E. MacPhee, reprinted by permission of Kluwer Academic Publishers (Stuart, A.J. 1999); Figure 6.5 photo reprinted by permission of the National Museum of Wales; Figure 6.6 redrawn from Pleistocene extinction of *Genyornis newtoni*: human impact on Australian megafauna, *Science*, 283, reprinted by permission of the American Association for the Advancement of Science (Miller, G.H. *et al.*, 1999a); Figure 6.7b courtesy of Jennifer Foster; Figure 6.8 redrawn from A provisional map of forest types for the British Isles 5000 years ago, *Journal of Quaternary Science*, 4, Copyright 1989, © John Wiley & Sons Ltd., reproduced with permission (Bennett, K.D. 1989); Figure 6.9 from Persistent places in the Mesolithic landscape: an example from the Black Mountain Uplands of South Wales, *Proceedings of the Prehistoric Society*, 61, reproduced by permission of The Prehistoric Society (Barton, R.N.E. *et al.*, 1995); Figure 6.10 modified and redrawn from *The Environmental Impact of Later Mesolithic Cultures*, reprinted by permission of Edinburgh University Press, www.eup.ed.ac.uk (Simmons, I.G. 1996); Figure 6.11 redrawn from The development of Denmark's nature since the last glacial, *Danmarks Geologiske Undersøgelse, V Raekke,* 7-C, reprinted by permission of the Geological Survey of Denmark and Greenland (Iversen, J. 1973); Figure 6.12b redrawn from The mid-Holocene *Ulmus* decline at Diss Mere, Norfolk: a year-by-year pollen stratigraphy from annual laminations, *The Holocene*, 3, reprinted by permission of Hodder Arnold

(Peglar, S.M. 1993b); Figure 6.12c redrawn from The mid-Holocene *Ulmus* fall at Diss Mere, south-east England – disease and human impact?, *Vegetation History and Archaeobotany*, 2, pp. 61–8, Fig. 5, © Springer-Verlag, Berlin, Heidelberg, 1996, reprinted by permission of Springer-Verlag GmbH & Co. KG (Peglar, S.M. and Birks, H.J.B. 1993); Figure 6.13a redrawn from *Pfahlbauten rund um die Alpen*, pub Konrad Theiss Verlag, reprinted by permission of Landesdenkmalamt Baden-Württemberg and A. Kalkowski (Schlichtherle, H. 1997); Figure 6.13c modified and redrawn from Reconstructing the Neolithic Landscape at Western Lake Constance in *Estuarine Archaeology: The Severn and Beyond; Archaeology in the Severn Estuary*, 11 edited by S. Rippon, reprinted by permission of Archaeology in the Severn Estuary/Severn Estuary and Levels Research Committee (Maier, U. and Vogt, R. 2001); Figure 6.13d adapted from *Enlarging the Past*, reprinted by permission of the Society of Anitquaries of Scotland (Coles, B. and Coles, J. 1996); Figure 6.14 photo courtesy of Somerset Levels Project; Figure 6.15 redrawn from *The Making of the American Landscape* edited by M.P. Conzen, reprinted by permission of Routledge/Taylor & Francis Books, Inc. (Williams, M. 1990); Figure 6.16 redrawn from *Grazing Ecology and Forest History*, reprinted by permission of CABI Publishing, CAB International (Vera, F.W.M. 2000); Figure 6.17 redrawn from *Thorne Moors: a palaeoecological study of a Bronze Age site*, University of Birmingham Occasional Publication 8, reprinted by permission of the author (Buckland, P.C. 1979); Figure 6.18 redrawn from 10,000 years of change: The Holocene entofauna of the British Isles in *Holocene Environments of Prehistoric Britain* edited by K. Edwards and J. Sadler, *Quaternary Research Association Proceedings*, 7, Copyright 1999, © John Wiley & Sons Ltd., reproduced with permission (Dinnin, M. and Sadler, J. 1999); Figure 6.19 after *Atlas of the Land and Freshwater Molluscs of the British Isles*, pub Harley Books (Kerney, M.P. 1999), map produced by the Biological Records Centre, CEH Monks Wood, from records supplied by the Non-marine Mollusc Recording Scheme; Figure 6.20 redrawn from *The Environment of Early Man in the British Isles*, pub Paul Elek, reprinted by permission of the author (Evans, J.G. 1975); Figure 6.22b modified and redrawn from Palaeoecological investigations towards the reconstruction of environment and the land-use changes during prehistory at Céide Fields, western Ireland, *Probleme der Küstenforschung im südlichen Nordseegebeit*, 23, reprinted by permission of Niedersächsisches Insitut für historische Küstenforschung (Molloy, K. and O'Connell, M. 1995); Figure 6.23 photo courtesy of Barry Raftery, University College, Dublin; Figure 6.25 redrawn from *The Dartmoor Reeves*, pub B.T. Batsford, reprinted by permission of Chrysalis Books Group (Fleming, A. 1998); Figure 6.26 redrawn from *Land Snails in Archaeology*, pub Seminar Press, reprinted by permission of the author (Evans, J.G. 1972); Plate 7.1 and Figure 7.2 photos courtesy of Brenda Westley; Plate 7.2 photo courtesy of Jodi Davidson and Shaun Buckley; Figure 7.3 photo by Harold D. Walter, courtesy Museum of New Mexico, neg. no. 128725; Figure 7.7 photo courtesy of J. Boardman; Figure 7.8 redrawn from Modelling long-term anthropogenic erosion of a loess cover: South Downs, UK, *The Holocene*, 7, 1, reprinted by permission of Hodder Arnold (Favis-Mortlock, D. *et al.*, 1997); Figure 7.9 photo courtesy of Susan Ripper, University of Leicester Archaeological Services; Figure 7.10 after River sediments, great floods and centennial-scale Holocene climate change, *Journal of Quaternary Science*, 8, 2, Copyright 2003, © John Wiley & Sons Ltd., reproduced with permission (Macklin, M.G. and Lewin, J. 2003) and *Catena*, 42, Edwards, K.J. and Whittington, G., Lake sediments, erosion and landscape change during the Holocene in Britain and Ireland, pp. 143–73, Copyright 2001, with permission from Elsevier (Edwards, K.J. and Whittington, G. 2001); Figure 7.11 photo Amsterdams Archaeologisch Centrum (AAC), University of Amsterdam, courtesy of H.A. Heidinga; Figure 7.12 redrawn from Recent and long-term records of soil erosion from southern Sweden in *Soil Erosion on Agricultural Land* edited by J. Boardman *et al.*, Copyright 1990, © John Wiley & Sons Ltd., reproduced with permission (Dearing, J.A. *et al.*, 1990); Figures 7.13a and 7.13c reprinted from *Catena*, 42, Edwards, K.J. and Whittington, G., Lake sediments, erosion and landscape change during the Holocene in Britain and Ireland, pp. 143–73, Copyright 2001, with permission from Elsevier (Edwards, K.J. and Whittington, G. 2001); Figure 7.13b redrawn from Radiocarbon and palaeoenvironmental evidence for changing rates of erosion at a Flandrian stage site in Scotland in *Timescales in Geomorphology* edited by R.A. Cullingford *et al.*, Copyright 1980, © John Wiley & Sons Ltd., reproduced with permission (Edwards, K.J. and Rowntree, K.M. 1980); Figure 8.1 photo copyright reserved Cambridge University Collection of Air Photographs; Plate 8.1 from *Pfahlbauten rund um die Alpen*, pub Konrad Theiss Verlag, reprinted by permission of Landesdenkmalamt Baden-Württemberg and O. Braasch (Schlichtherle, H. 1997); Figure 8.2 redrawn from Economic development in Denmark since agrarian reform in *Archaeological Formation Processes* edited by K. Kristiansen, pub Nationalmuseet, Copenhagen, reprinted by permission of the author (Kristiansen, K. 1985); Figure 8.3 photo courtesy of Ed Yorath; Figure 8.4 photo M. Rasmussen, courtesy of Historik-Arkæologisk Forsøgscenter, Lejre; Figure 8.7 Crown Copyright – Royal Commission on the Ancient and Historic Monuments of Wales. Reproduced by permission. Figures 9.1, 9.2, 9.3, 9.5, 9.6, 9.8 and Table 9.2 after *Climate Change 2001: The Scientific Basis*, pub Cambridge University Press, reprinted by permission of the Intergovernmental Panel on Climate Change (Houghton, J.T. *et al.*, 2001); Table 9.1 after The Kyoto negotiations on climate change: a scientific perspective, *Science*, 279, reprinted by permission of the American Association for the Advancement of Science (Bolin, B. 1998); Figure 9.4 reprinted from *Quaternary Science Reviews*, 20, Ruddiman, W.F. and Thomson, J.S., The case for human causes of increased atmospheric CH4 over the last 5000 years, pp. 1769–77, Copyright 2001, with permission from Elsevier (Ruddiman, W.F. and Thomson, J.S. 2001); Figure 9.7 redrawn from Observed climatic variability and change in *Climate Change 2001: The Scientific Basis* edited by J.T. Houghton *et al.*, pub Cambridge University Press, reprinted by permission of the Intergovernmental Panel on Climate Change (Folland, C.K. *et al.*, 2001); Figure 9.9 redrawn from Calculating regional climatic time-series for temperature and precipitation: methods and illustrations, *International Journal of Climatology*, 16, Copyright 1996, © John Wiley & Sons Ltd., reproduced with permission (Jones, P.D. and Hulme, M. 1996); Figures 9.10 and 9.11 redrawn from Changes in Sea Level in *Climate Change 2001: The Scientific Basis* edited by J.T. Houghton *et al.*, pub Cambridge University Press, reprinted by permission of the Intergovernmental Panel on Climate Change (Church, J.A. and Gregory, J.M. 2001); Figure 9.12 redrawn from A greenhouse warming connection, *Nature*, 392, reprinted by permission of Macmillan Magazines Ltd. (Salawitch, R.J. 1998).

In some instances we have been unable to trace the owners of copyright material, and we would appreciate any information that would enable us to do so.

1 Environmental change and human activity

Introduction

Twenty-five thousand years ago the world was in the grip of the last ice age. The 14 ka[1] which followed saw some of the most dramatic climate changes in the recent history of the earth. Documentation of the rapid nature of some of those changes is one of the great achievements of Quaternary science, and the evidence makes a persuasive case for the relevance of research on past environments to contemporary environmental concerns such as global warming. The climatic shift from a regime of arctic severity to one of relative warmth that began around 15 ka BP led to the virtual disappearance of the continental ice sheets, to contraction of the mountain glaciers, and to the replacement of barren tundra by mixed woodland over large areas of Europe and North America. Meltwater from the wasting ice sheets raised global sea level by over 120 m, while a combination of climatic and vegetational changes exerted a major influence on a range of other environmental processes such as weathering rates, soil formation and the activity of rivers.

The end of the last ice age at 11.5 ka BP was rapidly followed by the earliest agriculture and then by the first large settlements and increasingly complex societies. In some areas, human activity had significant environmental effects even early in the post-glacial, but with the transition from hunter-gatherers to sedentary agriculturalists, to urban and then to industrial communities, people have had an increasingly profound effect on landscape. Indeed, over the last five millennia, anthropogenic activity in the temperate mid-latitude zones has become almost as important as natural agencies in determining the direction and nature of environmental change. Moreover, with the increased burning of fossil fuels and other forms of atmospheric pollution, human activity may be beginning to dictate the course of future climate changes for the first time in the history of the earth.

Landscape, people and climate are three variables which are inextricably linked (Figure 1.1), and an understanding of the course of recent environmental change requires an analysis not only of the elements themselves, but also of the way each influences the other in the broader context of earth systems science. The purpose of this book is to examine the interactions between people and the natural environment against a background of climate change. This reflects an increasing recognition by scientists and politicians alike of the importance of integrating scientific and social perspectives. Together they enable us to understand how natural environments have been transformed as human landscapes. It is also increasingly recognised that this integrated perspective is an essential part of planning for a sustainable future. The main focus of the book is on the northern temperate zone of Europe and North America where the effects of environmental change have been particularly marked and where the evidence for both natural and anthropogenic past processes is especially well preserved. Examples are also drawn from other geographical areas, however, where these help to illustrate the diversity of past people–environment relationships.

The book also seeks to draw on and integrate the differing academic traditions in the study of

(a)

(b)

Figure 1.1 The Merveilles Valley in the high Alps of Mont Bégo on the French–Italian border: (a) glacially striated rock surfaces where Bronze Age communities have pecked art showing weapons and animals; (b) a plough scene (Barfield and Chippindale, 1997). This landscape was made cultural by human agency, and we may speculate that seasonal pastoralists in the high Alps attached particular significance to this dramatic landscape, or the route across the Alps on which the art lies (photos Martin Bell)

environmentally focused archaeological science: from North America, anthropological and earth science perspectives; from Scandinavia, ethnohistoric approaches to cultural landscape; and from Britain and western Europe, environmental archaeology and new social and perceptual dimensions.

The time frame of the book covers the transition from the last cold stage in the Northern Hemisphere (**Late Weichselian** in Europe; **Late Devensian** in Britain; **Late Wisconsinan** in North America), to the present warm episode (interglacial) which began around 11.5 ka BP (**Holocene** in Europe and North America; **Flandrian / Postglacial** in Britain). The database is broad ranging, drawing on material from geology, geomorphology, geography, biology, archaeology, anthropology, history and social theory. However, the approach to the material is firmly rooted in geography and archaeology, in that the emphasis throughout is on landscape as the home for the human race. In this introduction many of the key concepts and terms (in bold) are introduced and defined and, in particular, we consider the integration of the social and scientific perspectives.

Earth science, geography and archaeology

The relationship between earth science, including physical geography, and archaeology has been long standing and productive. Discovery in ancient sedimentary contexts, such as caves and river gravels, of human bones and stone tools accompanying the bones of extinct animals led, in the mid-nineteenth century, to recognition of the antiquity of humanity (Grayson, 1986). In this way the foundations were laid both for Darwinian evolution and the development of archaeology as an academic discipline. Archaeological sites and finds are preserved within sediments, so that a full contextual understanding generally requires the application of geological, pedological or geomorphological approaches. This has led to the development of **geoarchaeology**: archaeological research which draws on the methods, techniques and concepts of the earth sciences, and which has been a particularly influential strand of archaeological science in North America (Herz and Garrison, 1998; Rapp and Hill, 1998), where many archaeological sites are stratified in riverine sedimentary contexts. Earth science approaches are becoming increasingly important in north-west Europe (Brown, 1997), the Mediterranean and South-west Asia (French, 2003; Wilkinson, 2003) and in other areas of the world as well.

The disciplines of geography and archaeology have much in common, being concerned respectively with the spatial and temporal dimensions of the human condition. The prime concern of

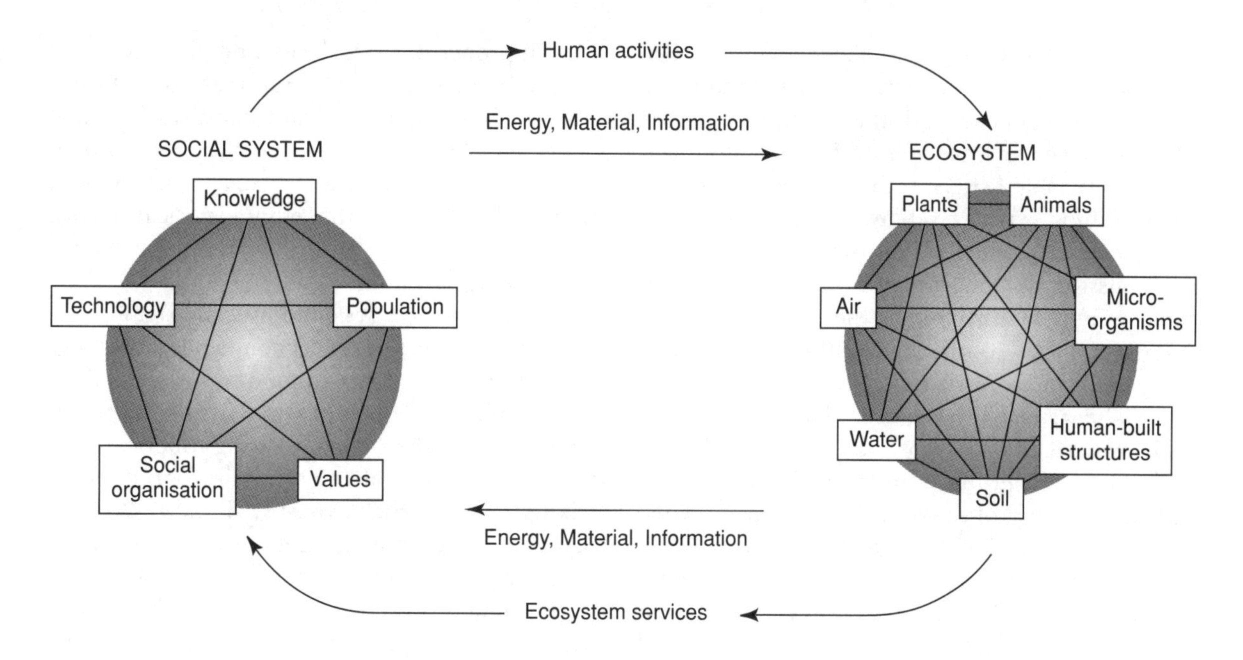

Figure 1.2 The interactive relationship between an ecosystem and a human social system (after Marten, 2001)

geography is to understand the processes that operate within the natural environment (physical geography) and to evaluate the ways in which people interact both with their environment and with each other (human geography). Archaeology deals with those aspects of the human past which are mainly elucidated using material remains (including environmental evidence) rather than written sources. Both physical geography and archaeology have been profoundly influenced by the science of **ecology**, which is concerned with the interactive relationship of organisms to each other and to their environment.

The components of an **ecosystem** (a living community and its environment) and its relationship to a human social system are shown in Figure 1.2. This includes two-way exchanges of energy, materials and information and the interactive effects of each factor on others. The diagram also includes the notion of **ecosystem services**, those commodities and benefits that the environment provides for people (Daily, 1997). This concept, recently developed in the USA, aims to quantify the full range of benefits which may be secured by environmental relationships which are **sustainable**, i.e. meet present needs without compromising the ability of future generations to meet their own needs (Marten, 2001). Each of the various elements (climate, geology, soils, flora, fauna, disease and people) are interlinked so that impact on one factor can have repercussions throughout the system, including a feedback effect on the original factor (Butzer, 1982).

Ecological concepts have been highly influential in the development of the subdiscipline of **environmental archaeology:** the study of the ecological relationships of past human communities, and the interactions between people and environment through time (Evans and O'Connor, 1999). Palaeoecological investigation using biological evidence has been a major aspect of archaeological science in north-west Europe, particularly Denmark (Kristiansen, 2002) and Britain, for more than a century, and is an approach that is increasingly being adopted in the Americas (Reitz *et al.*, 1996; Dincauze, 2000).

Not only does archaeology benefit from these relationships with earth and biological science but the benefits are reciprocal. Archaeological sites preserve dated contexts containing information about past environments, and these help to provide a past dimension (**time-depth**) for studies of environmental processes. Such contexts also contain evidence of the interaction between past human communities and the environment, the effect that people have had on their environment through time and the spectrum of environmental relationships experienced by societies, including those very different from our own. In this way the core mission of archaeology can be seen as complementary to **anthropology**, the science of people (Gosden, 1999). Both explore the diversity and richness of human existence.

It is increasingly recognised that there are few truly natural environments (i.e. those unaffected by humans). People have contributed to the present condition of most of the world's environment types, even in some of the most remote areas such as Pacific islands and tropical rainforest long considered pristine. In most areas of the world, what we see, what environmental scientists analyse and what conservationists seek to preserve is, to varying degrees, a human creation. For this essential reason an understanding of past human activity should often be part of effective conservation strategies (Chapter 8).

Landscapes are a product of the interaction between humans and environment which creates distinctive mosaics on varying scales reflecting particular ways of life, such as agricultural systems (Crumley, 1994). Landforms, soils, plants and animals have been modified by people who, in many areas, have also created a socially constructed landscape marked by particular arrangements of sacred places, wild places, settlements, fields, tracks, tombs, woodland, etc. This is the concept of the **cultural landscape**, an approach pioneered most notably in Scandinavia, where there has been a close relationship between **ethnohistorical** research (work on historically attested folk practice) and palaeoenvironmental science (Birks *et al.*, 1988; Berglund, 1991). By definition, therefore, landscapes are the product of human agency. Furthermore, environments do not have a neutral and independent existence, they exist in relation to organisms whose environments they are (Ingold, 1986, 1990, 2000). The materials which people use have physical properties, which make them useful, and they also possess attributed social significance (e.g. high or low status, female associations, magical properties). The heathland plant gorse, or furze (*Ulex europaeus*), illustrates the point (Evans, 1999: 105). It has a practical value as fuel, animal fodder, etc., but its gathering from the heath would also have played a part in articulating gender and social roles because the ethnohistorical record shows that particular groups were responsible for this activity.

Physical properties and social significance together contribute to the economic role that things play and thus to the articulation of social relationships. A place may be attractive for the resources it offers and the food-gathering opportunities it affords, because of the symbolism attached to striking forms of rock exposure (Bradley, 2000b), their colours (Cummings, 2000) or a combination of factors. These approaches draw, for instance, on **phenomenology**, i.e. the way in which landscapes were encountered and perceived (Tilley, 1994). Breton tomb locations (Figure 1.3) demonstrate the interrelationships between these perspectives; the example shown is a very rich and diverse coastal estuarine environment, tombs were located on what, in the time of lower Neolithic sea level, were rocky rises, or in the case of Guennoc an island. Rocky landforms are likely to have contributed to the significance of place and tomb passages are oriented on landscape features (Scarre, 2002). Such considerations highlight the need for an interdisciplinary approach, integrating environmental science and social perspectives thereby combining a landscape ecological approach (Forman, 1995) and phenomenology.

Increasingly, cross-fertilisation is taking place between those concerned with environmental and social perspectives (e.g. Edwards and Sadler, 1999; Evans, 2003). Simple explanations of the impact of environmental change on people, and of people on environment, are often not adequate. There is a need to move from an emphasis on the false dichotomy of people versus nature to a more integrated perspective. This may, for example, involve communities and their environments in a process of **coevolution**, in which interactive relationships have mutual influence (Rindos, 1989; Redman, 1999). Such an approach has proved particularly valuable in developing a better understanding of agricultural origins (p. 151).

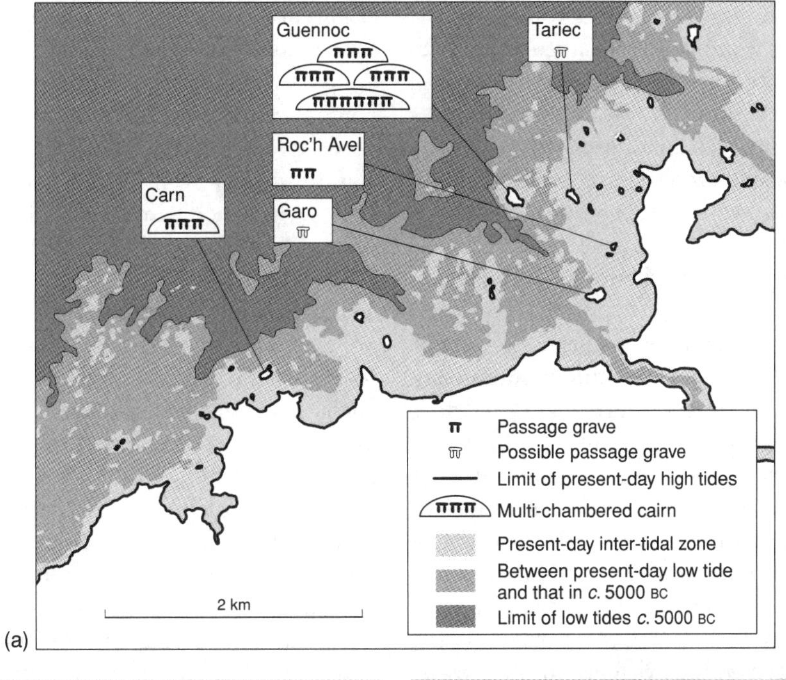

Figure 1.3 Aber Benoit, Brittany, France: (a) the locations of Neolithic tombs on what at a time of lower sea level were rocky rises or islands above a now drowned coastal plain, some tombs show evidence of orientation on topographic features (after Scarre, 2002, Figure 6.6); (b) one of the four tombs on the small island of Île Guennoc; (c) one of the reasons special importance was attached to this island may have been the natural giant perched boulder which dominates the approach to the island from the estuary to the south (photos Martin Bell)

Changing perspectives in science

Scientific techniques are the basis for our understanding of environmental change. They have provided major advances in dating and contribute in an ever-increasing number of ways to archaeology (Brothwell and Pollard, 2001). In parallel, the opening up of new perspectives in science is creating a much wider relevance for research on people and environmental interaction over extended timescales. The preoccupation of science has traditionally been a search for order, linear progress and predictability leading to law-like statements. Increasingly, however, the applicability of this approach has been questioned in the historical sciences (Gould, 1999). In subjects such as geology, evolutionary biology, environmental science and archaeology, traditional approaches have been criticised as **atemporal**, in so far as they give insufficient emphasis to temporal context and episodic events. Recognition of this is especially significant for archaeology because human behaviour is manifestly not predictable and law-like.

In the earth sciences there has been a challenge to the literal interpretation of **uniformitarianism**: the idea that present processes are the key to those operating in the past (pp. 21 and 50). In evolutionary biology the notion of uniformity of rates of evolutionary change (**gradualism**) has been challenged by Gould and Eldridge's (1993) concept of **punctuated equilibrium,** in which episodes of **stasis**, or little change, are interrupted by episodic, more rapid changes. Important in this approach is the notion of **contingency:** the existence of unique historical configurations of phenomena at particular points in time (Bintliff, 1999, 2004). Outcomes are not predictable and chains of events may impact in different ways, sending history cascading in various possible directions (Gould, 1999; see p. 20 for discussion of chaos). Parallel developments are seen in ecology, where the classical science which developed between the 1950s and 1970s emphasised pattern and regularity. Key concepts were **succession**, an orderly process of community change, leading to the creation of a **climax community**, the most stable community which could exist under given conditions. The quest was for natural ecological laws unaffected by the perturbing influence of human agency.

Palaeoenvironmental scientists have become progressively dissatisfied with this framework. In many parts of the world, there is no stage in the Holocene when conditions were sufficiently stable for the development of a climax (Simmons, 1999: 120). Moreover, as noted above, there are very few environments unaffected by human activity. Even in areas where human impact was minimal, research in population biology has provided little evidence of the predicted stability of the ecological climax and has revealed major population oscillations (Pickett and White, 1985). Furthermore, it is apparent that succession does not necessarily lead to a predictable outcome, and several end results are possible, creating communities which are more patchy and diverse in character than the climax model predicted (Forman, 1995; Simmons, 1999). The result has been the development of **dynamic ecology** in which emphasis is placed, in particular, on the roles of disturbance, disharmony and chaos (Worster, 1990).

Increasingly, therefore, scientific approaches are concerned with specific temporal contexts, the role of chance factors and non-linear processes. Such changing perspectives in both environmental science (Simmons, 1993b) and archaeology (McGlade, 1995, 1999; McGlade and van der Leeuw, 1997a) provide a framework for both a fuller appreciation of the complexities of environmental change through time, and an environmental science which is fully integrated with an understanding of the role of human agency. No longer are human environments marginalised as aberrations: human agency can now take its place as one of a spectrum of disturbance factors.

The causes of environmental change: cycles, pattern and chance

Natural environmental change comes about in many ways and on a wide range of timescales. Within this general framework, however, it is possible to distinguish those **processes** which are long term and gradual from those **events** which are sudden and frequently catastrophic. The former category would include such diverse phenomena as mountain building, the movement of the great lithostratigraphic plates, climate change, soil formation and ecological successions in biotic communities. Examples of events include major storms, outbreaks of disease, earthquakes and volcanic eruptions, and tidal waves (tsunamis). Events may be seen as pressure points in the relationship between people and nature, and are of special interest in that they help to define the nature of those relationships because many of them, for example volcanism, are precisely locatable in space and time. The distinction between events and processes serves to highlight the importance of timescale, but it must also be acknowledged that long-term processes may themselves be made up of multiple superimposed events. Hence long-term changes in sea level may, in reality, and certainly in terms of human perception, take the form of a number of discrete coastal inundation events.

The effects of people on landscape also operate at a range of different scales. Some can be considered as 'events', such as burning, warfare, or the failure of built structures such as dams. More significant, however, has been the impact of gradual processes over time, such as vegetation changes brought about by grazing of domestic stock, or the effects of irrigation, drainage and ploughing. In many parts of the northern temperate zone, the most far-reaching impact of human activity during the prehistoric period was the creation of an agricultural landscape which, although a gradual process, was likely to have comprised a series of clearance and burning events.

Many natural changes follow a regular or cyclical pattern and are astronomically determined. Figure 1.4 shows examples of these cycles, ranging from twice-daily tidal cycles, cycles of day and night, two-weekly spring tidal cycles, to the phases of the moon, seasons and years. There are also much longer astronomically determined cycles with wave lengths of *c*.100 ka (eccentricity cycle), 41 ka (obliquity cycle) and 23 ka (precessional cycle) which are related to the earth's orbital and axial parameters (Chapter 3).

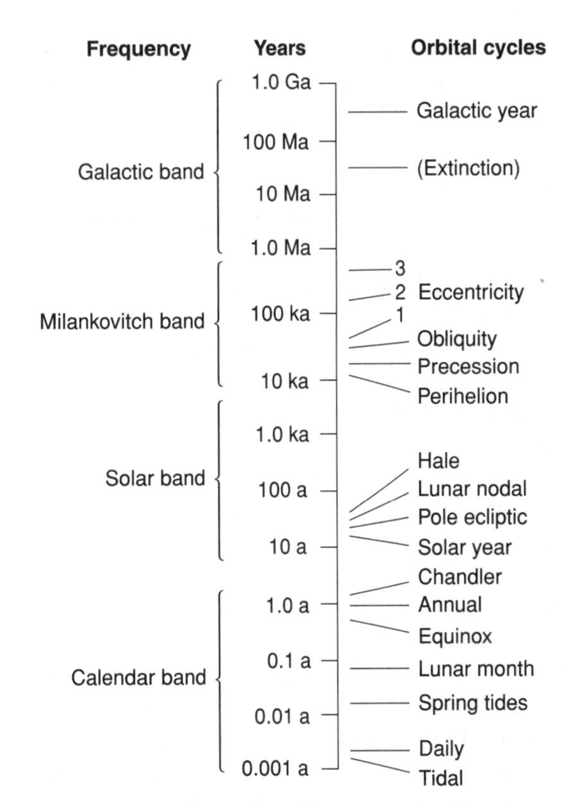

Figure 1.4 Examples of cyclical environmental phenomena on timescales varying from hours to millennia (adapted from House, 1999, Figure 2.5)

Regularities are apparent throughout the recent geological record. On the longer timescale, glacial and interglacial episodes have occurred at regular or quasi-regular intervals; cycles of vegetation change are detectable within interglacial episodes; and global sea levels have fluctuated in response to repeated expansions and contractions of the great continental ice sheets (Chapter 3). Short-term natural changes are also apparent over the last few thousand years: advances and retreats of mountain glaciers have occurred at intervals of 1–1.5 ka (Grove, 2002), variations in Scottish peat sequences are detectable over a *c*.210-year cycle (Chambers *et al.*, 1997), while fluctuations in solar activity (sunspots) are evident over cycles of 11 and 22 years. In the North Atlantic sediment record, oceanographic changes are apparent at periodicities of *c*.550, 1 ka and 1.5 ka years (Bond *et al.*, 1997; Chapman and Shackleton, 2000). On much shorter timescales, annual cycles from laminated sediments and tree rings are increasingly important as sources of chronological precision in studies of environmental change and tidal and seasonal cycles of molluscan growth have been used to detect patterns of seasonal exploitation in coastal midden sites.

Natural environmental clocks form the basis for human timekeeping, for the calendar and for the prediction of astronomical phenomena. Many natural cycles are superimposed on others, and when trends on different wavelengths happen to coincide mutual reinforcement can lead to pronounced change. Other **stochastic**, or chance, processes are of an unpredictable nature. Examples include the geomorphological and climatic consequences of volcanic eruptions (p. 169), and perturbations within the atmospheric and oceanographic systems which give rise to hurricanes, storms, floods, tidal surges, blizzards, droughts, etc. These events can sometimes be detected in the palaeoenvironmental record. Extreme climatic events with devastating human consequences are a distinctive feature of tropical and subtropical areas. In the temperate zone extreme events also occur and are more frequent during some **secular climatic episodes** (major periods of distinctive climate). There is also growing evidence that they are especially concentrated in unstable phases of climatic transition (Knox, 2000; Starkel, 2002). Destructive storms were a feature of the climate of north-west Europe during the Little Ice Age (p. 94), and this area has also been affected by a succession of intense storms during the late 1980s to the present. Examples are the hurricane which devastated parts of southern Britain on 16 October 1987 felling 15 million trees (Figure 1.5) and the storm on 26 December 1999 in France which felled an estimated 360 million trees. Events of this kind have made palaeoecologists increasingly aware of the effects of stochastic events. These recent storms, and similar extreme weather events that have affected areas such as the interior United States, and the devastating consequences of successive droughts in the Horn of Africa (Pearce, 1989), have fuelled alarm about global climatic change arising from industrial activity since the nineteenth century (Chapter 9).

Figure 1.5 A wood north of Brighton, England, devastated by a storm on 16 October 1987 (photo *The Argus*, Brighton)

Increasing awareness of the possibility of rapid, and potentially catastrophic, environmental changes comes about because of a growing recognition of the interrelationships that exist between the components of ecosystems and the implications of **feedback** effects involving the relationships between the components of an ecosystem illustrated in Figure 1.2. **Negative feedback** reduces the effects of the originally induced change, thus leading to a resumption of stable equilibrium conditions. **Positive feedback** reinforces the consequences of change. In human communities, systems for the exchange of environmental information (p. 140) are examples of feedback mechanisms; these are often negative, and reduce the effects of change. The effects of positive feedback are illustrated by the clearance of woodland by prehistoric communities, as this may have serious consequences for geomorphological processes (e.g. water retention in the soil, runoff, erosion) and ultimately for the human communities themselves. The concept of environmental **sensitivity** is the likelihood that a given change in conditions will produce an identifiable response. The degree of stability can therefore be seen as relating to the temporal and spatial distribution of resisting (negative feedback) and disturbing (positive feedback) forces (Brunsden, 2001). These concepts are important because human agency, and other disturbance factors, play a key role in sensitising environments to the effects of factors such as climate change.

Figure 1.6 shows schematic representations of various types of environmental change and equilibrium. Negative feedback often has a self-limiting effect on change and ensures that various forms of metastable equilibrium are maintained (Figure 1.6b). This is the condition also known as **homeostasis** in which change occurs within certain defined limits as in Figure 1.6e. Human activity patterns are set within the observed and predicted limits of environmental change. Every so often, however, those limits will be exceeded, either by a chance (stochastic) event (Figure 1.6c/d), or because a long-term trend (**dynamic equilibrium**) has crossed a critical threshold

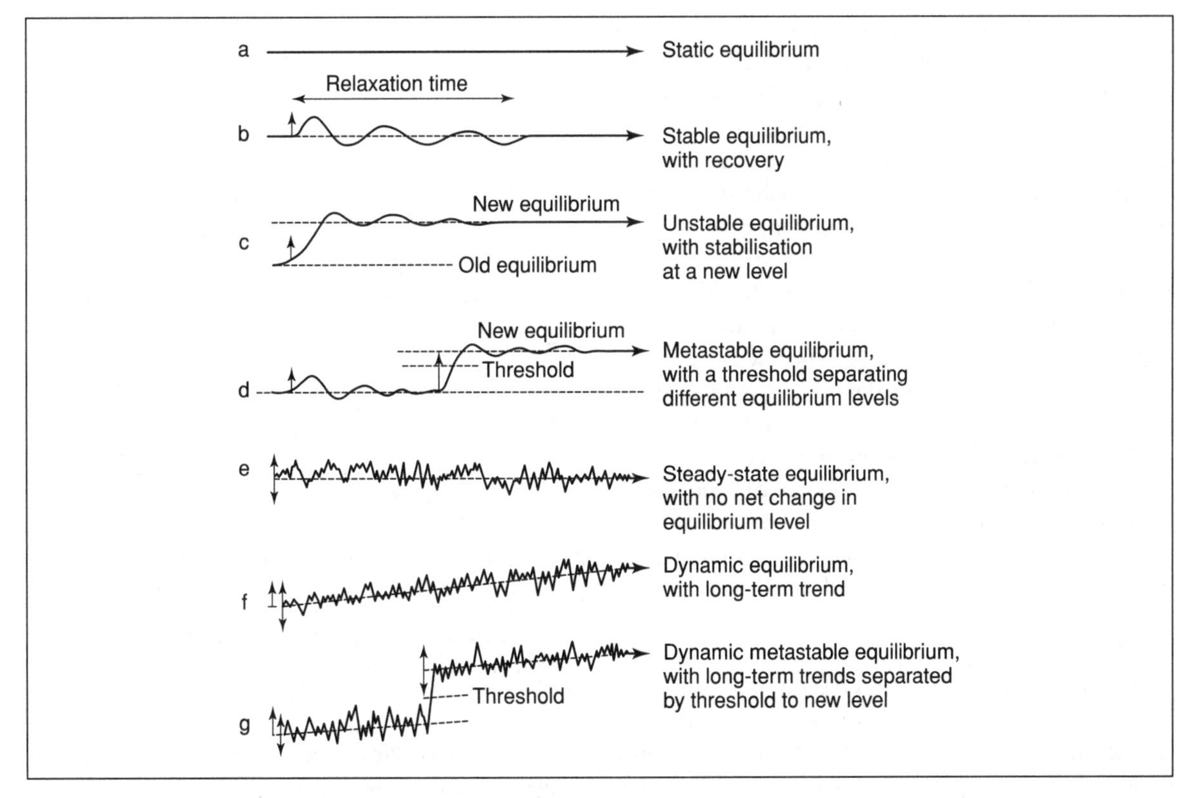

Figure 1.6 Diagrammatic representation of various forms of environmental change and equilibrium. The horizontal axis is time and the vertical arrows indicate disturbance factors affecting controlling variables (modified from Butzer, 1982, Figure 2.3)

(Figure 1.6g), following which a new state of equilibrium becomes established.

If stochastic phenomena occur sufficiently frequently over a period for which historical records are available, it may be possible to achieve some understanding of how often events of a specific magnitude are likely to recur (their **recurrence interval** or **return period**). This approach to hazard prediction is widely used in the field of civil engineering, for example in the design and construction of sea defences or flood prevention works (Handmer, 1987; Shaw, 1988). Perceived recurrence interval will also have influenced the response of past communities to particular hazards. Those which occurred frequently were most likely to have given rise to major adaptive responses. In the case of infrequent hazards, people would have been more likely to have taken a chance, as indeed millions do today in the San Andreas Fault earthquake zone of California, or the volcano-dominated, yet densely settled, Bay of Naples. It must be stressed, however, that estimates of recurrence interval are based entirely on recorded frequency over a particular time period, and that over other time periods (both past and future) the frequency of hazards may be very different. Indeed the new science of chaos theory (Chapter 2) carries the implication that reliable prediction of stochastic environmental phenomena is a virtual impossibility!

Questions of scale: space and time

Approaching questions at an appropriate scale of temporal and spatial resolution is an issue requiring particular consideration since Quaternary science and archaeology often operate on rather different scales. It is particularly important for archaeological investigations, for example, to focus on scales which are relevant from a human perspective (Stein, 1993). Many of the changes considered in the following pages operated on a global scale, such as the transition from glacial to interglacial conditions. More restricted in their effects were macroscale changes which are reflected in the catchment-wide development of distinctive climatic/vegetational zones. At an even more restricted spatial scale were mesoscale effects relating to regions, and microscale changes affecting individual localities such as the specific sites which are often the focus of detailed archaeological inquiry (Dincauze, 2000). Time and space are directly related: all movement in space has a temporal dimension which is the essence of **time–space geography**. The timescales of environmental change vary greatly from glacial episodes spanning perhaps 100 ka to earthquakes whose duration may be measured in minutes.

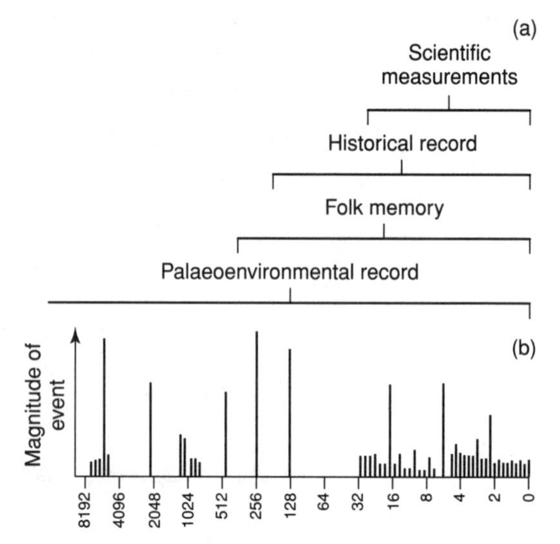

Figure 1.7 Sources of evidence for environmental phenomena against time (shown on an arithmetic scale). The vertical bars represent the magnitude of recorded environmental events using different sources (adapted from Bell, 1992)

The chronological precision at which environmental changes can be investigated is highly variable. In some deep ocean cores, for example (p. 42), it may be impossible to attain a resolution of less than 1000 years, due partly to slow rates of sediment accumulation on the seabed and partly to biogenic mixing (Lowe and Walker, 1997). Dating based on artefacts (e.g. pots and metalwork) or on the radiocarbon technique (p. 53) can provide an age within 100–300 years and, under ideal conditions, to a precision of a few tens of years. In some cases even better temporal resolution at an annual scale can be achieved as a result of recent advances in dendrochronology (p. 57) and work on annually laminated lake sediments (p. 58).

Sources of information for environmental change in the context of different timescales (Driver and Chapman, 1996), are shown on Figure 1.7. In many areas of scientific inquiry **instrumental records** (of temperature, rainfall, erosion, etc.) are often only available for short timescales. Weather records in Britain go back to AD 1659 (Chapter 2), while many other environmental phenomena have instrumental records going back only a few decades. Written history and folk memory may provide some information for earlier centuries, but for the vast majority of time we are reliant on the palaeoenvironmental record. The horizontal scale in Figure 1.7 is arithmetic, highlighting the greater range of sources we have in more recent times. Events of a range of magnitudes will be recorded in the recent record but in many cases this will not include the highest magnitude events, which are of low frequency and may not therefore be included in the period of instrumental records. Historic and oral records are selective in that they are likely to record only high-magnitude events. Preservation of the palaeoenvironmental record also varies temporally and spatially: at some times and places the circumstances of preservation mean that events of a range of magnitudes will be recorded, whereas during others there will be over-representation of high-magnitude events. The recent record is of particular importance because of the wider range of sources of evidence and the greater opportunity to compare them. Longer palaeoenvironmental records may be calibrated by reference to the shorter periods represented by historical and instrumental records. In this way Barber *et al.* (1994), for example, have created bog surface wetness curves providing evidence for past climate from European peat sequences (p. 32). The recent palaeoenvironmental record is also important in increasing our understanding of how environmental change impacts on societies, including the evidence from folk memory and the ethnographic record of societies very different from our own.

In some areas historical records of environmental change exist over extended timescales. Writing appears in Mesopotamia *c*.3300 BC and the earliest historic records in Egypt are from *c*.3100 BC. The first recorded eclipse occurred in China at *c*.1876 BC. Early historical records of environmental phenomena, or environmental disasters, are of great interest but demand the careful and critical analysis of sources, using the approaches of the historical or classical scholar and including consideration of the context in which they were written and the original purpose of the records.

Time itself cannot simply be conceived of as a fixed calendrical concept as we may be tempted to assume from our extensive use of scientific dating techniques, such as radiocarbon. Time is a cultural construct which differs from one social context to another; in some situations cyclical aspects may be emphasised whereas in others linear aspects may assume greater importance (Bradley, 1991; Gosden, 1994; Murray, 1999). Such complexities are apparent in the ethnohistorical record. The Luo of western Kenya, for example, recognise daily cycles measured in terms of natural phenomena, the cycles of sun and moon, the seasonal cycles of climatic episodes. Deeper historic time is measured in terms of cycles of generations and a linear sequence of unique events/disasters: famines, epidemics, locust plagues, etc. (Dietler and Herbich, 1991). In that particular case calibration of oral history is possible using colonial records which demonstrate that the linear time that events cover is about a century. Other examples discussed below (pp. 171 and 243) demonstrate that oral histories can preserve highly accurate accounts of environmental phenomena over timescales of at least 200–400 years. Far longer, but less precise records, have also been claimed. Australian aboriginal art, it has been argued, may preserve memories of sea-level rise occurring between 9 and 7 ka BP (Flood, 1983: 143).

The reconstruction of environmental change over the past 20 ka, therefore, involves analysis at a range of spatial and temporal scales within a chronological framework of variable precision. In the following pages natural environmental changes are considered principally at the macro- or mesoscale, whereas human activity is generally evaluated over much smaller scales, sometimes even in the context of short-term events within the lives of particular communities. In these cases it is not possible, nor indeed desirable, to seek law-like generalisations regarding the effects of environmental change on people, or of their reciprocal effects on the environment. Much depends on the social context within which particular changes occur and effects are likely to have varied greatly in time and space. A climatic change such as that experienced in north-west Europe during the Little Ice Age (from AD 1550 to 1850) may have crippled some communities but acted as a stimulus to others (p. 177). Hence, those sections of the book dealing with the human perspective take the form of a series of case studies arranged in a broadly chronological sequence.

Environmental change and human perception

Research into past human environments has often shown a tendency towards **environmental determinism**, the view that a particular set of environmental parameters would give rise to only one human response. This tendency has been reinforced by an emphasis on linear evolutionary progression and human adaptation. Determinists have argued that many cultural changes such as the collapse of civilisations were a direct result of environmental changes. Determinism remains a significant strand of thought in archaeological writing, for instance in the recent work of Baillie (1995 and 1999) which postulates the effects of volcanic eruptions and comet impact on past societies (p. 174). Those examples deserve particular consideration because they are underpinned by chronological precision made possible by recent advances in dendrochronology (pp. 57 and 169). That many forms of environmental change can have dramatic social implications is beyond doubt. Many writers today adopt a less dogmatic position than the determinists. This position is more akin to **possibilism**: the view that environments may limit, but

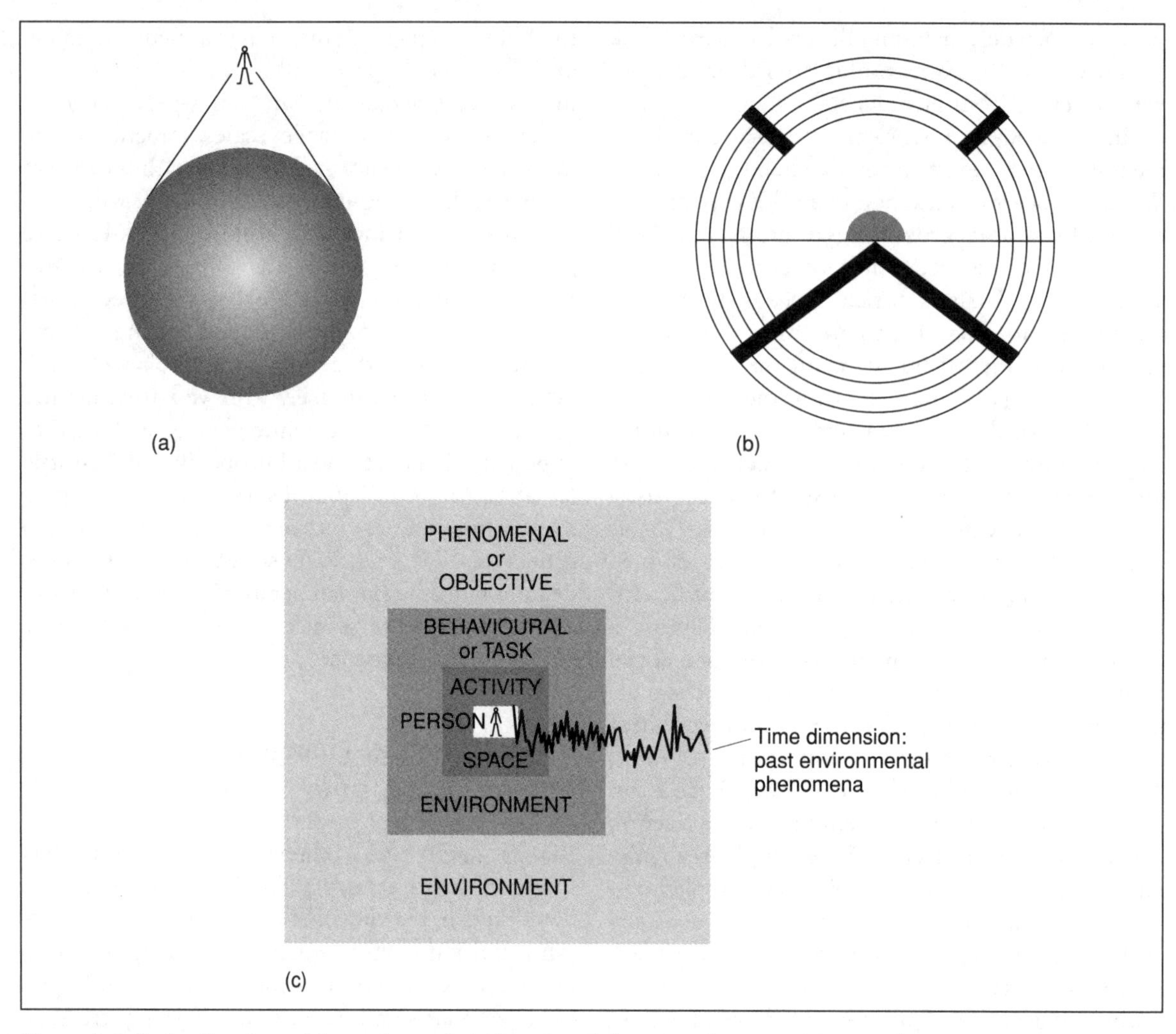

Figure 1.8 Contrasting views of the environment: (a) later twentieth-century view as a globe seen from space (after Ingold, 2000); (b) as seen in Yupik Eskimo cosmology, with possible routes (paths) shaded (after Fienup-Riordan, 1990); (c) as seen by a behavioural geographer (modified after Small and Witherick, 1995)

not necessarily cause, human behaviour and biology (Hardesty, 1977). The challenge for the future is to move beyond assertions of effect to a more sophisticated understanding of the relationship between environmental changes and people.

In the human sciences there is increasing recognition of the need to consider human perception of the environment. This issue is being addressed from the complementary perspectives of social anthropology (Ingold, 2000) and geography (Simmons, 1993b), while in archaeology there is growing emphasis on **cognition** (thought processes used in acquisition and organisation of knowledge (Renfrew, 1993; Mithen, 1998). **Cognitive–processual archaeology** is concerned with the processes of cultural change and, in particular, with the interaction of people and environment with due regard to the role of human perception and symbolism (Renfrew and Bahn, 2000: 491). Perception of the environment by other communities may be very different from our own, as an anthropological perspective so clearly shows. A detached modern view (Figure 1.8a) of the globe, as seen in high-technology images from outer space, is in total contrast to the views of Yup'ik Eskimos (Figure 1.8b) in whose cosmology the dwelling is at the centre and routes lead in various directions through concentric environmental zones (Ingold, 2000).

The routes or paths in Figure 1.8b are important but frequently overlooked elements in archaeology, which has a tendency towards preoccupation with dots on the map rather than the ways in which they articulated together as parts of a living landscape (Bell, 2003). Routes represent the communication and information-gathering system. They will often have played a pivotal role in structuring landscape and, together with topographic features (particularly passes), ancient routes will have helped to determine the encounters that successive peoples had with specific landscapes over extended timescales. Ingold (1993a) notes that 'movement is the essence of perception' and 'paths and tasks impose a habitual pattern of movement on people'. Routes, areas of abundant resources, or topographic locations of particular social significance, will have contributed to the creation of **persistent places**, those sites frequented over extended timescales (Barton *et al.*, 1995). Concentrations of human activity at such places would have been one of the factors contributing to the patch dynamic of environments (p. 182).

The third environmental representation in Figure 1.8c is that of the **behavioural geographer**, who is concerned with the ways in which people perceive, respond to and affect their environments. The individual is at the centre surrounded by the day-to-day environment of activity space. Beyond that is the wider behavioural or task environment, that part of the perceived environment which influences individual decision making. Beyond this is the phenomenal environment including the other physical, biotic and human elements. The sawtooth line is representative of the time dimension of past environmental phenomena, some of which lies within the individual's memory, or the folk memory of the community.

The reliability of decisions made by an individual about the environment is limited by factors such as perception and time: the concept of **bounded rationality.** Particular limiting factors include the relationship between timescale of observation and occurrence, i.e. the wavelength or recurrence interval of phenomena, and the spatial scale of individual or group environmental experience. In pre-literate societies much will depend on the veracity and timescale of individual and folk memory and on the communication systems for transmitting knowledge both spatially (by interaction with other individuals and groups) and temporally (from one generation to the next). The anthropological record provides some idea of the diversity of communication systems: myth and oral history, song, dance and art must in many societies all have played a key role in communicating environmental knowledge vital to a group's survival.

It follows that environments cannot be defined as abstract entities but are defined in relation to a subject. They are 'not independently given but are constituted in relation to organisms (including human beings) whose environments they are' (Ingold, 1986, 1990). We cannot simply reconstruct the resources of an area and regard these as static elements of the landscape, simply there for the utilisation by human communities. Resources are essentially socially and culturally defined, some may be favoured while others are avoided for social, religious or other reasons. In the same way people cannot be regarded simply as responding to changes in the 'actual' environment which can be reconstructed (with more or less accuracy) from the palaeoenvironmental evidence; rather, their response was to their perceived environment which reflects their own lived world of experience (**lifeworld**) (Butzer, 1982; Brandt and van der Leeuw, 1987).

Cultural ecology acknowledges that human beings have the capacity to adapt in terms both of their biology (i.e. genetically) and in terms of culture (the learned pattern of behaviour and understanding). These two facets of the human condition, though complementary, operate at very different timescales, cultural change being a more rapid and flexible response than biological evolution (Durham, 1978). Culture represents a system of thought linked to habitual action which organises and explains the natural world through routine actions, cosmology, science, religion, etc. (Whittle, 2003). Hence culture has the capacity to play a mediating role between people and environment, a form of buffering against the risks and uncertainties resulting from both cyclical environmental change and the effects of chance events (Butzer, 1982). Factors which can be described as 'cultural' account for the much greater adaptability of people by comparison with other elements of the biosphere. Human beings do not simply respond to natural factors as determinists assume, they possess the capacity of free will and for planned long-term independent action. Environmental change may, for instance, create the opportunity, or necessity, for change in human society but not determine the character, trajectory or the timescale of that change.

As Watts (1983) has written, 'we interpret the world within the limits of a historically conditioned imaginative vision'. We need to give special consideration to the effects of environmental variables on human groups and individuals. Our understanding of the range of ways in which human communities perceive environmental change and hazards can be informed by hazards research in geography (Slaymaker, 1996; Smith, 2001) and by the perspectives of ecological anthropology (Hardesty, 1977; Vayda and McCay, 1978; Halstead and O'Shea, 1989). These will include consideration of how environmental phenomena influenced the thought processes of individual decision makers in prehistory (Mithen, 1990, 1998). Hazards are not neutral entities and their effects depend on the social context (Hewitt, 1983), in particular on the coping or hazard-buffering strategies which the specific community has evolved to overcome the effects of environmental changes. Such strategies may be highly diverse, as we outline at the beginning of Chapter 5, and frequently they take the form of changing the environment itself as outlined in Chapter 6. The existence of these relationships significantly undermines assumptions of simple determinism. The ability of communities to adapt to environmental change will vary according to the context and nature of a society, the rigidity or adaptability of its social organisation and economy, and its technology and population structure.

Perception and response, therefore, inevitably embody a historical dimension, while ideas and ethics regarding the environmental past inevitably influence decision making for the future. This constitutes a dialogue between past and present which is part of the *raison d'être* for this book.

Scope and structure of the book

The book has been divided into nine chapters, each of Chapters 2–8 dealing with a particular aspect of landscape, climate and society over the last 25 ka. It also falls naturally into three parts. The discussion begins with natural environmental changes providing the context in which the drama of human evolution and social development has been (and still is being) played out. In Chapter 2, the different types of evidence that form the bases for environmental reconstruction are introduced and the ways in which these data sources can be obtained and interpreted are assessed. This chapter also considers the methods by which increasingly precise chronological frameworks for recent environmental change are being established. Building on that information, Chapter 3 describes the pattern of both long-term and short-term climatic change in north-west Europe and North America, and considers the causes of climatic change over a range of temporal scales. Chapter 4 then examines the impact of climatic events of the past 25 ka on the biotic and abiotic components of northern temperate zone environments.

The second part leads into a discussion of people in the landscape, a theme which is developed largely through a series of case studies based on the archaeological and ethnohistorical records. Figure 1.9 shows the relationship between climatic stages and selected archaeological periods which are discussed in the book. A more detailed review of worldwide archaeological chronologies is given

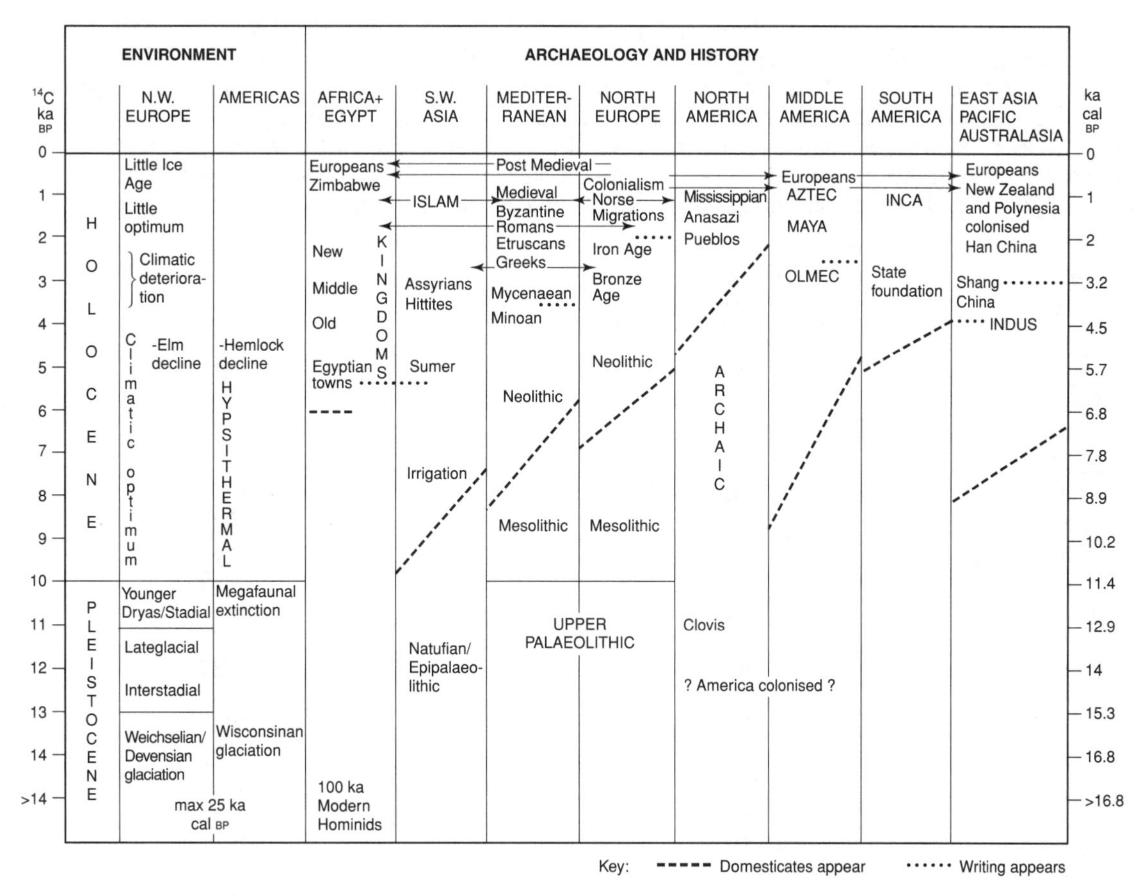

Figure 1.9 Chronological chart for the last 17 ka showing climatic stages and archaeological periods in selected areas

in Scarre (1993). Chapter 5 takes a critical look at the effects of environmental change on people and the coping strategies developed by human communities, while Chapter 6 shows how people have created diverse cultural landscapes over the last 11 millennia and Chapter 7 considers the relationship between people, climate and erosion, all with examples drawn especially from mid-latitude regions of the Northern Hemisphere.

The third part of the book deals with the present and looks to the future. Preserved within the contemporary cultural landscape is a rich record of the natural and cultural processes that have combined to produce that landscape. It is argued that an understanding of this past dimension is an important element in the development of sustainable strategies for the future. The challenges of nature conservation and heritage management are considered in Chapter 8. Chapter 9 examines the impact of nineteenth- and twentieth-century industrial activity on contemporary climate, and discusses the possible consequences of these anthropogenically induced climatic changes for global climatic and terrestrial environment. From an initial consideration of the effects of climate on people, therefore, the analysis has moved to an examination of the impact of human society on both present and future climate. To what extent climatic changes arising from human activity will amplify or modulate the natural climatic rhythms discussed in Chapter 3 is one of the most intriguing problems confronting historic scientists, and one to which there are, as yet, no conclusive answers.

Table 1.1 Calibrated radiocarbon ages 500–15,000 yrs BP based on the calibration program CALIB 4.3 (after Stuiver *et al.*, 1998). In this program the calibrated radiocarbon ages are expressed as *age ranges* at 1σ and 2σ, reflecting uncertainties inherent in the calibration calculations. In order to provide an estimate of likely calibrated ages, we have taken the mid point of the age range (2σ) for the highest probability distribution for each radiocarbon date as given by the calibration program, assuming a standard error of ± 50 yrs. As such, the calibrated ages above should be considered only as **indicative age values.**

^{14}C *age (yrs BP)*	*Indicative calibrated age (yrs BP)*	*Indicative calibrated age (yrs AD/BC)*
500	516	1435 AD
1000	890	1061 AD
1500	1369	582 AD
2000	1964	14 BC
2500	2594	644
3000	3167	1217
3500	3785	1835
4000	4463	2513
4500	5169	3219
5000	5708	3758
5500	6279	4329
6000	6837	4887
6500	7402	5452
7000	7797	5847
7500	8288	6339
8000	8856	6906
8500	9508	7558
9000	10,192	8242
9500	10,863	8913
10,000	11,451	9500
10,500	12,582	10,632
11,000	13,016	11,066
11,500	13,548	11,598
12,000	13,988	12,038
12,500	14,835	12,885
13,000	15,365	13,415
13,500	16,224	14,274
14,000	16,807	14,857
14,500	17,383	15,433
15,000	17,962	16,012

Note on dates

1. **Dates.** Throughout this book, the shorthand form is used for years before present (BP): ka – thousand years; ma – million years. Present is taken as 1950 calendar years AD. Wherever possible radiocarbon dates, which form the basis for much of the chronology of the past 40 ka, have been calibrated to calendar years BP using INTCAL 98 (Stuiver *et al.*, 1998), (see Chapter 2). For practical purposes calibrated radiocarbon years, ice-core years, varve years and dendrochronological dates are considered to be equivalent. Where ages have been produced in dendrochronological years the convention BC/AD is retained. We also use dates BC/AD in periods when dating is mainly based on historical sources. See Table 1.1 for a comparison of radiocarbon and calibrated dates.

2 Evidence for environmental change

Introduction

Evidence of past environments and of climatic change comes in a variety of forms. For the relatively recent past, instrumental data on temperature and precipitation are available for limited areas of the world, although the timescale for these records seldom exceeds 300 years. Historical evidence, in the form of old diaries, annals, ships' logs, woodcuts, pictures, etc., may yield additional information, although these tend to provide 'snapshots' of former climatic conditions. Some historical records, such as harvest dates and crop yields however, constitute time-series data of climatic change. By far the most widely used bases for environmental reconstructions are **proxy** records. The term 'proxy' is used to refer to any line of evidence that provides an *indirect* measure of former climates or environments (Ingram *et al.*, 1981), and can include material as diverse as pollen grains, insect remains, glacial sediments and tree rings, as well as data on crop yields, harvest dates and parish records. Some of the more widely used proxy records, along with historical evidence and meteorological data, are introduced in the following pages. The difficulties of interpreting these various lines of evidence in the context of environmental and climatic change are also considered. The chapter concludes with an examination of the principal dating techniques that are currently employed in the establishment of timescales of recent environmental change. We begin the discussion, however, with some methodological considerations.

Scientific methodology

Any attempt to unlock the secrets of the past inevitably involves the palaeoenvironmentalist in **scientific** inquiry. Precisely what constitutes a science has been a matter of debate among philosophers for many years, but a general view might be that this is a branch of study that is concerned with the search for truths and the establishment of general laws relating to the natural world, using as a basis the systematic collection of observed facts. Opinion differs, however, over what constitutes the most appropriate **scientific method**, i.e. the procedures employed by scientists to solve problems and hence to search for truths. In this section, two popular views of science, **inductivism** and **falsification**, are examined, for both approaches have underpinned investigations of climatic and landscape change (Haines-Young and Petch, 1986). The related approach of **multiple working hypotheses** is also discussed, and the 'new science' of **chaos theory** is briefly introduced. Finally, because palaeoenvironmental reconstructions using proxy data rely heavily on modern analogues ('the present is the key to the past'), it is necessary to consider the assumptions that underlie what has become known as the **uniformitarian** approach to earth history.

Inductivism

Inductivism is widely regarded as the essence of scientific inquiry, for it appears to provide science with a sound, logical (rational) methodological basis. It still underpins most work in the earth sciences while the quantitative approach, which has characterised a considerable proportion of the research output of geographers, geologists, environmental scientists and archaeologists over the past 25 years, is firmly rooted in inductivism. It rests upon the notion that the accumulation of knowledge relies on experience, and hence scientific statements acquire meaning by virtue of their **empirical** (i.e. experimental or observational, as opposed to theoretical) basis. In other words, generalisations that form the basis for scientific laws are derived from observations of reality. Moreover, once laws or theories are established in this way, there is a basis for explanation or prediction by means of **deduction**. The inductive approach therefore can be seen as a stepwise ascent in science from observation to theory (O'Hear, 1989) and could, perhaps, be summarised along the following lines: collection of data by observation (experience) → ordering of facts (measurement, classification, definition) → generalisations (induction) → law/theory construction → explanation (prediction via deduction). This line of reasoning, which argues that real-world phenomena can be explained by showing them to be instances of repeated and predictable regularities, underpins much of classical science extending back to the ancient Greeks (Chalmers, 1999), and has frequently been referred to as **classical rationalism**.

Despite its widespread adoption by scientists, inductivism has been the subject of penetrating critical scrutiny (e.g. Russell, 1961; Popper, 1972). A particular problem concerns **verification**, for no number of apparently confirmatory statements can ever show a general proposition to be true. Scientific knowledge can never be more than partial, and hence there must always be the possibility that anomalies exist which remain to be discovered, and which may refute a general statement or law derived by inductive reasoning. It is for this reason that many scientists speak in terms of **probabilities** rather than absolutes. A further difficulty concerns the objective nature of facts. It is implicit in inductive reasoning that observation precedes theory development, that there is a clear distinction between fact and theory and, moreover, that facts obtained by this form of inquiry are 'objective'. However, all observations or measurements are inevitably made in the context of prevailing theory. The ways in which data are collected, the processes involved in excavation and in the development of sampling strategies, and the assumptions underlying the techniques that are used, all have a significant effect on the evidence obtained. A considerable body of knowledge has been generated about the way in which the physical or human world is structured and hence that corpus of information will inevitably exercise an influence on new observations, experiments, etc. The 'objective' basis of facts from such observations must therefore be called into question. A related problem is that observations and measurements will inevitably be influenced by the level of technology that is available at any particular time. In the early years of the twentieth century, for example, it was widely accepted that four separate glacial episodes had occurred in the mid-latitude regions of the Northern Hemisphere during the course of the Quaternary. By the 1950s, technological advances in ocean coring led to the discovery of the oxygen isotope signal in deep ocean sediments (Chapter 3), which now suggests that as many as 50 cold or glacial stages occurred during the course of the Quaternary (Shackleton *et al.*, 1990). The question arises, therefore, as to whether the truth of a fact can ever be satisfactorily demonstrated.

Most contemporary inductivists would acknowledge these difficulties, and would accept that science does not begin with unbiased and unprejudiced statements about reality (Chalmers, 1999). Indeed, as Adams (1988) has observed, 'facts are now understood as compelling interpretative statements reached by comparing the results of more or less precise measurements obtained within a theoretical scheme'. Hence, such 'facts' should be regarded not as 'proved', but rather as 'accepted' pending further critical inquiry (see below). Theories may be conceived of by a variety of routes (accident, inspiration, creative acts, etc.), all of which precede observation but defy logical analysis. This **sophisticated inductivism** lies at the heart of many areas of current scientific inquiry, and is characteristic of much of the research that has been undertaken on environmental change. Such lines of thinking also accord with the approach to archaeological interpretation advocated by Hodder (1999), with its emphasis on **self-reflexivity,** in other words a mode of inquiry which is critically aware of the effects of scientific and archaeological assumptions on the data which are obtained. Hodder further argues for an approach which is **multivocal** (literally many-voiced) and which values a diversity of forms of investigation. This can be seen as an inherently logical stance for both archaeology and Quaternary science, subject areas in which research teams are becoming increasingly interdisciplinary.

Falsification

The most sustained challenge to inductivism is to be found in the writings of Karl Popper (1972, 1974). He accepted the empirical basis of scientific inquiry and also that observations will inevitably be guided by existing theory. He argued that although the truth of a proposition can never be conclusively demonstrated, statements can be rejected. In other words, while hypotheses cannot be **verified** they can, in fact, be **falsified.** Popper's view of scientific investigation, therefore, is a **deductive** one based on conjecture and refutation. Theories are viewed as tentative conjectures which are tested by observation, experiment, measurement, etc.; those that fail are rejected and replaced by further conjectures. In this way, according to Popper, science proceeds by trial and error, with the strongest theories being those that are clear, precise, detailed and broad-ranging. Theories that contain more detail, however, are potentially more falsifiable; hence, a theory or hypothesis gains strength the more wide-ranging and precise it is, the more falsifiable it is, but most importantly the more it resists falsification and hence constitutes a challenge to science. Because falsification attempts to provide a sound, rational basis for deciding between the relative merits of different theories by testing them critically, the term **critical rationalism** has been applied to this form of scientific reasoning.

Two further aspects of this line of thinking merit consideration. First, failure of a hypothesis or conjecture will lead to outright rejection, for theories can be modified to enable them to resist falsification. Second, although a theory can never be verified, it can be **confirmed.** The essential difference between verification and confirmation is that the former implies that a theory or hypothesis has been shown to be true, whereas confirmation merely implies that a theory has resisted falsification and hence has been accepted **for the time being.** An example might be the theory of plate tectonics, which has been supported by a range of geological evidence that has emerged over recent decades (Frankel, 1988). From the perspective of the critical rationalist, this is a high-order theory in so far as it is potentially highly falsifiable, but as it has so far resisted falsification, it can be regarded as being confirmed by the evidence that is currently available.

Critics of falsification have argued that this form of scientific inquiry is too inflexible and does not conform with what scientists actually do. Indeed, it has been suggested that had falsification been rigorously applied, many currently accepted scientific ideas would never have survived, simply because they appeared to conflict with prevailing observations (Chalmers, 1999). Moreover, in so far as observations are both theory-dependent and fallible, where an observation conflicts with a theory, there is no logical reason why it should not be the former that is in error rather than the latter. Hence it becomes possible to confirm a theory by testing it with a fallible observation. During the nineteenth century, for example, the occurrence in many inland areas of Britain of unconsolidated deposits containing marine shells (Figure 2.1) was used to corroborate the theory of a major marine inundation widely believed to have been the biblical Flood. Following the adoption of the glacial theory, however, it became clear that these shelly deposits were not marine in origin as originally thought, but were glacially transported (Sutherland and Gordon, 1993). The notion of the Great Flood achieved almost universal acceptance, but unfortunately it was based on a contemporary interpretation now known to be false.

Popper's ideas have been adopted by many natural scientists, and Haines-Young and Petch (1980, 1986) have presented compelling arguments for the more widespread adoption of critical rationalism within physical geography. Indeed, a falsificationist approach has found favour in

Figure 2.1 An exposure of glacial till containing marine shells at South Shian, western Scotland. It was this type of evidence that was used to infer the former marine inundation or 'Great Flood' (photo Mike Walker)

certain areas of Quaternary science (e.g. Birks, 1986; Peglar and Birks, 1993; Wagner *et al.*, 1999). However, the majority of earth scientists continue (perhaps, albeit, unconsciously) to follow an inductivist approach to scientific inquiry, and it seems that the 'challenge of critical rationalism', described some 20 years ago by Haines-Young and Petch (1980), has yet to be sustained within the historical earth sciences.

Multiple working hypotheses

The method of multiple working hypotheses was outlined initially by Chamberlin (1897, reprinted 1965) and involves the formulation of as many hypotheses as possible in an attempt to explain the same phenomenon. Weaker or mistaken theories are progressively eliminated as the hypotheses are tested critically against each other. The aim is to achieve an explanation that is more nearly correct than would have been the case if only a single hypothesis had been considered. Although the method was a precursor to falsification, it has much in common with it (Haines-Young and Petch, 1983, 1986) as decisions have to be made between competing theories, and scientists are encouraged to find evidence that will lead to the elimination of all but one of the working hypotheses. Applications of the approach can be found in geomorphology (Baker and Payne, 1978) and palaeolimnology (Battarbee *et al.*, 1985), while the method of multiple working hypotheses has been viewed as one of the fundamental philosophical principles of palaeoecology (Birks and Birks, 1980). It is also particularly appropriate in archaeology, where interdisciplinary research teams formulate and evaluate working hypotheses from a range of scientific perspectives.

Chaos theory

Much of what has been said so far rests upon the assumption that science is essentially **reductionist** in its approach to the gathering of knowledge. In other words, we can reach an understanding of the complexities of the modern world by analysing them into simpler constituents from which we derive relatively straightforward rules that we call the 'laws of nature'. These physical laws govern the operation of global systems at a range of spatial and temporal scales. In seeking to formulate and to understand these laws, scientists have traditionally looked for **regularities** (or order) in real-world phenomena, and have attempted to use this knowledge to make statements about the future behaviour of natural systems. Classical science therefore has both a deterministic and a predictive basis (see Inductivism above). Over the last three decades, however, a conceptual revolution has occurred, particularly in mathematics and physics, which has raised fundamental questions about the philosophical and methodological basis of classical science. This has been the recognition of **chaos**.

Scientists now realise that small adjustments in the variables of natural systems can have far-reaching consequences. Tiny differences in input may lead to overwhelming differences in output – a phenomenon which has been referred to as 'sensitivity to initial conditions'. The most famous example of this is what has become known as the **butterfly effect**, i.e. in weather patterns, the notion that a butterfly stirring its wings in Beijing can transform storm systems next month in New York (Gleick, 1987). It is also now apparent that in deterministic systems, random (stochastic) behaviour can exist side by side with order, and hence although systems may obey immutable and precise laws they do not always act in predictable and regular ways. Simple laws may not produce simple behaviour; rather, deterministic scientific laws can produce behaviour that appears random. Hence, order can breed its own kind of chaos (Stewart, 1990). By the same token, however, while chaos theory emphasises the randomising forces that cause systems to become disorderly, there are occasions where disordered systems can revert to a high degree of order, a phenomenon that has been referred to as **antichaos** (Kauffman, 1991). Indeed, Cohen and Stewart (1994) have taken this notion one stage further by introducing the concepts of **simplexity,** the tendency of simple rules to emerge from underlying disorder and complexity, and **complicity**, where interacting systems coevolve in a manner that changes both, leading to a growth in complexity from simple beginnings. This complexity is unpredictable in detail, but its general course is comprehensible and foreseeable.

These discoveries have transformed the face of contemporary science, for fundamental questions are now being raised about measurement, experimentation and predictability, and about the verification or falsification of theories. In the earth and environmental sciences, it is now acknowledged that many geomorphic systems show clear evidence of chaotic dynamics ('non-linear dynamical systems', or 'NDS') and deterministic complexity (Phillips, 1996). This poses particular problems for the historical earth scientist who employs the end product (i.e. landform or sedimentary evidence) as a starting point in the explanation of landscape evolution. The fact that chaotic behaviour is characteristic of geomorphic systems means that it may not, in fact, be possible to reconstruct the initial conditions from an analysis of the end product, and this will inevitably have an effect on the validity and accuracy of the hypotheses that can be constructed to explain historical landscape development (Harrison, 1999). The ramifications of chaos theory, or other aspects of NDS theory, in the earth and environmental sciences are, therefore, potentially far-reaching, not only for the way in which research proceeds, but also in the search for explanations of patterns or trends in historical data (Phillips, 1999). Indeed, the ripples of chaos theory have spread far beyond the analysis of physical and biological systems, for demographers, military historians, sociologists and economists (among others) have found non-linear dynamics valuable in re-evaluating (and often restructuring) old theories and creating new ones (Shermer, 1995). In archaeology, the growing recognition of the importance of chance (contingency: Chapter 1) enables human agency to take its place as one of a series of perturbing influences on the environment, and attempts are increasingly being made to model human non-linear dynamic systems as one way of understanding the role of chance factors in human–environment relationships (McGlade and van der Leeuw, 1997b).

Uniformitarianism

Early attempts to reconstruct patterns and processes of environmental change were rooted in a philosophy that became known as **catastrophism.** Hence landscape change was seen as being brought about by earthquakes, floods, volcanic eruptions and other cataclysmic events (Chorley *et al.*, 1964). The biblical Flood was considered to have been of widespread significance (the **diluvial** view) and strongly influenced interpretations of the geological record. Underlying this line of reasoning was the almost universally accepted view of Ussher (1650–54) that the Creation had occurred in 4004 BC. This clearly allowed only a limited timescale for geological and geomorphological processes. In addition, the influence of the church on scientific inquiry was pervasive and divine intervention was frequently invoked to account for otherwise inexplicable phenomena.

The development of the **uniformitarian** approach to earth history was a radical departure from the catastrophist school of thought. It was first proposed by the geologist James Hutton (1788, 1795) who argued not only for a geological timescale extended beyond that allowed for by Ussher, but also for continuity in the operation of geological processes through time. His views are summarised in the well-known aphorism 'the present is the key to the past', in other words, that former changes of the earth's surface may be explained in terms of those processes observed to operate at the present day. Preternatural or catastrophic forces were rejected. This radical reinterpretation of geological history was espoused by Hutton's co-worker John Playfair (1802), and particularly by Charles Lyell (1830–33) who is often regarded as the founder of modern geology. This new approach to earth history came to be known as the **fluvialist** school, because of the emphasis on river erosion as a process in the shaping of the earth's surface. Elements of uniformitarian reasoning, notably **actualism** (what exists now also existed in the past) and **gradualism** (geological processes operate at slow rates and in small increments) are implicit in Darwin's work on evolution and natural selection (Stoddart, 1986). In the twentieth century, the sciences of palaeontology (the study of fossil remains) and palaeoecology (the study of the ecological relationships of past organisms) were firmly underpinned by uniformitarian principles (Rymer, 1978).

Although uniformitarian reasoning is now implicit in most studies of earth history, debate about the validity of the approach continues (Frodeman, 1995). A major difficulty is that in using the present to interpret the past, environmental scientists are employing **analogy** as their main interpretative argument (Delcourt and Delcourt, 1991). However, every context could be regarded as being unique, no analogy is exact, and hence no argument from analogy is certain. This problem is further compounded in landscape reconstructions by the nonlinear nature of many geomorphic systems, which makes it difficult to reconstruct initial conditions from end products (see above). Furthermore, it is becoming clear that many former plant or animal distributions and, indeed, certain depositional environments, have no modern analogues, in which case uniformitarian reasoning cannot be applied. Equally, it cannot be assumed that processes in the geological and biological past have operated at a constant rate, as demanded by the early uniformitarianists. In the same way it cannot be argued that certain plant species occupied *precisely* the same ecological niche or covered *exactly* the same geographical range in the past as at present. Such strict adherence to uniformitarian principles (**substantive uniformitarianism**) finds little favour with contemporary earth scientists. Furthermore, a rigid uniformitarianism is manifestly unsuited to studies involving human agency, since it is becoming increasingly apparent that past societies were very different from those of today. Hence, most historical scientists tend to incline to the view that although currently observable geological and biological processes must have operated throughout history, they would have done so on varying timescales (Chapter 1). Catastrophic events such as floods, earthquakes and volcanic eruptions, which have profound effects on modern landscapes, must therefore have been equally effective in the past. Ironically, while uniformitarianism initially replaced catastrophism as a methodological basis for earth history, with a better understanding of the contrasting timescales of environmental change, cataclysmic events can now be reconciled within a uniformitarian framework. This concept of uniformitarianism has been referred to as **actualism** (see above) or **methodological uniformitarianism** and, despite its practical and philosophical limitations, continues to underpin most work in the historical earth sciences and in palaeoecology.

Fossil evidence

The term **fossil** is used to describe any organism or part of an organism that is buried by natural processes and subsequently permanently preserved. It includes skeletal material, plant remains as well as trace fossils, impressions of organisms, trails of organisms, tracks and borings (Bromley, 1996). Human artefacts, however, are not regarded as fossil material. The fossilisation process involves chemical and/or physical changes to the organic material which can result in often delicate structures being preserved. Where little or no chemical change occurs subsequent to death (e.g. shell, wood) the term **subfossil** rather than fossil is applied. Fossils are divided into larger **macrofossils** and smaller **microfossils**, this somewhat arbitrary distinction being based on whether or not a microscope is required for study. Detailed examination of macrofossils (e.g. animal bones, wood, molluscs) is often made under a microscope or scanner, but they can be seen with the naked eye. By contrast, microfossils can only be detected using microscopy and their study requires the use of a microscope throughout.

Fossil evidence is a central element in environmental reconstruction (Lowe and Walker, 1997). A considerable database has been assembled over the years on the ecological requirements of present-day plants and animals, although it must be stressed that for a number of groups of biota, there are still major gaps in our knowledge of their contemporary ecology. Nevertheless, sufficient is known about the ecological affinities and associations of many modern species to make inferences about former climatic and environmental conditions by means of uniformitarian reasoning. The following is a sample of some of the fossil records currently employed in palaeoenvironmental research. Note, however, that although the different forms of evidence are described individually, they are frequently used in combination as part of **multi-proxy** investigations, i.e. where evidence from a range of different fossils or other proxies is combined to form a basis for climatic or environmental reconstruction (Oldfield *et al.*, 2003; Walker *et al.*, 2003).

Figure 2.2 An exposure of **tufa** (calcium carbonate precipitated from carbonate-saturated springwaters in limestone regions) at Caerwys, North Wales, showing preserved plant macrofossils, largely reeds in growth position (photo Mike Walker)

Macrofossils

Plant remains

Macroscopic plant remains found in Late Quaternary deposits include fruits, seeds, wood and other parts of plants including leaves, buds, scales and spines (Grosse-Brauckmann, 1986; Wasylikowa, 1986). They are best preserved in lake sediments and in peat deposits where anaerobic conditions obtain, but they also occur in riverine sediments, in cave sediments, in buried soils, and in middens and pits on archaeological sites from which carbonised remains of fruits and seeds are often recovered (Jones and Colledge, 2001). More unusual contexts in which plant macrofossils have been found include tufa deposits (Figure 2.2), volcanic ashfalls and animal (e.g. packrat) middens (Warner, 1990a). Plant macrofossils are frequently deposited close to the original point of growth, and in fen and bog sites in particular they provide important data on local vegetation communities. They have been employed to reconstruct, *inter alia*, Holocene fluctuations in the altitudinal limits of the treeline (Figure 2.3), regional patterns of Holocene vegetation (Baker, 2000), vegetation and climate changes during the Lateglacial (Birks, 2003), and the nature of prehistoric economies (Moore *et al.*, 2000). Plant macrofossils have proved to be particularly valuable in investigations of Quaternary cold stage floras, as pollen data from these periods are often sparse (West, 2000).

Molecular and genetic evidence has also been recovered from plant macrofossils as well as from bone (see below). **Lipids**, which are the plant oils, resins and waxes of the natural world, can provide valuable archaeological information, for example, on plant foods (Evershed *et al.*, 1999, 2001), while studies of ancient **DNA** in plant

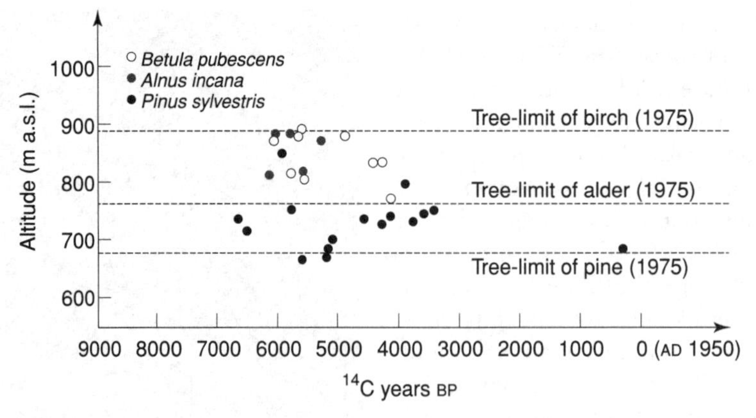

Figure 2.3 Radiocarbon dates for subfossil *Betula pubescens, Alnus incana* and *Pinus sylvestris* in relationship to altitude in the Scandes Mountains, central Sweden. The upper limits for birch, pine and alder in 1975 are also shown (after Kullman, 1988, Figure 3)

remains have provided valuable new insights into the origins and spread of agriculture (Jones and Brown, 2000; Jones, 2001).

Mollusca

Terrestrial and freshwater Mollusca (Figure 2.4) are preserved in a range of sediments where there are high concentrations of calcium carbonate. These include tufa, colluvial, fluvial and lacustrine deposits, cave sediments, aeolian sediments and buried soils. Under acidic conditions, in areas of calcium-deficient bedrock, Mollusca are rapidly leached and are usually absent (Lŏzek, 1986). Land Mollusca provide information about local environments, being particularly responsive to the amount of vegetation cover and shade (Evans, 1972; Preece, 2001). They have been used to infer land-use changes associated with prehistoric human communities (Taylor *et al.*, 1998: Chapter 6), and also to reconstruct patterns of climate change (Rousseau *et al.*, 1993, 1994). Good reviews of the applications of non-marine Mollusca in palaeoenvironmental reconstructions can be found in Goodfriend (1992) and Preece (2001).

Marine Mollusca are preserved in Late Pleistocene and Holocene contexts, including boreholes taken from the seabed, in beach gravels and estuarine sediments, and in localities inland where they have been transported by glacier ice (Merritt, 1992). They provide evidence of former sea-surface temperatures (Peacock, 1989, 1993), and they have also proved useful as a medium for dating Late Quaternary events, including sea-level changes and glacier advances (Lowe and Walker, 1997). In addition, midden dumps of marine shells around contemporary shorelines provide information on prehistoric coastal exploitation (Stein, 1992).

Fossil insects

Insects are extraordinarily successful animals comprising more than half of the total number of plant and animal species known today (Coope, 1986), and their fossil remains have been used in a diverse range of Quaternary environmental reconstructions (Ashworth *et al.*, 1997; Robinson, 2001). The largest order is the Coleoptera (beetles), which have colonised almost every terrestrial and freshwater habitat on earth. Yet many are **stenotopic**, tolerating a narrow range of environmental and climatic conditions, and it is this particular characteristic that makes Coleoptera such valuable palaeoecological and palaeoclimatic indicators (Elias, 1994). Their remains, which are extremely robust, are preserved in almost any sediment that contains plant macrofossils and can frequently be identified to the species level. Moreover, the considerable body of information that is available

Figure 2.4 Land molluscs (a) *Discus rotundatus* (diameter 5.4 mm); (b) *Vallonia costata* (diameter 2.4 mm); (c) *Carychium tridentatum* (length 1.7 mm); (d) *Truncatellina cylindrica* (length 1.7 mm); (e) *Pupilla muscorum* (length 3.5 mm); (f) *Clausilia bidentata* (Length 11.6 mm); (g) *Acicula fusca* (length 2.2 mm) (photo Ian Clewes)

on their present-day ecological associations and distributions means that uniformitarian reasoning can readily be employed to make inferences about Quaternary environments (Buckland and Coope, 1991).

Coleoptera have been most widely, and successfully, employed in the reconstruction of Quaternary climates, and have provided a basis for the quantitative reconstruction of summer, winter and mean annual temperatures. Using the

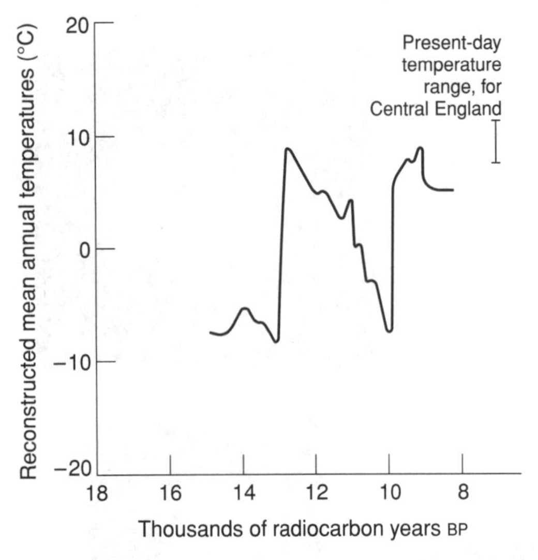

Figure 2.5 Reconstructed mean annual temperatures in England for the period 14.5–8 ka BP based on the MCR analysis of fossil beetle assemblages (after Lowe and Walker, 1997, Figure 4.22)

mutual climatic range method[1] (Atkinson *et al.*, 1987; Elias, 1997), it has proved possible to make inferences about temperature conditions during the last interglacial (Coope, 2000a), during the last cold stage (Coope, 2000b) and, in particular, during the transition from the last cold stage into the Holocene (Coope *et al.*, 1998; Miller and Elias, 2000; Figure 2.5). Coleoptera have proved valuable in other areas of Late Quaternary research, for example as indicators of local habitat change (Elias, 1994) and regional vegetation cover (Ponel and Lowe, 1992). They are also valuable in archaeology, providing evidence for past human landscapes (Dinnin and Sadler, 1999), and as indicators of living conditions and economic activities in urban contexts (Kenward and Hall, 1995).

Mammalian remains

Animals bones occur in many Quaternary deposits including cave sediments (Figure 2.6), fluvial and colluvial sediments, lacustrine and marine deposits, peats and soils, and in burial chambers, middens, hunting sites and other contexts associated with human activity (Stuart, 1982; O'Connor, 2000). The bones range in size from those of large mammals which, in an archaeological context, have frequently been exploited and deposited by people, to the remains of small mammals, birds, amphibians and reptiles which are often more valuable as ecological indicators. In some sediments the bones become **permineralised** as salts from circulating groundwaters are deposited in the vacant

Figure 2.6 Charterhouse Warren Swallet, Mendip Hills, England, showing speleothems and animal bones on the cave floor (photo Peter Smart)

pore spaces, while in acid peats where much of the mineral fraction has been lost by decalcification, only the flexible collagen fraction remains. Remarkably preserved human bodies such as 'Grauballe' and 'Tollund Man' from Denmark, and 'Lindow Man' from England, occur in peat bogs where decay has been inhibited by anaerobic conditions and the chemistry of the peat (Turner and Scaife, 1995; van der Sanden, 1996). Late Quaternary vertebrates reflect former local habitats (grassland, woodland, heathland) and hence can provide useful information on landscape history (Woodman *et al.*, 1997); they can be used to infer climatic conditions (Yalden, 2001); and they provide evidence for the process and diffusion of animal domestication (Clutton-Brock, 1989).

Vertebrate remains have also proved to be extremely valuable in Quaternary biostratigraphy (Lister, 1992; Currant and Jacobi, 2001). In addition, recent advances in molecular biology have enabled ancient DNA sequences to be obtained from Quaternary vertebrate remains. This offers an exciting new tool for the testing of hypotheses about human evolutionary history, such as the relationships between Neanderthals and modern humans (Krings *et al.*, 1997; Ovchinnikov *et al.*, 2000), and the expansion of early humans out of Africa (Templeton, 2002). Other applications of ancient DNA include studies of palaeodisease, kinship and population studies, and the origins of animal domestication (Brown, 2001).

Microfossils

Pollen and spores

Pollen grains (Figure 2.7) are derived from the seed-producing plants (Angiosperms and Gymnosperms) and are disseminated over wide areas by wind, water, animals or insects. Spores from the lower plants (Cryptogams) are entirely wind-dispersed. The grains become incorporated into peats, lake sediments and soils where they are well preserved, providing that anaerobic conditions obtain. In so far as the composition of the pollen rain is a reflection of regional vegetation cover, fossil pollen and spores obtained from stratified sequences of sediment will provide a record of vegetational (and hence environmental) change through time (Moore *et al.*, 1991). Pollen and spores constitute one of the most important proxy data sources for inferring former environmental conditions, and the range of applications of the technique is extensive. These include the reconstruction of local (i.e. site-specific) vegetation histories (Smith and Cloutman, 1988; Charman, 1994), the evolution of regional vegetation patterns at a variety of temporal scales (Huntley and Webb, 1988), the reconstruction of glacial–interglacial sequences (Reille *et al.*, 2000), studies of plant migration and forest history (Birks, 1989), and investigations of sea-level change (Shennan *et al.*, 1994). Pollen analytical data have been widely employed to reconstruct past climatic conditions using, for example, the occurrence in a fossil assemblage of particular 'indicator species' whose climatic requirements can be quantified (Zagwijn, 1994; Isarin and Bohncke, 1999), or multivariate statistical methods to generate quantified temperature estimates from fossil assemblages based on data from modern climate–plant relationships (Bartlein and Whitlock, 1993; Birks, 1995). These, in turn, have provided a basis for the development of models of past global climates (Wright *et al.*, 1993). Finally, much of what is known about the patterns of Holocene woodland clearance and early farming practices has been obtained from pollen analytical evidence (Birks *et al.*, 1988; Gaillard *et al.*, 1992).

Rhizopods or testate amoebae

These are protozoa that are found in a variety of freshwater habitats as well as in soils (Charman *et al.*, 2000; Charman, 2001). They are often well preserved in *Sphagnum* peats and have proved to be valuable indicators of past hydrological changes, including variations in the chemistry of mire water (Tolonen, 1986) and soil moisture content (Warner and Charman, 1994). They have also been employed in studies of sea-level change (Charman *et al.*, 1998), but have proved to be most valuable as indicators of changes in surface wetness on peat bog surfaces. As mire surface wetness variations are often a reflection of regional changes in precipitation, testate amoebae, in association with other proxies from peat sequences (see below), provide a basis for the construction of Holocene precipitation records (Woodland *et al.*, 1998; Charman *et al.*, 1999).

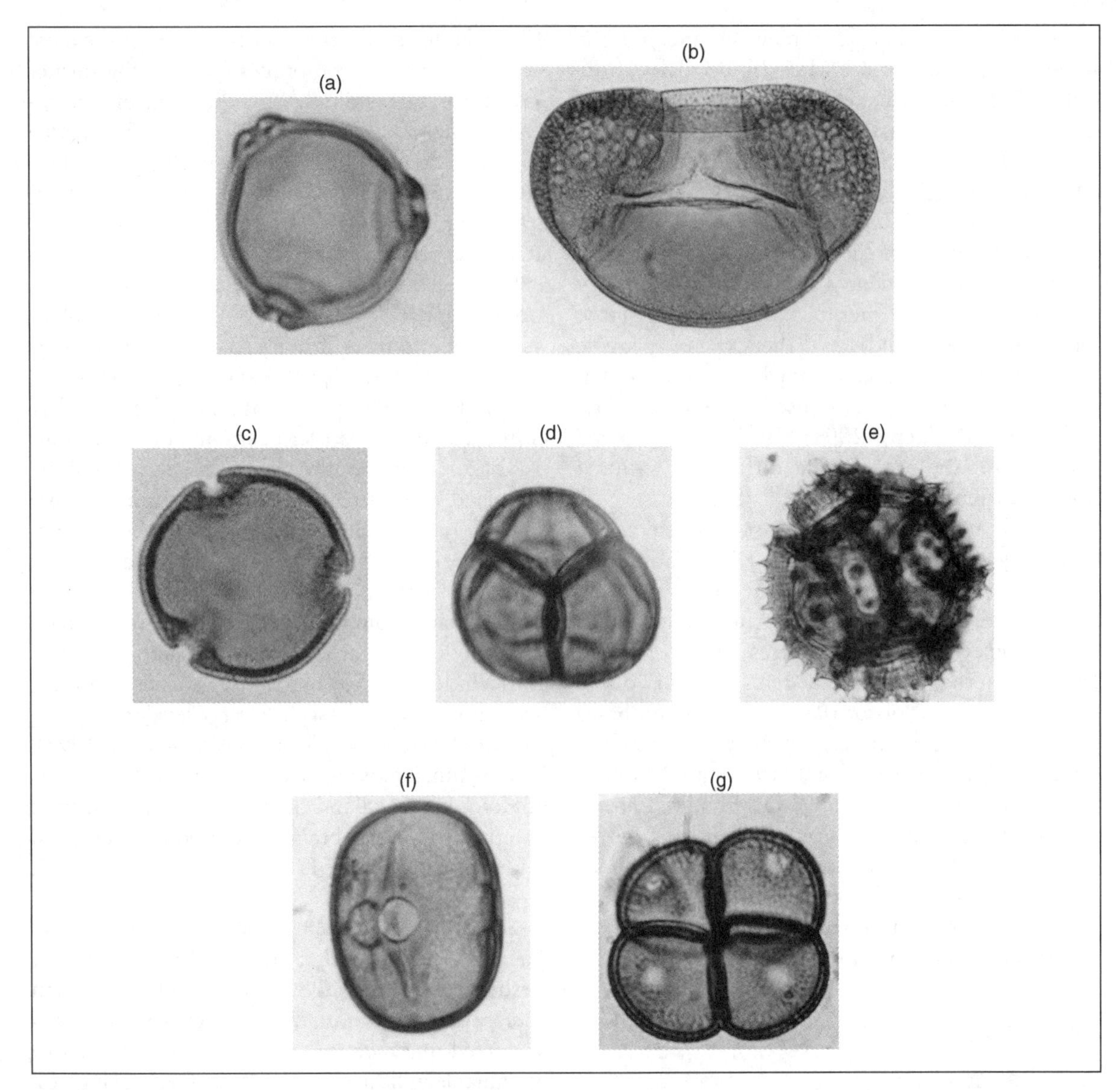

Figure 2.7 Pollen grains: (a) *Betula pendula* (diameter 35 μm); (b) *Picea excelsa* (80 × 70 μm); (c) *Tilia cordata* (35 μm); (d) *Erica cinerea* (35 μm); (e) *Taraxacum officinale* (35 μm); (f) *Vicia sativa* (35 × 30 μm); (g) *Typha latifolia* (40 μm) (photo Ian Clewes)

Diatoms

Diatoms are microscopic unicellular algae (Figure 2.8) that live in ponds and lakes, in estuaries and in the sea, their distribution being controlled by a range of environmental variables including acidity, degree of oxygenation of the water, mineral concentration, and especially water temperature and salinity (Stoermer and Smol, 1999). Fossil diatoms are found in many aqueous sediments and have been used to provide evidence, *inter alia*, of changes in lake water depth (Brugam *et al.*, 1998), recent lake acidification (Sullivan *et al.*, 1992), historic and prehistoric land-use changes (Renberg *et al.*, 1993), and regional climate and environmental change (Pienitz *et al.*, 1999). Diatom analysis has also been employed in the study of sea-level change to isolate

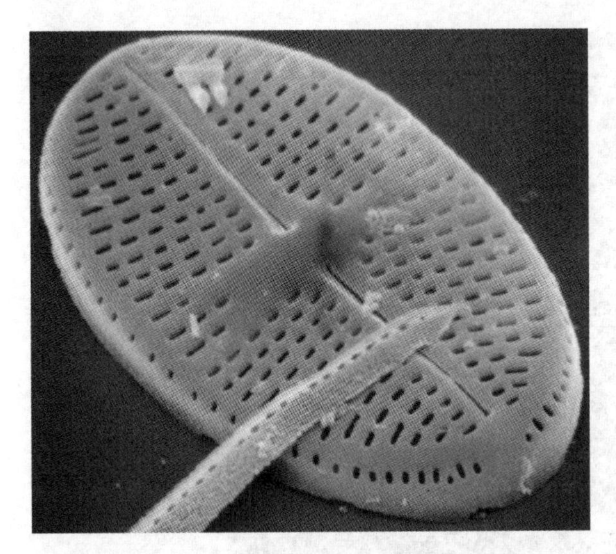

Figure 2.8 Scanning electron micrograph (SEM) of a diatom *Psammothidium subatomoides*. This species is a common diatom in acid lakes where it lives attached to a variety of substrates (photo R.W. Battarbee)

marine 'transgressions' and 'regressions' in coastal sediment sequences (Long *et al.*, 1998). In addition, diatoms have been extracted from deep-ocean cores and have been used, in association with other marine micro-organisms (Foraminifera, radiolaria, coccoliths), as a basis for palaeo-oceanographical and palaeoclimatic reconstructions (de Vernal and Hillaire-Marcel, 2000: Jiang *et al.*, 2002).

Chironomids

Chironomidae (chironomids), the non-biting midges, are valuable proxy climate indicators, as their distribution and abundance are closely related to summer lake surface water temperatures (Walker *et al.*, 1991). Head capsules of the larval stage are well preserved in freshwater sediments, and possess sufficient diagnostic characteristics to enable specific identifications to be made (Hofmann, 1986). Statistical comparisons between modern chironomid assemblages and contemporary July temperatures have provided a basis for quantitative estimates of past summer temperatures for both north-west Europe and North America during the Lateglacial and Holocene periods (Lotter *et al.*, 1999; Figure 2.9).

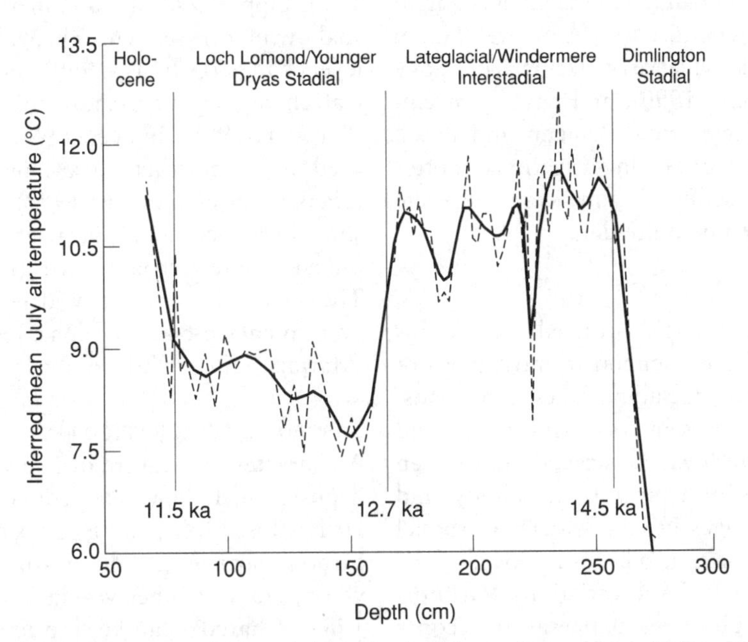

Figure 2.9 Temperature fluctuations during the Lateglacial and early Holocene (approximately 14.7 to 11.5 ka BP) based on the fossil chironomid record from Whitrig Bog, southern Scotland. The dates are in Greenland ice-core years BP (after Brooks and Birks, 2000, Figure 3)

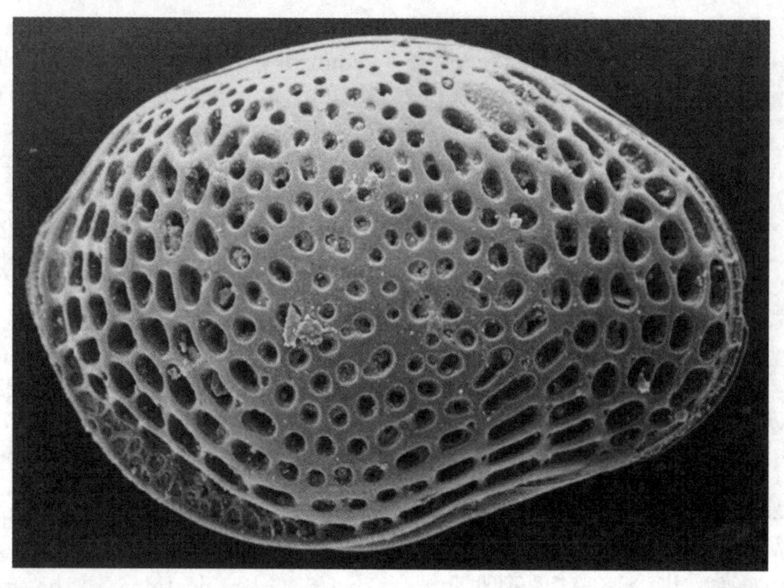

Figure 2.10 Scanning electron micrograph of the right valve of *Loxoconcha* sp., a marine inshore species of ostracod (length c.0.6 mm) (photo Eric Robinson)

Cladocera

Cladocera (water fleas) form a major component of the microcrustacean fauna of freshwater lakes and ponds. The skeletal fragments of cladoceran fossils are often abundant in lake sediments and in many cases can be identified to species level (Hann, 1990). They provide evidence of lake palaeoecology (Nilssen and Sandøy, 1990), and have been employed in both single-proxy (Duigan and Birks, 2000) and in multi-proxy investigations (Lotter *et al.*, 1997) of lake sediment sequences to generate quantified palaeotemperature data.

Ostracods

Ostracods are microscopic bivalved crustaceans (Figure 2.10), and are common in most types of aquatic environment, including lakes and ponds, streams, rivers, estuaries and oceans (Griffiths and Holmes, 2000). Freshwater ostracods have been used as indicators of temperature, salinity and eutrophication changes in lake waters (Carbonel *et al.*, 1988). They have also been employed in the reconstruction of lake-level variations (Griffiths *et al.*, 1994) and Holocene temperature records (Forester, 1987), while both non-marine and marine ostracods have been used in studies of sea-level change (Penney, 1987).

Foraminifera

Foraminifera (Figure 2.11) are marine protozoans that occupy habitats ranging from salt marshes to the deep oceans of the world (Murray, 1991; Lipps, 1993). Foraminifera from estuaries or shallow marine sequences provide evidence of sea-level change (Gehrels, 1999) and of oceanographical changes in near-shore marine environments (Bergsten, 1994). In deeper waters, they have been used to reconstruct variations in sea-ice cover (Haake and Pflaumann, 1989), and sequences of short-lived oceanographical changes that can be linked with terrestrial records (Asioli *et al.*, 1999). They have been most widely employed, however, in palaeo-oceanographic reconstructions (see 'Marine sediments', below).

Charred particles (charcoal)

A characteristic feature of many Late Quaternary deposits, including lake sediments and peats, as well as fossil soils, is the inclusion of microscopic carbon particles resulting from the burning of wood, grass or other vegetation (Patterson *et al.*, 1987). Charcoal can be seen under a microscope, for example in pollen samples, but as burning episodes create magnetically enhanced mineral particles, its presence can also be detected in sediment

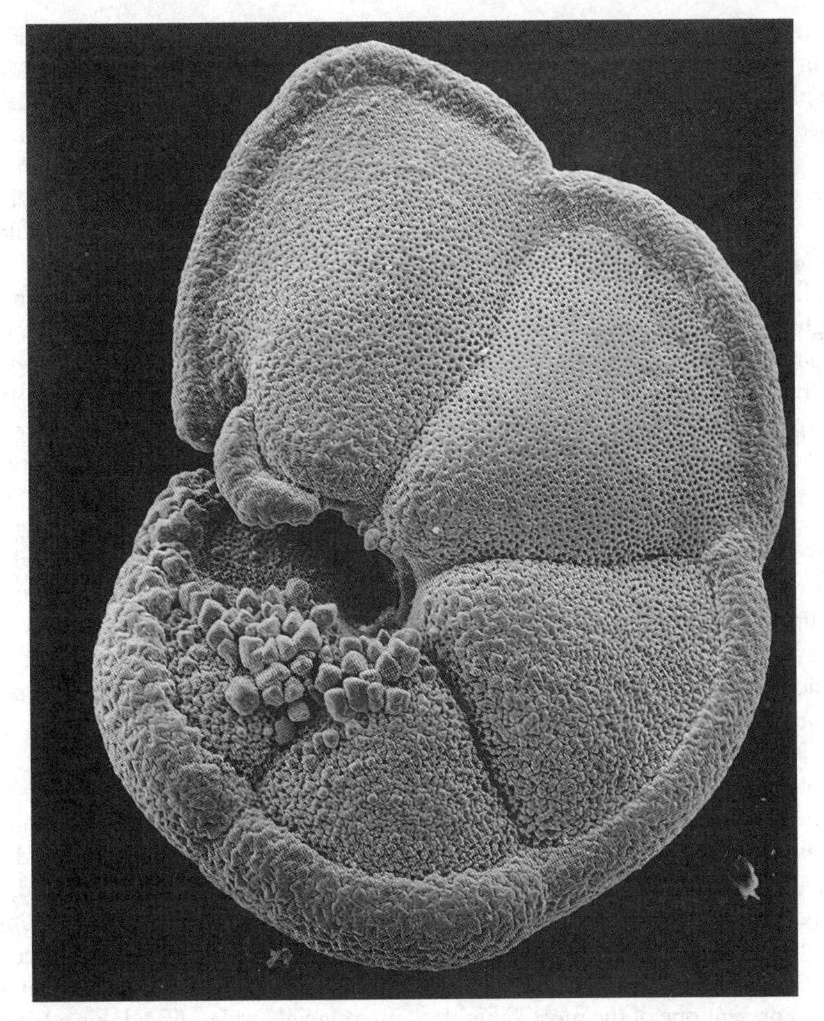

Figure 2.11 Scanning electron micrograph of *Globorotalia menardii*, a planktonic foraminifer characteristic of tropical waters (diameter *c*.700 μm) (photo Brian Funnell)

cores by the measurement of magnetic properties (Rummery, 1983). Most charred particles in Holocene sediments reflect burning by people, but some fires may have begun during droughts or may have been caused by lightning (Tolonen, 1986). When combined with other forms of proxy data (e.g. pollen, plant macrofossil or dendrochronological evidence), the analysis of charred particles adds a valuable additional dimension in studies of land-use history (Mellars and Dark, 1998; Pitkänen *et al.*, 1999).

Sedimentary evidence

Although interest in Late Quaternary sediments has often been stimulated by their fossil content, valuable information about former climatic and environmental conditions can frequently be derived from the nature of the sediments themselves and from the landforms that they comprise. Sedimentary evidence also constitutes a key element in the expanding research area of **geoarchaeology** (Rapp and Hill, 1998). Physical, chemical and biological properties of sediments can be used to make inferences about the environment of deposition, while stratigraphic relationships and contrasts provide evidence of depositional changes through time. Temporal and spatial variations in rates of sediment accumulation may be governed by climate, and hence may constitute a proxy record of climatic change. However, in some situations (e.g. lakes or valley floors) sediment accumulation has been strongly influenced by human activity, in which case the analysis of the sedimentary sequence provides a proxy record of anthropogenically induced landscape change (Chapter 7).

Peat

Peat accumulates in waterlogged localities where the breakdown of vegetal material is reduced by anaerobic conditions. Such areas are known as **mires**, some of which form where drainage is impeded (e.g. enclosed basins or river floodplains), whereas others are initiated and maintained by high atmospheric moisture levels. The latter are termed **ombrogenous mires** and occur as **raised bogs**, domed-shaped accumulations of peat that develop in lowland areas often following the infilling of a lake or pond, and **blanket mires**, namely extensive spreads of peat which cover the landscape in upland areas where rainfall is high (Lowe and Walker, 1997).

Peat deposits represent one of the most valuable terrestrial 'archives' for palaeoecological research (Godwin, 1981), for not only does the peat constitute an ideal medium for the preservation of fossil evidence but, in so far as peat development is closely related to climatic conditions, the stratigraphy of ombrogenous peat profiles represents a proxy record of climatic change (Blackford, 2000; Barber *et al.*, 2000). For example, in many bogs of north-west Europe, distinctive horizons are found separating dark, well-humified peats from overlying light-coloured peats and less humified *Sphagnum* peats (Plate 2.1), these **recurrence surfaces** reflecting a shift from drier to wetter conditions. Peat humification changes (Blackford and Chambers, 1991) and variations in the nature and degree of preservation of plant macrofossil remains found in peat profiles (Barber *et al.*, 1994) reflect changes in mire-surface wetness over time. When such evidence has been obtained from ombrogenous mires, where surface water is purely derived from precipitation, these data form the basis for high-precision palaeoclimatic reconstructions (Chambers *et al.*, 1997; Barber *et al.*, 2003; Figure 2.12). In some peat profiles, evidence of human activity is also preserved. Lenses of minerogenic sediment provide indications of erosion on hillslopes, possibly associated with woodland clearance (Edwards *et al.*, 1991), while ash and dust particles in ombrogenous peats reflect erosion of arable fields by wind and therefore constitute evidence for early agricultural activity (Aaby, 1986). Geochemical data from peat profiles, including down-core variations in silicon and titanium (Hölzer and Hölzer, 1998), lead (Shotyk *et al.*, 1998) and mercury (Martínez-Cortizas *et al.*, 1999), provide further evidence of historic and prehistoric human activity. Blanket bogs and raised mires are considered further in Chapter 6, p. 216.

Lake sediments

Lake basins are natural sediment traps and frequently contain a history of deposition spanning thousands of years. Indeed, in some lake sequences, such as those that accumulated in deep tectonic basins or volcanic craters (**maars**), lake sediment records may extend over several glacial–interglacial cycles (Tzedakis *et al.*, 1997; Reille *et al.*, 2000). As with peat deposits, lake sediments are ideal media for preserving a range of macroscopic and microscopic fossils, but a considerable amount of palaeoenvironmental information can be derived from the nature of the lake sediments themselves. For example, in mid-latitude lake sequences, the climatic amelioration at the end of the last cold stage is represented by the transition from minerogenic to organic deposits. This lithostratigraphic change reflects increased organic productivity within the lake ecosystem and also a reduction in mineral inwash as the catchment slopes became stabilised by vegetation (Lowe and Walker, 1997). Reduced inwashing of soils from around the basin catchments is also reflected in the marked decline in concentration of chemical bases (e.g. Ca, Mg, Na and K) in early Holocene

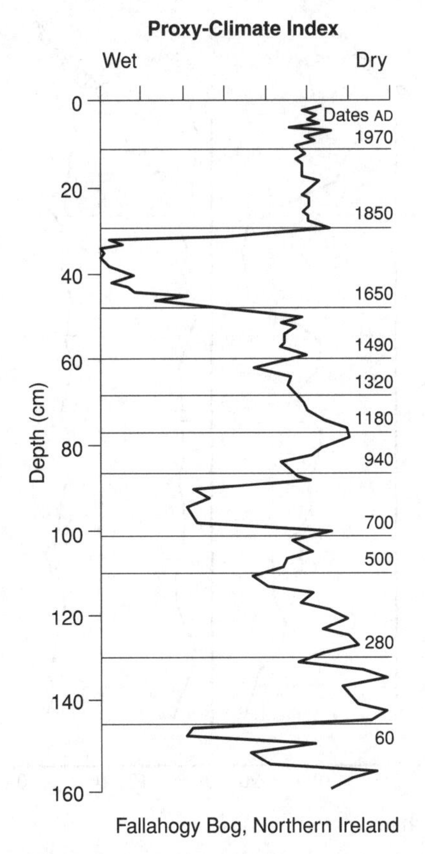

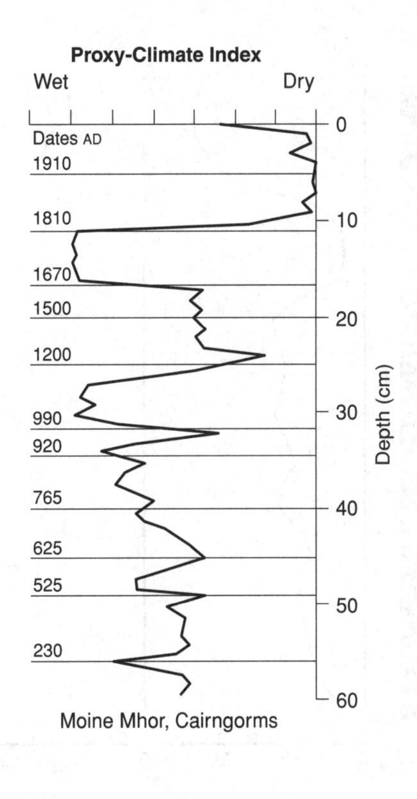

Figure 2.12 Proxy climatic indices based on plant macrofossil data from Fallahogy Bog, Northern Ireland and Moine Mhor, western Cairngorms, Scotland. The indices are based on the results of detrended correspondence analysis (DCA: see Chapter 3, note 6 for an explanation) and indicate relative wetness of the growing mire surfaces. The dates (AD) mark significance points of change in the plant macrofossil records (Barber *et al.*, 2000, Figure 5)

lake sediments. The curves for these bases (Figure 2.13) therefore provide proxy records of lake catchment stability and instability (Walker and Lowe, 1990). Fluctuating water levels in lakes are reflected in abrupt changes in sediment stratigraphy, and where such changes can be shown to be regionally synchronous, they provide a basis for reconstructing past rainfall regimes (Harrison and Digerfeldt, 1993). In mountain regions, variations in minerogenic content in lacustrine sequences from downstream glacial lakes have produced detailed evidence about former glacier activity (Nesje *et al.*, 2000). Evidence of anthropogenic activity is also preserved in many mid- and late Holocene lake sediment sequences with land-use changes, such as woodland clearance, reflected in variations in sediment flux into the lake basin (Dearing, 1991; David *et al.*, 1998). As with other sedimentary records, however, a multi-proxy approach which combines a range of different data sources provides the basis for the most detailed environmental reconstructions from lacustrine sequences (e.g. Walker *et al.*, 2003).

Cave sediments

Caves also form natural sediment traps and contain materials that originate within the caves (**autochthonous**) as well as sediments that are brought in from outside (**allochthonous**). The

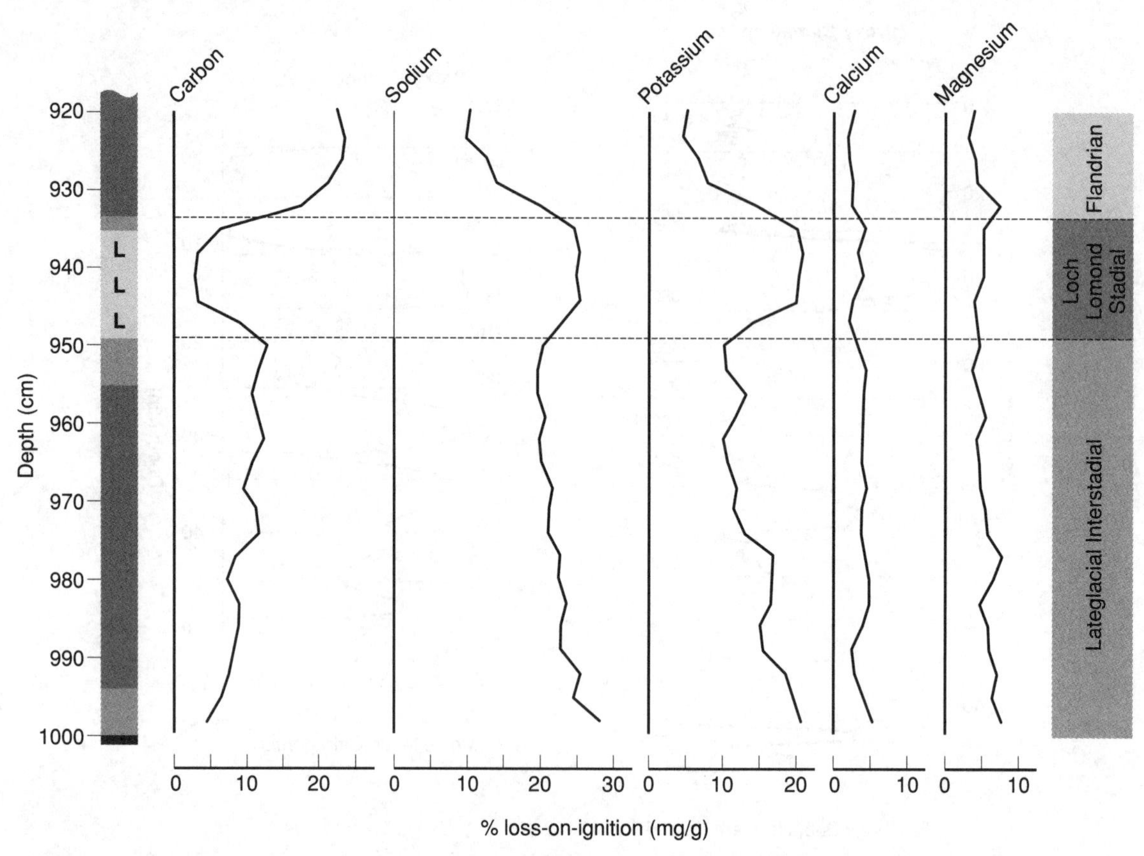

Figure 2.13 Variations in abundance of selected chemical elements in Lateglacial and early Holocene sediments (*c*.14.7–11.5 ka BP) from Druim Loch, Isle of Skye, Scotland (after Walker and Lowe, 1990, Figure 15)

former category includes scree, rock rubble and fine-grained materials (cave earth) derived from the weathering of the cave walls and, in limestone areas, secondary mineral deposits of calcium carbonate which are collectively known as **speleothems**, the most common forms of which are stalagmites, stalagtites, flowstones and tufas (Figure 2.14). Allochthonous materials include fluvial, glacial, colluvial, periglacial and aeolian deposits. Caves contain valuable information on environmental change. In many caves, a palaeoclimatic signal may be represented by distinctive suites of micromorphological features within the cave sediment record, by variations in the input of allochthonous materials, or by fluctuations in the mineral magnetic properties of the fine sediment fraction (Woodward and Goldberg, 2001). Speleothems (p. 43) are an additional palaeoenvironmental indicator, as speleothem formation appears to be associated with periods of warmer climate (Atkinson *et al.*, 1986). Caves were favoured by both wild animals and early humans, and so often contain abundant vertebrate remains and artefacts (Figure 2.6). In addition to vertebrates, other fossils have been recovered from cave contexts, including molluscs (Goodfriend and Mitterer, 1993) and pollen (Carrión *et al.*, 1999). Detailed climatic information can also be obtained from the isotopic record in cave speleothems (see below), while the stratigraphic record of cave deposits can provide valuable data on both natural environmental change (Valen *et al.*, 1996) and on the human occupation of cave sites (Wattez *et al.*, 1989).

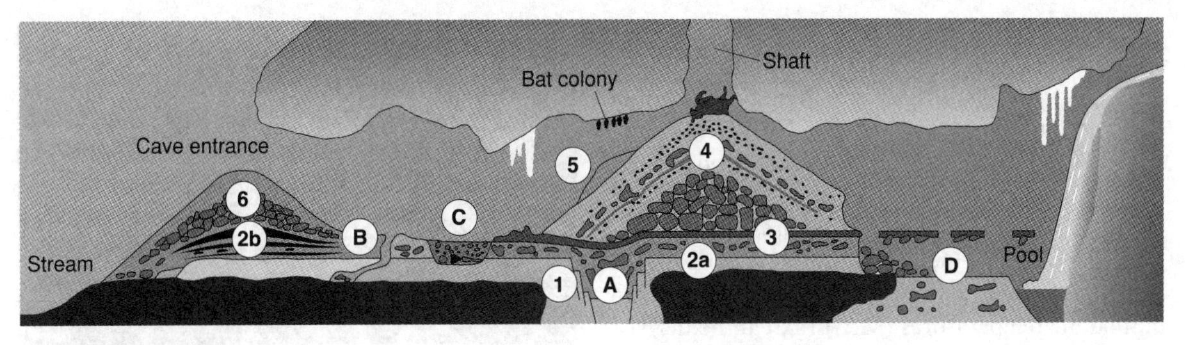

1. Water-laid sands and clays
2a. Deposits of an animal lair
2b. Hearths of fires made by people
3. Stalagmite floor
4. Talus cone with bones of animals which fell down a shaft
5. Bones and dung of bats
6. Second talus cone at cave mouth

A. Collapse pit
B. Burrow
C. Human burial
D. Washing out and redeposition by stream

Figure 2.14 Vertical section through an imaginary bone cave, illustrating some important types of cave deposit (after Sutcliffe, 1970)

Glacial sediments

Glacial sediments cover large areas of the mid-latitude regions of the world, forming an intermittent blanket over one-third of the land area of Europe and around half of the continent of North America. The geographical distribution of these glacially derived sediments and their landform assemblages provide evidence of the former extent of the great ice sheets and mountain glacier complexes that developed during the cold stages of the Quaternary (Benn and Evans, 1998). Moreover, patterns of former ice movement can be inferred from the physical and chemical properties of the sediments, and from the orientation or alignment of landforms produced by both glacial erosion and deposition (Lowe and Walker, 1997). Although around two-thirds of the global ice volume present during the Quaternary cold stages disappeared during the Holocene, active glaciers remain in many high-latitude and high-altitude regions of the world, and the distribution of glacigenic sediments and landforms (especially moraines) in those areas reflects glacier fluctuations during the course of the present interglacial (Davis and Osborn, 1988). Furthermore, an appreciation of the former extent of glacier ice, allied to glaciological principles derived from the study of contemporary ice sheets and glaciers, has enabled increasingly sophisticated modelling of former ice sheets and glaciers (Siegert, 2001). Such glaciological reconstructions not only provide evidence of the behaviour of Quaternary ice masses, but they allow inferences to be made about former climatic conditions (Dahl and Nesje, 1992; Ballantyne, 2002). They also provide a basis for predicting ice sheet behaviour under different scenarios of future climate change (Sugden and Hulton, 1994).

Periglacial deposits

The term 'periglacial' is widely used to refer to those high-latitude and high-altitude regions of the world where frost action constitutes the dominant geomorphological process. Cyclic freeze–thaw activity, the growth of ground ice, and the presence in many (but not all) periglacial environments of permanently frozen ground (**permafrost**), leads to the development of a suite of highly distinctive deposits, sedimentary structures and landforms (French, 1996). Moreover, the sparse vegetation cover that is characteristic of much of the periglacial domain means that aeolian and fluvial activity are also highly effective geomorphological processes. Relict periglacial phenomena from the cold stages of the Quaternary are found throughout the mid-latitude regions of the Northern Hemisphere, and constitute unequivocal evidence of climatic change (Ballantyne and Harris, 1994). By using modern analogues from present-day periglacial environments, quantitative estimates of former climatic conditions, particularly mean annual air temperatures (MAATs), can be derived from relict periglacial phenomena (Table 2.1). Of particular value in this respect are **ice wedge casts** (sedimentary infillings of thermal contraction cracks in the former permafrost surface: Figure 2.15), **sand-wedge casts** (aeolian infillings of thermal contraction cracks: Plate 2.2), **frost cracks, cryoturbation structures** (including **involutions**, contortions in sediments produced by the action of ground ice: Plate 2.3), and **frost mound remnants** (Vandenberghe and Pissart, 1993). These have been used to infer former MAATs as well as permafrost distributions (Huijzer and Vandenberghe, 1998). As with other lines of evidence, however, periglacial features have proved to be most useful when combined with other lines of data in multi-proxy approaches to palaeoclimate reconstruction (Isarin *et al.*, 1997; Huijzer and Vandenberghe, 1998).

Figure 2.15 Fossil ice wedge cast exposed in a road cutting near Aberystwyth, west Wales (photo Mike Walker)

Slope deposits

A range of sediments occurs on hillslopes and in valley bottom situations as a result of slope processes. These include **head** deposits that develop under periglacial conditions (French, 1996), **talus** or **scree** deposits which may also be periglacial in origin, landslide debris, and colluvial and solifluction deposits (Figure 2.16) that are more characteristic of erosion under temperate climatic regimes (Rice, 1988). Geomorphological processes on hillslopes are closely related to changes in vegetation cover, precipitation and temperatures, and hence a palaeoenvironmental record relating to climate and land-use change may be preserved in stratified sequences of hillslope sediments (Chapter 7; Bell and Boardman, 1992). Evidence from colluvial sequences, particularly of debris flows and avalanches, are increasingly being employed as indicators of short-lived Holocene climate changes (Frenzel *et al.*, 1993; Figure 2.17), and may provide a unique source of information on extreme climatic events such as heavy snowfall or intense rainfall (Matthews *et al.*, 1997). Colluvial sequences may also contain fossil materials (e.g. molluscs), which again form a basis for palaeoenvironmental reconstructions (Preece *et al.*, 1995).

Alluvial deposits

The investigation of lake sediments (see above) is one aspect of the science of **palaeohydrology**, i.e. the study of water and sediment dynamics in the past. The other element of palaeohydrological investigations is concerned with changes in river erosion and deposition, and with temporal fluctu-

Table 2.1 Climatic significance of periglacial evidence as expressed by the mean annual air temperature (MAAT) and the mean temperature of the coldest month (after Huijzer and Vandenberghe, 1998)

Periglacial phenomena	*Climate information*	
	Mean annual air temperature (MAAT) in °C	*Mean temperature of the coldest month in °C*
Thermal contraction cracks		
Ice-wedge cast, fossil sand wedge, composite-wedge cast	fine-grained substrate: ≤–4 coarse-grained substrate: ≤–8 (to –6)	≤–20
Seasonally frozen ground, soil wedge with primary (or secondary) infilling	≤–1 to 0	≤–8
Periglacial involutions		
Type 2 Large-scale (amplitude ≥0.6 m) down sinking or up doming forms	fine-grained substrate: ≤–4 coarse-grained substrate: ≤–8 (to –6)	
Type 3 Small-scale (amplitude <0.6 m) down sinking or up doming forms	≤–1	
Type 4 Solitary forms of variable amplitude in drops or diapirs	≤–1	
Perennial frost mounds		
Open-system pingo	≤–3 to ≤–1	
Closed-system pingo	≤–6 to ≤–4	
Palsa	organic: ≤–1 mineral: ≤–6 to –4	
Cryogenic microfabrics		
Banded fabrics	≤–1 to 0	
Lenticular platy microstructures	≤–1 to 0	
Cryogenic microfabrics in cave deposits	≤0	

ations in river regimes linked to climate change (Brown and Quine, 1999). Staircases of terraces in many river valleys of the temperate zone reflect episodes of aggradation (sediment accumulation) and incision (downcutting) in response to climate change (Vandenberghe *et al.*, 1994; Fuller *et al.*, 1998). These terrace sedimentary sequences frequently provide the stratigraphic context for archaeological finds of Palaeolithic (Wymer, 1999) and Holocene (Waters, 1996) age. In many areas, however, there are indications that land uplift has also stimulated incision and subsequent terrace development, and hence river terraces can provide useful data on crustal movements (Maddy, 1997). Over recent timescales, however, Late Holocene alluviation is often more closely related to patterns of human activity (Macklin, 1999; French, 2003) and, as was the case with lake sediments, alluvial sequences may also preserve a record of land-use changes (Chapter 7). Indeed, river valleys have long been centres of human occupation and their sediments frequently contain well-preserved artefactual evidence, along with other fossil remains such as pollen, plant macrofossils, molluscs and vertebrates (A.G. Brown, 1997; Bridgland, 2000).

Figure 2.16 Colluvial sediments of mostly prehistoric Bronze Age date in the Cuckmere Valley, England (photo Martin Bell)

Aeolian deposits

Aeolian activity in the temperate zone is now largely confined to coastal areas and to upland regions where wind erosion remains an important agent of geomorphological activity (Ballantyne and Harris, 1994). During the Quaternary cold stages, however, extensive spreads of **loess**[2] materials accumulated in the mid latitudes of the Northern Hemisphere from the 'Loess Plateau' of China, through the loess belt of Europe which extends from northern France to the Ukraine, to the Great Plains of North America. In addition, in both western Europe and interior North America, there are extensive spreads of **coversands**,[2] which form largely featureless surfaces which are occasionally fashioned into undulating dunes (Lowe and Walker, 1997). These wind-blown materials were derived from poorly vegetated periglacial regions, from glacial and outwash sediments, and from the continental shelves exposed by falling sea levels (Madole, 1995). In areas such as the Chinese Loess Plateau and central Europe, sequences of loess deposits and interbedded palaeosols (see below) provide valuable archives of long-term climate change, the fossil soils reflecting warmer (interglacial) episodes while the wind-blown materials accumulated during cold (glacial) stages (Derbyshire, 1995, 2003). However, climatic and environmental changes over much shorter timescales may also be detectable in aeolian sediment sequences (Olsen *et al.*, 1997; Vandenberghe *et al.*, 1998a). Aeolian sediments are also valuable indicators of former wind direction, and are important data sources for reconstructing and modelling past climates (Isarin and Renssen, 1999). In north-west Europe, coastal dunes which have continued to form throughout the Holocene are rich sources of

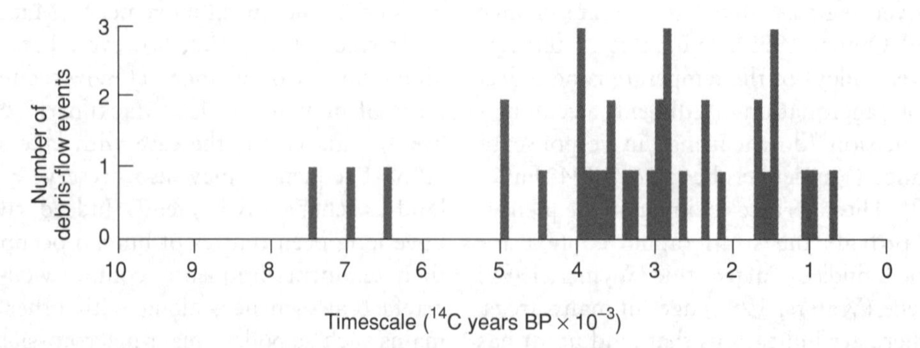

Figure 2.17 Temporal patterns of Holocene debris-flow events in southern Norway (after Matthews *et al.*, 1997, Figure 7c)

molluscs, bones and artefactual material (Huddart *et al.*, 1999), while well-preserved archaeological sites, sometimes with fields and cultivation marks, are found around many parts of coastal Europe (Figure 2.18).

Palaeosols

Palaeosols may be either **buried soils** below sediments (e.g. alluvium, colluvium, landslip deposits) or below archaeological monuments (Plate 2.4), or they may be currently exposed **relict soils** preserving some characteristics of former pedogenic regimes. A considerable amount of useful data has been obtained from buried soils in particular. By using modern soils as analogues, inferences can be made about former climatic and environmental conditions under which palaeosols developed (Bronger *et al.*, 1995). **Soil micromorphology**, examination in thin section of the distinctive arrangement of particles and voids that comprise the **soil fabric**, provides additional evidence of the history of soil development and related environmental changes (Kemp, 1998; French, 2003). This technique is increasingly being applied to the analysis of occupation horizons on archaeological sites, and is of particular value in the identification of microtraces of successive activity patterns, and thus in the social use of space (Matthews *et al.*, 1997). It can also offer insights into cultivation histories and other farming practices (Davidson and Simpson, 2001). Pedological data can be further augmented by palaeoecological evidence. Calcareous soils often contain fossil molluscs (Carter, 1990), while pollen grains are preserved in acid soils such as podzols or brown earths (Dimbleby, 1985). Soil pollen analysis can provide valuable information on regional vegetation changes (Caseldine, 1984), and although this technique has perhaps proved most useful in archaeological contexts where, for example, soils have been discovered beneath field monuments (Molloy and O'Connell, 1993), soil pollen analysis has considerable potential in broader-scale inquiries of settlement and the wider landscape (Whittington and Edwards, 1999). Pedogenic processes occurring during the Holocene are considered further in Chapter 6, p. 212.

Figure 2.18 Bronze Age site within wind-blown sand at Bjerre, North Jutland, Denmark, showing marks of ard furrows (photo Jens-Heinrich Bech, Museet for Thy og Vester Hanherred)

Coastal deposits and landforms

Relict coastal deposits and landforms both above and below the contemporary shoreline provide evidence of past changes in land and sea level (Figure 2.19). A record of former episodes of low sea level can be found in present offshore areas in the form of submerged coastal landforms (cliffs, caves, reefs, platforms, spits, shingle bars and river valleys), terrestrial landforms and deposits now covered by the sea (Stoker and Holmes, 1991), and the remains of terrestrial fauna and flora recovered from the seabed and from boreholes in the current sea floor (Ekman and Scourse, 1993). Evidence for former higher sea levels includes the presence of such erosional features as clifflines, caves and marine abrasion platforms at altitudes

Figure 2.19 Beaches above present day sea level in Ertebølle, Denmark. These have been raised above present sea level by glacio-isostatic uplift [Chapter 4] and therefore provide evidence of past land/sea changes. The photograph also shows the site of a Mesolithic shell midden: see also Figure 5.13 (photo E. Johansen, courtesy of S.H. Andersen)

higher than the contemporary shoreline, and also the occurrence of marine deposits (beach gravels, sand spits and marine clays) in situations above present-day sea level (Davies and Keen, 1985). A history of sea-level change can be reconstructed in those areas where a record of marine and terrestrial sedimentation has been preserved. Archaeological sites of many periods are found within coastal sedimentary sequences (Chapter 5), and these can make an important contribution to understanding the nature of past human coastal exploitation and the dating of coastal environmental change (Bell *et al.*, 2000; Pye and Allen, 2000). Where marine deposits (e.g. silts) are overlain by terrestrial sediments, a decrease in marine influence is apparent and this is described as a **negative sea-level tendency**. In the contrasting situation, where marine sediments overlie terrestrial deposits, the marine influence is increasing and a **positive sea-level tendency** is recorded (Shennan *et al.*, 1994). Such evidence was previously interpreted in terms of **marine regressions** and **transgressions** (falls and rises in sea level). However, these terms are now less widely used as they imply 'absolute' (i.e. worldwide or **eustatic**) changes in sea level, whereas only 'relative' sea-level tendencies, in other words changes in sea level arising from local land/sea-level movements, can really be inferred from these records (Lowe and Walker, 1997). The Holocene sequence of sea-level change around the coasts of the British Isles and north-west Europe (Chapter 3) has been reconstructed largely on the basis of this type of stratigraphic evidence (Shennan *et al.*, 2000b).

Marine sediments

In the deep oceans of the world, marine sediments have been accumulating in a relatively undisturbed manner for hundreds of thousands, or even millions, of years, and one of the major achievements of Quaternary science in the second part of the twentieth century was the development of a technology that enabled cores of ocean sediment to be recovered from the ocean floor, sometimes in water depths of more than 3 km. Microfossils within these sediments, principally Foraminifera, coccolithophores, radiolaria and diatoms, provide the basis for palaeo-oceanographic reconstructions of both surface and deep-water changes throughout the Quaternary (Haslett, 2002). These data provide an important input to global circulation models (GCMs) of long-term climate change (Wright *et al.*, 1993). In addition, Foraminifera contain a record of temporal variations in isotopic composition of ocean waters over successive glacial–interglacial cycles, which comprises an additional proxy record of long-term climate change (p. 42).

Ice cores

Of comparable importance in Quaternary science have been the technological advances that have made it possible to obtain a continuous core record from ice sheets and glaciers (Plate 2.5). Although valuable palaeoenvironmental and palaeoclimatic data have been obtained from low-latitude ice sheets (Thompson, 2000), the majority of research has been focused on the polar ice sheets. Key sites in Greenland include those on the ice sheet summit, the Greenland Ice Core Project (**GRIP** and **NorthGRIP**)[3] and the Greenland Ice Sheet Project (**GISP2**)[4], while in Antarctica perhaps the most important drill site is that at Vostok Station. Records from Greenland extend back through the last interglacial, in other words beyond 120 ka BP (Dansgaard *et al.*, 1993), while in Antarctica, a continuous 420 ka record has been obtained (Petit *et al.*, 1999). An even longer Antarctic core, extending back beyond 500 ka, is currently being drilled (**EPICA, Dome C**), and may eventually reach ice that is more than 800 ka old (Wolff, 2002). These ice core records, which can be dated by incremental and other methods (see below), provide evidence, *inter alia*, of changes in atmospheric trace gases (Figure 2.20), atmospheric dust and volcanic aerosols (Hammer *et al.*, 1997). They also contain evidence of human activity, ranging from pollution from Roman and medieval copper smelting

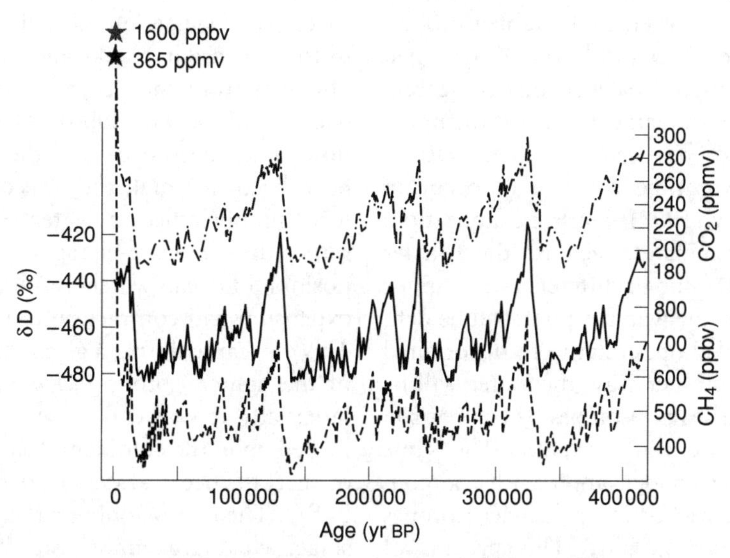

Figure 2.20 The Vostok ice core record showing fluctuations in greenhouse trace gases (CO_2 and CH_4), plotted along with variations in δD content (see below), over the last four climatic cycles. The present-day Antarctic CO_2 (365 ppmv) and CH_4 (1600 ppbv) values are also indicated (from Raynaud *et al.*, 2000, Figure 4)

(Hong *et al.*, 1996) to the dramatic changes in atmospheric greenhouse gas content resulting from industrial activities over the past 200 years (Raynaud *et al.*, 2000). Quantified estimates of former temperatures have been obtained from analysis of the gas content in the ice cores (Severinghaus *et al.*, 1998), and from oxygen isotope profiles (Johnsen *et al.*, 2001). In addition, the ice sheets contain an oxygen isotope record which can be linked to that obtained from sediments from the deep oceans, a topic that is considered in the next section.

Isotopic evidence

Isotopes are atoms of an element that are chemically similar but have different atomic weights. It now appears that almost all of the known elements have more than one isotope; hydrogen, for example, has two, whereas carbon and oxygen have three. Indeed, although there are 92 naturally occurring elements, some 270 naturally occurring isotopes of these elements are known to exist (Gray, 1981). Not all isotopes are chemically stable and from a number there is a spontaneous emission of particles. These are the **radioactive** isotopes, several of which are used in radiometric dating (pp. 53–7). However, some **stable** isotopes, most notably oxygen (^{18}O and ^{16}O) and hydrogen (H and D), are also used in palaeoclimatic research. These isotopes are fundamental constituents of water, but the ratios between them (i.e. $^{18}O/^{16}O$; H/D) will vary over time as changes occur in water state (Bradley, 1999). During evaporation, for example, discrimination against the heavier isotopes (^{18}O and D) due to differences in vapour pressure means that the water vapour will be deficient in the heavier isotopes relative to the original water source. Put another way, the water will be *relatively enriched* in heavier isotopes by comparison with the atmospheric water vapour. The same effect occurs during condensation with the condensate becoming enriched in the heavier isotopes relative to the water vapour.[5] This process of **isotopic fractionation** is controlled, *inter alia*, by temperature, and thus from an analysis of the isotopic ratios contained within fossils, peat, or glacier ice, it is possible to reconstruct the sequences of isotopic changes that have taken place, and thus derive a proxy record of former climatic conditions.

Microfossils in deep-ocean sediments

Micro-organisms that live in the oceans, such as the Foraminifera (see above), preserve in their skeletal remains a record of the isotopic composition of the ocean waters at the time they were alive. Hence, by analysing the isotopic content of microfossil remains down a core of deep-sea sediment (Figure 2.21), a record can be obtained of changes in the oxygen isotopic composition of ocean waters over time. Such records may extend back 2–3 million years (Shackleton *et al.*, 1990).

It was initially believed that down-core variations in the isotopic signal in foraminiferal remains reflected changes in ocean water temperatures and, indeed, some estimates of former sea-surface temperature changes have been obtained from **planktonic** (top 50 m of the water column) marine microfossils (Ruddiman *et al.*, 1986a; Duplessy *et al.*, 1992). However, although planktonic organisms are governed by a range of influences, including temperature and salinity variations (Patience and Kroon, 1991), data from different parts of the world's oceans show a broad similarity over successive glacial–interglacial cycles between the isotopic record in planktonic microfossils, and that obtained from the deeper water (**benthic**) forms (e.g. Cortijo *et al.*, 2000). As temperature variations in the deep oceans of the world are likely to have been minimal, these records do not, therefore, simply reflect a water temperature record; rather they indicate changes in the isotopic composition of ocean waters resulting largely from the expansion and contraction of the great ice sheets. This is because during a glacial stage, large amounts of the lighter isotope ^{16}O would have been preferentially removed from the ocean systems and locked up in the continental ice sheets. As a consequence, the ocean waters were relatively 'enriched' in ^{18}O. The reverse obtained during an interglacial stage, when large amounts of ^{16}O-rich meltwaters were released back into the world's oceans. Hence, rather than reflecting a temperature proxy, *sensu stricto*, the isotopic trace constitutes a **palaeoglacia-**

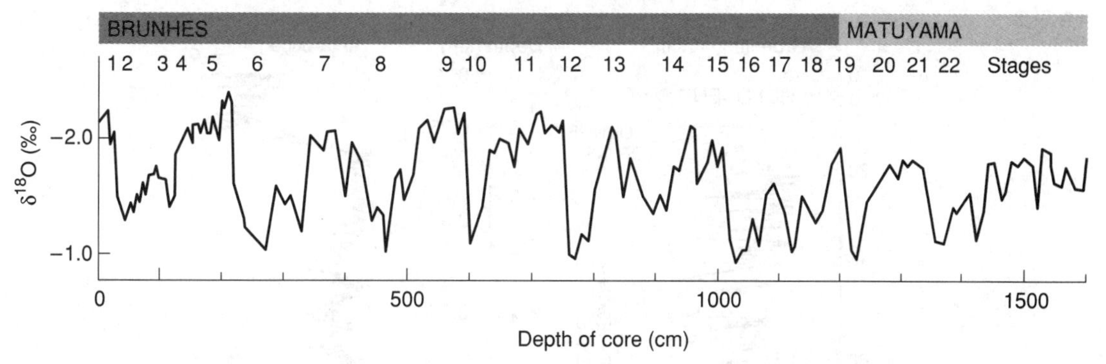

Figure 2.21 Oxygen isotope trace from deep-ocean sediment core V28–238. The Brunhes–Matuyama geomagnetic boundary has been dated to *c*.735 ka BP. Odd-numbered isotopic stages represent warm (interglacial) episodes, while even-numbered stages indicate cold (glacial) stages (after Shackleton and Opdyke, 1973, Figure 9). See Chapter 3 for explanation

tion record. In so far as it reflects the major changes in land-ice volume, however, the deep-ocean oxygen isotope signal represents a unique climate proxy for glacial–interglacial cycles. Moreover, as the marine isotopic trace can be read as an index of changing land-ice volumes, it is also a record of eustatic sea-level change, since an expansion of global ice must result in a concomitant fall in eustatic sea level (Shackleton, 1987).

Ice cores

Preserved within the ice sheets is a continuous record of changes in the ratios of oxygen and hydrogen isotopes in precipitation falling on the ice sheets (Figure 2.22). Because the fractionation of these isotopes is largely temperature dependent, down-core variation in the isotopic profiles reflects variations in global climate. These changes broadly parallel the isotopic shifts recorded in ocean core records (Bond *et al.*, 1993), but the level of stratigraphic resolution is markedly higher in the ice cores. Indeed, very high-resolution isotopic records are now available, particularly from Greenland, which enable climatic changes to be detected over time intervals measurable in individual years (Stuiver and Grootes, 2000). Although not without problems (Jouzel *et al.*, 1997), the oxygen isotope signal has also been used as a means of reconstructing past temperatures (Cuffey and Clow, 1997).

Speleothems

Speleothems are mineral deposits found in limestone caves and are composed largely of calcium carbonate that has been precipitated from cave waters. In deep caves speleothems tend to form in isotopic equilibrium with their parent seepage waters, and hence layers of calcium carbonate will contain a record of the isotopic composition of cave waters throughout the period of speleothem accumulation (Gascoyne, 1992). As the $\delta^{18}O$ ratio in cave waters is closely controlled by temperature, the isotopic trace through a section of cave speleothem provides a proxy record of climatic change (Lauritzen, 1995: Figure 2.23). The method has been used to reconstruct patterns of climatic change over timescales spanning thousands of years (Lauritzen and Lundberg, 1999a; McDermott *et al.*, 2001), but also offers the potential for detecting climatic variations in the recent past on an annual timescale (Baker *et al.*, 1993).

Tree rings

Variations have been recorded in the stable isotope ratios of oxygen, carbon and hydrogen in tree-ring cellulose. In part, these reflect fluctuations in the isotopic content of precipitation which are largely determined by former temperature levels. As a consequence, relationships can be established between $\delta^{18}O$, $\delta^{13}O$ and δD and climate, a field of

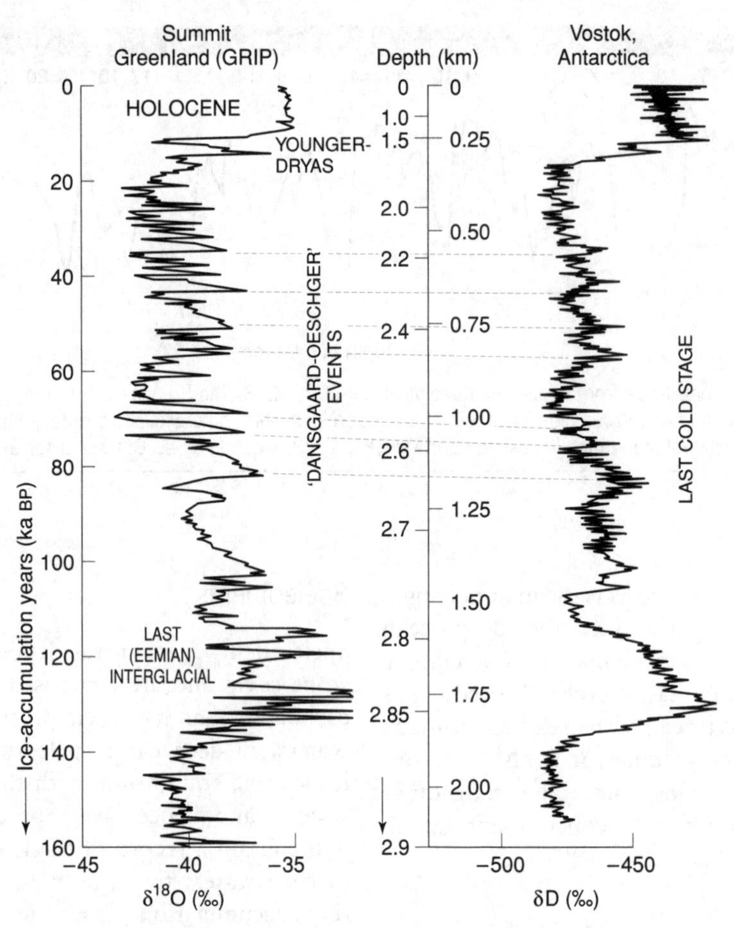

Figure 2.22 Stable isotope variations during the past 160 ka recorded in the GRIP Greenland Summit core (oxygen isotopes) and the Vostok core from Antarctica (deuterium ratios) (after Peel, 1994)

research known as **isotope dendroclimatology**. The technique has been used to infer changes in former temperatures (Sonninen and Jungner, 1995), precipitation regimes and relative humidities (Loader and Switsur, 1995), while high-resolution studies involving isotopic analysis across individual tree rings suggest that this approach can also be used to obtain seasonal climatic information, including extreme and short-lived climatic events (Loader *et al.*, 1995). However, isolating the principal controls on stable isotopes in tree-ring sequences is not always straightforward, and is perhaps the major obstacle to deriving reliable palaeoclimatic data from these records (McCarroll and Pawellek, 2001).

Peat

Plant cellulose material in ombrotrophic peat bogs contains a record of $\delta^{18}O$ and δD in former precipitation, which can be interpreted as a proxy record of climatic change (Brenninkmeijer *et al.*, 1982), and the method has been used to reconstruct variations in temperature and precipitation during the course of the Holocene (Dupont, 1986). Although problems have been encountered in isolating the temperature effects from other variables in the isotopic signal (van Geel and Middledorp, 1988), recent improvements in the measurement of oxygen isotope ratios in plant cellulose suggest that the method has considerable potential for

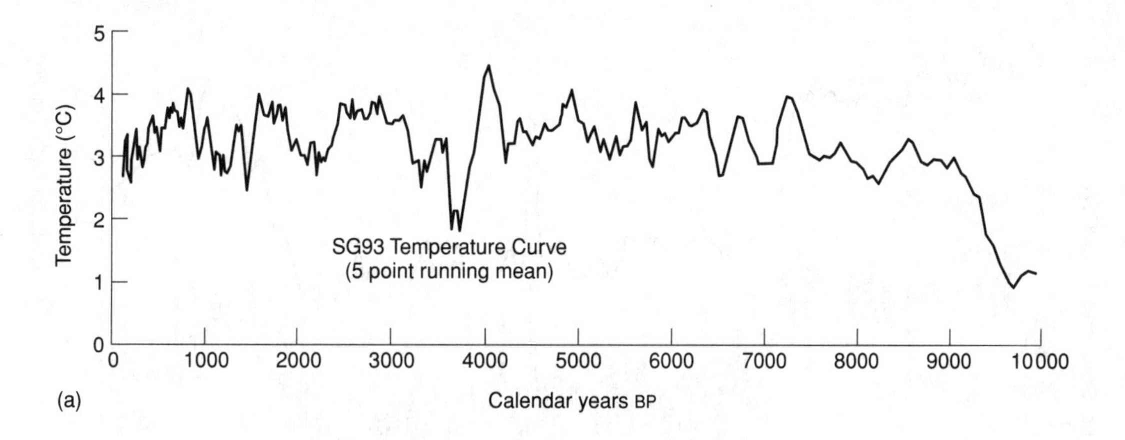

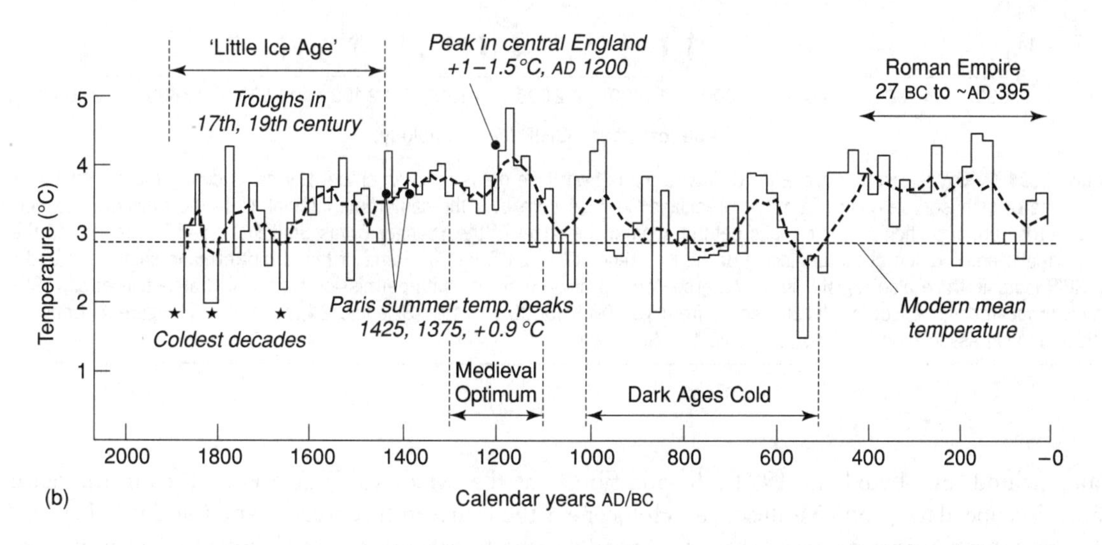

Figure 2.23 Temperature curves based on the oxygen isotope record in a speleothem from Dunderlandsdalen, northern Norway: (a) the curve for the last 10 ka; (b) the curve for the last 2 ka compared with known 'historic' events (after Lauritzen and Lundberg, 1999b, Figure 10. Reproduced by permission of Hodder Arnold)

reconstructing climate changes over the timescale of the Holocene (Saurer *et al.*, 1998; Hong *et al.*, 2000).

Lake sediments and other isotopic studies

In lakes where large numbers of submerged aquatic plants and algae use dissolved CO_2 for photosynthesis, an insoluble carbonate (marl) is precipitated, the oxygen isotope content of which is related to the isotopic composition of the lake waters at the time of precipitation. As the $\delta^{18}O$ variations of lake waters will be related to former precipitation and temperature levels, analysis of the isotopic record of lake carbonates provides a potential means of reconstructing past climates and environments, often at high levels of resolution (Jones *et al.*, 2002). Isotope/climate records can also be obtained from contained fossils, such as ostracods (Figure 2.24). The technique has also been employed in multi-proxy investigations of lake sediment records, in which environmental reconstructions have been based on both chemical and biological evidence (Lotter *et al.*, 1992;

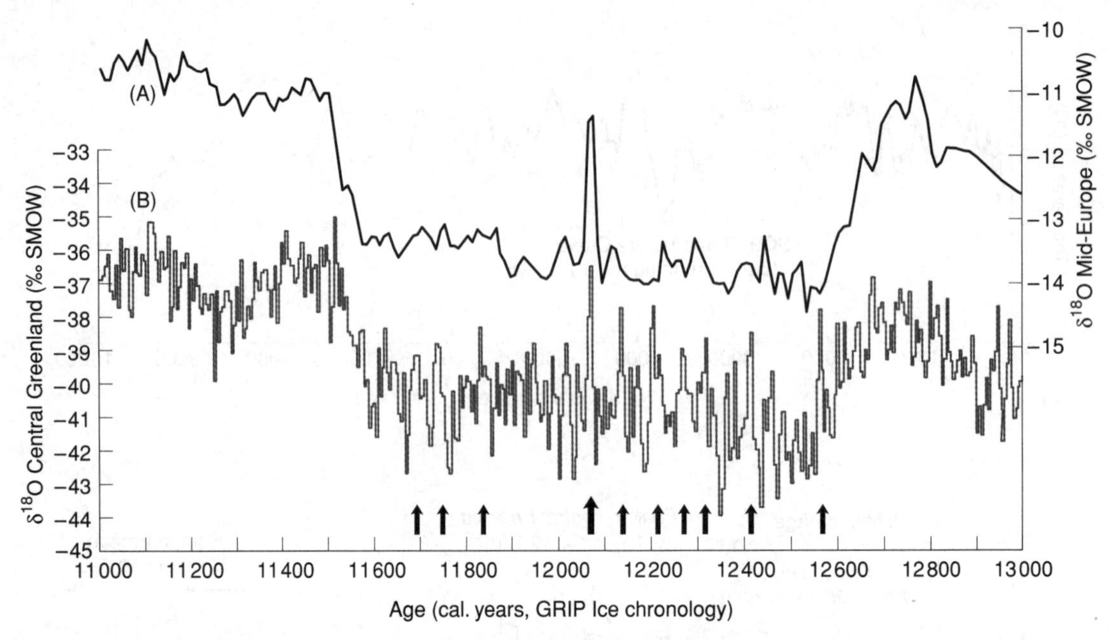

Figure 2.24 (A) Oxygen isotope trace ($\delta^{18}O$) in ostracods from lake marls from Amersee, Germany, during the period 13-11 ka BP. (B) The oxygen isotope record from the Greenland GRIP ice core over the same time interval. Note the similarity between the two records and, in particular, the climatic deterioration at the onset of the Younger Dryas Stadial (12.8–12.6 ka BP) and the rapid climatic amelioration at the beginning of the Holocene (11.6–11.5 ka BP). Some of the abrupt climatic shifts recorded in the GRIP isotopic trace also register in the Amersee $\delta^{18}O$ profile (Reprinted with permission from von Grafenstein *et al.*, 1999. A mid-European decadal isotope-climate record from 15,500 to 5000 years BP, *Science*, **24**, pp. 654–57, Figure 4. Copyright 1999 American Association for the Advancement of Science.)

Hammarlund and Lemdahl, 1994). In addition, stable isotope data from Mollusca in Holocene lakes reveal environmental trends that are broadly consistent with those reflected in pollen records from the same profile (Bonadonna and Leone, 1995). Isotopic evidence is also of great importance in the study of past diet and thus the history of plant and animal domestication (p. 150). Isotopic fractionation takes place within the food chain, giving rise to distinctive isotopic signatures for aspects of past diet: marine foods, terrestrial foods, plant foods, animals foods, etc. (Sealy, 2001; Richards *et al.*, 2003a).

Historical evidence

Documentary records constitute a valuable source of information on past climates, particularly for the last millennium during the episodes known as the 'Medieval Warm Period' from the tenth to the fourteenth centuries, and the 'Little Ice Age' of the fourteenth to the eighteenth centuries (Grove, 2002). Three types of data source have been employed: observations on weather phenomena, observations relating to natural phenomena closely controlled by weather, and phenological records (Bradley, 1999).

Weather records

These are observations made on specific weather phenomena such as unusually warm periods, exceptional snowfall, the duration of frosts, great storms, etc., and can be found in such diverse sources as diaries, annals, chronicles, letters, sagas, personal papers, and administrative and commercial records (Lamb, 1977). Some early Chinese documentary sources date back to the second century BC, intermittent records exist from Greece from about 500 BC, while in northern and western Europe documentary records become increasingly common from around AD 1100 (Ingram *et al.*, 1981; Pfister *et al.*, 1998). From the sixteenth century onwards, there is a marked increase in maritime records (ships' logs, etc.), while the papers of the great trading and exploration companies (East India companies, Hudson's Bay Company) also provide material for climatic reconstruction (Catchpole and Faurer, 1983; Wilson, 1985). Early newspaper reports on unusually severe winters or periods of heavy snow (e.g. Pearson, 1973) are also useful.

Documentary sources have been used to reconstruct climate (both in terms of temperature and precipitation) over large areas of Europe from the Middle Ages onwards (e.g. Ogilvie, 1992; Pfister, 1992). Where documentary sources are available over a number of years, it is possible to develop time-series of climate change. Individual months are allocated an index according to available information, these indices relating to maximum surfeits and deficits for temperature and precipitation (Pfister, 1992). Using this approach, temporal patterns of of historical climate change can often be reconstructed in considerable detail (e.g. Pfister *et al.*, 1998; Figure 2.25).

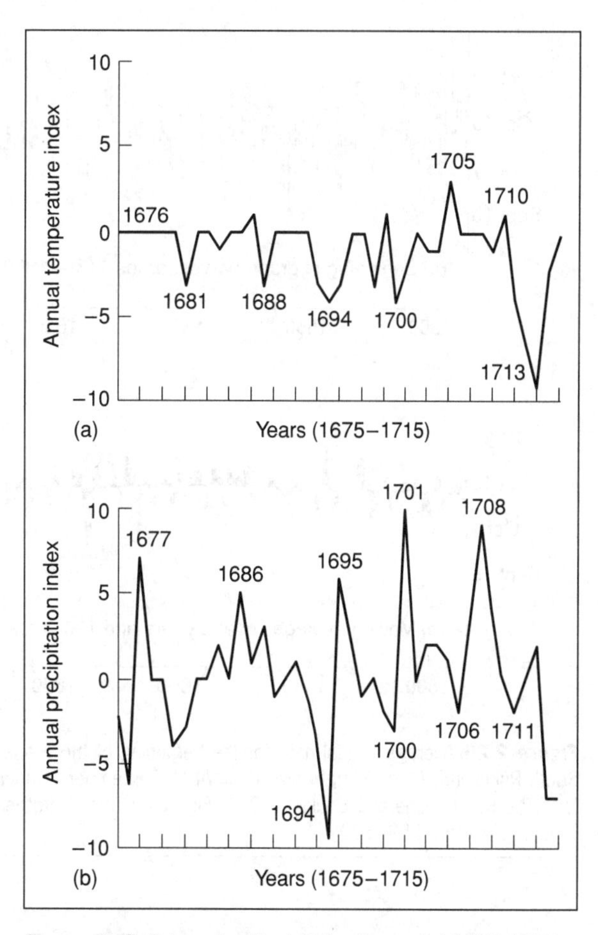

Figure 2.25 Palaeoclimatic data for the period 1675–1715 inferred from documentary sources from southern Portugal: (a) annual temperature index; (b) annual precipitation index (after Alcoforado *et al.*, 2000, Figure 2. Reproduced by permission of Hodder Arnold)

Weather-dependent natural phenomena

These include droughts, floods, duration of freezing of rivers or estuaries, the timing of ice breakup, and the movement of alpine glaciers. As all of these are governed by climate, they are sometimes called **parameteorological** phenomena (Bradley, 1999), and constitute a further proxy data source. Details of such phenomena are to be found in ancient inscriptions, documents and pictorial records. Examples include the remarkable 400-year record of the opening of the port of Riga which provides evidence of the extent of Baltic Sea ice (Lamb, 1995); the use of adminstrative records extending back to 1634 which detail the times of freezing on Dutch canals, and which therefore provide an index of winter severity (van den Dool *et al.*, 1978); and the employment of ecclesiastical sources to infer weather extremes (droughts, floods, etc.) during the later seventeenth century in southern Portugal (Alcoforado *et al.*, 2000).

Phenological records

These relate to the timing of recurrent biological phenomena, and include crop harvest records, flowering and fruiting of plants, and the timing of animal migrations. Where data are available over a number of years, such proxy records constitute time-series data of climatic change. For example, data on crops yields have been used to reconstruct climatic trends in Switzerland during the course of the Little Ice Age (Pfister, 1984), while variations in the date of the grape harvests formed the basis for a reconstruction of climatic change throughout north-west Europe back into the fifteenth century

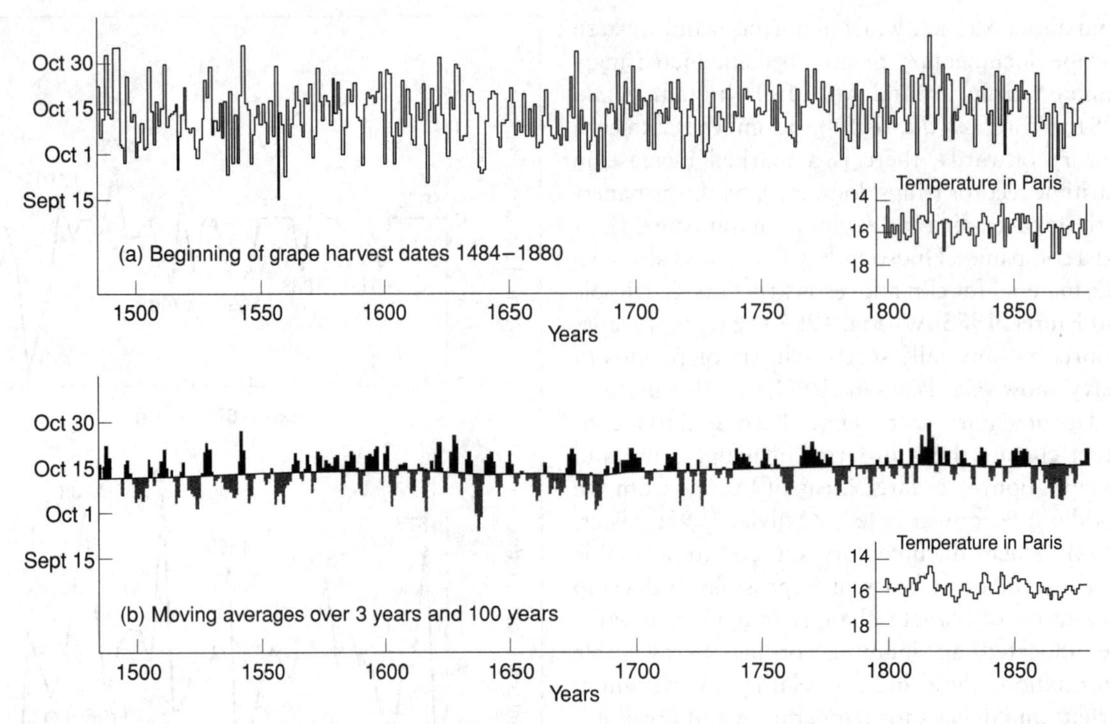

Figure 2.26 Average annual date for the beginning of the grape harvest in north-east France, French Switzerland, and the South Rhineland. At lower right are mean April–September temperatures (°C) in Paris during the period of instrumental records (after Le Roy Ladurie and Baulant, 1981, Figures 1 and 2. Rotberg, Robert I.: *Climate and History.* © 1981 Princeton University Press, Princeton, 259–69.)

(Figure 2.26). Significantly, a close relationship has been established between the proxy temperature record derived from vine harvest data and the pattern of Alpine glacier movement during the Little Ice Age obtained from other historical sources (Bray, 1982).

Instrumental records

Although measurements of rainfall had been reported from India as early as the fourth century BC, systematic recording of climatic data did not begin until the period of rapid scientific advancement in the seventeenth century AD (Shaw, 1985). The first temperature measurements, using the newly devised thermometer, date from around 1660, while barometric pressure and rainfall records, the latter obtained from carefully designed rain-gauges, began some 30–40 years later (Lamb, 1981). Throughout the early and middle years of the eighteenth century, various European weather stations were established and from about 1780 onwards, daily weather maps can be produced for large areas of north-west Europe based on these data sources (Lamb, 1995). Some of the longest climatic records using instrumental observations have been obtained from the British Isles and include the series of mean monthly temperatures for central England extending back to 1659 (Manley, 1974; Probert-Jones, 1984), and the continuous sequence of rainfall measurements for the East Midlands area from 1726 onwards (Craddock, 1976). In Europe, there is an instrumentally based temperature series from Paris beginning in 1665 (Pfister and Bareiss, 1994), while temperature records from Italy and the Netherlands extend back to the early years of the eighteenth century (Figure 2.27).

Although instrumental series provide ideal information for the reconstruction of former cli-

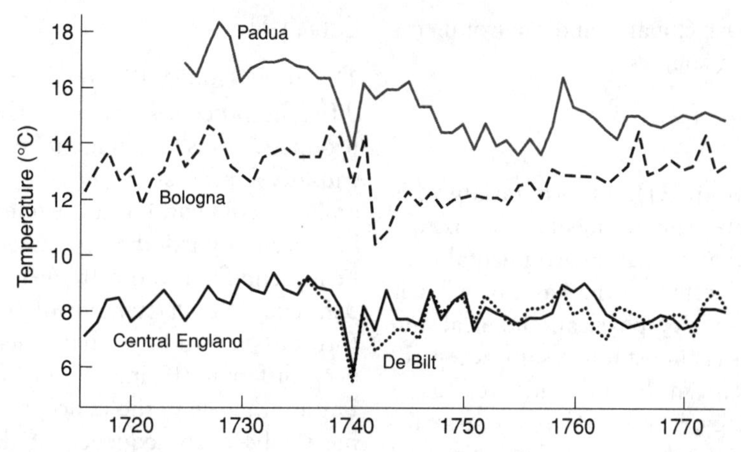

Figure 2.27 Comparison between historical temperature records from Bologna and Padua (Italy) and those from De Bilt (the Netherlands) and central England (after Comani, 1987, Figure 3)

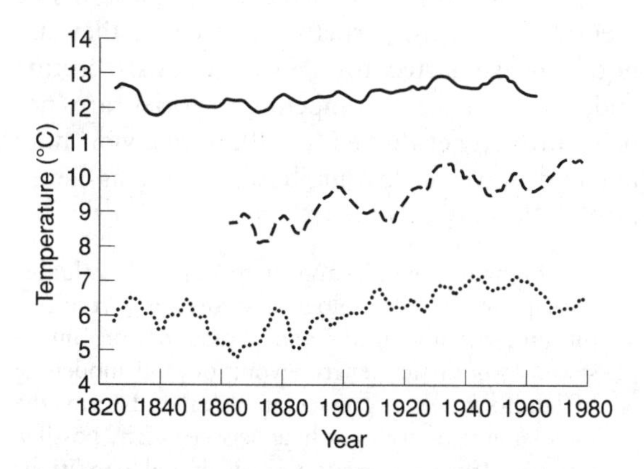

Figure 2.28 Time series of mean annual temperatures from the eastern United States (top), Iowa City (middle) and Minnesota (bottom). Data were smoothed by a normal curve-smoothing function, and values are plotted at the midpoint of the smoothing interval (after Baker *et al.*, 1985, Figure 1)

mate, they are restricted to the relatively recent past and the very early records are all from Europe. In Canada, the earliest continuous temperature record is from the Hudson Bay area and begins in 1760 (Ball and Kingsley, 1984), while in the USA the oldest temperature data are from towns on the east coast (New Haven, 1780; Baltimore 1817), although a continuous temperature series extending over 160 years is available for Minnesota (Figure 2.28).

While questions might be raised about the accuracy of some of these early instrumental records, both in terms of calibration with modern equivalents and reliability of recording practice, they are undoubtedly valuable sources of climatic data for the recent historical period. Moreover, where both instrumental and documentary data are available, the former can be used as an independent check on the reliability of the chronicled sources (e.g. Pfister *et al.*, 1994; Camuffo *et al.*, 2000).

Assessment of proxy data sources

Although each of the foregoing may be regarded as a tried and trusted means of reconstructing patterns of environmental change during the Late Quaternary, these different lines of evidence are all circumscribed by problems whose ramifications must be appreciated if meaningful inferences are to be made about former landscapes and climatic conditions. Some of these problems are explored in this section. It should be stressed, however, that the aim is not to undermine the methods themselves, nor to seek to invalidate the evidence upon which palaeoenvironmental reconstructions have been based. Rather, the intention is to encourage a critical and cautious approach when making inferences about past climates and environments based on proxy data sources.

Uniformitarianism

As was noted above (p. 21), uniformitarianism is fundamental to palaeoenvironmental reconstruction, and the idea that former environmental conditions can be interpreted on the basis of what is known about present-day physical, chemical and biological processes is almost universally accepted. Once uniformitarianism became adopted as a working methodology, it was a relatively straightforward matter to decipher the geological record in terms of contemporary earth-surface processes. However, quite apart from the philosophical problems associated with uniformitarian reasoning (see above), interpretation of the biological (i.e. fossil) record on the basis of uniformitarian reasoning is more complicated because of the assumptions that need to be made about both the fossil evidence and the present state of knowledge of plant and animal ecology (Lowe and Walker, 1997). If contemporary fauna and flora are to form the basis for the interpretation of the ecology of their fossil counterparts, it is axiomatic that the environmental parameters that influence the distribution of present-day plants and animals are known. Equally, it must be assumed that contemporary flora and fauna are in equilibrium with their environmental controls, and that the same was true of those elements preserved in fossil assemblages. There is also the presumption that the ecological affinities of plants and animals have not changed through time and, moreover, that former plant and animal communities have present-day analogues. It is increasingly recognised that this is often not the case. For example, the rapid climatic changes that occurred during the Lateglacial period (Chapter 4) produced plant and animal communities for which there do not seem to be obvious modern analogues. Furthermore, because later Holocene human activity in areas such as north-west Europe has impacted on so many natural ecosystems, there are few (if any) surviving analogues of ecological communities of the early Holocene. Hence, critical appraisal is necessary of the use of contemporary analogues in palaeoenvironmental reconstruction.

Equifinality

The term **equifinality** refers to the notion that different processes can give rise to similar end-products, thus making it difficult to deduce causative process from the geomorphological and/or sedimentological evidence. However, it has been argued that to describe landforms as being 'equifinal' more frequently reflects either a deficiency in understanding of process, or that landforms *appear* to be similar when, in fact, they are very different (Haines-Young and Petch, 1983). Put another way, the concept of equifinality may merely be a consequence of the poor ability of geomorphological classifications to resolve between differing forms (Harrison, 1999). In other words, the concept of equifinality has been employed as an excuse for methodological and experimental shortcomings. If this is correct, then questionable theories are being perpetuated because they are not being subjected to rigorous critical scrutiny and, in particular, competing theories are not being properly evaluated (p. 20). Such a view does not find universal favour, however, for as Beven (1996: 298) points out:

> . . . equifinality would appear to remain a valuable concept in geomorphological studies as a result of the inherent limitations and constraints on understanding both the genetic evolution and modelling of landforms. It expresses, in shorthand form, the impossibility of distinguishing between many possible histories from different possible initial conditions and different possible process mechanisms on the basis of the available evidence.

Indeed, even Haines-Young and Petch (1983) have conceded that it is logically possible for similar forms to be produced by different processes. Hence, although there may perhaps have been a tendency to acknowledge equifinality as a problem too readily in the historical earth sciences, awareness of the concept serves as a useful reminder of the difficulties that can arise when inferring process from form on the basis of proxy data sources.

Taphonomy

The term **taphonomy** refers to the analysis of the processes that have combined to produce a fossil assemblage, and understanding of the taphonomy of an assemblage is an essential prerequisite for palaeoenvironmental reconstruction (Lowe and Walker, 1997), or for inference about the use of biological resources (e.g. animals) by past human populations (Lyman, 1994). Interpretation of palaeoenvironmental evidence demands careful consideration of how the biota in question came to be incorporated into that context. Sometimes the taphonomy of an assemblage will be relatively simple, for example molluscs in a buried soil representing species that lived and died more or less on that site. The taphonomy of other assemblages will be more complicated, such as Mollusca which may be sorted by earthworms in soils (Carter, 1990), or have been eroded from soils and subsequently incorporated into valley sediments, or into archaeological contexts such as ditches. Similarly, pollen grains deposited in lake sediments will have followed a variety of depositional pathways (wind, animals, inflowing streams, slopewash, etc.) and will derive from different source areas (local, extra local, regional, long distance, etc.). In many cases, it may not be possible to quantify these different components and, as a consequence, the taphonomic complexity of this type of fossil assemblage will impose constraints on subsequent palaeoenvironmental interpretations.

Preservation and contamination

Fossil assemblages vary markedly in degree of preservation. Where anaerobic conditions have obtained since burial and where the sediment matrix is relatively fine, excellent assemblages of fossils which closely resemble the life assemblage may be recovered. Such conditions are rare, however, and it is more common to find fossil assemblages that reflect the post-burial operation of physical and chemical processes. Where the sediment matrix consists of sand-size material, in riverine deposits for example, the more delicate biological remains will frequently be abraded or differentially destroyed. Similarly, if completely anaerobic conditions are not achieved (e.g. in rapidly accumulating terrestrial peats), oxidation may affect the assemblage and again the more fragile and chemically susceptible elements, such as pollen grains, will be destroyed. As a consequence the fossil assemblage will be biased towards the stronger, more robust elements, and may bear little relationship to the original living assemblage.

A related problem is contamination by both older and younger material. In peat profiles, for example, percolating groundwaters may carry microfossils down the profile, thereby introducing younger pollen, insects and other fossils into older assemblages. Rivers frequently contain assemblages of both animal bones and Palaeolithic artefacts whose components are of markedly different age, reflecting successive episodes of fluvial erosion and deposition and the incorporation of earlier material into later sediments. Similarly, bioturbation (faunal disturbance) on a lake or sea bed may lead to admixtures of contemporaneous and previously deposited fossil assemblages. Problems of contamination are not confined to the interpretation of fossil assemblages, however, for they may equally affect the movement of archaeological artefacts. Sediments are also affected: for example, erosion around a lake shore may result in older material being reworked and subsequently incorporated into the sediment sequence, thereby posing problems for stratigraphic interpretation. Reworking also occurs in hillslope contexts where colluviation, gelifluction (under periglacial conditions) and the processes of sheetwash and rillwash lead to the mixing of older and younger materials and any archaeological artefacts they may contain.

Climatic inferences from historical data

A number of problems are encountered in the interpretation of historical data as evidence of past climate. With documentary sources, particular problems surround the reliability of subjective and impressionistic records, such as diaries and annals. The most serious errors are likely to arise from inaccurate dating; from spurious multiplication of events; from acceptance of distorted or amplified accounts; and from the inclusion of events for which there is no reliable evidence (Ingram *et al.*, 1981). As for phenological records, the major difficulty lies in establishing a cause-and-effect relationship between these data and climatic parameters. Variations in crop yield, for example, may be a reflection of climatic change, but they may also have been affected by socio-economic factors, including changing agricultural practices, fluctuations in market price, population change, disease or the effects of war. An unambiguous link between historical evidence and climatic change may therefore be difficult to establish (Parry, 1978). More positively, however, historical records can provide direct evidence for the ways in which climatic and environmental phenomena were perceived by past human populations whose outlook may have been very different from our own (Fagan, 2000).

Climatic inferences from other proxy data

Reconstructing former climatic conditions on the basis of proxy records from geological or biological sources is a far from straightforward process. Any climatic inference is two stages removed from the original evidence. Take, for example, the interpretation of climate from pollen analytical data. The first stage is to reconstruct former plant communities and vegetational patterns from the pollen evidence. Once that has been achieved, the second stage is to make a climatic inference based upon that reconstruction. In other words, an interpretation (of climate) is being made from an interpretation (of vegetation) and errors can be incurred at both stages in the analysis. In the case of pollen evidence, this problem is compounded by the fact that it is not always possible to identify pollen and spores to individual species. A pollen diagram comprises a data bank at a variety of taxonomic levels and this clearly imposes constraints on the reconstructions of former plant communities and hence on climatic inference. This problem of taxonomic imprecision also arises with other forms of biological evidence.

Climatic reconstructions from fossil evidence have often been based on particular **indicator species** whose present-day ecological affinities are reasonably well known (e.g. Zagwijn, 1994). However, problems relating to the taphonomy of fossil assemblages (see above), along with other factors such as differential migration responses of plant to climate, may result in erroneous palaeoclimatic inferences (Lowe and Walker, 1997). More recent approaches have therefore attempted to use **assemblages** (as opposed to individual species taken from the assemblages) as a basis for climatic reconstructions. These frequently involve the use of multivariate statistical techniques which first determine the relationships between plant or animal assemblages and prevailing climatic conditions, and then use that information to infer past climatic conditions based on the fossil data (Birks, 1995). These quantitative approaches are powerful new tools in climatic reconstruction, and have been applied to a range of fossil evidence including pollen, plant macrofossils, Coleoptera, chironomids, and deep-ocean microfossils.

A further difficulty in the interpretation of proxy data sources in the context of former climatic conditions concerns the timescale over which geological or biological response to climatic change takes place. Some proxy variables (e.g. Coleoptera) react swiftly to climatic change and the response may be measured over a matter of years. In other cases (e.g. glaciers and treeline fluctuations), there may be a lag in response, the duration of which may not only vary temporally and geographically, but also with the direction of change, i.e. whether the climate is ameliorating or deteriorating. The 'coarseness' of many proxy data records is an additional problem. Slow rates of lacustrine sedimentation, peat accumulation or colluviation, for example, mean that biological records such as pollen, plant macrofossils, diatoms and Mollusca will seldom be interpretable on a timescale of less than 50 years, and hence short-lived climatic variations may go undetected. Finally, there is the problem of distinguishing between the climatic and the anthropogenic signal in proxy records, a difficulty which becomes particularly acute during the mid- and late Holocene as natural environments were transformed into cultural landscapes through accelerating human impact.

Dating of proxy records

Four types of technique are currently employed in the dating of Late Quaternary proxy records: radiometric methods, incremental methods, methods that establish age equivalence, and artefact dating. The techniques are described here only briefly, but

Plate 2.1 A peat profile at Bolton Fell Moss, northern England showing a recurrence surface dividing lower, darker, well-humified peats from upper, lighter, less well-humified peats. This well-defined stratigraphic horizon reflects a marked increase in mire surface wetness (photo Mike Walker)

Plate 2.2 Fossil sand wedge of early Anglian age at Broomfield, Essex, England (photo Peter Allen)

Plate 2.3 Involution structure of Late Devensian age at Lackford, Suffolk, England (photo Mike Walker)

Plate 2.4 Buried soil beneath the bank of Avebury Neolithic henge, which was constructed around 4700 BP. (photo Martin Bell). For a land mollusc diagram from the Avebury soil see Figure 6.26

(a)

(b)

Plate 2.5 Drilling of the GRIP ice core at Greenland Summit: (a) the drill being raised into position; (b) an ice core under examination (photos Jorgen-Peder Steffensen, courtesy of Dept. of Geophysics, Niels Bohr Institute, University of Copenhagen)

Plate 2.6 A glaciolacustrine varve sequence from Nummi-Pusula, southern Finland (photo Jari Väätäinen, Geological Survey of Finland (GSF))

Plate 2.7 Annual layers of ice in the Quelcayya ice cap, Peru (photo Lonnie G. Thompson, The Ohio State University)

Plate 2.8 Mazama ash (*c*.6.8 k ^{14}C yrs BP) exposed in fluvial sediments near Lumby, British Columbia, Canada (photo Neville Alley)

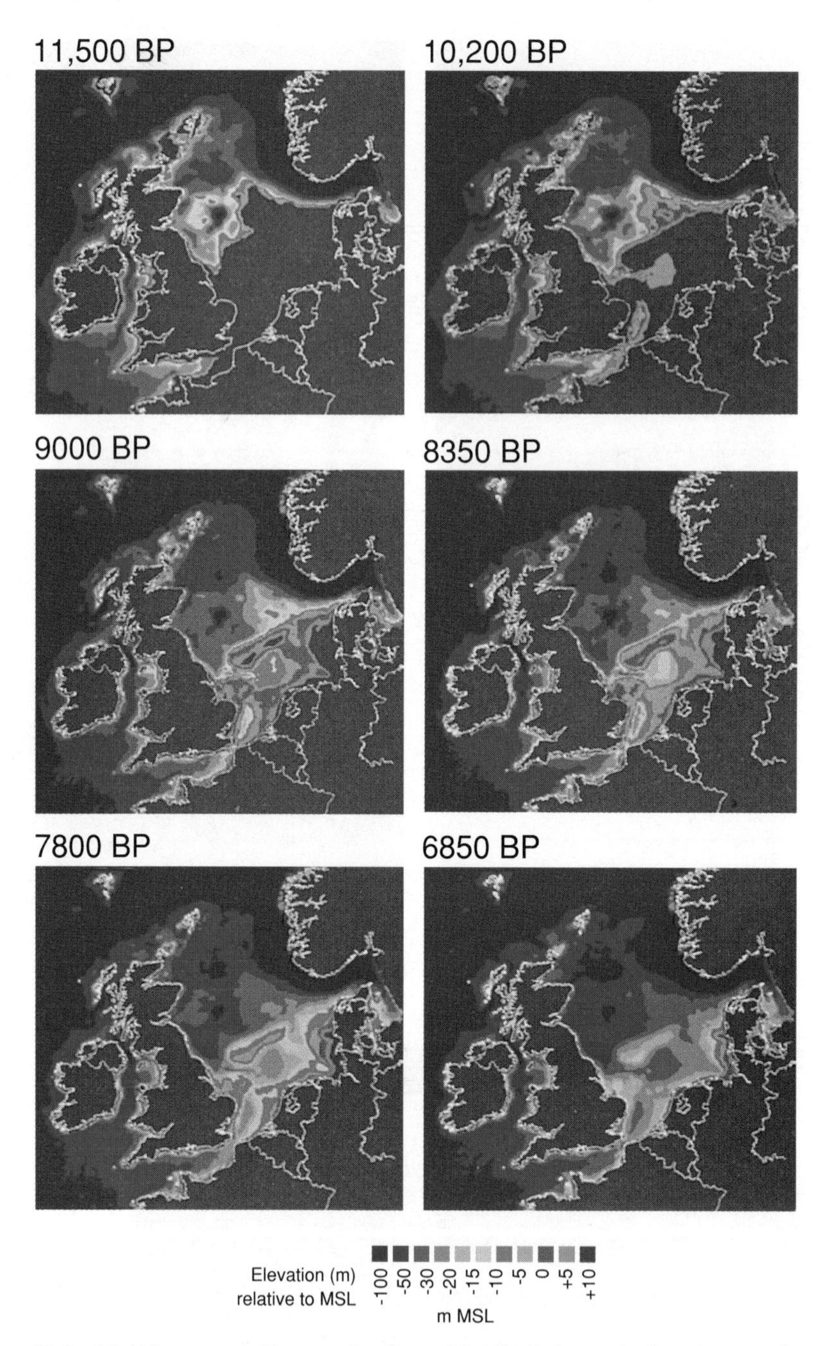

Plate 4.1 Palaeogeographic reconstructions of the North Sea and adjacent areas of north-west Europe during the early Holocene (after Shennan *et al.*, 2000b in *Holocene Land-Ocean Interaction and Environmental Change around the North Sea* edited by I. Shennan and J.E. Andrews, Geological Society of London, Specal Publication No. 166, pp. 299–319, Figure 5. Reprinted by permission of The Geological Society.)

a more in-depth discussion and evaluation of Quaternary dating methods can be found in Smart and Frances (1991), Lowe and Walker (1997), Taylor and Aitken (1997), Wagner (1998) and Bradley (1999).

Radiometric dating

All radiometric dating methods are based on the radioactive decay of unstable chemical elements or **isotopes** which emit atomic particles in order to achieve a more stable atomic form. A number of dating techniques involving radioactive isotopes are routinely used in the earth sciences, but the three that are most appropriate to the Late Quaternary are **radiocarbon, uranium-series** and **optical dating.**

Radiocarbon dating

Atoms of **radiocarbon** (^{14}C) form in the upper atmosphere through the interaction between cosmic-ray neutrons and nitrogen, and these are subsequently absorbed by all living organisms through the carbon dioxide cycle. Decay of ^{14}C occurs but there is constant replenishment in the living organisms (e.g. the outer rings of a tree) from the atmospheric reservoir. Following death of the plant or animal, ^{14}C decay continues, but no replacement can take place. Hence, measurement of the ^{14}C remaining in a fossil will provide an age for the death of that organism. The method is applicable to all organic materials and the time range normally extends back to *c.*45 ka BP (Bowman, 1990). However, in laboratories which specialise in dating older materials, and where relatively large samples are available, the dating limit may be 55 ka or even beyond (Taylor, 2001). Because of technical difficulties in the measurement of radiocarbon in the laboratory, dates are always given as a mean age with one standard deviation. Hence a date of 5000 ± 50 BP (before present) indicates that there is a 68 per cent chance that the date lies in the range 4950–5050 years. There is, of course a one in three chance that the true age might lie outside that range. The chance of a true age falling within the range of double the standard deviation (100 years in the above example) is 95.4 per cent.

There are two ways in which ^{14}C in fossil material is measured. Conventional radiocarbon dating involves the counting of the decay products (beta particles) over a period of time. Once the rate of emission of those particles has been established, the residual ^{14}C activity of the sample can be determined. The alternative approach, accelerator mass spectrometry (AMS), uses particle accelerators as mass spectrometers to determine the number of ^{14}C atoms in a sample of material. The advantages of AMS are speed of operation (actual measurements may take only a matters of hours by contrast with conventional procedures which may take days, or even months), and the fact that only very small samples of material (1 mg of carbon or less compared with 5–10 g of carbon for conventional samples) are required for dating. This means that dates can be obtained on, for example, individual seeds, fragments of Coleoptera (Walker *et al.*, 2001) or even pollen grains (Brown *et al.*, 1989). The technique is particularly valuable in archaeology, enabling ages to be obtained, for example, on tiny samples of museum objects of special value, such as human remains; on animal bones with cut marks which can be linked to human activity; and on bone artefacts, including Palaeolithic mobile art. The disadvantages of AMS are the costs of installing the facilities (in excess of £3 m. at 2004 prices) and the lower levels of analytical precision that are achievable by comparison with conventional dating. Most beta-counting laboratories can generate dates with smaller errors than AMS laboratories, although the gap is closing quite fast (Hedges, 2001). There is no doubt, however, that AMS ^{14}C dating has been a major chronological development. In addition to dating samples of peat, charcoal and wood, it has provided dates on important artefacts where only tiny samples can be taken, such as the Turin Shroud, the Dead Sea Scrolls, and prehistoric rock paintings (Gove, 2000; Valladas *et al.*, 2001). Using the technique, specific plant macrofossils such as seeds can now be dated and this has had major implications for the study of plant domestication (p. 150; Smith, 1997d; Hillman *et al.*, 2001). It has also been employed in the dating of key Late Quaternary events, including the initial peopling of Australasia and the Americas (Turney *et al.*, 2001; Fiedel, 2002); climatic and environmental changes during the glacial–Holocene transition

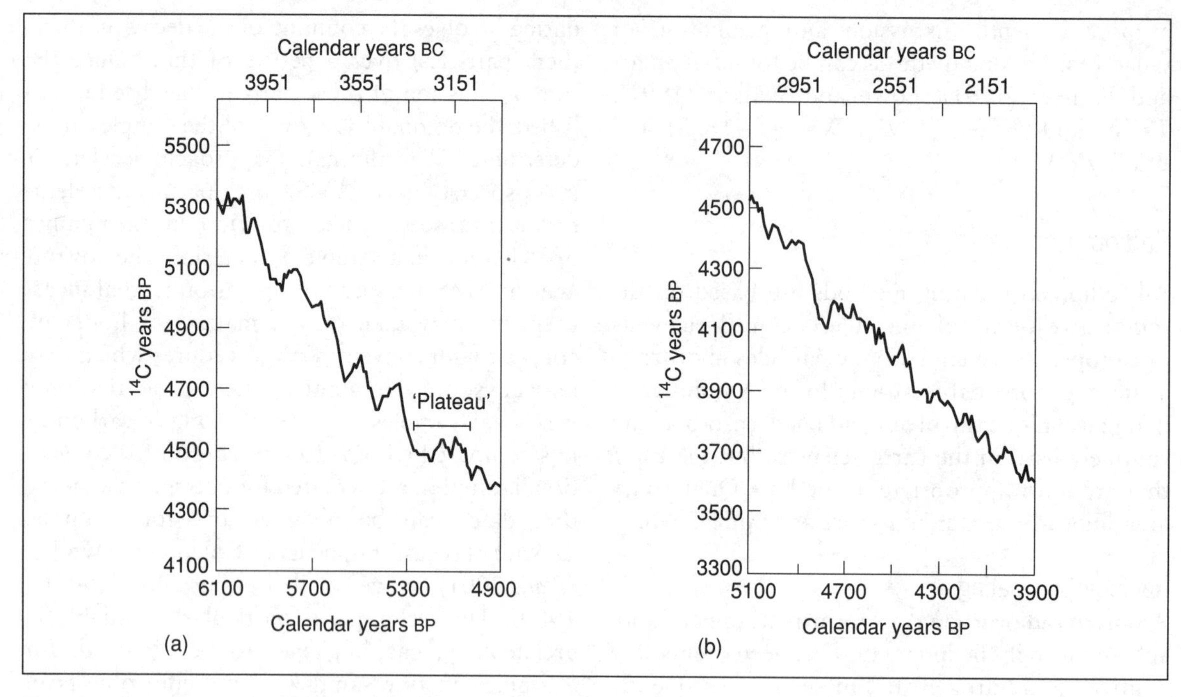

Figure 2.29 High-precision calibration curves covering the periods (a) 3.5–4.5 k ^{14}C yrs BP; (b) 4.4–5.3 k ^{14}C yrs BP, based on the INTCAL98 program of Stuiver *et al.* (1998). Note the 'plateau' of apparent ^{14}C age in (a), see text for further explanation

(Birks and Ammann, 2000); and oceanographical changes in the North Atlantic during the last deglaciation (Eiriksson *et al.*, 2000).

One difficulty with radiocarbon dating is that not only have the amounts of ^{14}C in the atmosphere fluctuated during the course of the Holocene, but there is an increasing divergence between radiocarbon and calendar ages over time, to the extent that radiocarbon determinations underestimate true age by as much as 3.5 ka at 20 ka BP (Bard *et al.*, 1998). However, radiocarbon years can be **calibrated** to calendar years.[6] For samples of up to *c.*12.0 ka in age, ^{14}C dates on samples of fossil material can be compared with ^{14}C age determinations on independently dated **dendrochronological** sequences (Figure 2.29; p. 57), while for materials older than 12.0 ka BP, calibrations are based on samples of fossil coral which have been dated by both ^{14}C and **uranium series** (see below), with the latter providing an independent check on the ^{14}C age (Figure 2.30). It is now generally accepted practice to calibrate radiocarbon dates in this way so that radiocarbon determinations can be presented as calendar years (Stuiver *et al.*, 1998; van der Plicht, 2002). Throughout this book radiocarbon dates are quoted in a calibrated form unless otherwise specified (see Chapter 1, note 1, p. 16).

Where the calibration curve is steep, calibration may reduce the original standard deviation, where it levels a **plateau** will be created (Figure 2.29) and the standard deviation will be greater than that of the original date. When a 'stratified' sequence of datable contexts is present as in peat stratigraphy, tree rings or annually laminated sediments, dating precision can be significantly increased by taking a sequence of AMS samples, and then **wiggle-matching** the sequence of dates to the relevant portion of the calibration curve (see Table 1.1). This can provide dates with a precision of two to four decades which has proved of great value in pinpointing the dates for environmental or cultural changes (van Geel *et al.*, 1996; Mellars and Dark, 1998).

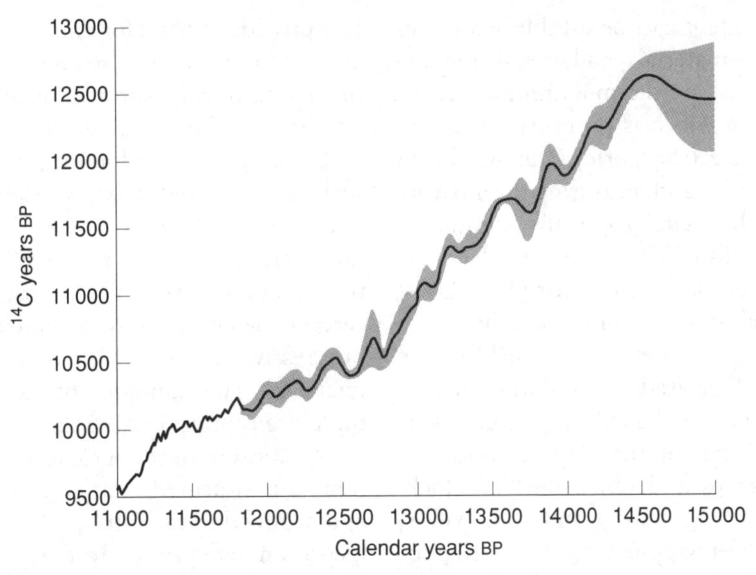

Figure 2.30 Radiocarbon calibration curve for the Lateglacial and early Holocene periods (*c*.9.5–13.0 k ^{14}C yrs BP). From 9.5 to 11.8 k ^{14}C yrs BP, calibrations are based on the dendrochronological record; the earlier part of the curve is based on paired ^{14}C/U series dates on fossil corals (after Stuiver *et al.*, 1998)

Uranium-series dating

The isotopes of uranium, ^{238}U, ^{235}U and ^{232}Th, decay to stable lead isotopes through a complex decay chain of intermediate nuclides with widely differing half-lives[7] (Ivanovich and Harmon, 1995; Bourdon *et al.*, 2003). Uranium and some products of the decay series are soluble, whereas others such as thorium (^{230}Th) and protactinium (^{231}Pa) are not. The latter therefore precipitate out and accumulate in sediments on lake floors or on the seabed. Hence the age of those sediments can be established by measuring the extent to which these daughter nuclides have decayed. By contrast, the ages of carbonate fossils of creatures that absorbed uranium from sea or lake waters during their lifetime (e.g. corals and molluscs) can be calculated from the extent to which the decay products of uranium (such as thorium and protactinium) have accumulated (or reappeared) in their carbonate shells. The age range of the method varies with the nuclides employed, with the ^{230}Th/^{234}U technique capable of dating samples up to 350 ka in age, while ^{231}Pa/^{235}U and ^{231}Pa/^{230}Th have upper age limits of *c*.200 and 250 ka respectively. The ^{234}U/^{238}U method, however, has a potential age range of up to 1.5 Ma (Smart, 1991). Recent developments using thermal ionisation mass spectrometry (TIMS) and AMS have reduced the lower limits of the ^{230}Th/^{234}U method to less than 100 years. Uranium-series dating is applicable to a range of materials including bone, teeth, cave speleothems and corals (Latham, 2001). It can also be used to date lake marls (e.g. Rowe *et al.*, 1999) and peats (Heijnis and van der Plicht, 1992). In addition uranium-series provides a means of calibrating the radiocarbon timescale beyond the time range of dendrochronology (see above).

Optical dating

When certain minerals (mainly quartz and feldspars) are exposed to the weak flux of nuclear radiation from radioactive elements (potassium, thorium or uranium), electrons are detached (*ionised*) from their parent nuclei and become 'trapped' within the crystal lattice of those minerals. The number of trapped electrons reflects the interval that has elapsed since ionisation began, and therefore forms a basis for dating (Aitken, 1998; Grün,

2001). Electron content can be established either by heating sample materials and measuring the resulting light emission (**thermoluminescence** or **TL**), the intensity of which is proportional to the number of trapped electrons, or by shining a beam of light onto the sample and measuring the intensity of the emitted light signal (**optically stimulated luminescence** or **OSL**). The latter involves the use of either a green light source (or green laser) or, in the case of feldspars, an infrared light source (**infrared stimulated luminescence** or **IRSL**). The TL technique was first used in the dating of pottery and other forms of baked clay (e.g. bricks and tiles) where firing of the clays empties the original electron traps and resets the 'TL clock' to zero (Aitken, 1985). Subsequently, however, the method has been applied to the dating of water- and wind-borne sediments, for exposure to sunlight also empties (bleaches) many of the electron traps, and hence the date of burial (i.e. last exposure to sunlight) can be determined by measuring the luminescence signal from the constituent mineral grains. The advantage of OSL over its predecessor TL is that OSL is associated only with electron traps that are relatively easily bleached, and hence any residual signal after deposition of sediment (i.e. 'older electrons') is at least an order of magnitude less than that for TL. Much younger ages (in some cases within the last 1000 years) can therefore be obtained using OSL dating (Duller, 2000). Optical dating techniques are now being applied to a range of Late Quaternary sediments, including coastal dunes (Bailey *et al.*, 2001), alluvial deposits (Rittenour *et al.*, 2003), glaciofluvial deposits (Gemmell, 1999), marine and freshwater deposits (Strickertsson and Murray, 1999) and loess sequences (Roberts and Wintle, 2001).

Other radiometric methods

Other radiometric methods for dating the recent past involve the use of both long-lived and short-lived radioactive isotopes. **Long-lived isotopes** are produced in the natural environment by the bombardment of the earth by high-energy cosmic rays, and include beryllium (^{10}Be), aluminium (^{26}Al) and chlorine (^{36}Cl). The production of these rare isotopes in lithological materials at the earth's surface provides a basis for dating (Gosse and Phillips, 2001; Stuart, 2001). One application has been in the dating of rock surfaces, principally using ^{36}Cl. The basis of the technique has some parallels with OSL dating described above, in that while a rock surface is covered it is protected from the direct effects of cosmic rays. Once a surface is exposed, however, the 'cosmic ray clock' is activated and the reaction between the cosmic ray flux and certain elements in rock minerals leads to the progressive accumulation of certain cosmogenic nuclides. The amount of cosmogenic ^{36}Cl is therefore proportional to the time that has elapsed since exposure of the rock surface. ^{36}Cl concentrations are measured relative to stable Cl by AMS, and production rates are calibrated from measurements on independently dated surfaces (Phillips *et al.*, 1996). Examples of the use of ^{36}Cl in this way include the dating of glacial landforms (Phillips *et al.*, 1990) and Holocene landsliding activity (Ballantyne *et al.*, 1998a) and estimates of denudation rates (Granger *et al.*, 1996). Cosmogenic isotopes can be employed in other geomorphological and environmental contexts, however, such as in the dating of lacustrine and fluvial sediments (Jannik *et al.*, 1991; Hancock *et al.*, 1999), riverine histories (Seidl *et al.*, 1997) and in studies of long-term landscape evolution (Fleming *et al.*, 1999). ^{10}Be has also been analysed in ice cores (Finkel and Nishiizumi, 1998), where it provides corroborative evidence for long-term changes in atmospheric ^{14}C concentration (see above), as well as a basis for correlating Greenland and Antarctic ice-core records (Beer *et al.*, 1992).

Short-lived isotopes include ^{210}Pb (range 1–150 years) and ^{137}Cs (1–55 years). The former employs the isotope of lead, ^{210}Pb, a decay product of radon, which is the naturally occurring radioactive gas that escapes from the earth. The latter involves the use of ^{137}Cs, an isotope produced artificially as a consequence of nuclear weapons testing since 1945. Caesium and lead isotopes have been successfully used in the dating of lake sediments and ombrogenous peats (Gale *et al.*, 1995; Oldfield *et al.*, 1995). Other short-lived isotopes that may offer a basis for dating are silicon (^{32}Si), a cosmogenic isotope, and argon (^{39}Ar) which is produced by chemical reactions in the stratosphere.

Potential applications include the dating of groundwaters, marine sediments and glacier ice (Olsson, 1986), although a number of technical problems need to be resolved before these can be used as reliable Late Quaternary chronometers (Stauffer, 1989).

Incremental dating

This group of dating techniques is based on the regular additions of material to organic tissue or to sedimentary sequences, and includes **dendrochronology, lichenometry** and **annually laminated lake sediments. Annual layers in glacier ice** also form a basis for dating.

Figure 2.31 Bristlecone pines (*Pinus longaeva*) growing on the semi-arid slopes of the White Mountains, Colorado (photo Mike Walker)

Dendrochronology

Dendrochronology (tree-ring dating) involves the measurement of increments of wood that are added annually to the outer perimeter of tree trunks. Years unfavourable for growth produce narrow rings, favourable years wide rings. The pattern of wide and narrow rings over a century or so will generally be distinctive of that period. By using distinctive rings or groups of rings (marker rings), it is possible to correlate wood from different trees and to establish long sequences linking living trees, building timbers, buried timbers from peats or archaeological sites, and other sub-fossil wood (Kuniholm, 2001). This technique of **cross-dating** enables tree-ring series to be established spanning hundreds and, in some cases, thousands of years (Schweingruber, 1988). The longest continuous tree-ring series from a single species is that obtained from the south-west United States where work on the bristlecone pine (*Pinus longaeva*: Figure 2.31) has produced a chronology extending back over 8 ka years (Ferguson and Graybill, 1983). In Ireland, dendrochronological work on oak has produced a continuous tree-ring record extending back to 7402 BP, while in Germany the oak record reaches back to 10.43 ka BP (Spurk *et al.*, 1998). The latter has now been linked to a 'floating' pine chronology[8] which extends this dendrochronological series back to *c.*12.0 ka BP (Friedrich *et al.*, 1999). In due course it may prove possible to extend this even further, as an older floating pine chronology is now available from Germany, the earliest parts of which date to *c.*14.3 ka BP (Friedrich *et al.*, 2001). As noted above, as well as providing a valuable basis for dating, these long dendrochronological records provide one of the principal means of calibrating the radiocarbon timescale.

Annual tree-ring thickness varies with climate, and it is often possible to reconstruct the history of climate change, particularly for the Late Holocene, from ring-width variations (**dendroclimatology**: Figure 2.32). These reconstructions enable inferences to be made about both temperature (Barclay *et al.*, 1999) and precipitation variations (Szeicz and MacDonald, 1996). Climatic inferences may also be possible from variations in wood density, and these often reveal a more sensitive annual or even seasonal climatic signal, such as summer temperature (Briffa *et al.*, 1990). Dendrochronological records have also been linked to other data-sets to make inferences, *inter alia*, about glacier behaviour and climate (Scuderi, 1987), Holocene humidity changes (Eronen *et al.*, 1999), Holocene temperature changes (Wilson and Luckman, 2003) and historical hydrological fluctuations (Cleaveland, 2000). Dendroclimatological records can also provide important information about atmospheric circulation dynamics linked, for example, to the El Niño/Southern Oscillation or to the North Atlantic Oscillation (Briffa, 2000), and about hemispherical-scale climate changes (Briffa and Matthews, 2002).

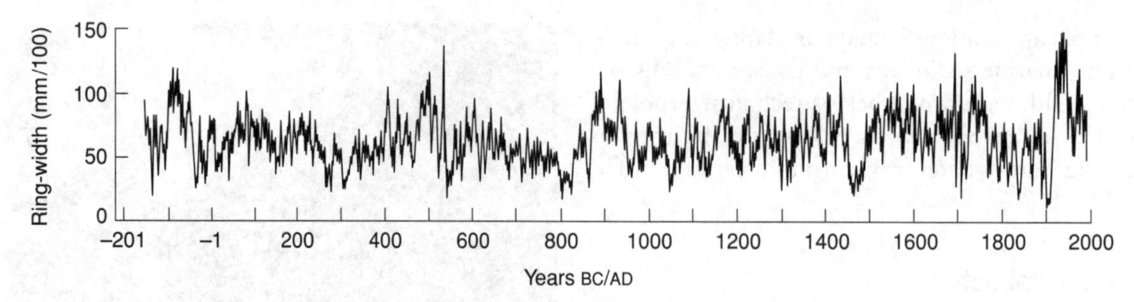

Figure 2.32 The Finnish Lapland pine chronology extending from the present time to the year 165 BC (after Eronen *et al.*, 1999, Figure 9. Reproduced by permission of Hodder Arnold)

Dendrochronology has become an increasingly important archaeological method in both Europe and North America (Baillie, 1995). Where suitable timbers survive, events can be dated to the year and sometimes to the season. Prehistoric wooden trackways have been dated in England and Ireland (Hillam *et al.*, 1990; Raftery, 1996), while in the Alpine lake region and in the pueblo area of the American south-west, settlement establishment, repair, expansion and abandonment have been dated with great accuracy (Dean, 1986; Coles and Coles, 1996). Such precision is especially valuable in the comparison of archaeological and palaeoenvironmental sequences, and in considerations of the effects of environmental change on human communities (Chapter 5).

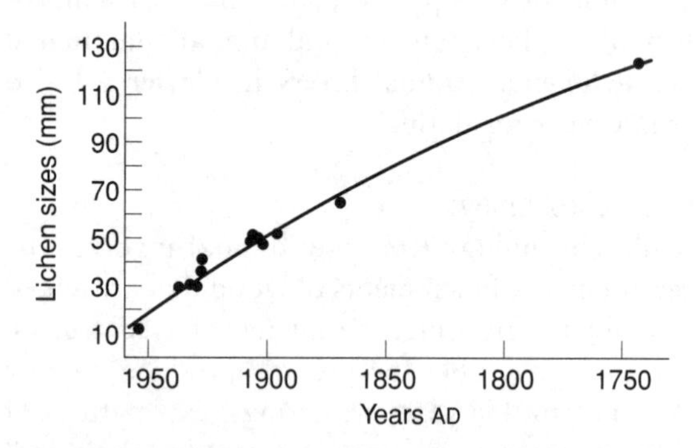

Figure 2.33 Lichenometric growth curve for western Norway (reprinted from Erikstad, L. and Sollid, L., 1986, Neoglaciation in South Norway using lichenometric methods, from *Norsk Geografisk Tiddskrift*, **40**, pp. 85–105, www.tandf.no/ngeog, by permission of Taylor & Francis As). The curve is based on lichen size measurements on newly deglaciated surfaces of known age (fixed points). The age of unknown surfaces can then be obtained by measuring average lichen size on those surfaces (*y* axis) and reading the appropriate age on the *x* axis

Lichenometry

Lichenometry is the technique of dating newly exposed surfaces using variations in lichen size. Measurements of lichen size on substrates of known age enables a growth-rate curve to be constructed for a particular locality (Figure 2.33). Reference to this curve enables unknown age to be constructed on the basis of lichen size. The method has been most widely employed in studies of Late Holocene glacier fluctuations (Bickerton and Matthews, 1993; Evans *et al.*, 1994), but it has also been used in the dating of plant colonisation (Matthews, 1992), proglacial river terraces (Thompson and Jones, 1986) and recent earthquake activity (Smironova and Nikonov, 1990). An additional application has been in the dating of archaeological features in coastal areas, for example around the Bothnian and Baltic coasts (Broadbent and Bergqvist, 1986).

Laminated lake sediments

Laminations in lake sediments (Plate 2.6) are usually referred to as **rhythmites** or, where the laminations develop because of annual variations in sedimentation, they are known as **varves** (O'Sullivan, 1983; Hicks *et al.*, 1994). Annual laminations in lacustrine sediments are formed as a consequence of seasonal, rhythmical changes in biogenic production, water chemistry and the inflow of mineral matter (Saarnisto, 1986). Seasonal contrasts may be reflected, for example, in the accumulation of diatom remains (Simola *et al.*, 1981), spring and summer calcareous precipitation (Peglar *et al.*, 1984) or variations in iron precipitation, all causing annual laminations (Renberg, 1981). Glaciolacustrine varves (alternating layers

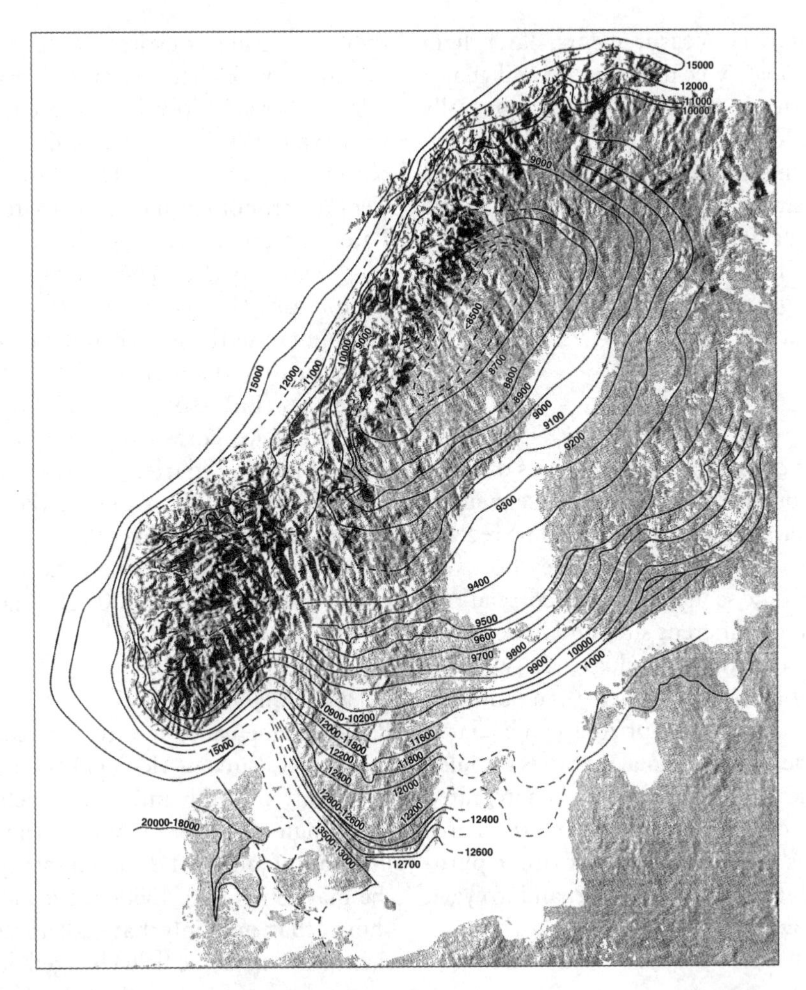

Figure 2.34 The pattern of retreat of the Scandinavian ice sheet at the end of the last cold stage. The dates from northern Sweden and southwards across the Baltic are based on clay-varve chronology and should be corrected by +365 years. Other dates are based on ^{14}C determinations (after Lundqvist, 1986)

of coarse and fine sediment) develop in proglacial lakes due to summer and winter contrasts in sediment input. Perhaps the best-known dating application of varved sediments has been in the development of a chronology for the retreat of the Fennoscandian ice sheet (Lundqvist, 1986: Figure 2.34), although laminated lake sediments have also been used as a chronological basis for studies of vegetational history (Lotter and Kienast, 1990), land-use history and human impact (Litt, 2003), glaciolacustrine environments (Ringberg and Erlström, 1999) and climatic change (Ridge and Toll, 1999).

Annual layers in glacier ice

Within the upper levels of the polar ice sheets, clearly defined layers of snow and ice are often visible (Plate 2.7), and these annual increments can be counted to provide a chronology for ice cores (Alley *et al.*, 1998). In the deeper parts of the ice sheet, however, the annual layers become increasingly deformed and diffuse and age estimates have to be inferred from theoretical flow models based on ice dynamics (Johnsen *et al.*, 1992). In addition to changes in the visual stratigraphy of the ice, other parameters can be used to date ice cores, including oxygen isotope ratios of the ice, electrical conductivity measurements, laser light scattering from dust, volcanic signals and major ion chemistry (Meese *et al.*, 1998). Incrementally based ice-core chronologies are particularly important for establishing a timescale for climate change during the transition from the last cold stage to the present interglacial (Kapsner *et al.*, 1995; Alley, 2000), and for the dating of Holocene climatic and other (e.g. volcanic) events (Dansgaard *et al.*, 1993; Zielinski *et al.*, 1997).

Age equivalence

This approach to dating the Holocene uses distinctive **marker** horizons within sedimentary sequences, or other characteristics of the stratigraphic record, to establish a chronology of events. Key horizons within a stratigraphic sequence enable correlations to be made between different sites and hence form the basis for a **relative chronology**, i.e. whether one deposit is older or younger than another. Moreover, where distinctive horizons can be dated by the radiometric or incremental methods outlined above, the sedimentary sequence may be integrated into an absolute chronological framework. Three examples of this type of approach are **tephrochronology, palaeomagnetic dating** and **oxygen isotope chronology**.

Tephrochronology

The term **tephrochronology** refers to the use of volcanic ash (tephra) in the establishment of a chronology of events. Following a volcanic eruption, volcanic ash is widely dispersed and becomes incorporated into terrestrial and marine sediments and also into glacier ice. Some tephras are visible in stratigraphic sequences (Plate 2.8), whereas others (microtephras) can only be found by detailed laboratory analysis of peat or lake sediment cores (Pilcher *et al.*, 1996; Turney, 1998). Individual tephra units can be distinguished by their mineral composition and geochemical properties (Hall and Pilcher, 2002), and can be dated either by radiocarbon determinations on interbedded organic material, or by their stratigraphic position in a sequence of annually laminated lake sediments or in glacier ice. Because the tephra layers are the product of a single, short-lived volcanic eruption, they effectively constitute time-planes in a body of sediment. Hence they provide a basis for correlation between localities where the ash layers are preserved (Davies *et al.*, 2002), as well as for a chronology of events (Haflidason *et al.*, 2000). Tephrochronology has been useful in a range of Late Quaternary investigations, including studies of fluvial erosion (Thorarinsson, 1981), glacier fluctuations (Dugmore, 1989), vegetation change and human settlement (Bennett *et al.*, 1992), landscape history (Hall *et al.*, 1993) and ocean-floor sediments (Eiriksson *et al.*, 2000). Tephras in the Greenland ice cores provide a record of historic and prehistoric glacier eruptions (Clausen *et al.*, 1997; Figure 3.27), while in the North Atlantic region, tephrochronology has been used to develop a time-stratigraphic framework for environmental events over the past 400 ka (Haflidason *et al.*, 2000).

Palaeomagnetism

Palaeomagnetism refers to the record of changes in the earth's magnetic field that may be preserved in a body of sediment. These include **declination** (the angle between magnetic and true north), **inclination** (angle of dip) and **intensity** (strength of the magnetic field). Evidence from lake sediments shows that not only have all three properties of the earth's magnetic field changed during the course of the Holocene, but that these changes are regular and have regional application (Thompson and Oldfield, 1986). The construction of master geomagnetic curves for particular areas, which can be dated by radiometric or incremental methods, therefore provides a means by which other lake sediment sequences can be dated and correlated (Figure 2.35). Although the technique has been mostly applied to lacustrine sediments (e.g. Snowball and Thompson, 1992), there is considerable potential for using magnetic parameters of other materials as a basis for dating and correlation, including glacial sediments (Walden *et al.*, 1992), aeolian/palaeosol sequences (Thompson and Maher, 1995), marine deposits (Robinson *et al.*, 1995) and a range of archaeological features and deposits (Sternberg, 2001). Practical aspects of environmental magnetism are described in Walden *et al.* (1999), while wider aspects of magnetic mineral applications in

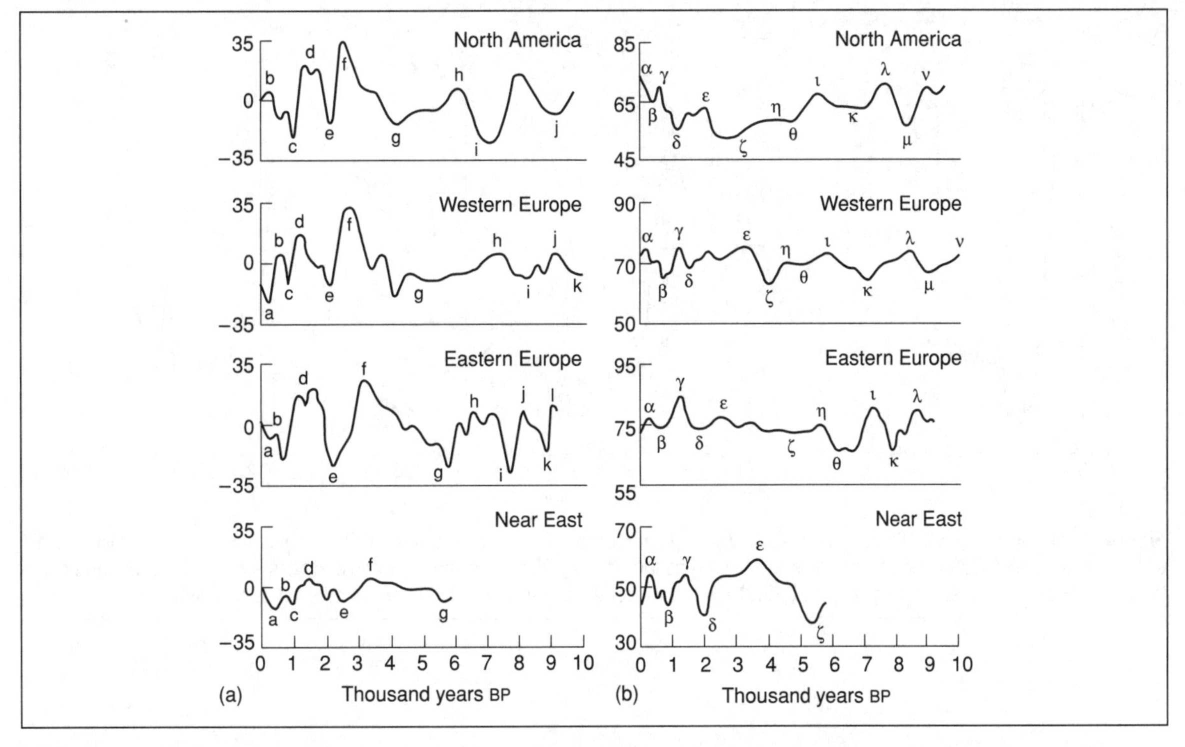

Figure 2.35 Regional declination (a) and inclination (b) master geomagnetic curves. Tree-ring calibrated timescale in calendar years BP (after Thompson, R. and Oldfield, F. (1986) *Environmental Magnetism*, Allen and Unwin, London)

Quaternary science can be found in Maher and Thompson (1999).

Oxygen isotope chronology

As the oxygen isotope signal in microfossils contained in deep-ocean sediments represents a global response to changes in land-ice volumes (see above), the inflections in the isotopic profiles obtained from cores from different parts of the world's oceans are essentially time-parallel events, and therefore constitute key marker horizons (Lowe and Walker, 1997). Not only do these provide a basis for inter-core correlations, but they also define the boundaries of isotopic stages and substages (Figure 2.21). As will be seen in Chapter 3, long-term climatic changes that are reflected in these isotopic records appear to be driven principally by changes in the earth's orbit and axis (the so-called **astronomical** or **Milankovitch** variables). Since these orbital variations are believed to be constant and their frequency is known, they provide a basis for timing the cycles in the marine oxygen isotope records, for the age of each cycle represented by the isotopic stages can be calculated by extrapolating back from the present day (Imbrie *et al.*, 1984). This technique, known as **orbital tuning**, has produced what is known as the **SPECMAP timescale** (Figure 2.36). This is based on the amalgamation of several marine isotopic ('stacked') records, provides a high-resolution chronology for the past 300 ka (Martinson *et al.*, 1987), and constitutes the global geochronological framework for recent environmental change and human history (Lowe, 2001).

Artefact dating

The occurrence of humanly produced objects (**artefacts**) of pottery, flint, metal, etc. in Quaternary sediment sequences has also provided a basis for

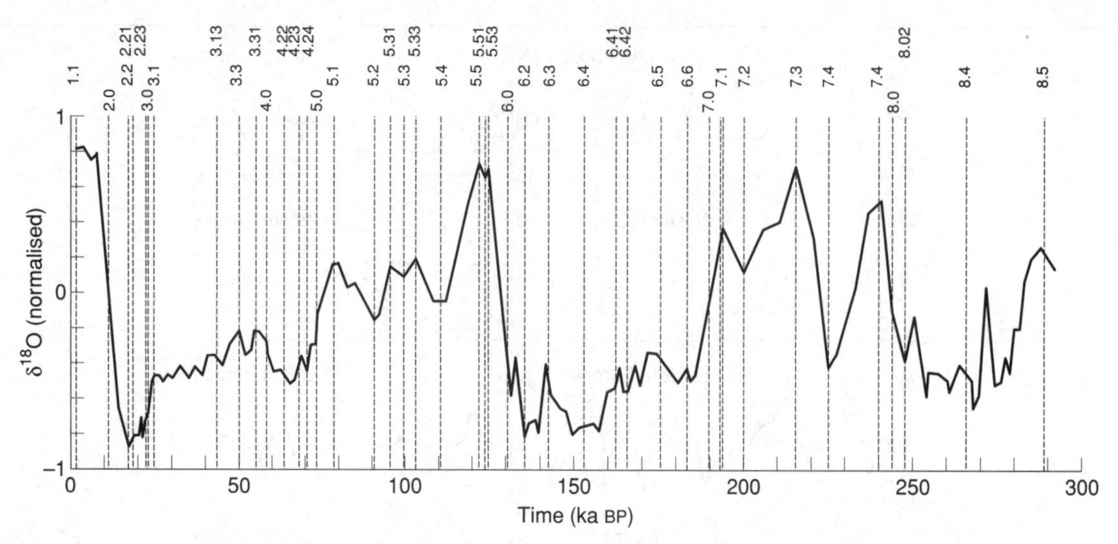

Figure 2.36 The composite ('stacked') deep ocean oxygen isotope record for the past 300 ka which has been 'orbitally tuned' to the astronomical (Milankovitch) variables (after Martinsson *et al.*, 1987). The major isotopic stages are indicated by integers (e.g. 3.0, 5.0) and the substages (e.g. 3.1, 5.3) and sub-substages (e.g. 3.13, 5.31) are numbered accordingly

dating. Archaeological periods are characterised by specific assemblages of objects. The ages of stone, bronze and iron, first defined in 1836 on the basis of Danish evidence by Christian Thomsen, have subsequently been considerably refined to form the basic framework for prehistory. Types and styles change through time as a result of technological innovation, improved skill levels, and because people wish to convey different meanings, for instance by the form and decoration of pots. Where several archaeological horizons are superimposed on one site, they often show marked changes in artefact type. The relative position of types within a sequence establishes a relative chronology. Where artefacts of the same type occur elsewhere in contexts which are historically or scientifically dated, **cross-dating** is possible. Examples of artefact dating include the use of flint, bone and stone tools to establish an Upper Palaeolithic chronology for parts of France and North America (Gamble, 1993), and the well-defined metalwork typologies of Europe which form the basis for continent-wide correlations during the Bronze and Iron Ages (Champion *et al.*, 1984). The discovery of metal artefacts in north European peat bogs (Coles and Coles, 1989) provides a valuable link between human activity and the broader picture of environmental and climatic change inferred from the peat bog records (p. 32).

The accuracy with which an artefact can be dated will be constrained by the overall chronological precision in the period and area concerned, and also by the timescale of innovation. In contexts from the last two millennia in Europe, or since European contact in North America, some artefacts may be dated to within a few decades. Coins, for example, can offer particular precision. In prehistory, however, artefact dating is more usually only accurate to within a few centuries. Moreover, there is always the possibility that objects, such as coins or pots, may have been in circulation for some considerable time before deposition. Hence, for much of the archaeological record, artefact dating is far less accurate than dendrochronology, varve chronology and, for older time periods, radiometric dating. Nevertheless, it remains an important technique for the dating of many archaeological sites, and is particularly important in the context of this book in so far as it links environmental sequences and human activity.

Notes

1. In the mutual climatic range method as applied to coleopteran data, modern distribution maps are first obtained for all of the species in a fossil assemblage, and the climatic range of each beetle type is then established using contemporary meteorological data. The two most important climatic parameters governing beetle distribution appear to be the temperature of the warmest month and the range between the warmest and coldest months. Using these two variables, the geographical range of each species can be plotted in 'climate space'. The overlap between these 'climate envelopes' for each coleopteran species will produce a *mutual climatic range* from which values of maximum temperature, annual temperature range and, by implication, minimum temperature can be obtained. These consitute the best estimates of the 'mutual climatic conditions' within which the particular mix of fossils existed (Lowe and Walker, 1997).
2. Loess is distinguished from coversand on the basis of grain size, the former typically falling in the size range 2–64 µm, while the latter is somewhat coarser (64 µm–2 mm).
3. For further information on the Greenland ice core project (GRIP) see: www.ngdc.noaa.gov
4. For further information on the Greenland ice sheet project (GISP2) see: www.gisp2.sr.unh.edu
5. Ratios of oxygen and hydrogen isotopes are measured not in absolute terms but as relative deviations (per ml) from the mean ratios of a standard, and are expressed in the form $\delta^{18}O$. Hence a $\delta^{18}O$ value of –30 per mille indicates that the sample is 0.3 per mille or 3 parts per millilitre deficient in ^{18}O relative to the standard. Standards are **PDB** (fossil belemnite shell) for carbonate (e.g. marine microfossils or speleothems) and **SMOW** (standard mean ocean water) for water, ice and snow.
6. For further information on radiocarbon calibration see: www.calib.org
7. The 'half-life' of a radioactive isotope is the period of time (t) that it takes for a given quantity of a radioactive isotope to reduce by half. Hence, if a parent nuclide is left to decay, after $t_{0.5}$ only 50 per cent of the original remains. It will then take the same period of time to reduce that 50 per cent to 25 per cent, and to reduce that 25 per cent to 12.5 per cent, and so on. This means that radioactive decay is not linear, but *exponential*. The half-life of radiocarbon is 5730 ±30 years, and the normal practical dating limit is 8 half-lives.
8. A 'floating chronology' is one in which tree-ring sequences are not tied *directly* to historically dated or living wood, but which can be related to the master chronology by other means, such as by high-precision and wiggle-match ^{14}C dating (p. 54).

3 Natural environmental change

Introduction

The environmental impact of climatic change can be observed at a range of spatial and temporal scales. Of fundamental significance are the long-term global climatic shifts that operate over time-scales of 10^4–10^6 years, and whose consequences are most spectacularly demonstrated by the repeated expansion and contraction during the course of the last 2–3 million years of the great continental ice sheets. Superimposed upon these macroscale climatic changes are short-term climatic fluctuations that occur over timescales of 10^1–10^3 years, and which are most notable for their effects on the vegetation cover of the early and mid-Holocene, on rates of operation of geomorphological processes, and on prehistoric and historic anthropogenic activity. Although this book is concerned primarily with environmental changes during the relatively recent past, the climatic regimes of the present interglacial reflect the operation of climatic processes over both long and short timescales. A proper understanding of recent climatic and environmental change can therefore only be achieved by examining the way in which climate has fluctuated over a range of temporal scales. Hence, the patterns of both long- and short-term climatic changes, their causal mechanisms and their impact on the landscape of the mid-latitude regions of the Northern Hemisphere, are examined in this chapter.

Patterns of long-term climatic change

A range of proxy data sources show that throughout the Tertiary period, the earth's climate has gradually cooled (Andrews, 1979), with this trend being particularly marked during the Miocene and Pliocene (from *c.*15 ma onwards). Moreover, pronounced oscillations increasingly became a feature of the global climatic pattern. Hence, although the prevailing climatic mode of the past 2.5 ma or so in the mid- and high-latitude regions of the world has been one of almost unremitting cold, markedly warmer climatic episodes have occurred at quasi-regular intervals. Thus, despite the popular conception of the Quaternary period as the Ice Age, it is now apparent that the **glacial** episodes, marked by expansion of the great continental ice sheets, were interspersed with **interglacial** phases during which global temperatures rose to be as high or even higher than those of the present day. There is also evidence to suggest that relatively short-lived climatic fluctuations occurred within the glacials and interglacials. A short cold episode which resulted in the local expansion of glaciers is usually termed a **stadial**, whereas a period of thermal improvement during a glacial episode when temperatures did not achieve levels comparable with those of the present day is referred to as an **interstadial**. Consequently, the pattern of long-term climatic change over the timescale of the past 2–3 million years is one of major oscillations between glacials and interglacials, superimposed upon which are minor climatic fluctuations involving stadial and interstadial episodes.

In low-latitude areas and in mid-latitude continental interior regions beyond the influence of glacier ice, cyclic changes in precipitation regimes occurred, with phases of higher rainfall (**pluvials**) being interspersed with drier intervals or **interpluvials.** In very general terms, there appears to be a relationship between these climatic oscillations and the glacial/interglacial cycles of the mid- and high latitudes, the glacial episodes usually being equated with interpluvial intervals and the interglacials with pluvial phases (Lowe and Walker, 1997).

Evidence for long-term climatic change

For many years, the most widely used proxy evidence for long-term climatic changes was the sequence of glacial and other cold-climate deposits found throughout Europe and North America. Early investigations of these glacial sediments led to the development of the classical scheme of four glacial stages, Gunz, Mindel, Riss and Würm in Europe, and Nebraskan, Kansan, Illinoian and Wisconsinan in North America, while analysis of organic sediments often found interbedded with the glacial deposits provided evidence of intervening interglacial episodes. The problem, of course, with this type of data is that as the ice sheets covered more or less similar areas during successive cold stages, evidence for earlier glacial episodes has been largely removed by succeeding ice advances. Hence, the terrestrial record from the mid- and high-latitude regions is unlikely to constitute anything other than a partial record of long-term climatic change. Moreover, the fragmentary nature of the evidence presents difficulties both in correlating glacial deposits and in the recognition of glacial stages (Lowe and Walker, 1997).

The shortcomings of glacial evidence as a proxy record for Quaternary climatic change have been thrown sharply into focus by data that have emerged over the course of the last three decades from the deep-ocean floors. By contrast with the terrestrial environment, sediments have been accumulating on the ocean bed in a slow but relatively uninterrupted manner throughout the Quaternary. Indeed, some marine depositional records extend as far back as 50 ma (Lear *et al.*, 2000). Technological developments in deep-sea coring in the years immediately following the Second World War enabled undisturbed cores of sediment to be raised from the ocean bed in water depths sometimes exceeding 3 km (Imbrie and Imbrie, 1979). The oxygen isotope trace obtained from microfossils taken from successive levels within these cores (see Figure 2.22) represents a continuous record of the changing isotopic composition of the ocean waters over time (see Chapter 2). As the isotopic signal from organisms that formerly inhabited the open oceans is largely a reflection of fluctuations in volumes of land ice (Shackleton and Opdyke, 1973), the isotopic trace provides a continuous record of changes in global ice volume which, by implication, can also be read as a record of glacial/interglacial fluctuations. Working from the top of the sequence, each isotopic stage has been given a number, even numbers denoting 'cold' (glacial) episodes while the 'warm' (interglacial) phases are denoted by odd numbers. Twenty-two isotopic stages can be recognised in the record from approximately the past 880 ka (Bassinot *et al.*, 1994), indicating something of the order of 10 glacials and 10 interglacials (or near interglacials) during that time period. Inferred ages for the boundaries of these isotopic stages are shown in Table 3.1. The total number of isotopic stages formally identified in Quaternary deep-ocean cores now extends down to marine oxygen isotope stage 116, which is dated to *c.*2.73 ma BP (Patience and Kroon, 1991). The most impressive feature of the deep-sea oxygen isotope record is that the pattern revealed by the isotopic trace is geographically consistent, and can be replicated in cores taken from different sectors of the world's oceans (Figure 3.1). Clearly, therefore, the marine oxygen isotope record provides a climatic signal of global significance.

In addition to oxygen isotopes, other indicators of long-term climatic change have also been

Table 3.1 Ages of marine isotopic stage boundaries over the past 880 ka (after Bassinot *et al.*, 1994)

Stage boundary	*Age (ka)*
1–2	11
2–3	24
3–4	57
4–5	71
5–6	127
6–7	186
7–8	242
8–9	301
9–10	334
10–11	364
11–12	427
12–13	474
13–14	528
14–15	568
15–16	621
16–17	659
17–18	712
18–19	760
19–20	787
20–21	820
21–22	865

obtained from ocean sediments. Using data on the modern ecological affinities of planktonic Foraminifera and other marine micro-organisms, former sea-surface temperatures (SSTs) can be inferred, and glacial–interglacial changes reconstructed (Figure 3.2). Temperature changes in deep-ocean waters can also be determined using the ratio between Mg and Ca in foraminiferal calcite (Lear *et al.*, 2000), while Cd/Ca ratios in foraminferal tests can be used to infer past variations in oceanic productivity and, in particular, changes in nutrient levels. These, in turn, can be related to variations in meltwater flux into the oceans and deep-water mass formation (Keigwin *et al.*, 1991). Deep-water circulation changes can also be inferred from carbon isotopes ^{13}C and ^{12}C ($\delta^{13}C$ ratios) in fossil Foraminifera (Saarnthein *et al.*, 1994), and from variations in grain size in ocean sediments (McCave *et al.*, 1995). A 20 ka multi-proxy record from the north-west North Atlantic from which not only surface temperature, but also salinity and sea-ice cover variations have been reconstructed, is shown in Figure 3.3.

Although the evidence from the deep-ocean floors provides the touchstone for reconstructions of climatic change during the Quaternary, proxy data from certain terrestrial contexts can provide a record of long-term climatic and environmental change that can often be correlated with the marine sequence. These include the lithostratigraphic and biostratigraphic evidence from deep tectonic basins, such as those beneath the Hungarian Plain or the southern Netherlands (Cooke, 1981; Zagwijn, 1996); long sedimentary sequences from deep lake basins including those in Australia (Kershaw, 1991), Colombia (Hooghiemstra *et al.*, 1993) and Japan (Fuji, 1988); shorter lacustrine sequences in volcanic craters (maars) or deep lakes spanning one or more glacial–interglacial cycles, for example in France (Reille *et al.*, 2000), Italy (Allen *et al.*, 1999) and the south-western United States (Lao and Benson, 1988); and the successions of loess and interbedded palaeosols that are found in many mid-latitude regions, but which are especially well developed on the Loess Plateau of China where loess/palaeosol sequences extend back *c.*2.4 ma (Kukla and An, 1989; Ding *et al.*, 1993). Detailed records of environmental change can frequently be obtained from these depositional contexts, including vegetational and climatic histories, with quantitative climatic reconstructions possible in a number of cases (Figure 3.4). Ice-core data, especially from the polar regions of Greenland and Antarctica (See Figure 2.20), but also from uplands in the tropics (Thompson, 2000), provide a further important archive of palaeoenvironmental data which can also be linked to the marine oxygen isotope sequence. The analysis of glacigenic sediments in both terrestrial and offshore areas, coupled with technological advances both in the investigation of subsurface stratigraphy and in techniques of dating, have led to the development of regional glacial stratigraphies which again can be related to the deep-sea oxygen isotope record (Elverhøi *et al.*, 1998; Sejrup *et al.*, 2000).

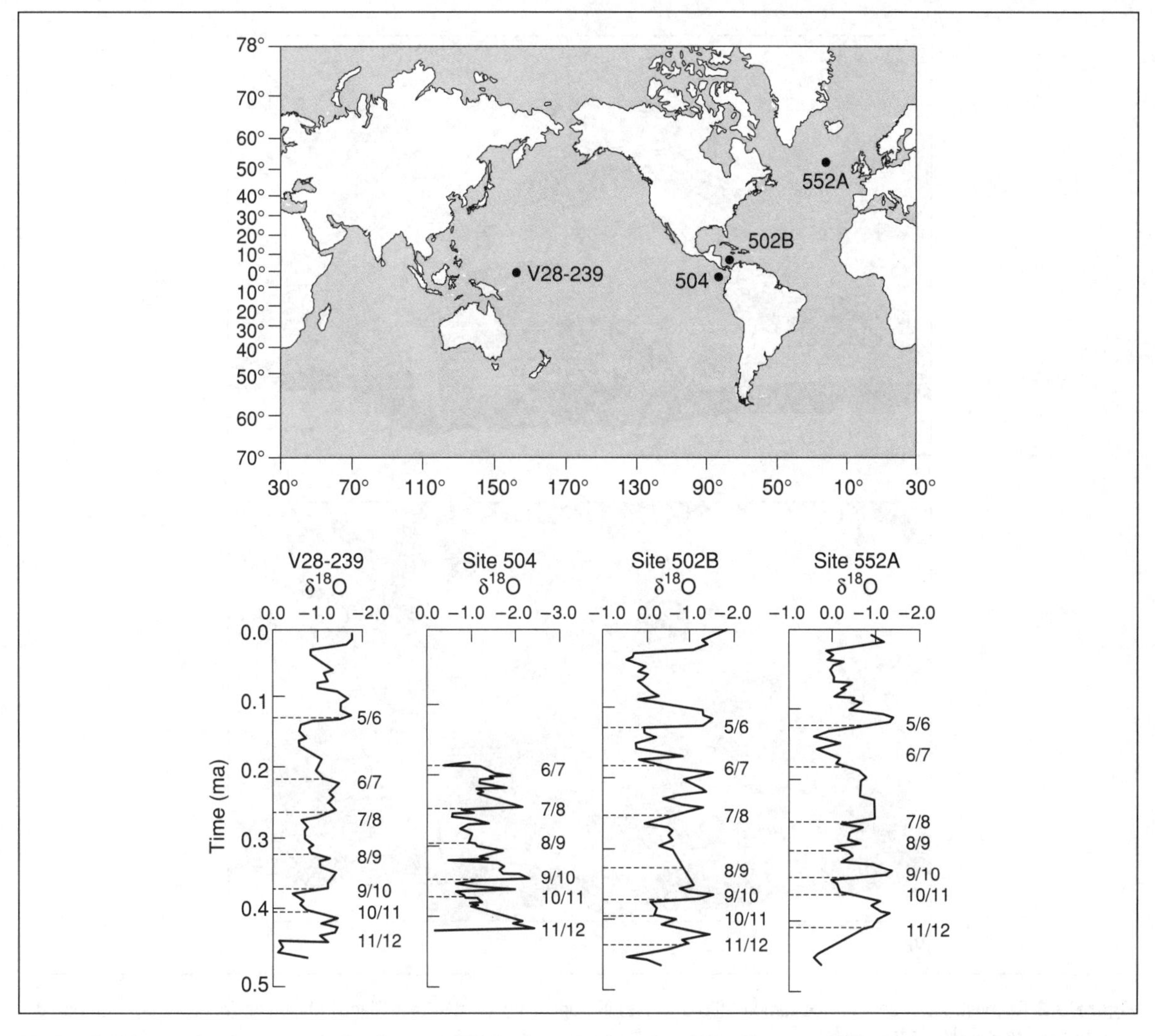

Figure 3.1 Oxygen isotope records for the past 470 ka from the Atlantic and Pacific oceans. The horizontal lines mark the boundaries between the marine oxygen isotope stages (after Williams *et al.*, 1988, Figures 1 and 5)

The nature of long-term climatic change

The oxygen isotope and other proxy records from deep-ocean sediments provide the most detailed indications of the nature and timing of long-term climatic change. The evidence suggests that over the past 800 ka or so, the global climate has been fluctuating in a rhythmical manner in a series of cycles, ranging in length from around 80 to 120 ka. Approximately 90 per cent of each cycle was characterised by glacial conditions in the mid- and high latitudes, with conditions as warm as those of the present interglacial accounting for only around 10 per cent of each glacial/interglacial cycle. Contrasts between full glacial and interglacial conditions were pronounced, involving annual temperature changes in excess of 15 °C, and marked variations in annual precipitation. Prior to *c*.800 ka, the fluctuations in climate appear to have been

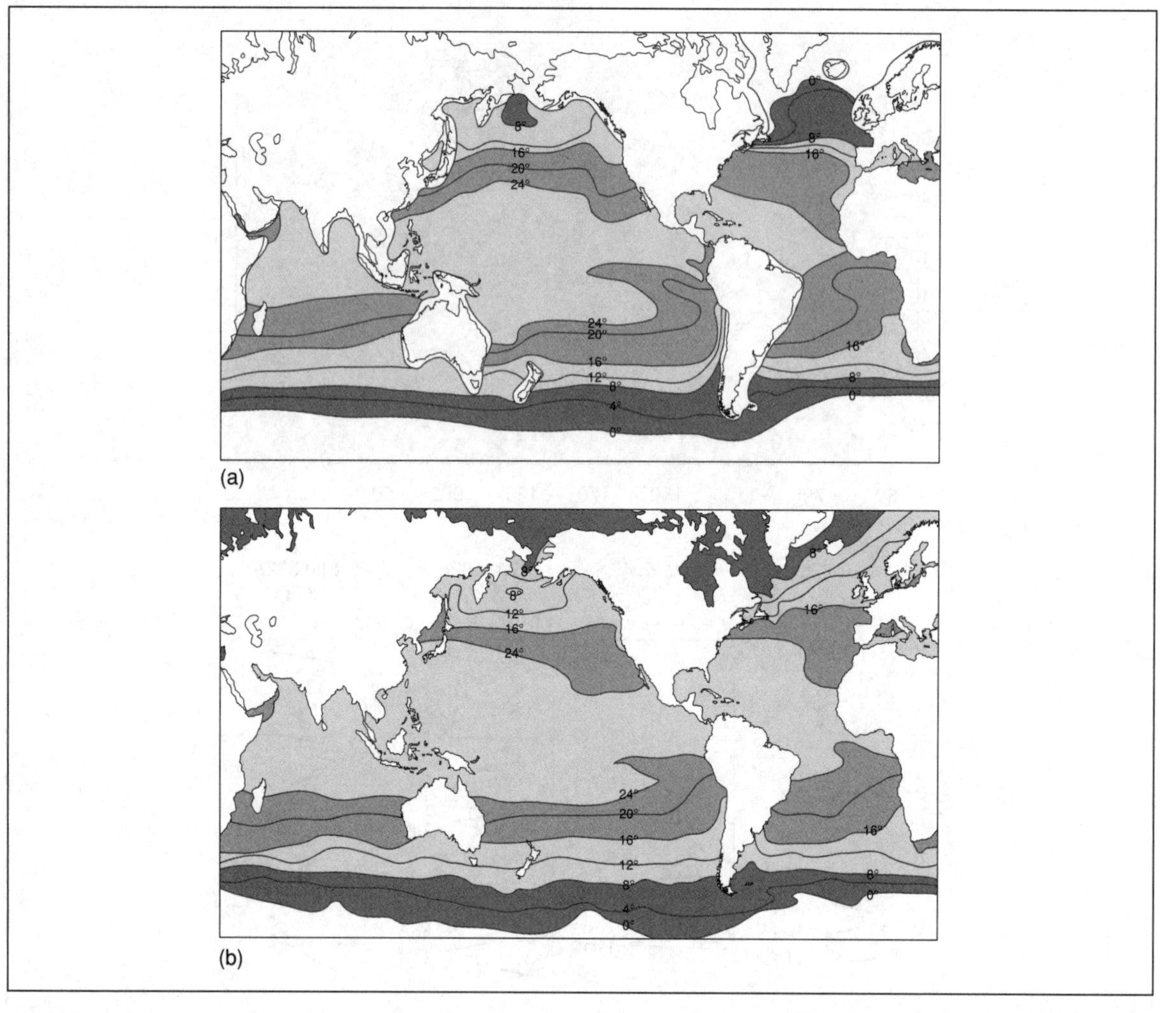

Figure 3.2 Reconstructed sea-surface temperatures (SSTs) in August for (a) the Last Glacial Maximum (*c.*22 ka BP), compared with (b) the present day (after CLIMAP, 1981)

of lower amplitude, however, with a periodicity of around 41 ka (Ruddiman *et al.*, 1986b). Possible reasons for this shift in climatic pattern are considered below.

A striking feature of the long-term climatic record is the relative rapidity of climatic change, something that has only become fully appreciated in recent years, with improvements in dating techniques and with the increasingly sophisticated nature of proxy climate records. In the oxygen isotope trace from the deep-ocean cores (Figure 3.1), the glacial stages typically end in abrupt **terminations,** i.e. sharp transitions in the isotopic profiles indicating an abrupt climatic amelioration. The magnitude and rate of change that occurred during these climatic shifts are remarkable, and are perhaps best reflected in the range of evidence from 'Termination 1' which marks the end of the last cold stage and the transition into the present (**Holocene**) interglacial. In Europe, for example, pollen, chironomid and cladoceran data from western Norway show a steep temperature rise of the order of ~6 °C in the first 500 years of the Holocene (~0.3 °C per 25 years), and a similar rate of

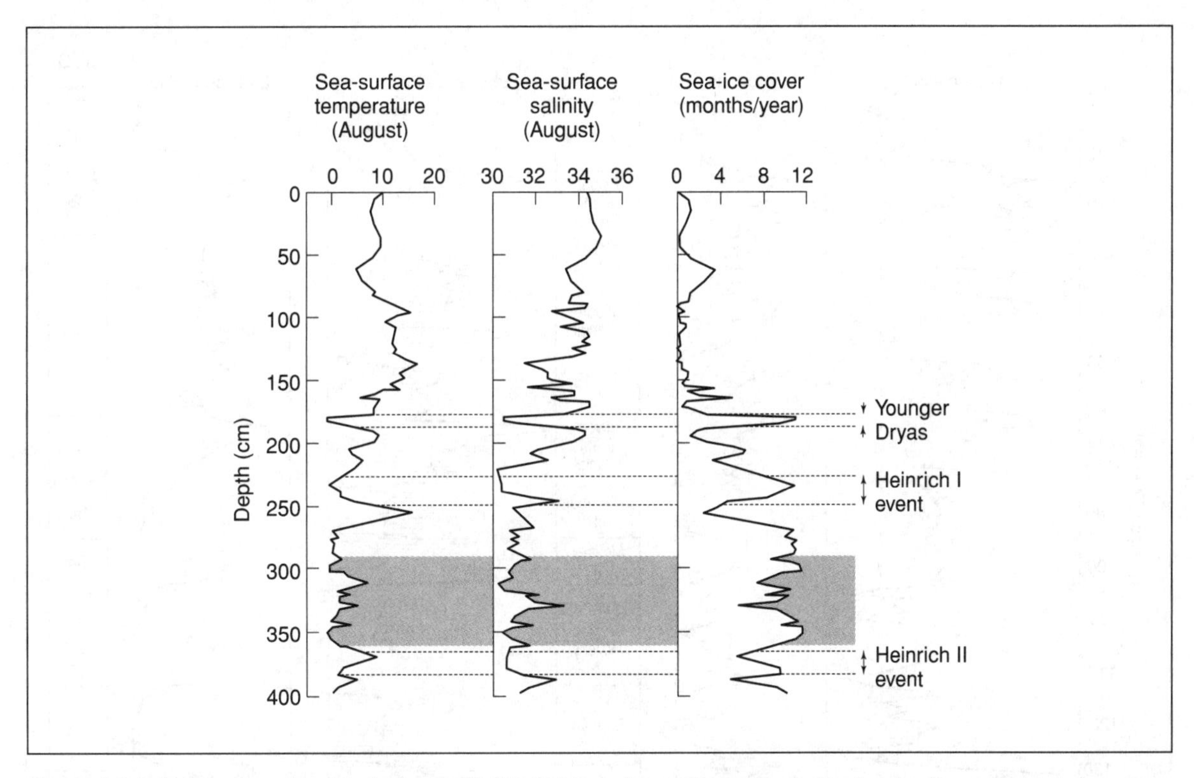

Figure 3.3 Palaeoceanographic data from core 91-045-094 from the north-west North Atlantic, showing reconstructed summer sea-surface temperatures, sea-surface salinity changes and variations in sea-ice cover. The shaded zone corresponds to the Last Glacial Maximum (LGM). Note the reduction in sea-surface temperatures and salinity, and the corresponding increase in sea-ice cover during the Younger Dryas, and during the Heinrich I and Heinrich II iceberg discharge events. These are explained in the text (after de Vernal and Hillaire-Marcel, 2000, Figure 11)

thermal improvement has been recorded from sites in Switzerland (Birks and Ammann, 2000). More spectacular rates of temperature change have been recorded in coleopteran evidence. Data from Britain suggest that the rise in mean annual temperature around 15 ka BP,[1] at the first warming at the end of the last cold stage (see below) may have been as much as 7.2 °C per century (Atkinson *et al.*, 1987), and comparable rates of climatic amelioration at the onset of the Holocene (around 11.5 ka BP) are indicated by coleopteran evidence from southern Sweden (Lemdahl, 1991). In the south-east Norwegian Sea, diatom evidence indicates an early Holocene rise in SSTs of around 9 °C within the space of half a century (Koç *et al.*, 1996), while in the Greenland ice-core record, a rise in temperature of 5–10 °C and a doubling of snow accumulation occurred at the beginning of the Holocene in 'less than a few decades, and possibly in less than a few years' (Alley, 2000). Such abrupt climatic changes must have had profound effects on the biosphere and, in particular, on human communities (p. 147).

Climatic changes in the North Atlantic region during the last cold stage

The chronology of the deep-ocean record suggests that the the last interglacial (the **Eemian** in Europe; **Ipswichian** in the British Isles; **Sangamon** in North America) ended around 115–120 ka BP and was followed by a general cooling trend, albeit with warmer interstadial episodes, which lasted until the beginning of the Holocene interglacial around 11.5 ka BP (Imbrie *et al.*, 1984).

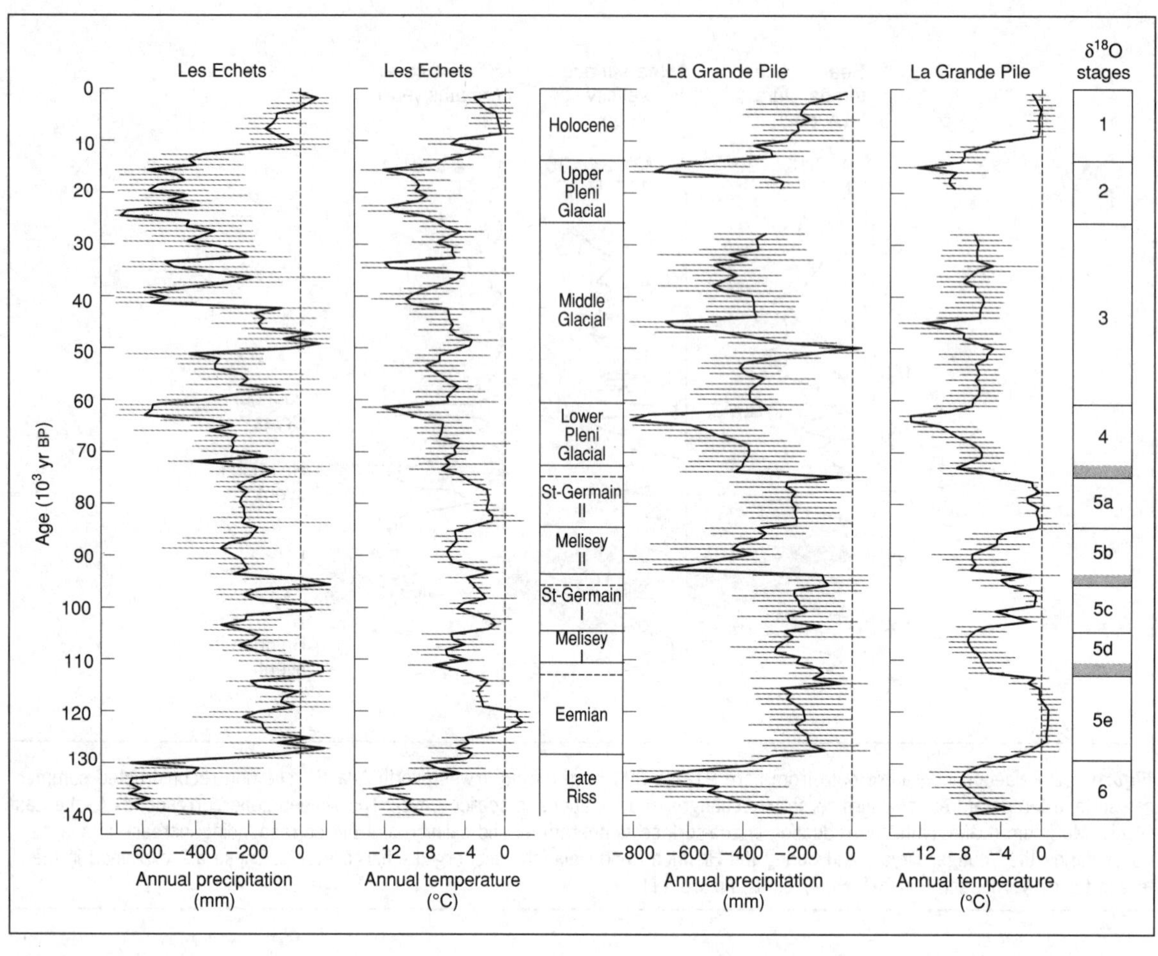

Figure 3.4 Reconstructions, based on pollen data, of variations in annual total precipitation and mean annual temperature (expressed as deviations from the modern values) over the past 140 ka in lake/mire sequences at Les Echets and La Grande Pile, eastern France. The climatic sequence is correlated with oxygen isotope stages 1–6 of the deep-ocean record (after Pons *et al.*, 1992, Figure 5)

This *c.*100 ka period of global cooling is also reflected in terrestrial sequences in both Europe and North America. By far the most detailed record of climate change during this period, however, is that contained within the Greenland ice cores, and this now constitutes the template against which climatic records from other parts of the North Atlantic region can be measured (Walker *et al.*, 1999).

Greenland

Data are now available from five major coring sites on the Greenland ice sheet, and drilling at a sixth (NorthGRIP) has recently been completed (Johnsen *et al.*, 2001). The $\delta^{18}O$ profiles are remarkably similar and can be regarded as proxies for regional climatic change (Figure 3.5). Greenland temperatures are estimated to have been *c.*20 °C colder at the last glacial maximum at around 20–25 ka BP. A distinctive feature of all of the records, however, is the recurring sequence of abrupt changes in oxygen isotope values over the course of the last cold stage (Johnsen *et al.*, 1992; Dansgaard *et al.*, 1993). These **Dansgaard–Oeschger events** lasted 500–2000 years and were characterised by temperature changes of up to 15 °C (Figure 3.6),

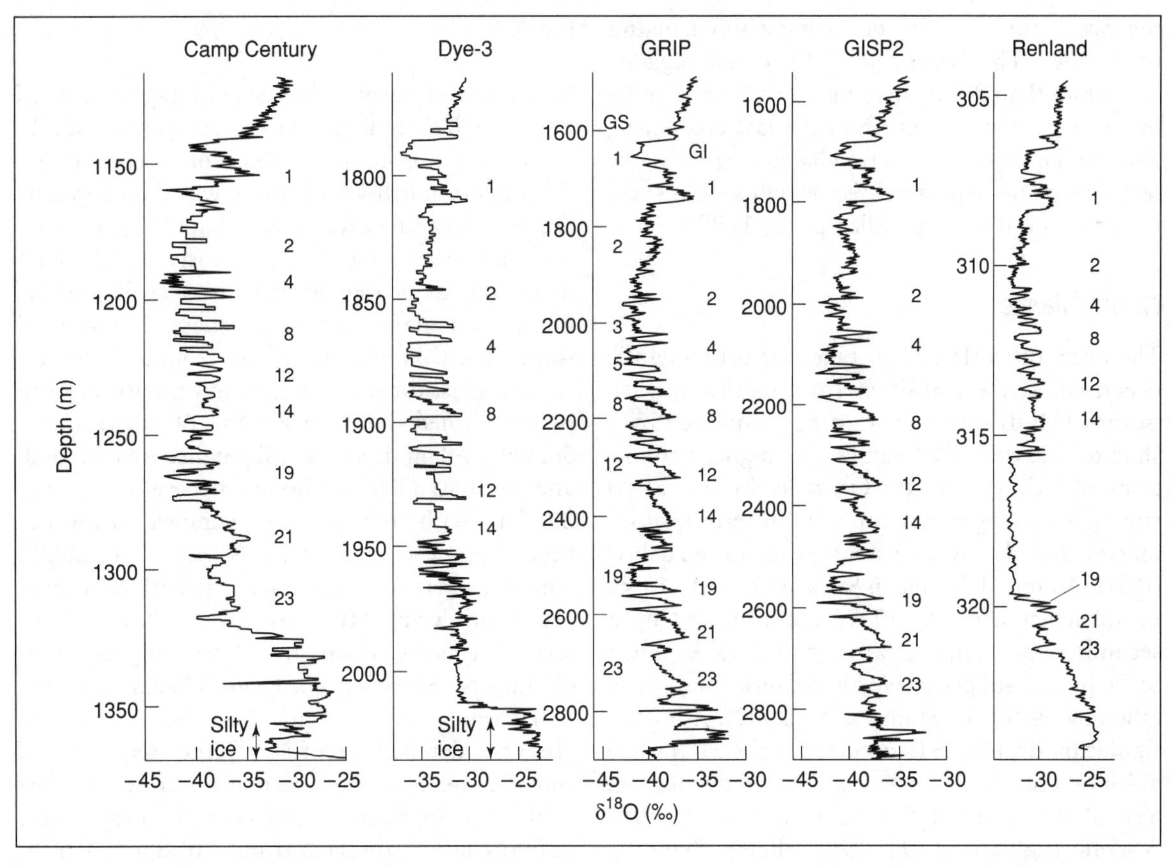

Figure 3.5 Continuous $\delta^{18}O$ profiles along sections of five Greenland deep ice cores. These are plotted on linear depth scales and extend back to before the beginning of the last cold stage. Some of the warm Greenland Interstadials (GI) of the last glacial cycle, defined initially in the GRIP ice core, are shown to the right of each profile (after Johnsen *et al.*, 2001, Figure 2)

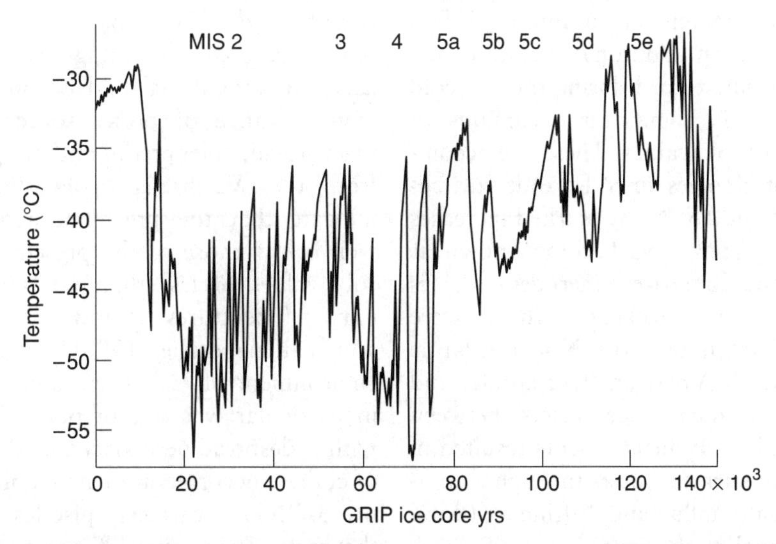

Figure 3.6 Past temperature changes for Summit, Greenland, reconstructed from the GRIP ice core. Temperatures at the Last Glacial Maximum are around 20 °C colder than present, while the amplitude of the rapid temperature shifts associated with the Dansgaard-Oeschger cycles between 15 and 80 ka BP are of the order of 12–15 °C (after Johnsen *et al.*, 2001, Figure 5)

the initial rapid warming occurring within a matter of decades. The Greenland $\delta^{18}O$ record suggests that more than 20 of these marked climatic shifts occurred during the course of the last cold stage, and similar abrupt climatic changes can also be seen in marine sequences (see below) and in some long terrestrial records (Allen *et al.*, 1999).

North Atlantic

The onset of the last cold stage in North Atlantic ocean cores, the transition from marine oxygen isotope (MOI) substage 5e to 5d, is marked by a shift to 'heavier' $\delta^{18}O$ values (i.e. higher proportions of ^{18}O relative to ^{16}O), reflecting increased storage of the lighter isotope ^{16}O in terrestrial ice masses, and also by an increase in ice-rafted debris (IRD). An initial decline in SSTs at around 115 ka BP (Shackleton *et al.*, 2002) was followed by a second cooling event at around 107 ka BP when SSTs in the subpolar North Atlantic may have fallen by ~4 °C (McManus *et al.*, 2002). A further significant drop in SSTs occurred at the MOI stage 5/4 boundary (*c.*70–75 ka BP), with reductions in SST of the order of 5–6 °C in the low-latitude Atlantic (Eglinton *et al.*, 1992), but perhaps by as much as 10 °C or more further north. The Dansgaard–Oeschger events recorded in the Greenland ice cores also appear to register in deep-ocean sediments in the North Atlantic, for SST shifts on millennial timescales and with strong amplitudes (up to ~5 °C) are apparent in foraminiferal data (Bond *et al.*, 1993). An additional feature of the North Atlantic marine record during the last cold stage are at least six significant excursions to heavier oxygen isotopic values. These are accompanied by abrupt changes in SSTs and, particularly between 40° and 55° N, by marked increases in IRD. These horizons have become known as **Heinrich layers** and comprise minerogenic debris released from melting icebergs which were periodically discharged into the North Atlantic from both the North American (Laurentide) and European (Fennoscandian) ice sheets between *c.*14 and 70 ka. These **Heinrich events** resulted in extreme cooling of surface waters through a combination of meltwater influx and drifting ice (Bond *et al.*, 1992; Fronval *et al.*, 1995).

Europe

In north-west Europe, the last cold stage is referred to as the **Weichselian**, while in central and southern Europe it is known as the **Würm**. Throughout this period, north-west Europe was dominated by the Fennoscandian ice sheet (Chapter 4), with ice build-up beginning during substage 5d, with major retreat phases during substages 5c and 5a, but with a major expansion phase during MOI stages 4–2 (Mangerud, 1991). Around the southern margins, three major maximum positions can be established: between 80 and 60 ka (**Karmøy Stadial**), 50 and 37 ka (**Skjonghelleren Stadial**) and the Last Glacial Maximum between 22 and 28 ka, with the shorter **Tampen readvance** occurring around 18–15 ka (Sejrup *et al.*, 2000). Similar patterns of glacier activity, with maximum glaciation during MOI stages 4 and 2, have been identified in the mountains of central and southern Europe (Sibrava *et al.*, 1986; Schlüchter, 2000).

Beyond the limits of these ice sheets and glaciers, detailed records of the sequence of climatic changes that occurred during the last cold stage in western and central Europe have been obtained from lacustrine, fluvial and aeolian deposits, and from their associated periglacial structures (Vandenberghe *et al.*, 1998b). In addition, long sedimentary sequences in deep lake sites, for example in north-eastern France (Beaulieu and Reille, 1984, 1992) and in the Massif Central (Reille *et al.*, 2000), have produced continuous pollen records extending back into the last (Eemian) interglacial and beyond. Climatic data from a range of proxy sources (pollen, plant macrofossil, coleopteran and periglacial evidence) from the Weichselian cold stage (Figure 3.7) indicate that the prevailing temperatures were well below those of the present day, with mean July values of less than 10 °C and mean January temperatures as low as –25 °C (Huijzer and Vandenberghe, 1998). However, while the environment in northern and western Europe in particular was one of polar desert conditions with widespread permafrost, marked ameliorations of climate occurred at quasi-regular intervals (Figure 3.7). The warmest episodes occurred during the early Weichselian/Würm in the Brörup and

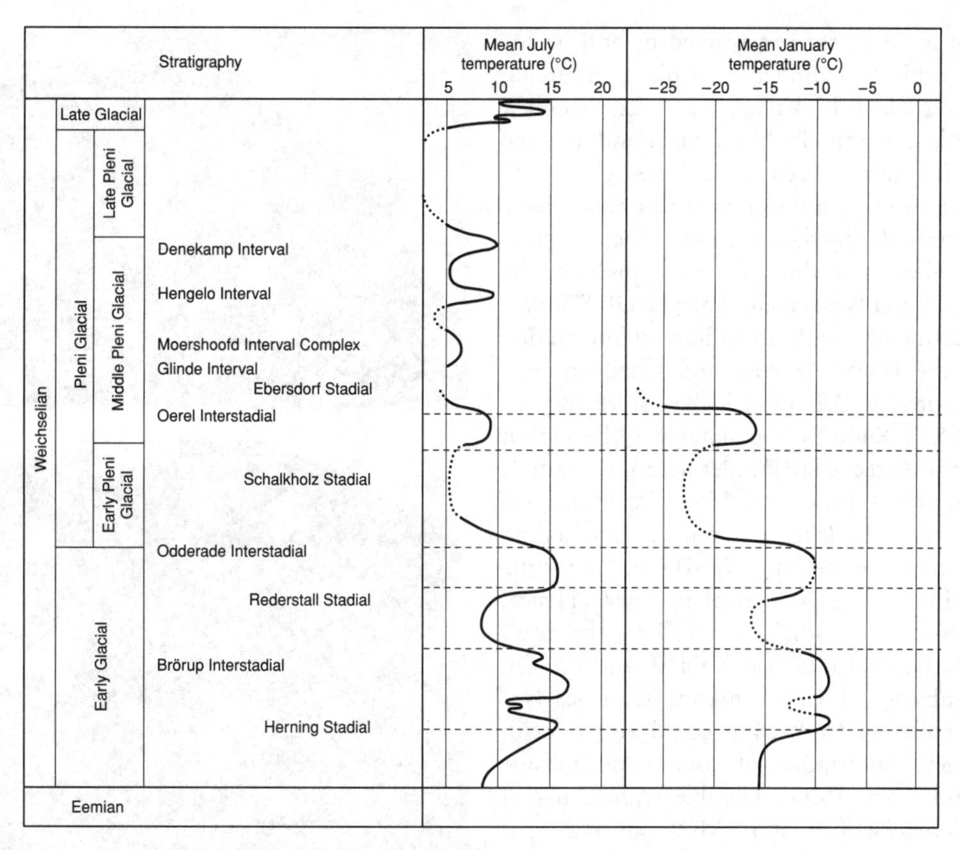

Figure 3.7 Reconstructed Weichselian summer and winter temperatures for north-west Germany based on palynological data, plant macrofossils and fossil Coleoptera (after Caspers and Freund, 2001, Figure 13)

Odderade interstadials of the Netherlands and northern Germany, when summer temperatures appear to have been only a few degrees below those of the present day (Caspers and Freund, 2001). These warm intervals appear to be the equivalent of the St Germain I and II interstadials (the former including two warm episodes separated by a brief deterioration) in the French pollen records (Guiot *et al.*, 1989). Correlation with the deep-sea record suggests ages of *c.*100 ka BP for the Brörup/St Germain I and *c.*80 ka BP for the Odderade/St Germain II warm episodes. Well-documented later interstadials (in the Netherlands and northern Germany) include those at **Oerel** (58–54 ka BP), **Glinde** (51–48 ka BP), **Moershoofd** (46–44 ka BP), **Hengelo** (39–36 ka BP) and **Denekamp** (32–28 ka BP: Behre and van der Plicht, 1992). By far the most widely recorded episode of climatic warming, however, is the **Bølling-Allerød Interstadial**, which has been recognised throughout northern and western Europe, and which has been dated to *c.*14.7–12.8 ka BP.

The British Isles

For much of the last (**Devensian**) cold stage, large areas of the British Isles resembled arctic tundra, with only the northern and western regions being directly affected by glacier ice. Ice sheets almost certainly developed in Highland Britain during the early Devensian, but while some glacial deposits have been tentatively attributed to this period, the extent of glacier ice and the timing of glacier events remain to be established. The Late Devensian glaciation is far better constrained in both space and time, with the main ice-sheet phase occurring between *c.*28 and 15 ka BP, when ice extended southwards down the Irish Sea basin and into the English Midlands (Bowen *et al.*, 2002).

As in Europe, a number of interstadials have been identified during the Devensian. There is widespread evidence for three warmer episodes, the **Chelford, Upton Warren** and **Lateglacial/Windermere Interstadials**, while an additional interstadial between the Upton Warren and Chelford may be represented at **Brimpton** in Berkshire (Bryant *et al.*, 1983). Radiocarbon dates for the Upton Warren and Lateglacial/Windermere interstadials suggest ages of *c.*46 ka and 15–13 ka BP respectively, while TL (thermoluminescence) dating indicates an age in the range 90–100 ka BP for the Chelford Interstadial (Rendall *et al.*, 1991). Hence, the Chelford Interstadial may well be the equivalent of the Brörup/St Germain II and oxygen isotope substage 5c of the marine sequence (see above), while Upton Warren could equate with the Hengelo Interstadial of continental Europe (Jones and Keen, 1993). Possible equivalents of the Odderade/St Germain (MOI substage 5a) interstadials have been found at sites in North Wales and in the upper Thames Valley (Chambers *et al.*, 1995; Maddy *et al.*, 1998), while in eastern Scotland and in the Outer Hebrides (Figure 3.8) there are records of short-lived interstadial episodes that may be correlated with the Glinde and Denekamp interstadials of continental Europe (Whittington, 1994; Whittington and Hall, 2002). Coleopteran evidence points to summer temperatures of around 15 °C during the Chelford Interstadial (1–2 °C lower than the present day), but reaching 18 °C in the Upton Warren and Lateglacial/Windermere interstadials. During the intervening cold periods, summer temperatures remained below 10 °C, with winter temperatures as low as –25 to –30 °C and mean annual temperatures typically in the range –8 to –12 °C.

Figure 3.8 Photographs of the interstadial deposits at Tolsta Head, Outer Hebrides, Scotland (photos Mike Walker)

Northern United States, Canada and the Arctic

In North America, the period corresponding to MOI substages 5d to 5a is termed the **Eowisconsinan**, while the MOI stages 4 to 2 are referred to as the **Wisconsinan**. Early in the Eowisconsinan, perhaps around 115 ka BP (~MOI substage 5d), the Laurentide ice sheet began to develop over Keewatin, Labrador and Baffin Island, and reached its maximum extent around its northern margins possibly during substage 5b (Clark *et al.*, 1993). However, large areas of northern Canada were subsequently deglaciated during MOI substage 5a (~85 ka BP: Miller *et al.*, 1999b), before the ice sheet expanded once again during MOI stage 4. A marked retreat phase in MOI stage 3 was followed by a major advance along the north-western, southern and north-eastern margins between ~27 and 24 ka BP. Around most of the ice margin, the Late Wisconsinan maximum ice extent either exceeded the extent of earlier Wisconsinan advances, or was similar to the Early Wisconsinan advance (Dyke *et al.*, 2002). During the Late Wisconsinan also, a major ice sheet (the Innuitian ice sheet) developed over the islands of the Canadian Arctic, which may have coalesced with the Greenland ice sheet to the east (England, 1999).

Important interstadial episodes during the Wisconsinan were the **St Pierre** at around 85 ka BP (~MOI 5a), the long **Port Talbot** dated to between *c.*64 and 40 ka BP, and the **Plum Point** (*c.*35–32 ka BP). Further south, at least five significant climatic shifts (from wetter to drier episodes) between 50 and 20 ka BP have been recorded in a pollen record from Florida (Grimm *et al.*, 1993). Interstadial episodes during the retreat of the Laurentide ice sheet include the **Erie Interstadial** at *c.*15.5–15 k ^{14}C years BP (*c.*18–17.5 ka BP), during which a significant retreat of the southern Laurentide margin occurred (Clark, 1994), and the **Two Creeks Interstadial** *c.*12.1–11.8 k ^{14}C years BP/14.1–13.8 ka BP (Kaiser, 1994). Pollen data from Baffin Island suggest that summer temperatures higher than at any time during the Holocene prevailed around 85 ka BP (Miller *et al.*, 1999b), while temperatures in the eastern USA comparable with those of the present day have been inferred for the early Port Talbot Interstadial (Berti, 1975). At the last glacial maximum, around 19–22 ka BP, temperatures in the eastern and central United States, as reconstructed from pollen and plant macrofossil evidence, were of the order of 10 °C below modern values for the summer months and 15–20 °C cooler during the winter months (Figure 3.9), with precipitation up to 40 mm below modern levels (Jackson *et al.*, 2000). Climatic warming following deglaciation during the period 15–13 ka BP saw summer temperatures rise to 16 °C throughout the north-eastern seaboard region, while to the south of the Great Lakes region, warming approached modern July conditions, but January temperatures remained well below those of the present day (Shane and Anderson, 1993).

Causes of long-term climatic change

The cause of global climatic changes over timescales of 10^4–10^6 years has been one of the longest-standing and most intractable of scientific problems. With the introduction, and ultimate acceptance, of the glacial theory in the middle years of the nineteenth century, added impetus was given to the search for an explanation of long-term climatic changes of such magnitude as to carry the temperate regions of the world from full glacial to interglacial conditions within what, in geological terms at least, appeared to be a relatively short timespan.

Numerous hypotheses that seek to explain long-term climatic change have been proposed over the years. Some theories relate to geographical changes on the earth's surface (**terrestrial theories**), including mountain-building episodes, changes in the patterns of oceanic circulation and changes in the disposition of the continental land masses (plate tectonics). Such geographical changes will inevitably affect the distribution of heat across the surface of the earth and hence, it has been argued, could lead to climatic change. Other theories involve possible changes in the earth's atmosphere (**atmospheric theories**), such as variations in the content of carbon dioxide and other trace gases, changes in the amounts of water vapour, and the occurrence of volcanic aerosols and particulates. These will influence the transmission of solar radiation through the atmosphere, and hence the amount of radiant energy received at the earth's surface. A third group of theories (**solar theories**) involves structural changes in the sun itself, perhaps leading to variations in the output of radiant heat and hence temperature fluctuations on the earth.

Most of these theories were developed before 1960, and their essentially speculative nature

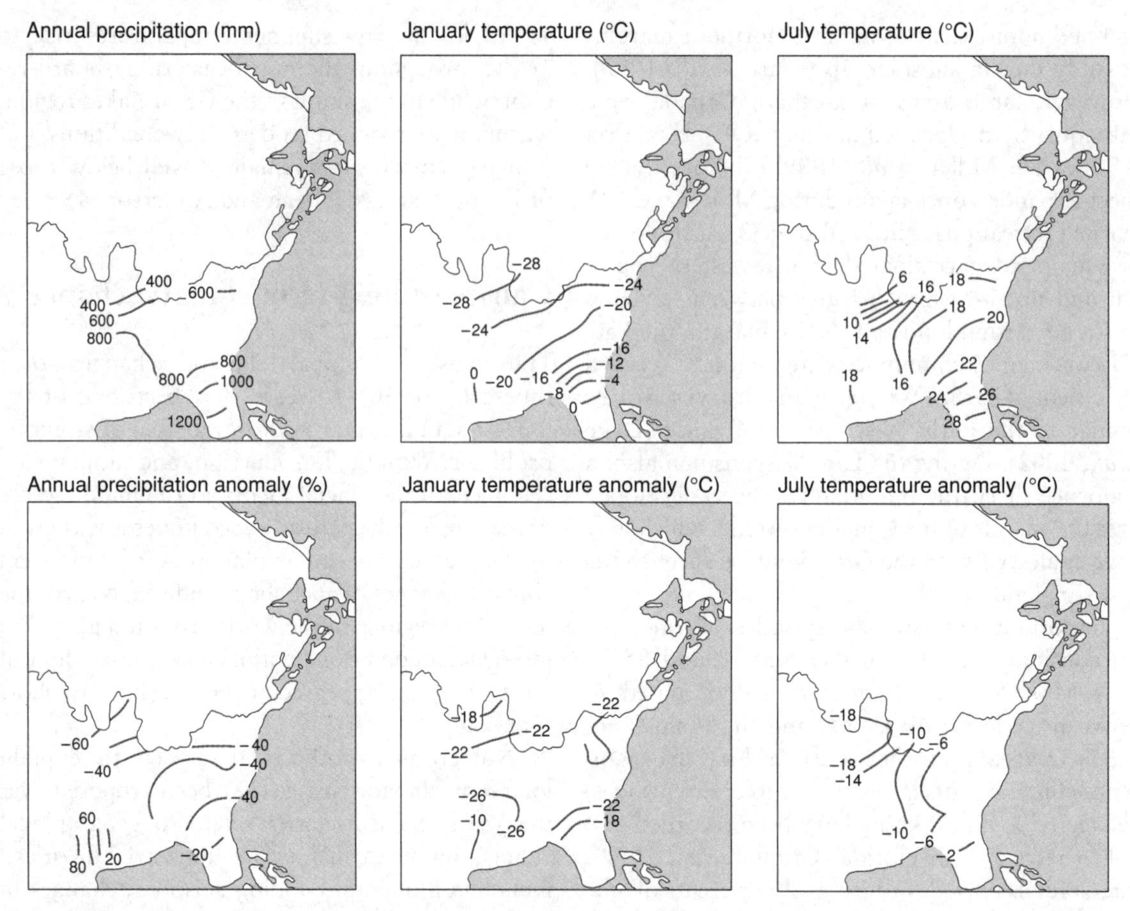

Figure 3.9 Maps of the Last Glacial Maximum climate and climatic anomalies (differences from modern climate) in eastern North America reconstructed from pollen and plant macrofossil data (after Jackson *et al.*, 2000, Figure 7)

reflects the fact that prior to that date, little firm evidence could be adduced in support of any of them. The chronology of long-term climatic change was uncertain, few reliable long proxy records existed, and the state of knowledge of atmosphere–ocean–terrestrial linkages was relatively rudimentary. Indeed the database was so limited that it was almost impossible to test the large number of theories that were then proliferating. During the 1960s and 1970s, however, technological improvements and innovations in a number of different areas of Quaternary research prompted a re-evaluation of the nature and causal mechanisms of long-term climatic change. Of particular significance was the development of coring equipment which, for the first time, allowed long sedimentary records to be recovered from terrestrial sites such as deep lake basins and, more importantly, from the ocean floors where undisturbed sediment sequences spanning the entire Quaternary period are to be found. The discovery of oxygen isotope variations in the microfossils of deep-ocean sediments (Chapter 2) provided a continuous proxy climatic record, while technical developments in dating techniques enabled a chronology of long-term climatic change to be established. The deep-sea record suggested a periodicity in long-term climatic change, i.e. cyclical fluctuations in climate over a long time period, that had been suspected but which, hitherto, had never been unequivocally demonstrated. This has led to a resurgence of interest in the notion that long-term climatic change might be a consequence of the ways in which the earth moves around the sun, what has become known as the **astronomical theory** of climatic change.

The astronomical theory

The idea that climatic changes may be triggered by variations in the earth's orbit and axis was first suggested by J.F. Adhemar in 1842 (Imbrie and Imbrie, 1979). The theory was further developed by James Croll in the 1870s, but was really given substance by the Serbian geophysicist Milutin Milankovitch who, in the 1920s, produced a substantial body of data to show that orbital and axial changes (which could be detected by astronomical observations and which had, in fact, been known for some time) would lead to variations in the amounts of radiant energy received at the earth's surface. Moreover, that these data could be depicted graphically to show how insolation at different latitudes had varied over time (Figure 3.10). Milankovitch also suggested that a link might reasonably be postulated between such fluctuations and glacial episodes. Until the early 1970s, however, the astronomical theory was largely disputed, principally because the quasi-periodicities of the earth's orbital elements could not be identified in the relatively fragmentary geological records then available, but also because of widespread uncertainty within the scientific community about the reliability of the long-term variations of the earth's orbital elements and the predicted insolation effects. Doubts were also expressed about whether a correlation could be established between the insolation curves and geological data, and also whether the predicted insolation changes could have induced climatic changes of the magnitude of those evident in terrestrial geological and biological records (Berger, 1980).

In a seminal paper in 1976, however, Hays *et al.* convincingly demonstrated that the astronomical frequencies were present in the isotopic record from the deep oceans, and it has subsequently become apparent that variations in the earth's orbit and axis are also reflected in other geological and biological data (see below). In essence, a scientific revolution has occurred to the extent that other theories of long-term climatic change have been relegated to a secondary role, with variations in the earth's orbit and axis now widely accepted as the 'Pacemaker of the Ice Ages' (Berger, 1992; Imbrie *et al.*, 1993a).

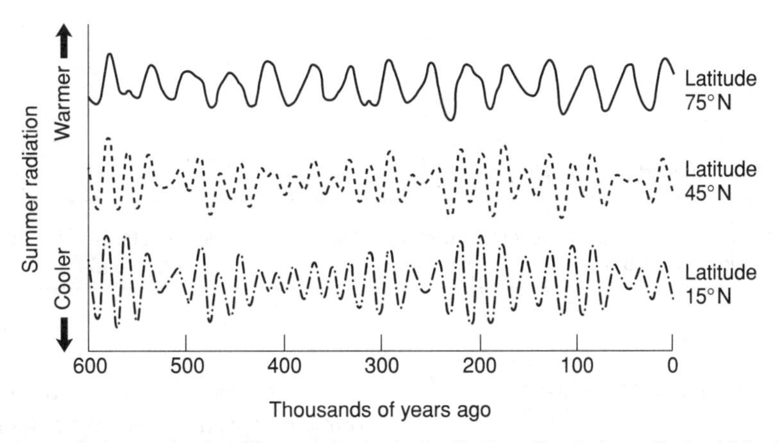

Figure 3.10 Milankovitch radiation curves for different latitudes in the Northern Hemisphere. First published in 1938, these show changes in summer radiation at 75° N, 45° N and 15° N (after Imbrie, J. and Imbrie, K.P. (1979) *Ice Ages: Solving the Mystery*, Macmillian, London)

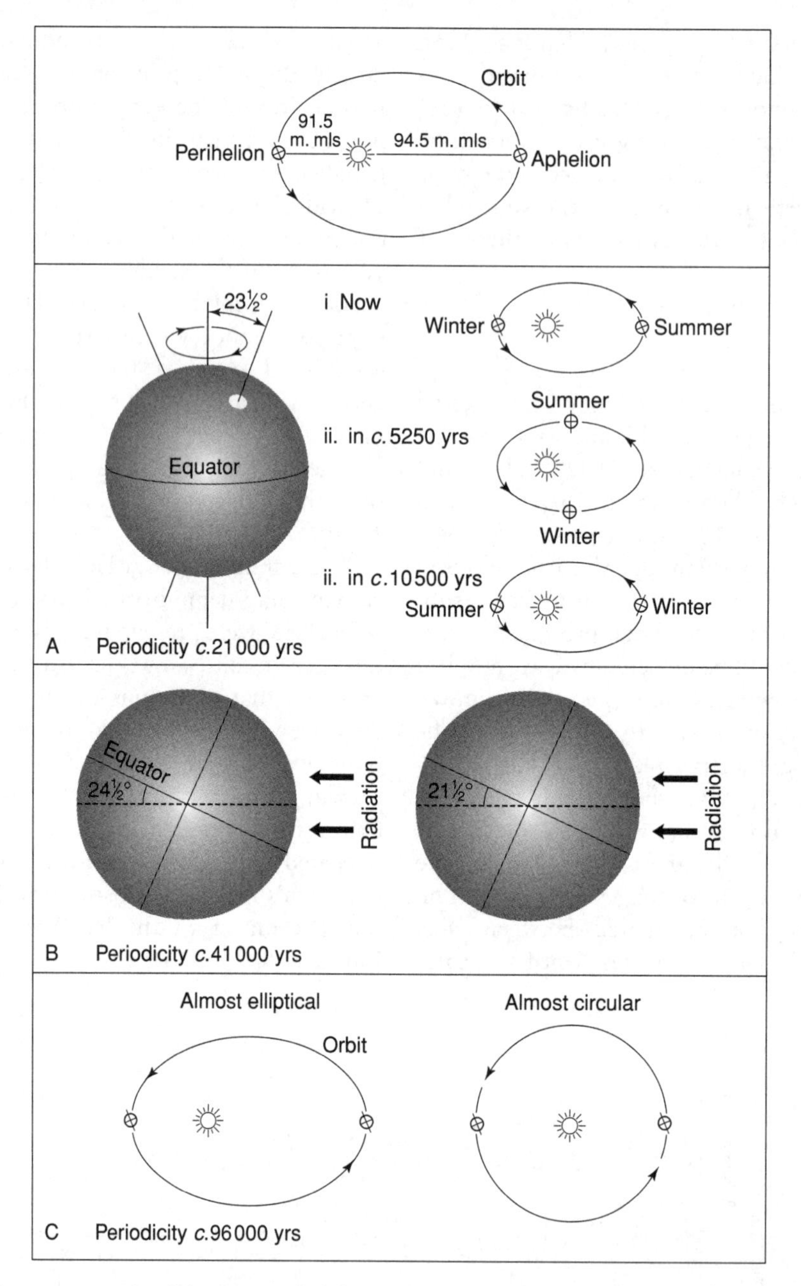

Figure 3.11 The three components of the astronomical theory of climate change: (a) precession of the equinoxes; (b) obliquity of the ecliptic; (c) eccentricity of the orbit

Elements of the astronomical theory

The precession of the equinoxes

The earth revolves around the sun in an elliptical orbit (Figure 3.11a) and each complete revolution takes one year. At times, therefore, during the course of a year the earth passes closer to the sun than at others. The period of closest passage is known as **perihelion** and the time of year when the earth is furthest from the sun is referred to as

aphelion. Clearly, when the earth is in perihelion (presently *c*.147 million km from the sun) there will be a greater intensity of solar radiation receipt than in aphelion (152 million km). However, because the earth is tilted on its axis (currently 23.5°), the heat received by the earth will not be uniformly distributed, but will vary depending on whether the Northern or Southern Hemisphere is tilted towards or away from the sun. At present, the Northern Hemisphere summer occurs in aphelion and the Southern Hemisphere summer in perihelion. Hence, Southern Hemisphere summers will be hotter than those in the Northern Hemisphere. By contrast, Northern Hemisphere winters (perihelion) should be milder than those experienced in the Southern Hemisphere (aphelion).

This situation is not immutable, however, for because of the gravitational pull exerted on the earth's equatorial bulge by the sun and the moon, the earth wobbles on its axis like a top (Figure 3.11a). As a consequence, the direction of tilt varies over time to the extent that after *c*.10.5 ka it has become completely reversed. The effect of this axial change is that the Northern Hemisphere summer now occurs in perihelion and the Southern Hemisphere summer in aphelion. Seen from above (Figure 3.11a), the seasons seem to move around the sun in a slow, regular fashion, hence the term **precession of the equinoxes** or **precession of the solstices**.[2] Hence, there is a regular cyclical change in patterns of heating and cooling across the globe. After 21 ka or so, the cycle is complete and a new one begins. In recent years, however, it has been established that although the precessional cycle averages around 21 ka, there are in fact two separate and interlocked cycles, a major one at around 23 ka and a minor one at *c*.19 ka.

The obliquity of the ecliptic

The second element in the astronomical theory (**obliquity of the ecliptic**) involves changes in the angle of tilt of the earth relative to the plane of the ecliptic (Figure 3.11b). Over time, the tilt will vary from 21° 39′ to 24° 36′ and any change in tilt will amplify or reduce seasonal contrasts. The greater the tilt, for example, the more pronounced will be the differences between summer and winter. Changes in axial tilt occur with a periodicity of around 41 ka. In other words, it takes around 41 ka for the tilt of the earth to change from *c*.24.5° to 21.5° and back to 24.5° once more.

The eccentricity of the orbit

As already noted, the earth's orbit around the sun is not circular but elliptical, a major consequence of which is that summers and winters in the two hemispheres will be of unequal length. Over time, however, due to planetary gravitational influences, the shape of the orbit will change from being markedly elliptical to less so (Figure 3.11c). The greater the eccentricity, the more pronounced will be the differences in solar radiation receipt and hence summer/winter contrasts. The **eccentricity of the orbit** is therefore the third element in the astronomical theory. The periodicity, i.e. the time taken for the orbit to change from one point of greatest eccentricity to the next, varies from around 95 to 136 ka, although this is generally referred to as the 100 ka eccentricity cycle. Over a longer timescale, a more significant 413 ka eccentricity cycle is also apparent.

These variables, either singly or in combination, lead to marked variations over time in the latitudinal and seasonal distribution of solar insolation, the principal results of which are long-term climatic fluctuations between glacial and interglacial episodes. It is important to stress, however, that of the three astronomical variables, only eccentricity affects the total amount of solar radiation received by the earth. The other astronomical variables contribute largely to a redistribution of solar insolation at different latitudes (Berger, 1988). It has been calculated, for example, that for particular latitudinal positions, precession and obliquity effects can generate monthly insolation values up to 12 per cent different from those of the present day. By contrast, the effect on total insolation of eccentricity is relatively small, of the order of 0.1 per cent or less (Berger, 1989). Because the variables are interlinked, however, each will exert an effect on the others. The effects of precessional cycles, for example, on spatial and temporal variations in solar radiation receipts will be modulated by orbital changes, for winter/summer contrasts will

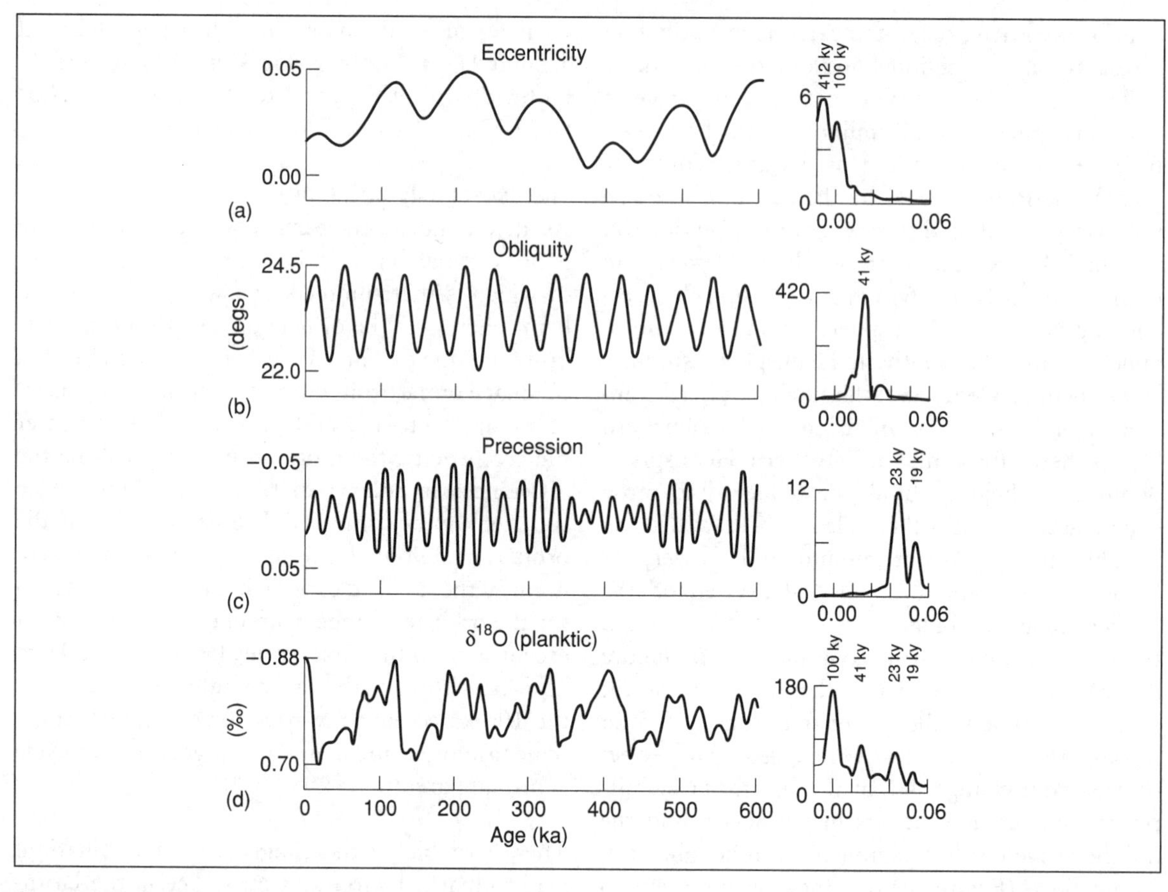

Figure 3.12 Orbital and climatic variations over the past 600 ka: (a) eccentricity; (b) obliquity; (c) precession; (d) variations in $\delta^{18}O$. Variance spectra on the right, with their dominant periods indicated in thousands of years (ka), are calculated from time series shown on the left (after Imbrie *et al.*, 1993a, Figure 17.1)

be more pronounced at times of maximum eccentricity than during episodes when the shape of the earth's orbit is less elliptical. Similarly, the greater the axial tilt, the more accentuated will be the seasonal contrasts and the more important will be the seasonal timing of perihelion. Overall, however, it would seem that the principal forcing factors of climate change are precession and obliquity, with obliquity in particular driving annual radiation changes in the higher-latitude regions (Imbrie *et al.*, 1993a).

Evidence in support of the astronomical theory

Over the course of the last 30 years or so, the astronomical variables have been detected in a range of proxy records. These include the following.

Deep-sea cores

Because long time-series data can be obtained from deep-ocean cores, a considerable amount of research on the linkages between the astronomical variables and terrestrial climatic changes has been focused on these proxy records (Imbrie and Imbrie, 1979). The data suggest that, for the past 1 ma, the earth's climatic rhythms have been dominated by the 100 ka eccentricity cycle, modulated or amplified by cycles at 19 and 23 ka (precession) and 41 ka (obliquity) intervals (Figure 3.12). Indeed, following a detailed statistical analyis of the oxygen isotope trace in five ocean cores from the Atlantic, Pacific and Indian oceans, Imbrie *et al.*

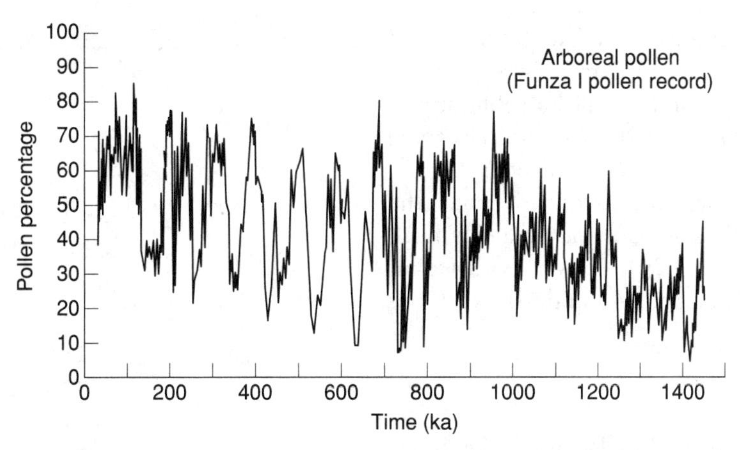

Figure 3.13 The 1.4 ma arboreal pollen record from Funza I, Eastern Cordillera, Colombia. Note the well-developed 100 ka periodicity in pollen percentage variations over the last 800 ka (after Hooghiemestra *et al.*, 1993, Figure 3A)

(1984) concluded that the Milankovitch variables would explain at least 77 per cent of the amplitude of the $\delta^{18}O$ variations observed in these cores. Other marine parameters, such as biological and lithological indicators of surface-ocean variability, have also been linked to astronomical influences (Ruddiman and Raymo, 1988). The marine data, therefore, would seem to confirm the hypothesis that the climatic fluctuations of the Late Quaternary are driven by changes in the earth's orbit and axis, in other words, what has become known as **orbital forcing** is the primary mechanism behind long-term climatic change (Imbrie *et al.*, 1992).

Coral reef sequences

In Barbados, coral reefs formed during high stands of sea level (global warming) were dated to around 82, 105 and 125 ka, and a possible link was inferred with precessional cycles (Mesolella *et al.*, 1969). Similar dates of 81, 108 and 130 ka have subsequently been obtained from raised coral terraces in Haiti (Dodge *et al.*, 1983). Data from New Guinea indicate successive reef terrace development at around 130, 107, 85, 60, 45, 40, 29 and 10 ka, the early and later parts of the record corresponding closely with enhanced seasonal insolation (higher summer and lower winter insolation than at present) predicted by the astronomical variables (Aharon, 1984).

Pollen data

Spectral analysis[3] of the 130 ka pollen record from Grande Pile in northern France suggests a close statistical relationship between fluctuations in the herb and pine pollen curves and precessional periodicities (23 and 19 ka). This, in turn, implies a response by continental vegetation to orbital forcing (Molfino *et al.*, 1984). In the long Funza pollen record from the Eastern Cordillera of Colombia which extends back over 1.4 ma, a 100 ka periodicity is evident in the sequence for the last 800 ka (Figure 3.13), while 23 and 40 ka periodicities occur throughout the profile (Hooghiemstra *et al.*, 1993).

Loess/palaeosol sequences

Detailed analysis of long sequences of loess and interbedded palaeosols on the Loess Plateau of China have revealed evidence of Milankovitch frequencies at periodicities of 41, 100 and 400 ka (Ding *et al.*, 1994; Liu *et al.*, 1999). Periodicities of *c.*20 ka and 40 ka have also been detected in loess/palaeosol sequences from Kashmir (Gupta *et al.*, 1991).

Ice-core data

Spectral analysis of $\delta^{18}O$ and δD profiles from Greenland (GRIP and GISP2) and Antarctic (Vostok) ice cores reveals the presence of both precessional and obliquity cycles, with the 19 and 23 ka precessional cycles more strongly represented in the Greenland data and the 41 ka (obliquity) a more stable component of the Antarctic isotopic profiles (Mayewski *et al.*, 1997; Yiou *et al.*, 1997). A temperature record reconstructed from the Vostok $\delta^{18}O$ and δD signals, which extends back over 420 ka (Figure 2.20), shows evidence of orbital periodicities at both 41 and 100 ka (Petit *et al.*, 1999).

Tropical lake data

Significant variations in low-latitude lake levels during the Late Quaternary are reflected in a range of geological evidence, and these have been linked to changes in precipitation regimes. An atmospheric general circulation model, which simulates climate over the past 18 ka, suggests that episodes of increased precipitation apparent in lake level records from tropical Africa are associated with periods of strengthened monsoonal circulation resulting from cyclical changes in the pattern of Northern Hemisphere insolation (Kutzbach and Street-Perrott, 1985; Street-Perrott *et al.*, 1990). This, in turn, implies a forcing of fluctuations in tropical lake levels by the astronomical variables. A 200 ka proxy rainfall record from South Africa, which shows a period of variation of *c.*23 ka (Figure 3.14), provides further evidence of orbital insolation forcing on African climate (Partridge *et al.*, 1997).

Lake sediment data

At Lake Baikal in continental Asia, Fourier analysis (which is a form of spectral analysis: see note 3 below) of the water content profile of the upper 500 ka of sediments reveals periodicities at 19, 23, 41 and 100 ka (Grachev *et al.*, 1998). Sedimentological data from a longer core record from the same lake which covers the past 12 ma showed evidence of periodic fluctuations in grain size. Spectral analysis indicated a periodicity at *c.*400 ka, which corresponds with the eccentricity parameter of solar insolation. Further cycles at periodicities of 600 and 1000 ka may also be related to the Milankovitch variables (Kashiwaya *et al.*, 2001).

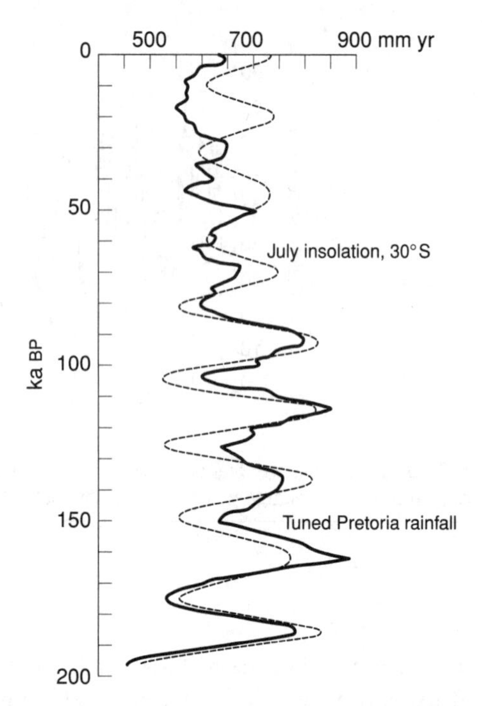

Figure 3.14 The Pretoria saltpan rainfall time series for the past 200 ka. Note the close relationship between the July insolation values based on the 23 ka precessional cycle and rainfall variability (after Partridge *et al.*, 1997, Figure 5)

Modifications to the astronomical theory

Although a considerable body of empirical evidence now exists to support the hypothesis that orbital forcing is a fundamental causal component of climatic change during the course of the Quaternary, it is now equally apparent that the astronomical variables are not the only elements in the equation and, indeed, a number of outstanding problems relating to the astronomical theory need to be resolved. For example, oceanographical evidence suggests that high-latitude cooling in the Northern Hemisphere began in late Miocene times (~10 ma BP) with the earliest land-ice accumulation in Greenland and in the Arctic (Wolf and Thiede, 1991). This long-term cooling cannot be explained by orbital forcing alone. Furthermore, geochemical and biological evidence from the North Atlantic indicates that during the course of the past million years, the climatic cycles of the earth have shifted from a periodicity of around

41 ka to a prevailing rhythm of *c.*100 ka (Ruddiman *et al.*, 1986b). This change, which became progressively marked after *c.*900 ka BP, was accompanied by an apparent intensification of glaciation with the growth of Northern Hemisphere ice sheets to maximum volumes, considerably larger than those attained during the previous 1.6–1.7 ma of the Quaternary. Again, these marked shifts in periodicity and intensity of climatic cycles cannot be accounted for solely by orbital forcing (Ruddiman and Raymo, 1988).

It would appear, therefore, that while long-term climatic changes are *largely* explicable in terms of the astrononomical variables, other factors serve to modulate or amplify their effects. A widely held view is that long-term global cooling (which can be traced back to the Cretaceous) results from changes in the distribution of the continental land masses accompanied by tectonic uplift, with tectonic activity in particular exerting a major influence on global climate during the course of the past 10 ma. For example, the closing of the Panama Isthmus could have prompted marked changes in the oceanic heat and moisture fluxes in the North Atlantic, with increased northward moisture transport being a major factor in ice sheet initiation and growth (Mikolajewicz *et al.*, 1993). Dating of the emergence of the isthmus has been contentious, however, with the latest estimates pointing to closure beginning around 4.6 ma and continuing until 2 ma BP (Maslin *et al.*, 1998). Of greater significance may have been the tectonic uplift during the Cenozoic[4] in areas such as the Alps, East Africa, North and South America (Raymo and Ruddiman, 1992). Climatic modelling suggests that such a marked increase in tracts of high land in the middle-latitude regions would enhance albedo-temperature feedback on a global scale and could have led to significant cooling especially during the autumn and winter seasons (Birchfield and Weertman, 1983). Land uplift may have also altered the wave structure in the airstreams of the upper atmosphere, the effects of which have been to cool the Eurasian and North American land masses and increase their sensitivity to orbitally driven insolation changes (Ruddiman and Raymo, 1988; Ruddiman and Kutzbach, 1991). Hence, in the late Tertiary (2.7–2.5 ma BP), tectonic changes could have brought the global climate to a critical threshold, after which orbital forcing led to the establishment of the Quaternary glacial–interglacial climatic regime (Maslin *et al.*, 1998). A similar tectonic threshold may have been reached during the mid-Quaternary, perhaps as a consequence of the uplift of the Tibetan Plateau, which may have contributed to the intensification of Northern Hemisphere glaciation, and the shift in periodicity from the obliquity-dominated (41 ka) cycles of the early/mid-Quaternary to the predominant 100 ka (eccentricity) cycle of the past ~900 ka (Ruddiman and Raymo, 1988). By the same token, uplift of submarine ridges may have led to significant changes in ocean circulation, with the formation Greenland–Scotland ridge possibly producing a coupled ocean–atmosphere climatic shift ~650–900 ka BP (Denton, 2000).

Tectonic uplift may affect climate in another way, however, for Raymo and Ruddiman (1992) have suggested that uplift of the Himalayas over the past 40 ma has produced a chain reaction in which more intense monsoonal circulation and increased rainfall has led to accelerated rates of chemical weathering. The weathering of silicate minerals involves CO_2 from the atmosphere, and the products are ultimately deposited on the deep-ocean floor, where they are lost from the global biogeochemical carbon cycle. As a consequence, the atmosphere is depleted in CO_2, which would contribute to the long-term climatic cooling described above. Changes in the atmospheric CO_2 content could also amplify or modulate the climatic effects induced by the astronomical variables. Evidence from deep-sea sediments and from polar ice cores indicates that during the last glacial period, atmospheric CO_2 levels were significantly lower than during the present interglacial (Shackleton *et al.*, 1992; Anklin *et al.*, 1997). The most detailed record has been obtained from the Vostok ice core in Antarctica (Figure 2.20) which shows dramatic shifts from around 180 to 280–300 ppmv during glacial–interglacial transitions. The high degree of correlation between the CO_2, CH_4 (methane) and isotopically derived temperature profiles from the Vostok core has prompted the suggestion that CO_2 and CH_4 may have contributed to the glacial–interglacial changes over the past 400 ka by amplifying the effects of orbital forcing (Petit *et al.*, 1999). The close phase relationship between the Vostok record and the 100 ka eccentricity cycle is also reflected in isotopically derived deep-ocean temperatures (Shackleton, 2000). The factors underlying these long-term variations in atmospheric CO_2 remain unclear, although they may be related to glacial/interglacial changes in patterns of deep-ocean circulation (Broecker *et al.*, 1985; Broecker and Denton, 1990). Whatever the causes, the data point very strongly to the importance of CO_2, and perhaps also other trace gases such as CH_4, in the system of climatic feedbacks that amplify the direct effects of insolation changes resulting from orbital forcing.

Amplification of orbital forcing may arise not only from the effects of atmospheric trace gases, however, but may also be generated by other feedback effects. This notion gains strong support from global atmospheric models. These have failed to simulate the onset of the last glaciation when forced by orbitally induced insolation changes alone, suggesting that feedback effects from changes in vegetation cover or ocean circulation are required (Mitchell, 1993). Such models involve major rearrangements of ocean water movements which would lead to cooling of high latitudes and increased moisture flux from equatorial to polar regions to generate glacier ice. Albedo changes resulting from vegetation cover and ice accumulation would amplify the climatic effects of weak orbital forcing (Khodri *et al.*, 2001). The build-up of large continental ice masses may be particularly significant in this context, and it has been suggested that the 100 ka climatic cycle of the last ~950 ka can be explained largely in terms of the operation of a coupled ocean–ice sheet mechanism (Denton, 2000). It is envisaged that ice sheets grow steadily over a 100 ka cycle, abstracting waters from the world's oceans and leading to a major reorganisation of deep-water circulation from an 'interglacial' to a 'glacial' mode. Seasonal insolation effects, resulting from the operation of the Milankovitch variables, will ultimately trigger ice-sheet collapse which then shifts ocean circulation back into an interglacial mode and the cycle begins again. Under this scenario, glacial to interglacial transitions (**terminations**: see above) reflect abrupt reorganisations of the ocean–atmosphere–cryosphere system that are simply triggered by orbital forcing. Such complex feedback loops are essential to explain the 100 ka cycles in particular, for it has long been recognised that variations in insolation forcing due to eccentricity changes, by themselves, are too small (of the order of 0.1 per cent) to be the direct cause of glacial–interglacial cycles (Imbrie *et al.*, 1993b).

Overall, therefore, while the orbital parameters, amplified or modulated by terrestrial and atmospheric feedback effects, appear to offer a coherent basis for explaining long-term climate change, a number of problems still remain to be resolved (Rial and Anaclerio, 2000). For example, although the longer 413 ka cycle is the largest component of eccentricity forcing, it does not register clearly in the marine oxygen isotope data. Why this is so is not at all clear. A further difficulty is that in the deep-ocean $\delta^{18}O$ sequence, although the predominant climatic mode of the last *c.*700 ka is around 100 ka, the duration of consecutive glacial periods actually varies from around 80 to 120 ka. The interval between the last two interglacials is 120 ka, while around 400 ka, the interval was *c.*80 ka, with three successive interglacials occuring in less than 200 ka (Raymo, 1997). These do not correlate closely with eccentricity-induced insolation changes. The search for a solution to these problems has led to an increasingly sophisticated analysis of the oxygen isotope signal, with power spectral analysis indicating that, rather than responding in a linear fashion to orbital forcing, the earth's climate system actually responds in a *non-linear* manner by frequency modulating[5] the eccentricity-related variations in insolation (Rial, 1999; Rial and Anaclerio, 2000). The consequence of this process is that the climate system switches periodically from an 80 ka mode to a 120 ka mode every 400 years or so. In other words, the 413 ka eccentricity cycle appears to be the principal component of the eccentricity signal, rather than the 100 ka cycle, despite the fact that it is the latter that manifests itself in terrestrial proxy records. Although this hypothesis potentially offers an exciting new insight into the operation of the Milankovitch variables, the precise cause of the frequency modulation itself remains to be established.

Patterns of short-term climatic change

Short-term climatic changes over timespans of thousands of years or less are most clearly marked at the end of the last cold stage and during the Holocene. This section examines the pattern and chronology of climatic change over the course of the present interglacial, beginning with the marked oscillation in climate recorded throughout the North Atlantic region at the close of the Weichselian/Devensian cold stage, and extending up to the climatic warming that followed the Little Ice Age of the medieval period.

The Lateglacial climatic oscillation

Throughout much of the last cold stage, cold surface waters occupied large areas of the North Atlantic, the oceanic polar front being situated off northern Portugal around latitude 40° N (Figure 3.15). By 15 ka BP, however, the polar front and the southern limit of winter sea ice had receded to a position near Iceland, allowing warmer waters to spread northwards around the coastline of western Europe (Ruddiman and McIntyre, 1981). The climatic impact of this oceanographic change was dramatic. Prior to 15 ka BP, for example, extensive areas of sea ice around the British coasts resulted in a climate of extreme continentality, with

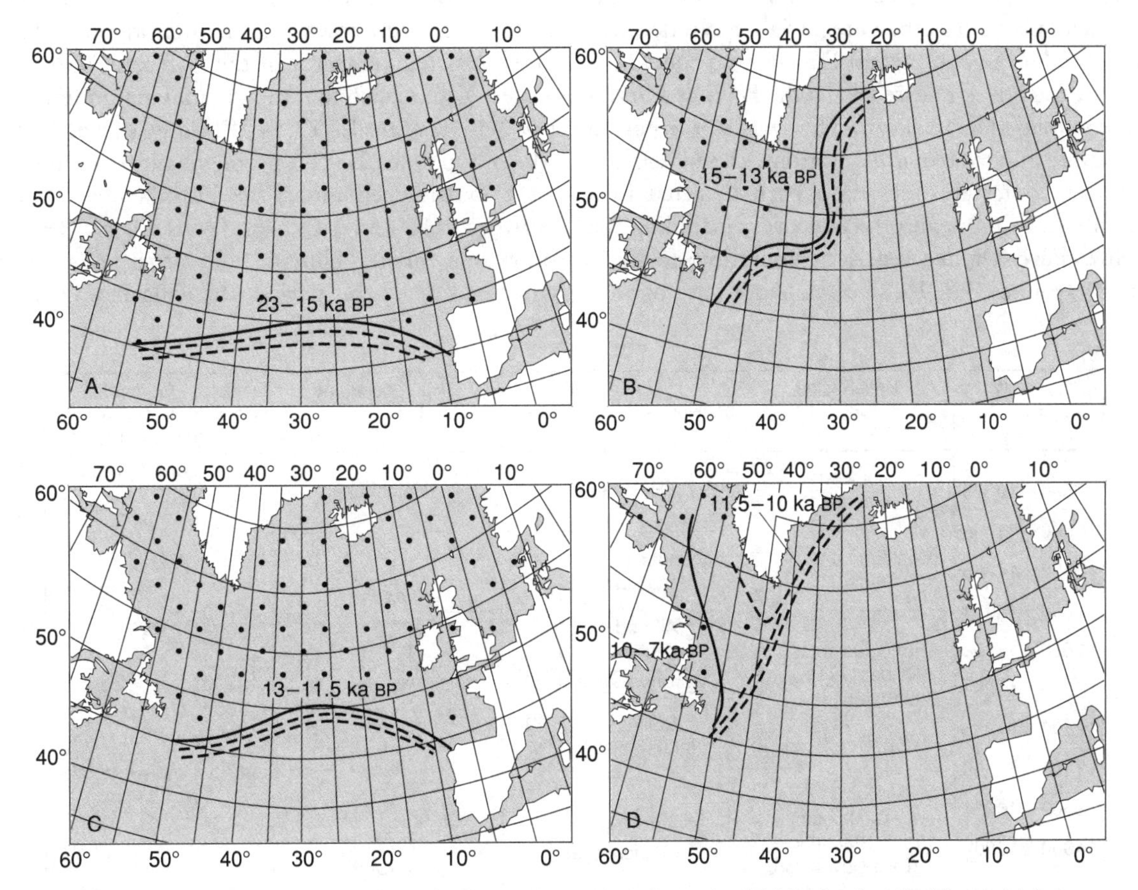

Figure 3.15 Movement of the North Atlantic polar front during the Late Devensian/Weichselian Lateglacial period (after Ruddiman and McIntyre, 1981)

temperatures of the coldest winter months in the range –20 to –25 °C and summer temperatures below 10 °C. By 14.5 ka BP, however, summer temperatures (as indicated by coleopteran evidence) were typically around 17 °C, with winter temperatures in the range 0–1 °C (Atkinson *et al.*, 1987). In other words, the climate of lowland Britain was as warm during the summer months as at the present day, with winter temperatures only a few degrees lower. Evidence for this climatic amelioration has been found throughout north-west Europe, with temperatures similar to those in Britain being experienced along the Atlantic seaboard (Lowe, 1994; Walker, 1995), although to the north-east, the chilling effect of the downwasting Scandinavian ice sheet kept temperatures much lower (Coope and Lemdahl, 1995). This episode of climatic warming, which began shortly after 15 ka BP and lasted for around 2000 years, is known in Britain as the **Lateglacial** or **Windermere Interstadial.** In continental Europe, the equivalent warm episode is referred to as the **Bølling-Allerød Interstadial** (Figure 3.16). This was not a period of uniformly warmer conditions, however, for shorter cooler episodes have been detected in a number of European proxy records, including the **Older/Early Dryas (Aegelsee Oscillation:** Switzerland) around 14.3–13.8 ka BP, and a later oscillation (**Gerzensee Oscillation:** Switzerland) in the period between 13.7 and 13.0 ka BP (Walker, 1995). A comparable warmer interstadial episode, again with two pronounced but short-lived colder intervals (Figure 3.17), is also reflected in climatic proxy records from the Greenland ice cores (e.g. Alley *et al.*, 1993; Stuiver and Grootes, 2000), where it has been designated **Greenland Interstadial (GI) 1** (Björck *et al.*, 1998; Walker *et al.*, 1999).

In North America, the Lateglacial Interstadial warmer episode is less well resolved both in terms of biostratigraphy and geochronology than is the case in western Europe, but it is reflected in pollen and other proxy records from the eastern United States. These indicate initial warming from *c.*19 ka BP in Kentucky and Virginia (Kneller and Peteet, 1993), but further north summer temperatures did not approach modern values until *c.*15–13 ka BP (Shane and Anderson, 1993). As in north-west Europe, the downwasting Laurentide ice sheet depressed temperatures in the northern USA and eastern Canada, so that the thermal maximum was not achieved until *c.*13.5–13 ka BP (Lowe, 1994; Miller and Elias, 2000). Although colder conditions were experienced along the Florida coast plain between *c.*16 and 14 ka BP (Watts *et al.*, 1992), only one major climatic downturn has been detected during this period, the **Killarney Oscilla-**

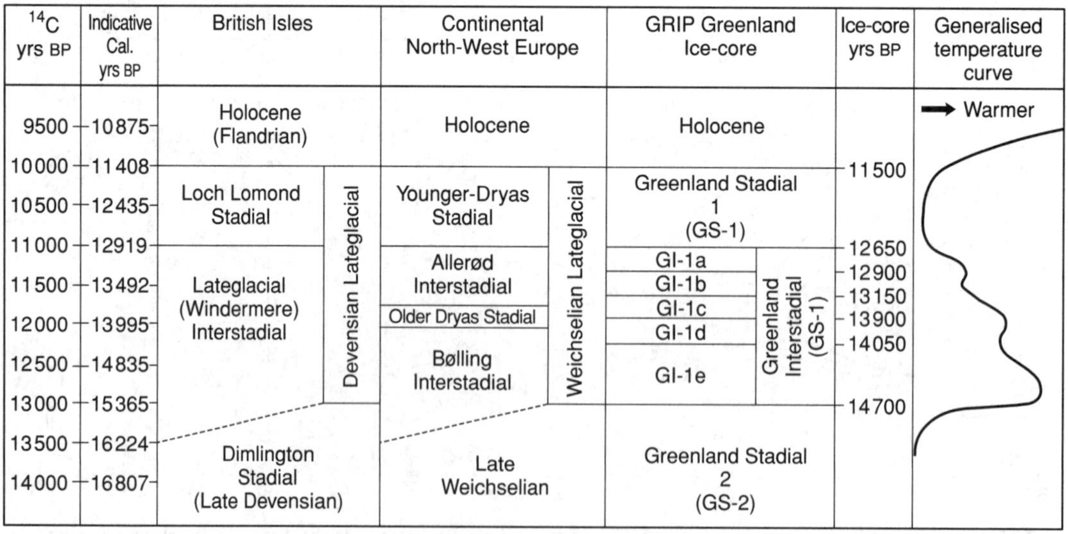

^{14}C yrs BP	Indicative Cal. yrs BP	British Isles	Continental North-West Europe	GRIP Greenland Ice-core	Ice-core yrs BP	Generalised temperature curve
9500	10875	Holocene (Flandrian)	Holocene	Holocene		→ Warmer
10000	11408				11500	
10500	12435	Loch Lomond Stadial (Devensian Lateglacial)	Younger-Dryas Stadial (Weichselian Lateglacial)	Greenland Stadial 1 (GS-1)		
11000	12919	Lateglacial (Windermere) Interstadial	Allerød Interstadial	GI-1a (Greenland Interstadial (GS-1))	12650	
11500	13492			GI-1b	12900	
12000	13995		Older Dryas Stadial	GI-1c	13150	
				GI-1d	13900	
12500	14835		Bølling Interstadial	GI-1e	14050	
13000	15365				14700	
13500	16224	Dimlington Stadial (Late Devensian)	Late Weichselian	Greenland Stadial 2 (GS-2)		
14000	16807					

Figure 3.16 The Lateglacial period (*c.*15–11.5 ka BP) in northwest Europe (after Lowe and Walker, 1997)

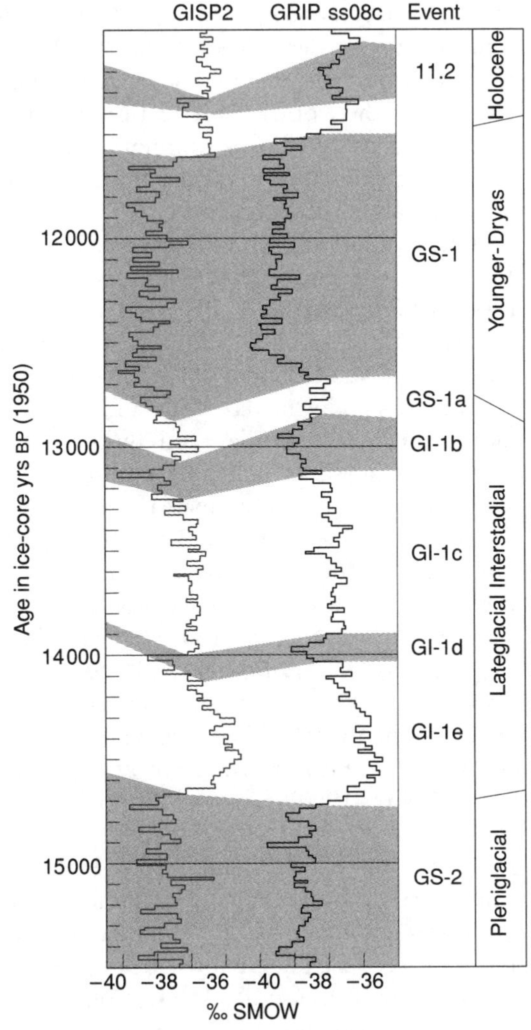

Figure 3.17 Oxygen isotope profiles ($\delta^{18}O$) from the Lateglacial and early Holocene in the GISP2 and GRIP Greenland ice cores (after Lowe *et al.*, 2001). The records show two major 'cold' events (shaded): Greenland Stadial 2 (GS-2) and Greenland Stadial 1 (GS-1), the latter broadly equating with the Younger Dryas; and two warmer phases: Greenland Interstadial 1 (GI-1), which equates with the Lateglacial Interstadial, and the early Holocene. Three short-lived colder episodes are also apparent: GI-1d, GI-1b during Greenland Interstadial 1, and the GH-11.2 event in the early Holocene. Note the rapid climatic ameliorations, reflected in the abrupt shifts in isotopic values, at the beginning of GI-1e (Lateglacial Interstadial) and at the end of GS-1 (Younger Dryas). For further explanation, see text. (After Lowe *et al.*, 2001, Figures 1 and 2)

tion (or '**Amphi-Atlantic Oscillation**') of Atlantic Canada and the north-east USA which has been dated to 13.2–12.9 ka BP (Levesque *et al.*, 1993; Cwynar and Levesque, 1995).

Between approximately 13 and 11.5 ka BP, polar waters once again spread southwards around the coastline of western Europe (Figure 3.15), with the oceanic polar front reaching its maximum southerly position off south-west Ireland (Ruddiman and McIntyre, 1981). The rate of movement of these polar waters appears to have been extremely rapid, with velocities in excess of 5 km yr^{-1} being inferred from evidence in cores from the eastern North Atlantic (Bard *et al.*, 1987). The climatic cooling that followed these oceanographic changes is referred to in Europe as the **Younger Dryas Stadial** and in Britain as the **Loch Lomond Stadial** and is dated to the period 12.8–11.5 ka BP. Evidence for this episode of colder climatic conditions is found from Finnish Karelia in the far north (Bondestam *et al.*, 1994) to the Iberian Peninsula in the south (Allen *et al.*, 1996), and from the west of Ireland (O'Connell *et al.*, 1999) to eastern Europe (Onac and Lauritzen, 1996). The terrestrial record for climatic cooling is impressive. There are indications of renewed glacier activity in mountain regions throughout Europe, while periglacial conditions returned to many lowland areas. The pollen records show that the woodland that had started to develop in response to climatic warming was replaced by scrub tundra in northern and western Europe, and by open-habitat, steppe vegetation further south (Lowe, 1994). Coleopteran evidence suggests that in Britain, summer temperatures fell below 10 °C, with winter temperatures in the range −15 to −20 °C (Atkinson *et al.*, 1987), figures that are in broad agreement with temperature reconstructions based on geomorphological evidence (Ballantyne and Harris, 1994). Elsewhere in Europe, an annual temperature reduction of up to 8 °C has been recorded during the Younger Dryas Stadial, the greatest temperature contrasts with the preceding Allerød Interstadial being experienced in areas adjacent to the western European seaboard (Walker, 1995). A marked climatic deterioration is also evident in a range of proxy records (oxygen isotope signatures, dust profiles, snow accumulation records) from Greenland ice cores (Figure 3.17), with temperatures over Greenland as much as 15 °C below modern values (Alley, 2000). This climatic downturn, which is comparable in terms of timing with the Younger Dryas of north-west Europe, has been termed **Greenland Stadial (GS) 1** (Björck *et al.*, 1998; Walker *et al.*, 1999).

The climatic signal of the Younger Dryas is less strongly marked in records from North America, and again registers most strongly along the eastern seaboard of the USA and in Atlantic Canada. Cooling of summer temperatures of the order of 3–4 °C is reflected in pollen records from New England (Peteet *et al.*, 1994), while further north in Maine and in Atlantic Canada chironomid data point to dramatic declines in summer pondwater temperatures, at some sites by as much as 20 °C (Cwynar and Levesque, 1995; Levesque *et al.*, 1997). In eastern and central North America, by contrast, coleopteran records show little evidence of climate change, with a 'plateau' of temperatures from around 15 to 11.5 ka BP (Elias *et al.*, 1996). In a global context, the Younger Dryas has proved to be something of an enigma, for although there are indications in some areas for a cold event around 11.5–12.5 ka, in other parts of the world evidence for cooling is either equivocal or absent (Peteet, 1993, 1995). A possible causal mechanism for the Younger Dryas in the North Atlantic region is outlined in a later section (p. 105).

The early Holocene amelioration

The global warming in climate that began at the beginning of the present interglacial around 11.5 ka is recorded in a wide range of marine and terrestrial evidence. Data from the deep-ocean floor, for example, suggest that by that time, polar waters had retreated to the north-west of Iceland, and that relatively warm surface waters (temperatures above 14 °C) were once more firmly established around the coasts of western Europe. A mean warming rate for this time period of almost 1 °C per century has been inferred from foraminiferal and stable isotope data for deep Atlantic waters off north-west Scotland (Austin and Kroon, 1996), while even more rapid rates of warming (9 °C in ~50 years) of surface waters are indicated by diatom data from the Greenland–Iceland–Norwegian seas (Koç *et al.*, 1993). The rapid wastage of the Laurentide and Fennoscandian ice sheets, and the worldwide retreat of mountain glaciers in response to climatic warming, is reflected in the abrupt shift in the isotopic trace in deep-ocean sediments at the marine oxygen isotope stage 2/1 boundary (Ruddiman and Duplessy, 1985).

In north-west Europe, isopollen maps (see Figures 4.17 and 4.18) show the rapid northward movement of the treeline and the widespread replacement of scrub tundra by birch woodland by 11 ka BP and by mixed woodland within a further 1000 years (Huntley and Birks, 1983). Multivariate analysis of pollen data from sites in France points to an abrupt increase in temperature after 10 ka BP of around 3–4 °C/500 years (Guiot, 1987a). More rapid rates of climatic amelioration (1.7–2.8 °C/century) have been suggested on the basis of Coleoptera from sites in Britain (see Figure 2.5) and France, the evidence indicating that by *c.*11 ka BP, both summer and winter temperatures had reached levels comparable with those of the present day (Atkinson *et al.*, 1987; Ponel and Coope, 1990). In eastern North America, palynological evidence for changing forest composition suggests that the beginning of the Holocene was characterised by a marked climatic shift, with temperatures in the period 11.5–11.0 ka BP rising rapidly to levels only a degree or so below those of the present day (Peteet *et al.*, 1994). Coleopteran data from sites in east-central North America also show that temperatures comparable with those of today were attained between 11 and 10 ka BP (Morgan, 1987). Although the waning Laurentide ice sheet continued to exert an influence on the early Holocene climate, pollen-based climatic parameters show that the modern north–south temperature gradients were established by 9 ka BP, with the east–west gradient in precipitation evident in the Midwest by around the same time (Webb *et al.*, 1993). This episode of climatic amelioration coincides with a precessionally induced increase in solar insolation, to the extent that by 10–9 ka BP, July solar insolation in the Northern Hemisphere was ~7 per cent above that of the present day.

By far the most impressive of all of the proxy records for climatic change at the beginning of the

Holocene are those from the Greenland ice sheet (e.g. Figure 3.17), where annual layer counting has enabled the climatic record to be resolved to within an accuracy of ~2 years (Taylor *et al.*, 1997). The data show that while the entire transition (of some 15 °C) from Younger Dryas to typical Holocene temperature values took around 1500 years, the main transition, which involved a temperature increase of the order of 5–10 °C, occurred over a much shorter timescale (Severinghaus *et al.*, 1998). The rise in temperature was accompanied by a doubling of snow accumulation, by a significant decline in wind-blown materials, and by a large increase in methane, indicating the rapid expansion of global wetlands. Remarkably, these major climatic and environmental changes occurred in less than a single human lifetime, over a few decades or, possibly, in less than a few years (Alley, 2000).

Striking though the climatic amelioration in the North Atlantic region at the beginning of the Holocene appears to have been, evidence from a range of proxy records suggests that the temperature rise was not unidirectional, but that a number of climatic oscillations occurred. The most pronounced of these were the **'Preboreal Oscillation'** at around 11.2 ka BP and a later oscillation around 8.2 ka BP (Björck *et al.*, 1996; Alley *et al.*, 1997). Both events register strongly in the GRIP ice core (Figure 3.17) where they have been designated the **Greenland Holocene** (GH) **11.2** and **8.2 events** (Walker *et al.*, 1999). The amplitude of these climatic oscillations appears to have been considerable, with the 8.2 event, for example, being marked by a temperature decline of 4–8 °C in central Greenland (Figure 3.18) and 1.5–3 °C at marine and terrestrial sites around the north-eastern North Atlantic. Although of a lesser amplitude and duration, there may be similarities in terms of causal mechanism between these events and the Younger Dryas, and again this is considered further below.

The Climatic Optimum (the Hypsithermal)

The Climatic Optimum of the present interglacial (the Hypsithermal in North America) occurred between *c.*9 ka and 4 ka BP. Again, the most impressive evidence for this climatic episode is that from the Greenland ice cores. The oxygen isotope signal from the NorthGRIP core (Figure 3.18), for example, shows warmer conditions between *c.*8 and 4 ka BP, with the 'thermal maximum' from *c.*7.5 to 4.5 ka BP (Johnsen *et al.*, 2001). Elsewhere in the North Atlantic region, the timing of the Holocene thermal maximum is both spatially and temporally variable. Pollen data, supported by climatic modelling experiments, show that

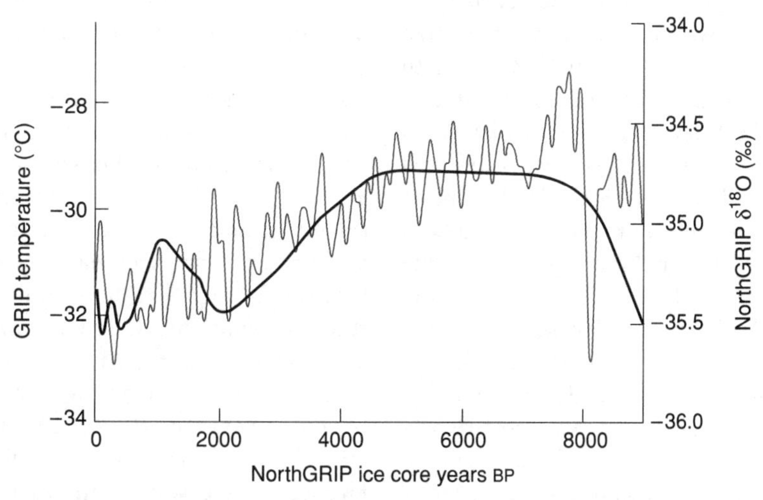

Figure 3.18 The NorthGRIP δ^{18} O record (thin line) plotted against the long-term temperature trend (thick line) obtained from the NorthGRIP profile (after Johnsen *et al.*, 2001, Figure 8). The trends are very similar suggesting that the δ^{18} O signal can be read as a temperature record. Note the pronounced isotopic shift which marks the short-lived 8.2 ka event.

around 9 ka BP, summer temperatures were higher than at present in Scandinavia and the Baltic region, whereas the British Isles and most of central and southern Europe were cooler. At 6 ka BP, by contrast, all of north, north-west and central Europe were experiencing higher July temperatures than at present (Huntley and Prentice, 1993). Higher temperatures than those of the present day around 9 ka BP have also been inferred for western former Soviet Union (Peterson, 1993). In parts of western and interior North America, pollen and other proxy data indicate that summer temperatures warmer than those of the present were attained as early as 9 ka BP (Thompson *et al.*, 1993), whereas in eastern North America, maximum warmth with summer temperatures above those of the present was not achieved until around 6 ka BP. By that time the Laurentide ice sheet had finally wasted away and summer insolation, due to precessional effects, was still 5 per cent greater than that of today (Webb *et al.*, 1993).

A range of proxies suggests that temperatures at the Holocene Climatic Optimum were of the order of 1–3 °C above those of the present day. In the GRIP ice core, for example, temperatures ~2.5 °C above present have been calculated for the period 8–5 ka BP (Dahl-Jensen *et al.*, 1998), while in the Nordic seas, surface waters may have been up to 6 °C warmer than present (Koç and Jansen, 1994). Pollen-based temperature reconstructions for 6 ka BP point to warmer summers over most of Europe by an average of around 2 °C, with summers warmer by 2 °C in mid-continent and in the far north (Huntley and Prentice, 1993). A similar temperature increase has been estimated for western Russia (Peterson, 1993). In the United States, pollen data again point to temperatures of the order of 1–2 °C higher than at present during the early and/or mid Holocene (Thompson *et al.*, 1993; Webb *et al.*, 1993), and these reconstructions are supported by other proxy data, including coleopteran records (Morgan, 1987). Further indications of increasing Holocene warmth are provided, *inter alia*, by the northward migration of woodland in both Europe and North America (Huntley and Birks, 1983; Jackson *et al.*, 1997), by the rise of the treeline in mountain regions to elevations well above current levels (Barnekow, 1999), and by the northward expansion of thermophilous organisms. The last-named include Mollusca (Rousseau *et al.*, 1993) and the European pond tortoise (*Emys orbicularis*), whose present-day breeding range is confined to the Mediterranean and eastern Europe, but whose remains have been found in deposits of mid-Holocene age in south-east England, Denmark and southern Sweden (Stuart, 1979). As *Emys* requires mean July temperatures well in excess of 18 °C to breed successfully, its fossil remains provide further evidence that temperatures were 1–2 °C above present-day levels during the Holocene Climatic Optimum in north-west Europe, an estimate which matches very closely that inferred from changes in treeline elevation (Barnekow, 1999).

Indications of former precipitation regimes can be obtained from a number of proxy sources, and especially from lake-level records. These data point to significant differences in precipitation regime between northern and southern Europe during the course of the Holocene, with a gradual trend towards wetter conditions in the Mediterranean in the early/mid Holocene, which was followed after *c.*6.5 ka BP by increasing aridity. In southern Sweden (Figure 3.19a), by contrast, drier conditions during the early Holocene were followed by a significant increase in effective moisture between *c.*9 and 7 ka BP, increased aridity between *c.*6 and 4 ka BP and a marked increase in effective moisture over the past 3.5 ka (Harrison and Digerfeldt, 1993). This reconstruction, based on lake-level fluctuations, is supported by data from lacustrine carbonates which indicate relatively dry and stable conditions between 8 and 4 ka BP, and a more humid and variable climate thereafter (Hammarlund *et al.*, 2003). In North America, a shift from dry to relatively moist conditions between 10 and 8 ka BP is indicated by lake-level records from north-eastern and south-eastern areas, whereas in the Midwest, the trend over the same period was from wetter to drier climatic regimes (Shuman *et al.*, 2002). In the eastern United States, the moister conditions of the early Holocene gave way to increasing aridity (Figure 3.19b), with the driest period centred on 6–7 ka BP (Webb *et al.*, 1993). This was followed by a rise in water levels from some time before 5 ka BP,

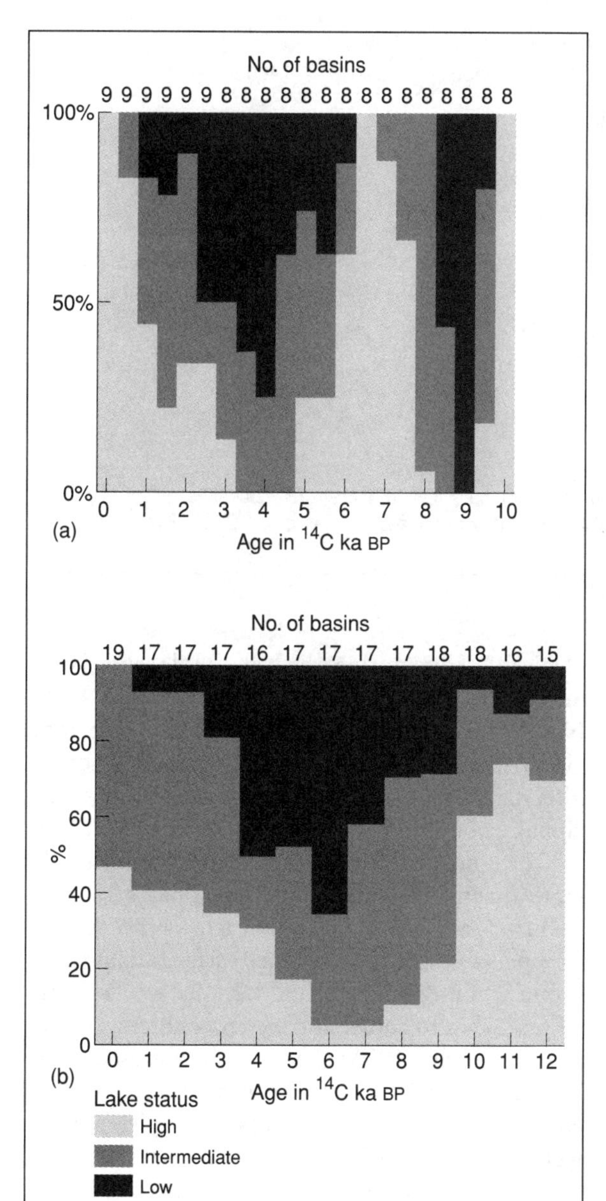

Figure 3.19 Holocene lake level records (a) for southern Sweden (after Harrison and Digerfeldt, 1993, Figure 11); and (b) from the eastern United States (after Webb *et al.*, 1993, Vegetation, lake levels, and climate in eastern Northern America for the past 18,000 years in *Global Climates since the Last Glacial Maximum* edited by H.E. Wright, Jr. *et al.*, published and reprinted by permission of the University of Minnesota Press, Figure 17.14.). Note that the timescales are in radiocarbon yrs BP (see Table 1.1 for calibration)

with a particularly pronounced increase between 3.5 and 3.0 ka BP (Almquist *et al.*, 2001).

Higher-resolution time-series data of Holocene precipitation regimes can be obtained from peat sequences where plant macrofossil and other data can be used to reconstruct surface wetness changes on mire surfaces (Chapter 2). In northern England, for example, a surface wetness record extending back almost 10 ka (Figure 3.20) shows that an early Holocene phase of relative dryness was followed by a pronounced shift to wetter conditions at 7.8 ka BP (Hughes *et al.*, 2000). More detailed palaeoprecipitation records from peat sequences have been obtained for the later Holocene, and these are discussed in the next section.

The late Holocene deterioration

During the later Holocene, from around 6 ka BP onwards, interpretation of terrestrial proxy records becomes more problematic, for the climatic signal is often difficult to disentangle from anthropogenic effects. Changes in forest composition, which had previously been widely employed as an index of climatic change, are now equally likely to reflect the activities of early human groups. The hydrological changes that accompanied woodland clearance, for example, in terms of runoff, surface water balance and lake sediment budgets, further exacerbate the problems of isolating the climatic signal in proxy records from this and succeeding periods. Despite these difficulties (which are considered further in Chapter 6), it has proved possible to gain some insight into the pattern of climatic change during the late Holocene in areas around the North Atlantic region. In addition, of course, there are records from the ice cores and from marine sequences, both of which are unaffected by anthropogenic activity.

Data from the Nordic seas point to a progressive decline in surface water temperatures from around 6 ka BP (Bauch *et al.*, 2001), while the Greenland ice core records indicate a decline in temperatures beginning around 4.5 ka BP (Figure 3.18). In North America, there was a southward movement, from about 4 ka BP onwards, of the boreal forest at the expense of mixed conifer–northern hardwood forest (Delcourt and Delcourt,

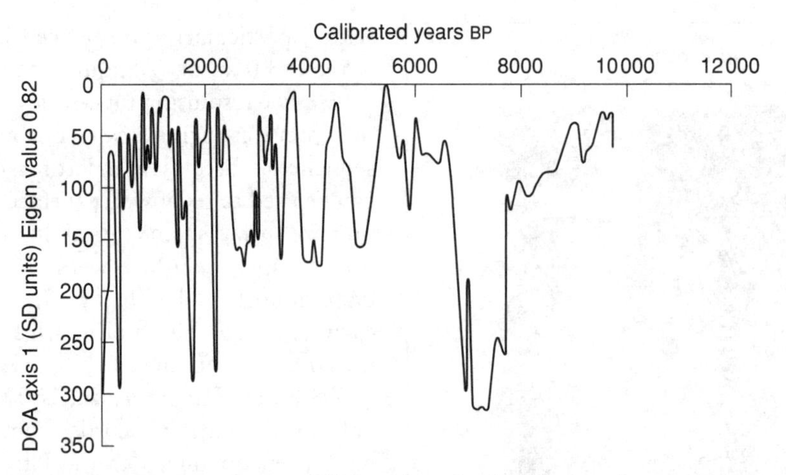

Figure 3.20 Reconstruction of mire surface wetness at Walton Moss, Cumbria, England. Increased wetness is reflected in the higher values of DCA (detrended correspondence analysis[6]) shown on the *y* axis (after Hughes *et al.*, 2000, Figure 10. Reproduced by permission of Hodder Arnold)

1984), while in parts of the American south-west a trend to increasingly arid conditions is detectable from around this time (Thompson *et al.*, 1993). In many parts of North America, and also in northern Europe, a significant decline in the upper limit of the treeline is recorded during the late Holocene, with data from northern Sweden suggesting a reduction in mean growing season temperature of 1.5 °C since 4.5 ka BP (Barnekow, 1999). A trend towards wetter (and probably also cooler conditions) is reflected in peat bog records and in lake-level data. A significant wet shift has been detected at around 4 ka BP on mire surfaces at sites in northern England (Barber *et al.*, 1994; Figure 3.20), while lake-level data from southern Sweden show a clear tendency towards higher water levels from *c.*3.8 ka BP (Harrison and Digerfeldt, 1993). In southern Finland, and in many other areas also, there is evidence of peat initiation from *c.*4.3 ka BP onwards (Korhola, 1995). Although there is evidence from many areas of north-west Europe that human activity and burning preceded peat growth (p. 216), the widespread evidence for renewed peat growth from around 4 ka BP is likely to be in part caused by climatic (higher rainfall, lower temperatures and increasing cloudiness, etc.) as well as anthropogenic factors. In both Europe and North America, there is abundant evidence for renewed glacier activity during the later Holocene (Nesje and Dahl, 2000), a period often referred to as the **Neoglacial**, while in the Arctic basin, driftwood evidence indicates a significant increase in sea ice after *c.*5 ka BP, again indicating increasingly severe climatic conditions (Stewart and England, 1983).

The most pronounced deterioration in climate during the late Holocene appears to have occurred around 2.8–2.5 ka BP. In north-west Europe, widespread recurrence surfaces (Plate 2.1), radiocarbon-dated to around 2.5 ka BP, first suggested a shift to cooler and wetter conditions at this time (Godwin, 1975), and this has subsequently been confirmed by data from a range of different sources, including increases in mire surface wetness (Figure 3.20), higher lake levels, lower treelines, glacier advances, snow avalanching and accelerated solifluction on upland slopes, and archaeological evidence for farm abandonment and a reduction in upland grazing (van Geel *et al.*, 1996; van Geel and Renssen, 1998). In North America, marked cooling at 3–2.5 ka BP in eastern Canada has been inferred from terrestrial evidence (Macpherson, 1985), while further north, dendroclimatological and palynological data from a network of sites extending from Alaska to Labrador reveal a progressive decline in summer temperatures from around 4 ka BP, the climatic deterioration being especially pronounced after *c.*2.4 ka BP (Diaz

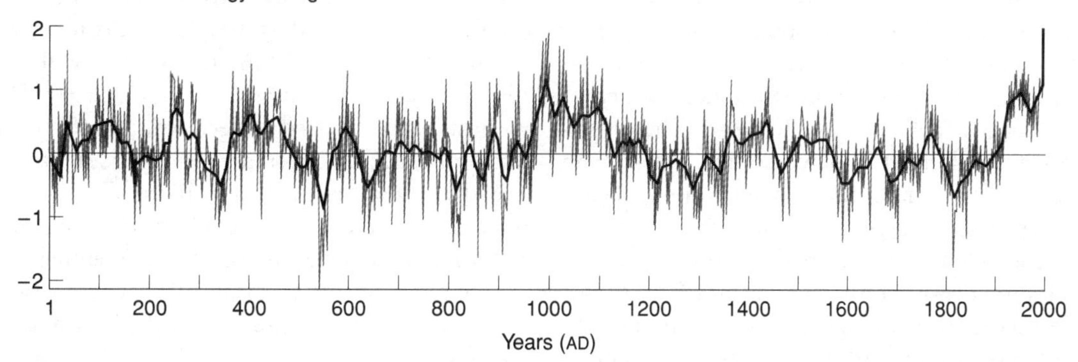

Figure 3.21 Northern 'high-latitude' temperature changes over the last 2000 years based on tree-ring data from sites in Mongolia, Siberia, Sweden and North America. The vertical scale shows departures (in °C) from the 2000-year mean (after Briffa, 2000, Figure 1). Note (a) the significant climatic downturn at c.AD 550; (b) the marked climatic improvement at c.AD 950; (c) the progressive climatic deterioration from c.AD 1100; (d) the coldest period of the 'Little Ice Age' after AD 1800

et al., 1989). There is also evidence for an increase in solifluction activity in many mountain regions from around 2.8 ka BP (Smith, 1993).

The historical period

A slight climatic recovery during the early historical period, with fluctuating, but generally warmer conditions, and with temperatures perhaps up to 1 °C above those of the present, appears to be reflected in proxy climate records from northwest Europe (e.g. Lauritzen and Lundberg, 1999b). However, a significant deterioration is evident around AD 500–600 and is reflected, *inter alia*, in dendrochronological records (Figure 3.21), in peat bog stratigraphies (Blackford and Chambers, 1995) and in increased mire surface wetness (Mauquoy and Barber, 1999). Documentary sources also lend support to the notion that, in Britain at least, a period of relatively warm and dry conditions was succeeded by a cooler and wetter climatic regime during the middle of the first millennium AD (Lamb, 1995).

Thereafter, many proxy records point to a warmer episode between *c.* AD 800 and AD 1200, which has been termed the **Little Optimum** or the **Medieval Warm Period** (Lamb, 1995). Dating of organic material associated with moraines indicates that there was widespread retreat of mountain glaciers during this period (Grove and Switsur, 1994), while a reduction in sea ice and generally warmer climatic conditions around the Arctic basin enabled Viking colonies to be established in Greenland and northern Newfoundland (Chapter 5). Ice-core data from Greenland point to temperatures around 1 °C above present-day levels, with thermal maximum around AD 900 (Dahl-Jensen *et al.*, 1998), while in Sweden dendrochronological and glacier records suggest a warmer climatic regime (~+1.2 °C) around AD 1150 (Karlén and Kuylenstierna, 1996: Figure 3.21). In southern England, documentary sources for the early medieval period (e.g. Domesday Book) provide evidence of extensive agriculture in the twelfth and thirteenth centuries which suggests an increase in both summer and winter warmth. Drier conditions in upland Britain are reflected in surface wetness curves from peat bogs (see Figure 2.12), and there is also evidence for a dry phase with aeolian activity in the Netherlands (Chapters 5 and 7). In the Sierra Nevada mountains of California, dendrochronological data suggest higher summer temperatures between AD 1100 and 1300 (Graumlich, 1993), although markedly colder conditions between AD 1200 and 1350 are indicated in dendrochronological and glacier records from western Canada (Luckman *et al.*, 1997).

From around AD 1400 onwards, summer temperatures over much of the northern temperate zone slowly cooled for ~200 years, reaching their lowest values early in the seventeenth century. The trend was not only towards cooler summers, however, but also towards colder and wetter winters and a significant increase in storminess, a period that is known as the **Little Ice Age**. The effects of this general climatic deterioration are seen in the readvance of mountain glaciers, in significant increases in geomorphological activity (landslides, avalanches, etc.) in upland areas, in flooding around the Atlantic coasts, in the expansion of Arctic sea ice, in vegetational changes in the mountains, and in widespread crop failure and abandonment of settlement (Grove, 2002). Detailed analysis of a range of proxy records, however, indicates that the single term 'Little Ice Age' may be something of a misnomer, for there appear to have been two significantly colder episodes, one during the seventeenth century, which was most severe in Eurasia, and a second during the nineteenth century (following a period of recovery in the eighteenth century) which was more severe over North America (Jones *et al.*, 1998). Temperatures in central Europe may have been 1–2 °C below normal during the late seventeenth century (Wanner *et al.*, 1995), while the nineteenth-century cooling in North America also saw temperatures falling to around 1.5 °C below those of the late twentieth century (Mann *et al.*, 2000). Globally, the seventeenth-century cold period appears to have been around 0.4 °C cooler than the twentieth-century warm period (Crowley, 2000), while the coldest decades of the nineteenth century were around 0.6–0.7 °C below the late twentieth-century mean (Mann *et al.*, 1998). These reconstructions, based on terrestrial proxy records (tree-ring records, documentary sources, etc.) are broadly supported by data from the GRIP ice core which also indicates two cold periods, one at AD 1550 and a second at AD 1850, when temperatures over Greenland were 0.5 and 0.7 °C respectively below those of the present (Dahl-Jensen *et al.*, 1998).

While there is perhaps stronger evidence for the Little Ice Age as a worldwide event, the global extent of the Medieval Warm Period is, perhaps, less convincing. In areas such as the south-east USA, southern Europe along the Mediterranean, and parts of South America, for example, climate during that period appears to have been little different from that of later times and, overall, the strongest evidence for warming comes from higher-elevation rather than from low-elevation records (Hughes and Diaz, 1994). However, although there may well have been temperature variations of 1–2 °C (or even more) in marginal situations at high latitude (Hughen *et al.*, 2000) or high altitude (Pfister, 1992), the reconstructed temperature record for the Northern Hemisphere as a whole (Figure 3.22) suggests a range of no more than ~0.5 °C, with Little Ice Age

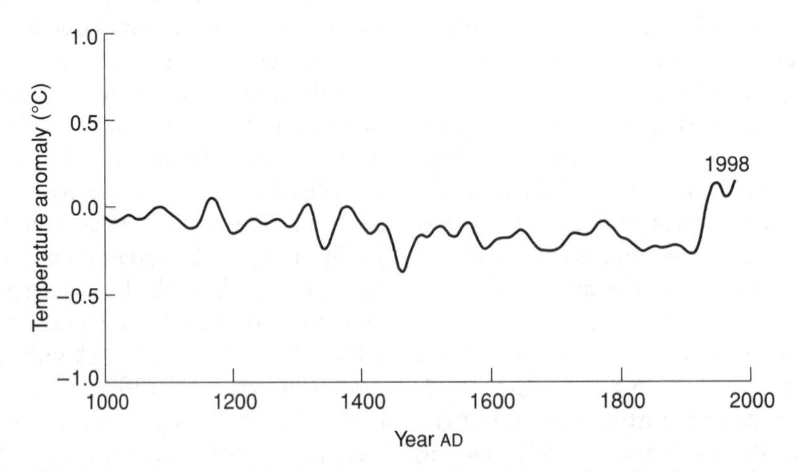

Figure 3.22 Northern Hemisphere mean annual temperatures reconstructed from palaeoclimatic records and calibrated against instrumental data for 1902–1980. The original data have been smoothed using a 40-year filter (after Bradley, 2000b, Figure 1)

temperatures about 0.45–0.5 °C lower than mid twentieth-century values, and temperatures around AD 1000–1200 only about 0.20 °C warmer than the Little Ice Age (Crowley and Lowery, 2000). Within that narrow envelope of variability, therefore, all of the documented environmental changes of the Medieval Warm Period and the Little Ice Age occurred. This puts into stark context the magnitude of projected future changes resulting from greenhouse gas and associated feedbacks, which involve projected global temperature changes of 1–2 °C by the end of the twenty-first century (Bradley, 2000b). The Little Ice Age is discussed in more detail on p. 177, while possible anthropogenic influence in future climatic scenarios is considered in Chapter 9.

Cyclical climatic change during the Holocene

While the foregoing provides a general overview of the sequence of climatic changes that occurred during the Lateglacial and Holocene, recent data suggest that the climatic history of the North Atlantic region may, in fact, be more complicated than this. In particular, a number of recent studies point to a cyclicity or periodicity in Holocene paleoclimatic records. For example, geochemical data from the Greenland GISP2 core suggest that cooler climatic conditions over Greenland occurred at ~2600-year intervals (O'Brien *et al.*, 1995), and a similar periodicity (~2400 years) has been detected in dendrochronological records from Ireland and California (Sonnett and Finney, 1990). Swedish dendrochronological records also show a 2–3 ka periodicity (Karlén and Kuylenstierna, 1996), a spacing similar to that between the Dansgaard–Oeschger events recorded in the ice cores (Chapter 2). In North Atlantic ocean cores, a pervasive ~1500-year cycle has been identified which can be traced back beyond the Holocene into records from the last cold stage (Bond *et al.*, 1997; Bianchi and McCave, 1999), while a 550 and 1000-year cycle has also been detected (Chapman and Shackleton, 2000). Interestingly, a *c.*600 and 1100-year cycle has also been found in mire surface wetness records from northern England (Hughes *et al.*, 2000) and Scotland (Langdon *et al.*, 2003: Figure 3.23), while a ~1450-year cycle is evident in the GISP2 geochemical record (Mayewski *et al.*, 1997). Much shorter amplitude cycles have also been noted. Century-scale temperature variability has been found in Holocene lake sequences in Greenland (Willemse and Törnqvist, 1999), while cyclical periodicities of 200 and 260 years have been recorded in mires in Denmark and in Scotland (Aaby, 1976; Chambers *et al.*, 1997). Even shorter frequency cycles are evident in the oxygen isotope signal in the GISP2 ice core, with periodicities as low as 2.7–6.3 years (Stuiver *et al.*, 1995). Possible causes of these quasi-regular changes in Holocene climate are considered in the following section.

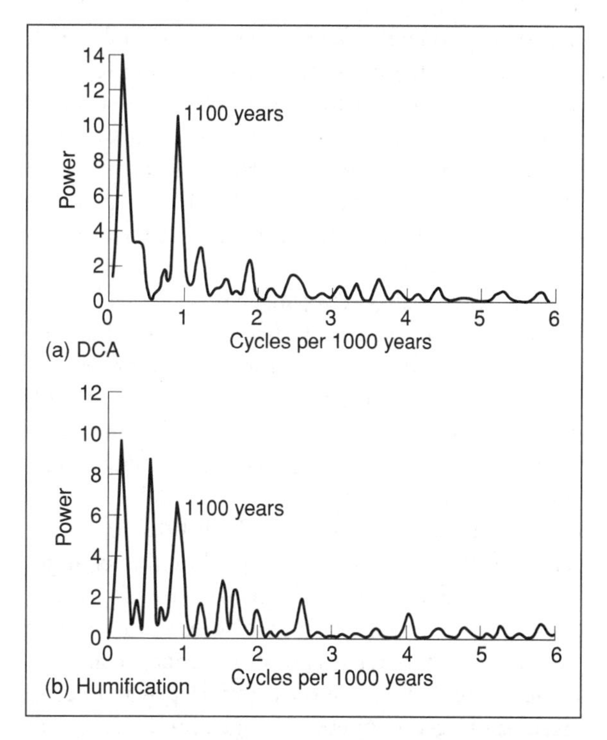

Figure 3.23 Periodograms showing 1100-year cyclicity over the past 7.5 ka in a Holocene peat sequence at Temple Hill Moss, south-east Scotland, as revealed by spectral analysis[3]: (a) plant macrofossil DCA[6] values; (b) detrended humification results (after Langdon *et al.*, 2003, Figure 10)

Causes of short-term climatic change

While the general pattern of climate change described above for the Holocene is consistent with that predicted by the operation of the Milankovitch variables, particularly precessional forcing, it is apparent from the various proxy records that rapid and often short-lived shifts in climate are superimposed upon the longer-term, orbitally driven climatic cycles (Clark *et al.*, 1999). The possible causes of these **sub-Milankovitch events** are discussed in this section.

The principal forcing mechanisms of climatic change over timespans of 10^1–10^3 years are generally considered to be variations in solar output and volcanic aerosols in the atmosphere. Indeed, comparisons between observational data and simulations from an energy balance climate model suggest that possibly as much as 64 per cent of pre-anthropogenic (i.e. pre-1850) decadal-scale temperature variations can be attributed to changes in solar irradiance and volcanism (Crowley, 2000). However, other factors may also force, modulate or amplify short-term climatic change. These include changes in both surface and deep-water ocean circulation, geomagnetic influences and changes in atmospheric gas content.

Solar output variations

Changes occur over time in the radiant heat emitted by the sun and involve both fluctuations in solar output as well as variations in the quality of solar energy. The former occurs as a consequence of **sunspot activity** and **solar flares**, while the latter reflects changes in the **ultraviolet** range of the solar spectrum.

Quantitative changes in solar output

It has long been considered that fluctuations in the solar constant (the amount of radiant heat emitted by the sun) are major factors in short-term climatic change, and general circulation model simulations suggest that a 2 per cent decline in solar output could reduce average surface temperatures by 4 °C (Hansen *et al.*, 1984). Satellite observations during recent solar cycles suggest a variation in solar irradiance of ~0.1 per cent (equivalent to a surface temperature reduction of ~0.2 °C) and, indeed, it has been suggested that many of the climatic variations of the past 11.5 ka may be explained by changes in solar irradiance of no more than several tenths of a per cent (Wigley and Kelly, 1990). Observations of the surface of the sun (the photosphere) indicate that the sun alternates between active and relatively quiescent phases, the most useful indications of these solar fluctuations being the growth and disappearance of **sunspots**. These are conspicuous dark patches that occur as shallow depressions in the general photospheric level and which may be the central parts of active or disturbed regions reflecting convectional activity within the photosphere. Hot areas (**plages**) have also been detected. In addition, violent eruptions, during which large amounts of energy are released over short timespans, are a regular feature of the outer parts of the sun, and are referrred to as **solar flares**. These short-lived disturbances, which occur from time to time in the sun's atmosphere, are superimposed on the steady output of radiant energy from the sun, and constitute a type of solar weather. Such short-lived variations in solar activity, reflected particularly in the appearance and disappearance of sunspots, are not random but appear to follow a cyclical pattern, with an 11-year periodicity especially prominent (the **Schwabe cycle**). Documentary records of sunspots (sunspot numbers) reflecting such a cyclical pattern extend back to about AD 1700 (Figure 3.24). Other elements of the solar weather also appear to reach maximum intensity at about 11-year intervals, although shorter-term fluctuations in solar activity have also been detected, for example cycles of solar flares with mean periods of 9 years, 2.25 years and 3 months (Landscheidt, 1984). The causes of these apparently cyclical variations in solar irradiance are uncertain, but they may relate to changes in the thermal structure of the sun under the influence of magnetic fields, or they may in some way be related to changes in solar diameter or convection (Lean, 1996).

Satellite-based measurements over the past two decades show a clear correlation between the solar irradiance and the 11-year sunspot cycle (Beer *et al.*, 2000), and patterns of solar activity over the 11-year cycle have frequently been linked with observed climatic and environmental changes (Hoyt and Schatten, 1997). Connections have been suggested, for example, with such diverse phenomena

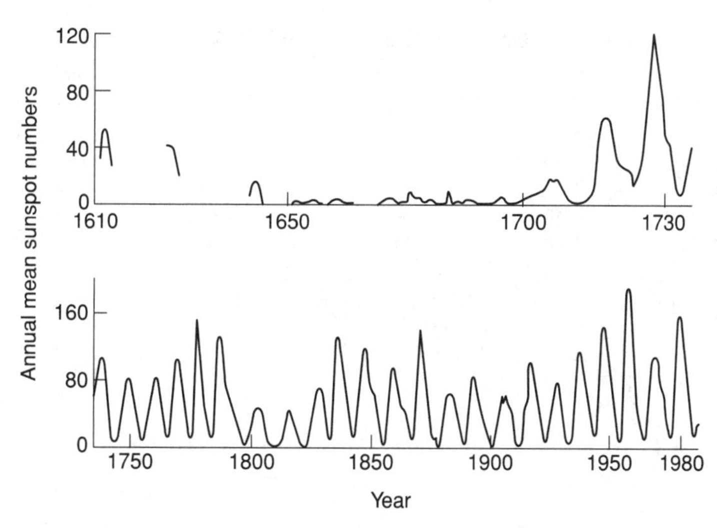

Figure 3.24 Annual mean sunspot numbers: AD 1610–1988 (after Stephenson, 1990, Figure 2)

as fluctuations in English temperature records, drought in the United States and floods on the River Nile, while solar flares and enhanced eruptional activity on the sun have been related to recent atmospheric circulation changes, rainfall patterns and incidence of thunderstorms (Landscheidt, 1984). The 11-year cycle has also been observed in SST variations over the course of the past century (White *et al.*, 1997), while a 22-year solar cycle (the 'double sunspot' or **Hale cycle**) has been noted in proxy climatic data from central Canada over the time period AD 1700–1979 (Guiot, 1987b).

Changes in solar irradiance have also been detected over the longer timescale of the Holocene. These longer periodicities cannot be observed directly, but are reflected in other data-sets (Table 3.2), Of particular significance are the records of **cosmogenic isotopes**, notably radiocarbon (^{14}C) and beryllium (^{10}Be), which form in the atmosphere as a result of the interaction between incoming galactic cosmic rays and certain constituents of the atmosphere. The rate of production of ^{14}C and ^{10}Be is modulated by the **solar wind**, a low-density stream of ionised gas which is ejected from the sun and which is related to overall levels of solar activity. The solar wind influences the magnetic field strength around the earth, so that at high levels of solar activity, incoming cosmic rays

Table 3.2 Postulated solar (and other) cycles and their possible reflection in Holocene proxy climate records (after Chambers and Blackford, 2001: 330)

Cycle	*Evidence*
Solar	
Schwabe: sunspot 10- to 11-yr	Dust, Greenland ice core; meteorological records
Hale: double-sunspot 22-yr	Storm tracks? tree rings? droughts
Gleissberg: 88-yr (or, 78-yr) (often cited as 80- to 90-yr)	Meteorological records; sea-surface temperatures? lake sediments
Suess: 211-yr (180- to 211-yr)	Peat humification? pollen data from peat bogs? carbon-14
Hallstattzeit: 2200-yr	Cosmogenic isotopes
Other	
1470-yr	Deep-sea cores; ice cores
c.2500-yr	Ice cores; tree rings; varves; lake levels
Interval between events c.2168 yr	Tree ring – narrow-ring events 1628 BC and c. AD 540

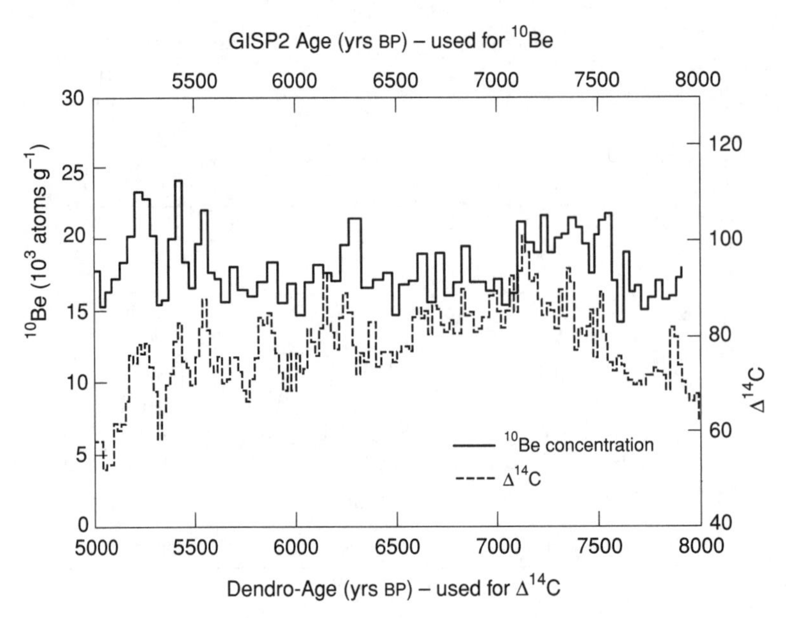

Figure 3.25 Beryllium 10 (^{10}Be) concentration in the GISP2 ice core compared with Δ^{14}C values from dendrochronological records. The ^{10}Be data are plotted on the GISP2 timescale whereas the Δ^{14}C record is plotted relative to the tree-ring timescale (after Finkel and Nishiizumi, 1997, Figure 3). Although the timescales are offset by ~100 yrs, the close similarity between the two data-sets supports the hypothesis that the two cosmogenic isotopes have been modulated by solar forcing

are increasingly deflected, resulting in a reduction of cosmogenic isotope production in the atmosphere. Long-term records of ^{14}C changes have been obtained from tree-ring records (Stuiver and Brazunias, 1993), and a continuous ^{10}Be record can be extracted from ice cores (Yiou *et al.*, 1997). The close relationship between these two data-sets (Figure 3.25) appears to confirm the hypothesis that short-term variations in ^{14}C and ^{10}Be reflect variations in their production rate due to solar modulation of cosmic radiation (Finkel and Nishiizumi, 1997).

Spectral analysis of the ^{14}C and ^{10}Be records shows evidence of a range of long-term solar periodicities, including the 88-year (**Geisberg**) cycle, a ~200-year cycle and a ~2500-year cycle (Rind and Overpeck, 1993). As noted above, a ~2500-year cycle has been detected in ice-core and oceanographical records, but it also is evident in varved sediment sequences, lake-level records and ocean core data (Lowe and Walker, 1997). Indeed, the similarity in timing between this long-term solar cycle and the Dansgaard–Oeschger cold events of the last cold stage (see above) has led to the suggestion that periodic marked increases in cloud cover, precipitation (snow) and declining temperatures as a result of solar/cosmic ray forcing, could have played a crucial role in the regularly occurring iceberg discharges (Heinrich events: see p. 72) recorded in North Atlantic deep-ocean cores during the last glaciation (van Geel *et al.*, 1998). In this respect, it may be significant that a statistically significant correlation has also been found between global glacier fluctuations over the course of the Holocene, and variations in atmospheric ^{14}C concentrations (Wigley and Kelly, 1990). It has also been suggested that solar forcing could have been responsible for the colder episode of the Younger Dryas (van Geel *et al.*, 2003).

During the later Holocene, discussions of the possible linkages between solar activity (as reflected in the ^{14}C and ^{10}Be records) and climate change have centred, in particular, on the period around 2600 BP and on the Little Ice Age. It has already been shown that proxy climatic records from many parts of the world point to a major climatic downturn between 2.5 and 2.8 ka BP, and this coincides with a significant increase in atmospheric

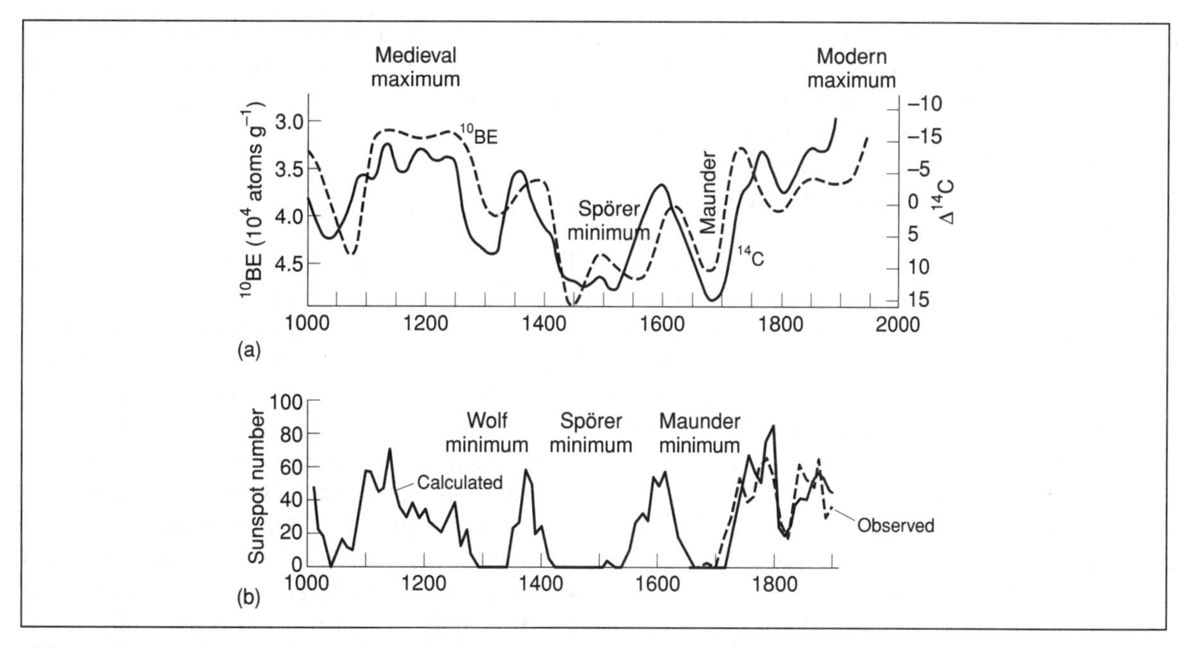

Figure 3.26 (a) ^{14}C and ^{10}Be records for the last millennium (after Lean *et al.*, 1996, Figure 2); (b) sunspot numbers for the same interval; the dashed line (a) gives the actual and observed averaged sunspot numbers (after Stuiver and Quay, 1980, Changes in atmospheric carbon-14 attributed to a variable sun, *Science*, **207**, pp. 11–19, Figure 12. Copyright 1980 American Association for the Advancement of Science). Note the close correspondence between sunspot numbers and the ^{14}C and ^{10}Be profiles

^{14}C (Δ^{14}C) and also in ^{10}Be. A marked reduction in solar activity has therefore been invoked to account for this climatic deterioration (van Geel *et al.*, 1996, 1998). Similarly, higher ^{14}C and ^{10}Be levels (Figure 3.26) correlate with the three sunspot minima of the medieval periods: the Wolf Minimum (AD 1281–1342), the Spörer Minimum (AD 1416–1534) and the Maunder Minimum (AD 1645–1715). A reduction in solar irradiance of the order of 0.25 per cent has been estimated for the Maunder Minimum (Lean *et al.*, 1992), with climatic modelling experiments suggesting that decrease of this order in solar irradiance would lead to a global temperature reduction of ~0.5 °C (Rind and Overpeck, 1993). Such a figure is in line with some estimates of temperature decline during the Little Ice Age (see above), although others are slightly higher (e.g. Crowley and North, 1991).

The foregoing suggests a strong linkage between solar activity and short-term climate change, and correlations between indicators of insolation change (such as ^{14}C and ^{10}Be) and proxy climate records appear to confirm the hypothesis of solar dominance of climate variability over the last several centuries. However, it is often difficult to translate these solar properties into absolute irradiance changes and, moreover, to prove the causal connection (Rind and Overpeck, 1993). Equally, much remains to be learned about precisely how the global climate system reacts to a solar forcing mechanism, while many aspects of the interrelationships and feedback loops between solar output variations and other sub-Milankovitch forcing factors have yet to be established. Some of these aspects are considered in the following sections.

Qualitative changes in solar output

Although the ultraviolet (UV) flux constitutes only a small fraction (about 1.05 per cent) of the total photon energy emitted from the sun, it is a significant factor affecting the heating of the upper atmosphere, principally through the formation of ozone, the concentration of which is important in controlling stratospheric temperature through UV absorption (Lean, 1984). Indeed, it has been suggested that UV variations may be a major component of the solar influence on climate, partly because solar spectral irradiance variations are proportionately larger at short wavelengths, but also because they carry a significant proportion of the total solar energy variability (Hoyt and Schatten, 1997). Measurements of atmospheric ozone content since the late 1960s indicate that concentrations vary with time, and there is evidence of an 11-year fluctuation almost in phase with the sunspot cycle which has been linked to variations in output of solar UV radiation (Keating, 1978). Satellite observations have confirmed these temporal changes in solar UV irradiance, and the data indicate that when the level of solar activity is high, as reflected by the sunspot number, so too is the sun's total UV radiative output (Lean, 1996). In addition, there is evidence of a shorter periodicity of UV irradiance, corresponding with the 27-day solar rotation period (Gérard, 1990).

Climatic modelling suggested that a 1 per cent increase in UV radiation at the maximum of the solar activity cycle generated 1–2 per cent more ozone in the stratosphere and that this would be accompanied by substantial stratospheric heating as more sunlight is absorbed by the increased ozone levels (Haigh, 1994). This, in turn, would lead to a strengthening of the stratospheric winds, and the poleward displacement of both the tropospheric subtropical jet streams and the mid-latitude storm tracks (Haigh, 1996). The result would be a significant short-term climatic change in the mid/high latitudes. Indeed, other models predict temperature variations of up to 2–5 °C at 50 km, especially in high-latitude summer regions (Gérard, 1990).

Although both the empirical data and modelling results are persuasive, once again testing the hypothesis that variations in the quality of solar output can cause climatic change is problematical. On the one hand, there is increasing observational evidence that the upper/middle atmosphere responds both chemically and in terms of temperature to variations in the solar UV flux and, moreover, that changes in both solar irradiance and UV flux follow a similar cyclical pattern. On the other hand, the present database is limited and only when long-term records over many solar cycles become available will it become possible to evaluate the effects of solar UV variations on the stratosphere, and to estimate precisely how such changes affect the troposphere and hence global climate.

Volcanic aerosols

The infusion into the atmosphere of large amounts of volcanic debris has long been regarded as a possible causal mechanism for producing short-term variations in climate (Lamb, 1977). It was originally believed that a reduction in temperature would result from the screening out of incoming solar radiation by atmospheric dust particles, and that this effect would be amplified by cloudiness as the particulate matter acted as foci for water droplets. Moreover, it has frequently been suggested that the effects of volcanic activity on climate would be relatively localised. However, it now appears that the amount of dust and ash that remains in the atmosphere for any length of time following a major episode of explosive volcanism is relatively small, and that the more significant climatic effects result from the release during an eruption of large volumes of sulphur volatiles (Bluth *et al.*, 1993). Once in the atmosphere, these are converted into sulphuric acid and become the dominant aerosol, the effect of which is to cool the lower troposphere by the backscattering of incoming solar radiation. Moreover, there is evidence to suggest that once injected into the atmosphere, these aerosols are disseminated globally (see below). The mean residence time of the sulphuric acid aerosols ranges from one to five years (Schönwiese, 1988).

Considerable empirical evidence exists to support the hypothesis of a link between short-term climatic change and recent volcanic activity. For example, the eruptions of the Tambora (1815), Krakatoa (1883) and Santa Maria (1902) volcanoes may have been associated with global temperature reductions of 0.2–0.5 °C for periods of up to five years (Rampino and Self, 1982), while a ~1 °C fall in tropical tropospheric temperature followed the eruption in 1963 of Mt Agung on Bali in Indonesia, the effects of this eruption being registered by a temperature decline of 0.8 °C as far north as latitude 60° N (Newell, 1981). More recently, the eruption in 1991 of Mt Pinatubo in the Philippines was followed, in 1992, by a global mean air temperature decline of ~0.5 °C at the surface and 0.6 °C in the troposphere (Parker *et al.*, 1996). Radiation receipt in the tropics following the Pinatubo eruption may have declined by as much as 10 per cent (Handler and Andsager, 1994). What appears critical in these data, however, is the location of the volcano itself, for equatorial eruptions appear to have a global impact whereas those in mid-latitude regions tend to induce cooling principally in the hemisphere in which the eruption occurred (Zielinski, 2000). Recent evidence also points to the complexity of the climate response to volcanic eruptions. For example, climate records indicate *warmer* winters in Eurasia (although colder summers) following the Pinatubo and El Chichón (1982) eruptions (Robock and Mao, 1992). Overall, however, these recent data appear to confirm the hypothesis that sulphur volatiles from volcanic eruptions can lead to a cooling of global climate by 0.2–0.3 °C for several years after the eruption (Zielinski, 2000).

Data on past episodes of volcanic activity have been obtained from polar ice cores. One method is to use tephra (i.e. silicate matter including volcanic glass) in the ice, and these materials can often be linked to the source of the eruption by geochemically matching the ice-core tephra with glass from the suspected eruption (Zielinski, 2000). Tephra is often relatively rare in ice cores, however, and hence an alternative approach which employs variations in acidity is more common. Acidity levels in annual ice layers can be established by electrical conductivity measurements which reflect the amount of sulphuric acid washed out of the atmosphere in precipitation over the course of a year. Hence down-core variations in acidity provide an indication of the fluctuations in the amount of atmospheric sulphuric acid and, by implication, of volcanic aerosols over time. These acidity profiles, therefore, constitute a proxy temporal record of volcanic eruptions. A 110 ka record of volcanic activity from the GISP2 ice core (Figure 3.27a), for example, shows clearly the major eruptions including Toba (Sumatra) around 73 ka which has been associated with the onset of marine oxygen isotope stage 4 (Rampino and Self, 1993) and the great Z2 eruption of around 53 ka which is believed to be of Icelandic origin. There is also a significant increase in volcanism between *c.*15 and 8 ka BP (see also Figure 3.27b), which has been linked with increased crustal fracturing and magma release in volcanically active areas such as Iceland following ice sheet wastage and isostatic recovery (Sejrup *et al.*, 1989).

The significance of these records lies in the fact that they provide a chronology of volcanic activity against which climate proxy records can be assessed. The Greenland ice-core volcanic records for the past 2000 years, for example, show a significant increase in volcanic activity during the course of the last 600 years compared with the previous 1500-year period (Table 3.3), and this matches well with a range of proxy data indicative of an overall cooling in Northern Hemisphere climate. There is also a close relationship between glacier activity over the last millennium and Greenland ice-core acidity, which led Porter (1986) to suggest that sulphur-rich aerosols generated by volcanic eruptions are a primary forcing mechanism of recent Northern Hemisphere glacier fluctations, and hence of climatic change. Other climate proxy records indicative of cooling have been linked with this recent episode of increased volcanism. Dendrochronological records from the Kola Peninsula, Russia, suggest a summer temperature decline of ~0.72 °C in the year immediately following an eruption (Gervais and MacDonald, 2001). Indeed, tree-ring data from a range of Northern Hemisphere sites show a very close correlation with the GISP2 ice-core record for the last 600 years, the lowest low tree-ring density values (reflecting cooler conditions) corresponding with episodes of explosive volcanism, and which are especially marked during several decades of the 1600s and in the early 1800s (Briffa *et al.*, 1998; Briffa, 2000). Temperature data for the Arctic for the last 400 years, reconstructed from a range of sources including tree-ring series, lake sediment records and ice-core evidence, also show markedly cooler conditions, especially in summer, which again may be linked to episodes of volcanism (Overpeck *et al.*, 1997). Of equal significance is the fact that the abrupt rise in Northern Hemisphere temperatures in the period 1920–50 (Figure 3.22) coincides with

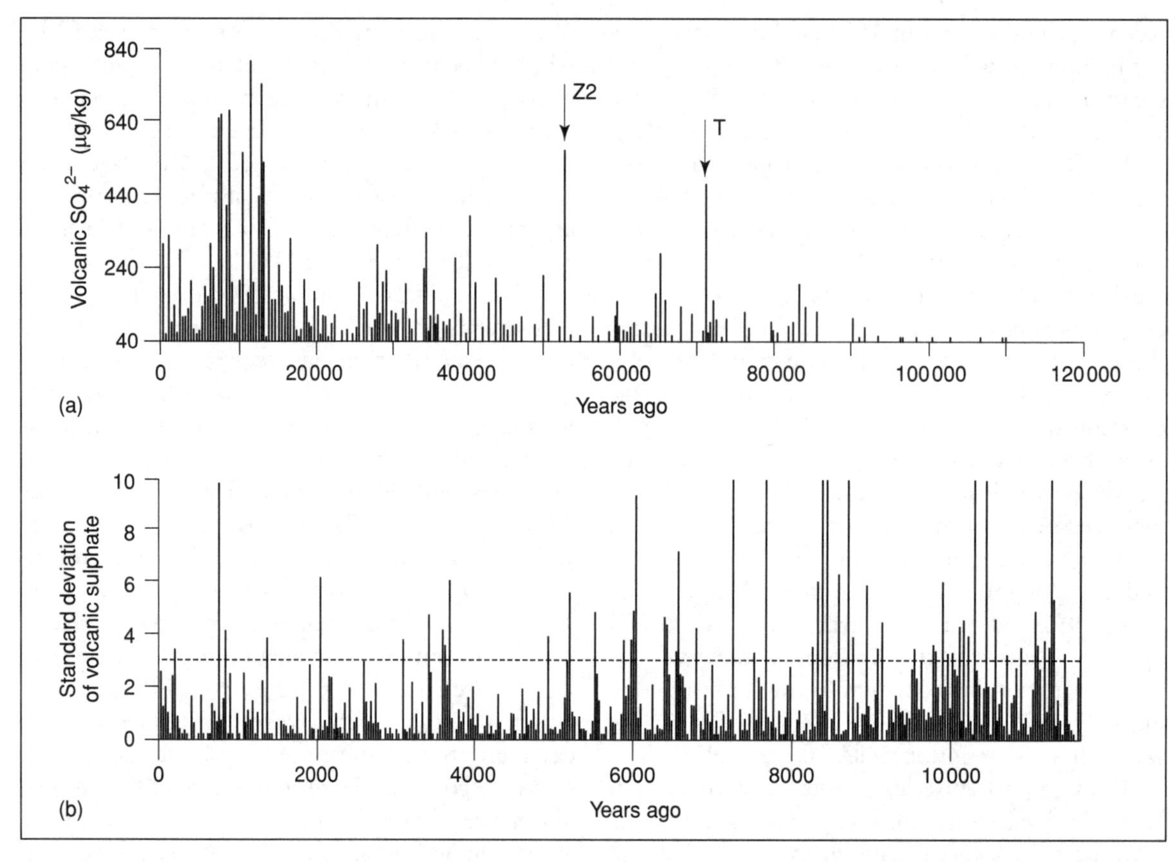

Figure 3.27 (a) Acidity profile (volcanic SO_4^{2-}) from the GISP2 ice core over the past 110 ka. Note the major Toba [T] and Z2 eruptions; (b) acidity profile (expressed as standard deviation of volcanic sulphate) from the GISP2 ice core for the Holocene. Note the increase in volcanic activity around 10 ka BP (after Zielinski *et al.*, 1997, Figures 1 and 3)

a distinct lack of evidence for volcanic activity (Zielinski, 2000).

There is now, therefore, a considerable body of confirmatory evidence to support the hypothesis of a relationship between volcanic activity and climatic change, albeit over very short timescales (decades or less) and, in particular, for the relatively recent past. Climatic modelling experiments lend further support to the hypothesis that volcanic activity can exert a significant impact on climate. For example, a recent simulation using an energy balance climatic model suggests that for the past 1000 years, 22–23 per cent of the reconstructed temperature variations for the pre-anthropogenic period can be explained by volcanism, and that for the Little Ice Age interval (AD 1400–1850), the volcanic contribution increases to 41–49 per cent (Crowley, 2000). What is apparent in both the empirical and modelling data, however, is that volcanic forcing alone is unlikely to account for all of the short-term changes that are manifest in the historical climate records, but rather that episodes of explosive volcanism tend to amplify climatic changes that are already occurring through a complex series of positive and negative feedbacks. Moreover, considerable additional research is required before the linkages between atmospheric volcanic aerosols, solar radiation balance and climatic change are firmly established.

Geomagnetism

Changes in the earth's magnetic field have frequently been linked to climatic change. During

Table 3.3 Major acidity signals in the Dye 3 and GRIP Greenland ice cores from the past 2000 years (after Clausen *et al*., 1998)

Age of event (AD)	*Dye 3 Magnitude of event*[1]	*GRIP Magnitude of event*[1]	*Possible source of eruption*
1912		64	Katmai, Alaska
1816	119	101	Tambora, Indonesia
1810		72	Equatorial region?
1783	272	212	Laki, Iceland
1601		78	High northern latitudes?
1477		51	Unknown
1259	305	464	Equatorial region: Mexico?
1179		65	Katka, Iceland
1106		77	Hekla, Iceland
934	50/82	128/91	Eldja, Iceland
898		53	Unknown
871		40	Iceland?
757		2/9	Iceland?
572		9/6	Iceland?
516	48/51		Iceland?

[1] The magnitude of the eruption (the amount of acidic gases released into the atmosphere) is expressed in megatons of acid (10^9 kg). Where two values are shown, these relate to different acids. The calculation of the magnitude is based on an eruption site in a high northern latitude.

the second half of the twentieth century, for example, in both central Europe and the USA there are close similarities between trends in geomagnetic activity and temperature (Figure 3.28), while longer-term variations in geomagnetic intensity have also been linked to patterns of climate change (Bradley, 1999). It has already been seen that variations in the solar wind can influence the magnetic field strength around the earth, and it been suggested that geomagnetic changes could influence, *inter alia*, the stratospheric greenhouse effect, the latitudinal extent of high-altitude cloud cover, and other elements that govern the general circulation (Fairbridge, 1983; Svensmark and Friis-Christensen, 1997). For example, the intensification of westerly zonal airflow and fluctuations of temperature in Europe, the USA and northern India, have been strongly linked to enhanced geomagnetic forcing (Bucha and Bucha Jr, 2002). However, although there is a certain amount of circumstantial evidence for a connection between changes in geomagnetism and climate, it is often difficult to distinguish between cause and effect, and it may be that both are reacting essentially to variations in solar activity (Goudie, 1993). Hence, the extent to which terrestrial geomagnetic fluctuations in themselves stimulate short-term climatic change remains an open question.

Ocean circulation

The importance of ocean water circulation in the development of global climatic systems is widely acknowledged, and hence any changes in the nature of ocean water movement are likely to have a climatic effect. Such changes have already been cited as a possible factor in long-term climatic change (e.g. the major oceanographical changes that would have followed the closing of the Gulf of Panama 4.6–2 Ma BP: see above), but there is now a substantial body of evidence to show that the changes in the pattern of oceanic circulation could have occurred over much shorter intervals (see above) and that these too may be a

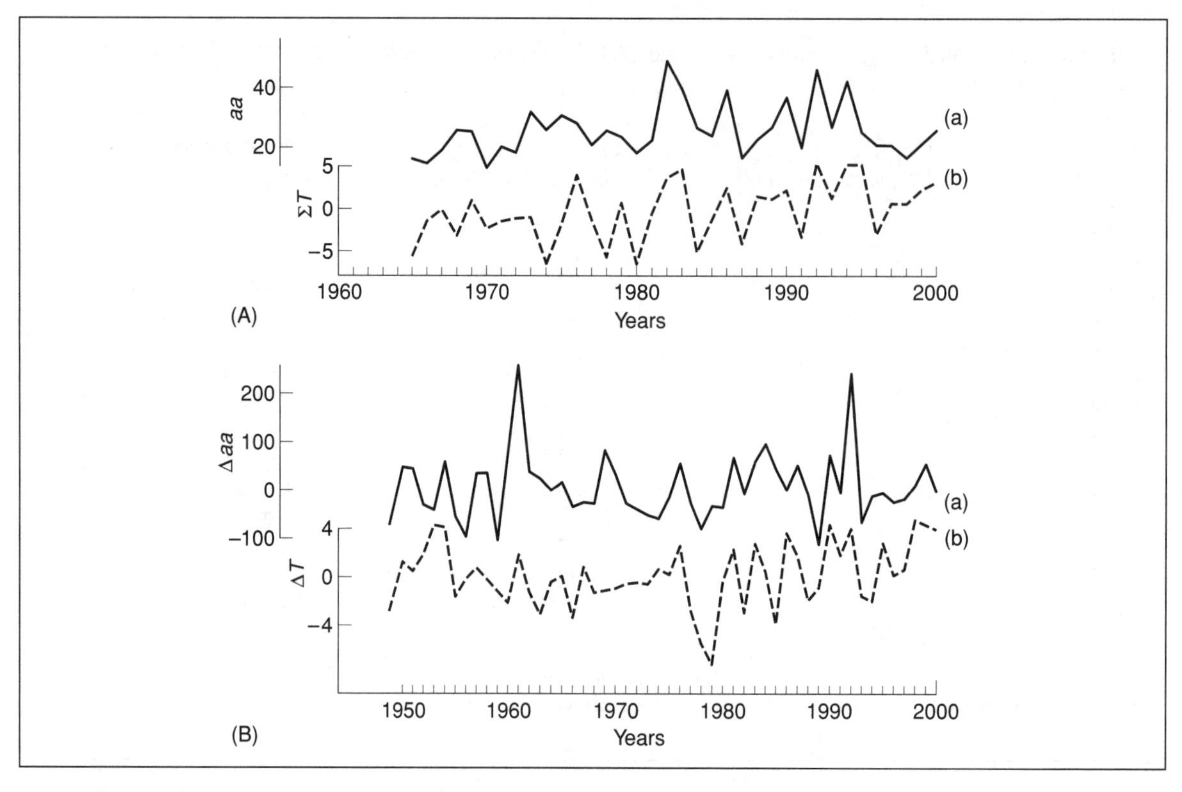

Figure 3.28 (A) Time series of (a) geomagnetic activity (aa index), and (b) air-surface temperature anomalies (May–July) in Prague; (B) Time series of (a) geomagnetic activity (aa index), and (b) mean US temperature anomalies (January and February) in North America (after Bucha and Bucha, Jr., 2002, Figures 3 and 6)

causal factor in climate change, this time at the millennial or sub-millennial scale (Rahmsdorf, 2002).

Ocean water movements are driven by a combination of interconnected factors. Some of these are internal, such as density, temperature and salinity variations, while others (frictional drag of the wind; inputs of fresh water from precipitation, rivers, glaciers, etc.) are external to the ocean system. Ocean core data (Chapter 2) can provide evidence of patterns of surface water movement, which can frequently be linked to climate change. Ocean core records also contain evidence of variations in the production of deep water, namely water masses with distinctive temperature and salinity characteristics that develop in the abyssal depths of the world's oceans. Deep-water circulation changes arise from variations in salinity (density) and temperature (**thermohaline circulation**) and are also important in terms of climate change. Within the North Atlantic, for example, a major **conveyor** system appears to operate in which water moves northwards in the upper levels of the ocean and sinks around 60° N to form a deep-water mass known as **North Atlantic Deep Water** (**NADW**). The return limb of the conveyor at depth transfers this deep water to the Southern Oceans (Broecker *et al.*, 1985). The climatic effects of this conveyor system are considerable, for the sinking (**ventilation**) of relatively warm surface waters in the northern North Atlantic releases enormous quantities of heat to the atmosphere. It has been estimated that an amount of heat equivalent to ~25 per cent of the solar heat reaching the surface of the North Atlantic is released to the atmosphere each year as a result of this ventilation process (Broecker and Denton, 1990). Any reduction in the effectiveness of the conveyor would clearly have major climatic consequences for climate around the North Atlantic.

The ways in which a combination of surface and deep-water ocean circulation changes can impact on climate, often over relatively short timescales, are illustrated by events in the North Atlantic at the end of the last cold stage where it has been suggested that episodic discharge of enormous volumes of meltwater from the wasting Laurentide ice sheet could have exerted a profound impact on ocean circulation, and hence on climate (Licciardi *et al.*, 1999). During the Younger Dryas, for example, NADW formation seems to have ceased, leading to a complete shutdown of the North Atlantic conveyor system (Broecker and Denton, 1990). This appears to have resulted from a decrease in density of North Atlantic waters, brought about partly by a major influx of icebergs and cold meltwater from disintegrating ice shelves or marine-based ice sheets in the Arctic basin, but also from the infusion into the North Atlantic of large quantities of meltwater from the wasting Laurentide ice sheet following the diversion of the Great Lakes drainage from the Mississippi to the St Lawrence system (Broecker *et al.*, 1989: Figure 3.29a). Computer modelling suggests that this involved the release of around 9500 km^3 of water into the North Atlantic from glacial Lake Agassiz (see Figure 4.22) via the Great Lakes and Gulf of St Lawrence (Leverington *et al.*, 2000). The result was a reduction in sinking of waters in the North Atlantic and cross-equatorial flow to feed that sinking, and a consequent decline in heat release to the atmosphere (Keigwin and Lehman, 1994). Rediversion of North American meltwater drainage and a reduction in iceberg influx into the North Atlantic led to a density increase of North Atlantic surface waters and reactivation of the conveyor. The associated heat release produced rapid climatic warming and brought about the end of the Younger Dryas event (Alley, 2000)[7]. Episodic discharge of meltwaters from proglacial lakes around the southern margin of the Laurentide ice sheet are also thought to be responsible for climatic changes during the early Holocene. It has been suggested, for example, that rapid drainage of glacial Lake Agassiz into the Arctic basin via the Mackenzie River system at *c.*11.3 ka BP (p. 133) may have increased arctic pack ice, freshened surface waters and impeded the formation of NADW, resulting in the hemispherical cooling known as the 'Preboreal Oscillation' (Fisher *et al.*, 2002), and which is recorded in the Greenland ice cores as the 'GH-11.2 event' (Figure 3.17). Similarly, catastrophic drainage of ice-dammed lakes during the later downwasting of the Laurentide ice sheet led to a chilling of ocean surface waters and caused the short-lived climatic deterioration that was experienced throughout the North Atlantic region at ~8.2 ka BP (Figure 3.29b).

The Younger Dryas and GH-8.2 events are also important because they emphasise the links not only between the oceans and climate, but also between the oceans, climate and the ice sheets. It was noted above, for example, that one of the features of the North Atlantic sedimentary record over the last glacial cycle is the presence of layers of ice-rafted debris (Heinrich layers), which reflect the episodic discharge of icebergs from the Laurentide and Fennoscandian ice sheets (p. 72). These **Heinrich events**, which appear to have led to ocean temperature changes of up to 5 °C in a matter of decades, find parallels in the Dansgaard–Oeschger cycles in the ice cores (Dansgaard *et al.*, 1993). It has been suggested that the Heinrich events result from periods of massive ice-sheet collapse, the subsequent iceberg and meltwater discharge leading to the chilling of surface waters, and significantly colder conditions on land areas around the North Atlantic. It seems unlikely that climate is the driving factor behind ice-sheet collapse. Rather, a combination of internal factors, including changes in ice-sheet mass balance and basal thermal regime, along with sea-level fluctuations around tidewater margins, appear to have prompted ice sheets to grow, reach a position of instability, and then collapse catastrophically (Andrews, 1998). Climatic forcing is then transmitted through the ocean system by abrupt SST [sea-surface temperature] changes. Hence, ice-sheet behaviour may not only be linked to climate change over longer timescales (e.g. the 100 ka cycle: see above), but possibly also at sub-Milankovitch periodicities.

During the course of the Holocene, periodicities in North Atlantic ocean circulation have been detected at ~550, 1100 and 1500-year intervals (Bond *et al.*, 1997; Chapman and Shackleton, 2000).

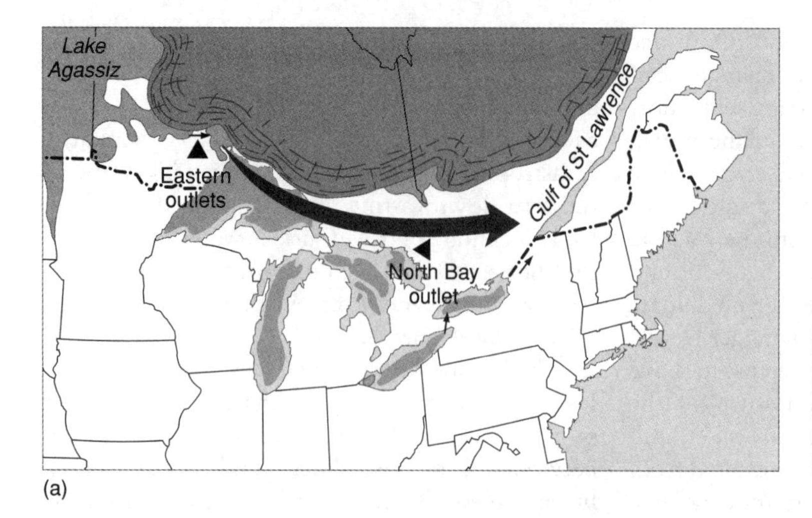

(a)

Source: Reprinted with permission from *Nature*, after Broecker, W.S. *et al.*, 1989, Routing of Meltwater from the Laurentide Ice Sheet during the Younger Dryas Cold episode, *Nature*, **341**, pp. 318–21, Figure 1. Copyright 1989 Macmillan Magazines Ltd.

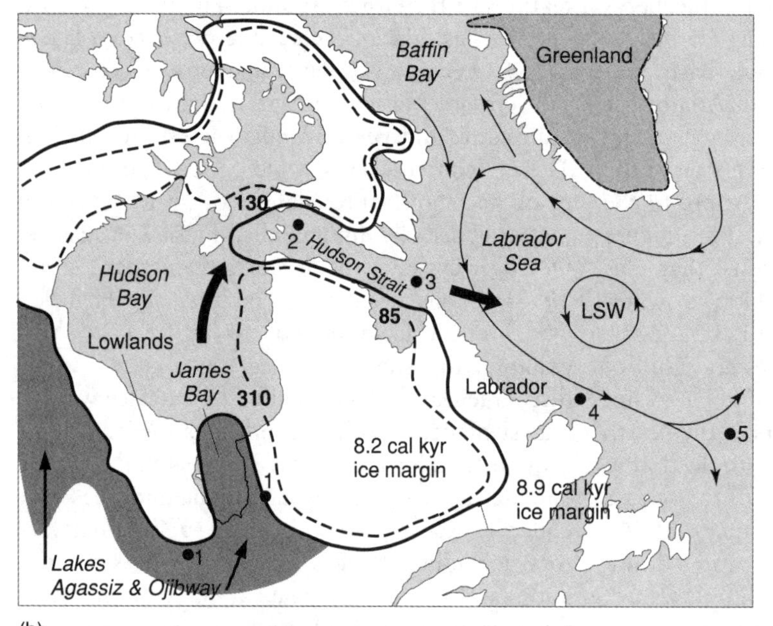

(b)

Source: Reprinted with permission from *Nature*, after Barber D.C. *et al.*, 1999, Forcing of the cold event of 8,200 years ago by catastrophic drainage of Laurentide lakes, *Nature*, **400**, pp. 344–48, Figure 1. Copyright 1999 Macmillan Magazines Ltd.

Figure 3.29 Routing of meltwaters in eastern North America during the wastage of the Laurentide ice sheet: (a) routing of Laurentide meltwaters via the St Lawrence system and into the North Atlantic, which may have been the trigger for the Younger Dryas event ~12.5-11.5 ka BP; (b) ice margins in north-eastern Canada at ~8.9 ka BP (thick black line) and at ~8.2 ka BP (hatched line) before and after disintegration of ice in central Hudson Bay. The opening of the ice barrier across the Hudson Strait resulted in the catastrophic drainage into the Labrador Sea of glacial lakes Agassiz and Ojibway and may have been the trigger for the GH-8.2 event

These data suggest an inherent oceanographical instablity over the course of the present interglacial, which could well have had a climatic impact. Indeed, changes in deep-water circulation in the North Atlantic have been linked to Holocene climatic events, for increased rates of deep-water flow to the south of Iceland appear to coincide with the Medieval Warm Epoch and with the episode of warmer climate around 2000 BP (Bianchi and McCave, 1999). By contrast, more sluggish flow coincides with the Little Ice Age and with the cooler climatic episode in the Dark Ages (after *c.* AD 500). Data covering the past 50 years indicate a close relationship between observed SST anomalies and changes in atmospheric circulation (Czaja and Frankignoul, 1999), while climatic modelling experiments suggest a close link between North Atlantic SSTs and European winter temperatures (Rodwell *et al.*, 1999).

The causes of these oceanographical changes during the course of the Holocene remain to be established, however. Links with ice-sheet oscillations, as has been suggested for the Heinrich events, seem unlikely, but it is possible that the cyclicities that have been detected in NADW circulation are driven, to some extent, by variations in solar activity (Chapman and Shackleton, 2000). Internal changes within the ocean system may also be a factor. It has been suggested, for example, that variations in freshwater flux through the terrestrial hydrological cycle could induce major thermohaline changes in the oceans, and that these could lead to marked surface temperature changes on timescales of only a few years (Rahmstorf, 1995). If this is so, then the Holocene oceans may not be in the relatively steady state that was once considered to be the case, and even if thermohaline circulation changes are not in themselves the trigger for short-term climate change, these could well have served to modulate or amplify the effects of other forcing factors (Keigwin and Boyle, 2000).

Atmospheric trace gases

A further factor in the equation may be the role played by the oceans in the global carbon cycle. We have already seen how long-term variations in CO_2 and other atmospheric trace gases may amplify or modulate the long-term global climatic changes induced by the Milankovitch variables (see above). However, changes in atmospheric CO_2 as reflected in the ice-core records (see Figure 2.20) may also influence climate over shorter timescales. The mechanisms behind the observed changes in atmospheric trace gas are not completely understood, but they appear to reflect the operation of sources and sinks of individual gases (Sundquist, 1993). In the case of CO_2, the oceans would have constituted a major sink during glacial periods at times of reduced vegetation cover. The ocean–atmosphere flux of CO_2 is governed by a range of interconnected factors (biological productivity, sea-ice cover, rate of circulation, etc.), and hence changes in any of these, for example at glacial–interglacial transitions, would have resulted in significant changes in the amounts of CO_2 released to the atmosphere. Hence, the dramatic rise in CO_2 and, subsequently, in CH_4 at the transition from the Younger Dryas to the Holocene (Severinghaus *et al.*, 1998) could have served to amplify the warming resulting from other forcing factors. On the other hand, the Holocene Antarctic CO_2 record, which shows an initial decline (8 ppmv) between 10.5 and 8.2 ka BP, and thereafter a gradual 25 ppmv increase over the period from 7 to 1 ka BP (Raynaud *et al.*, 2000), suggests that, prior to the anthropogenically induced increase in CO_2 of the last 300 years (p. 260), fluctuations in atmospheric trace-gas content have not been significant factors in short-term climate changes during the remainder of the Holocene.

Notes

1. Where ages within the past 20 ka are based on ^{14}C dating, these have been calibrated to calendar years (see Chapter 2).
2. The term **precession** describes the slow movement in the axis of rotation of a spinning body (e.g. a gyroscope) so as to describe a cone (Figure 3.11a).
3. **Spectral analysis** is a statistical technique which aims to identify **cycles** in time-series data. Cycles may be characterised in two related ways: the *period* of a cycle is the length of time between consecutive repeats, while the *frequency* is the number of cycles (i.e. repeats) that occur per unit of time (Green, 1995).
4. The **Cenozoic** is the fourth of the great geological eras, after the Proterozoic, the Palaeozoic and the Mesozoic. It began around 65 ma BP and includes the Tertiary and Quaternary periods.
5. The term **frequency modulation** as employed here is entirely similar to electronic modulation of a high-frequency carrier by a low-frequency modulating signal, such as is used in FM radio and television broadcasting (Rial, 1999).
6. Variations in plant macrofossil assemblages at different levels in a peat profile are likely to reflect a

range of environmental factors. One way in which the influence of these factors can be established is to employ **detrended correspondence analysis (DCA).** This is a standard multivariate statistical technique which assumes that any structure within the data (i.e. within the plant macrofossil record) is a response to changes in a few unknown environmental variables (wetness, dryness, competition, etc.). In numerous studies of ombrotrophic peats (e.g. Barber *et al.*, 1994; Hughes *et al.*, 2000; Langdon *et al.*, 2003), plotting of the first and second axes of the ordination has shown that the plant macrofossil taxa on axis 1 (the strongest axis in the ordination) tend to be arranged along a hydrological gradient from wet to dry. As this represents a water-level axis, the sample scores can therefore be used as an *index* of changes in bog-surface wetness. Moreover, because ombrotrophic mires are entirely rain-fed, the DCA scores also reflect former levels of precipitation.

7. However, Berger and Jansen (1995) specifically reject the 'conveyor-belt' hypothesis for the Younger Dryas, arguing that the event simply represents a pause during deglaciation when cold conditions were re-established for a short period. Others have suggested that solar forcing may be a more likely explanation (Renssen *et al.*, 2000; van Geel *et al.*, 2003).

4 Consequences of climatic change

Introduction

The global climatic changes discussed in the previous chapter had a profound impact on the landscapes of Europe and North America. The transition from a full glacial to an interglacial climatic regime which occurred between *c.*15 and 10 ka BP was reflected in the rapid wastage of the great ice sheets and mountain glaciers of the Northern Hemisphere, in a contraction of the periglacial domain, and in the initiation of a vegetation succession which resulted in a change from arctic tundra to woodland over much of Europe and North America within the timescale of a few thousand years. Dramatic changes also occurred in fluvial regimes, erosional activity and pedogenesis. Moreover, following the release into the oceans of enormous quantities of meltwater from the wasting ice sheets, global sea level rose by over 120 m, completely changing the configuration of coastal regions in many areas around the Atlantic basin. These geological, geomorphological and biological responses to climatic change, and their reflection in the landscapes of the mid-latitude regions of the Northern Hemisphere over the past 20 millennia, form the subject matter of this chapter.

The last glaciers in the northern temperate zone

Europe

At the height of the last glaciation, around 22 ka BP, there were two major centres of glaciation in mainland Europe, one located in the Baltic region and the other in the Alps, the two glacier complexes remaining as discrete entities throughout the Late Weichselian. Smaller glaciers also formed in the mountainous areas of central and southern Europe including the Pyrenees, the French Massif Central, the Apennines, the Dinaric Alps and on the island of Corsica (Sibrava *et al.*, 1986). In the Arctic, an ice sheet developed over the Barents and Kara seas, although whether this ice sheet extended southwards to cover mainland Russia remains uncertain (Grosswald and Hughes, 2002; Mangerud *et al.*, 2002). By far the largest ice sheet was the one that developed over Scandinavia (Figure 4.1) which, at its maximum, covered the whole of Norway, Sweden and Finland, and extended southwards into Denmark and Germany, and eastwards into Poland and the north-west USSR (Mangerud, 1991). The main ice dome lay over the Gulf of Bothnia where the ice may have been in excess of 2 km in thickness. The evidence suggests that the maximum extent of the Late Weichselian Fennoscandian ice sheet was reached earliest in the south-west (at around 28 ka BP) and later in the east (between 17 and 15 ka BP), and that the ice sheet was still advancing in the east while retreating in the west (Boulton *et al.*, 2001).

The British Isles

In the British Isles, the major ice accumulation and dispersal centres lay along an axis extending from the mountains of western Scotland to the uplands of Wales, and these fed a large ice stream that flowed southwards down the Irish Sea basin

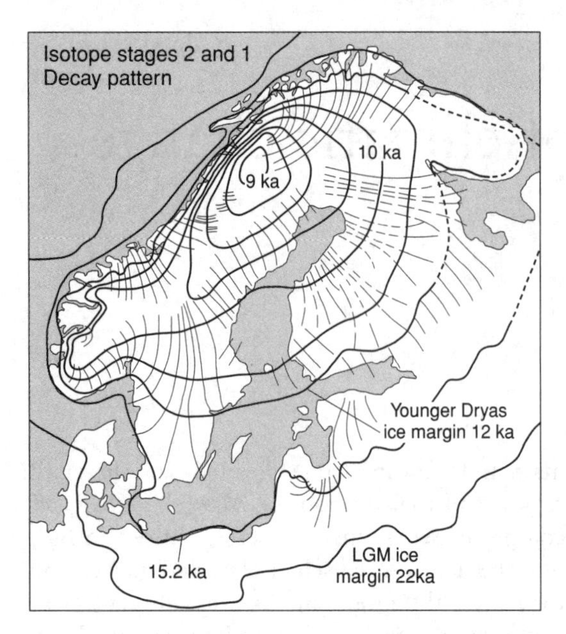

Figure 4.1 The Fennoscandian ice sheet at the Last Glacial Maximum, and the ice-frontal position during successive stages of ice-sheet wastage (after Siegert, 2001)

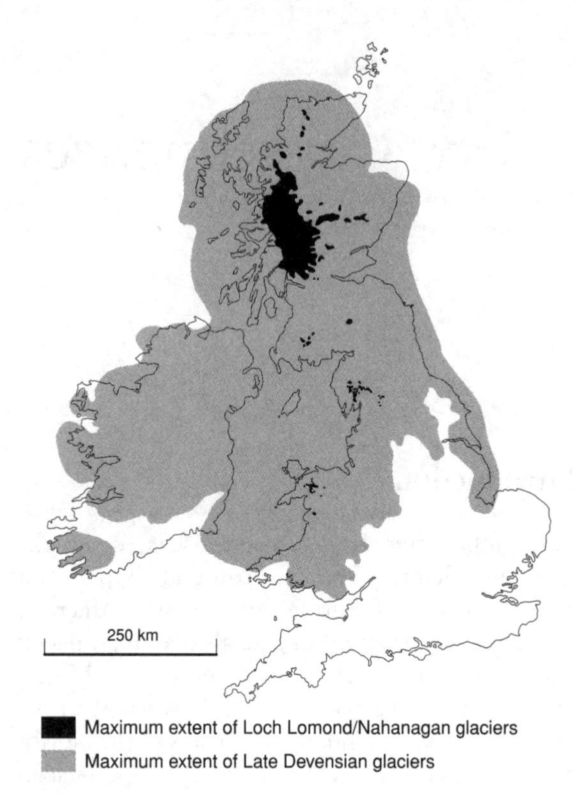

Figure 4.2 The last glaciers in Britain and Ireland (after Bowen *et al.*, 1986, 2002)

(McCarroll *et al.*, 2001). At the glacial maximum the ice sheet covered almost all of Scotland, Ireland, Wales and northern England, with only central and southern parts of England remaining ice-free (Figure 4.2). A combination of field evidence and glaciological modelling suggests an ice thickness of up to 1300–1400 m over the Grampian Highlands. At the glacial maximum, ice overtopped most of the mountainous areas of mainland Scotland, although the highest summits of some of the Hebridean islands, of the Lake District and of upland North Wales remained above the ice as nunataks (Ballantyne *et al.*, 1998b; McCarroll and Ballantyne, 2000). Whether the British and Scandinavian ice sheets were confluent at the Last Glacial Maximum is still a matter for debate, for while some have argued that the two ice sheets were confluent between *c.*28–22 ka BP (e.g. Sejrup *et al.*, 2000), other reconstructions (perhaps the majority) imply that there was no contact between the two ice masses during the Late Devensian/Late Weichselian (e.g. Bowen *et al.*, 2002).

North America

The largest glacier complex in the Northern Hemisphere was the Laurentide ice sheet of North America (Figure 4.3). Centred on Hudson Bay, the ice sheet extended southwards across the United States–Canada border into the Great Lakes region, westwards towards the Rocky Mountains, and eastwards into the Canadian maritime provinces. To the north, ice extended across the islands of the Canadian Arctic archipelago where the Innuitan ice sheet covered most of that region (Dyke *et al.*, 2002). The main accumulation and dispersal centre appears to have been to the east of the Hudson Bay throughout much of the glacial period (Veillette *et al.*, 1999), although the ice dome may have extended westwards across Keewatin with ice thicknesses perhaps in excess of 3 km. The maximum volume of ice, estimated at *c.*34 million km^3 (Marshall *et al.*, 2002), is over four times that of the Late Weichselian Scandinavian ice mass, and

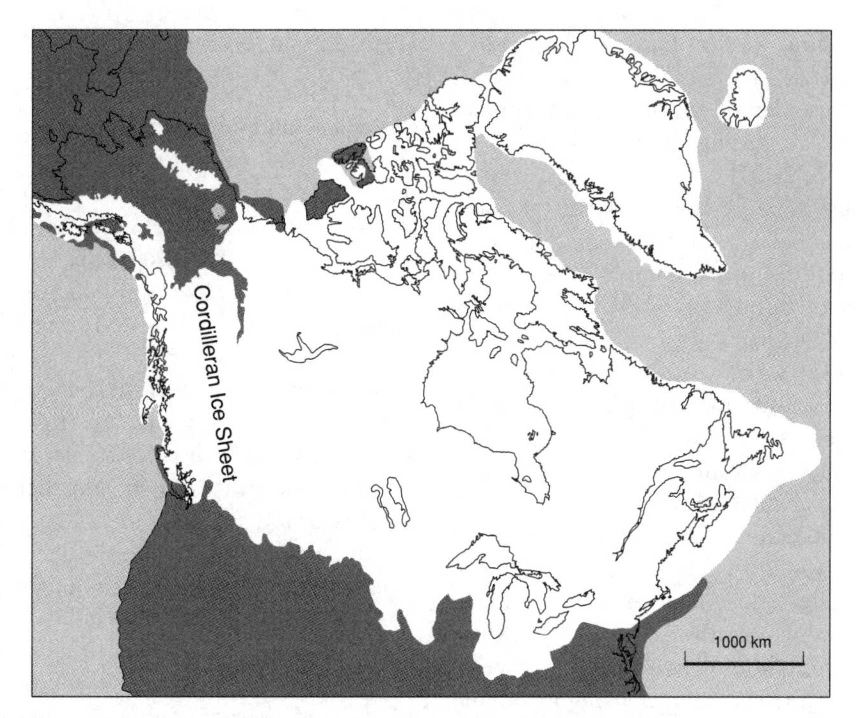

Figure 4.3 The Laurentide ice sheet at the Last Glacial Maximum (after Dyke *et al.*, 2002, Figure 1)

is significantly in excess of the estimated volume (*c*.24 million km^3) of the present-day Antarctic ice sheet (Denton and Hughes, 1981). As in Scandinavia, the maximum extent of the Laurentide ice sheet was time-transgressive, with the southernmost limit being reached ~24–21 ka BP, while along the northern margins in the Arctic the Late Wisconsinan maximum occurred some 12 ka later (Clark *et al.*, 1993). To the west, a major glacier complex extending from Alaska to southern British Columbia developed over the mountains. This Cordilleran ice sheet, which reached its maximum extent along the western seaboard ~17–18 ka BP (Clague and James, 2002), may have been coalescent at the Late Wisconsinan maximum with the Laurentide ice sheet to the east (Dyke *et al.*, 2002). In the western USA, smaller mountain glaciers and minor glacier complexes formed as far south as California (Osborn and Bevis, 2001).

Iceland and Greenland

Between western Europe and North America, an ice sheet covered almost all of Iceland, with ice thicknesses in the south of the island in excess of 1000 m (Ingólfson *et al.*, 1997), while the Greenland ice sheet was slightly more extensive than at the present day (Funder *et al.*, 1994).

Deglaciation: the ocean record

In the North Atlantic ocean core record, the first signs of deglaciation are evident from around 18.3–19 ka BP in a marked change in ocean circulation and, in particular, in an increase in low-density surface meltwaters released from the wasting ice sheets (Alley and Clark, 1999). By ~15 ka BP, it has been estimated that perhaps as much as half of the Northern Hemisphere ice had disappeared (Ruddiman and Duplessy, 1985). However, terrestrial proxy data indicate that cold and dry conditions continued to exist to the south of the ice sheets during this period of ice wastage, implying that a considerable volume of glacier ice disappeared while the climate of the Atlantic basin region remained cold. In Britain, for example, extensive wastage of the Late Devensian ice sheet began *before* the abrupt climatic warming that occurred throughout western Europe at around 15 ka BP (McCabe *et al.*, 1998), while a major retreat of the southern sector of the Fennoscandian ice sheet is also evident during the period 20–15 ka BP (Boulton *et al.*, 2001).

The reasons behind this apparent paradox of widespread deglaciation during continuing cold climate conditions are complex, but reflect a combination of ocean–atmosphere–cryosphere interactions and feedback effects. An increase in insolation from ~20 ka onwards, particularly in the high northern latitudes, may have been an initial trigger for deglaciation, but these Milankovitch effects would have been amplified by glacial and ocean circulation changes (Alley and Clark, 1999). The increasing meltwater flux from wasting ice sheets would have led to a progressive chilling of ocean surface waters, an extension of sea-ice cover and a reduction in North Atlantic deep-water (NADW) formation, leading to hemispherical cooling. During the same period, increasing instability around the fringes of the continental ice sheets due partly, perhaps, to rising sea levels, led to surging and/or collapse of some ice-sheet margins and a massive influx of icebergs and cold meltwaters into the North Atlantic (Bond and Lotti, 1995). The most recent of these **Heinrich events** (H-1) culminated in the North Atlantic around 16.5 ka BP, and was marked by a renewed fall in SSTs and an increase in sea-ice cover (see Figure 3.3). Dissipation of the iceberg-rich meltwaters following the H-1 event led to the re-establishment of NADW formation, a rise in sea-surface and atmospheric temperatures, and further ice-sheet wastage. In effect, the oceanographic changes prior to and during the H-1 event reversed the orbitally induced warming trend that began around 20 ka BP, and delayed the transition from full glacial to interglacial conditions by, perhaps, 2–3 ka (McCabe and Clark, 1998). Throughout the period 20–15 ka BP, the volume of the Northern Hemisphere ice sheets and glaciers was only partly reduced by an increase in summer radiation balance, but of equal importance was the decline in the poleward moisture flux across a chilled North Atlantic Ocean which resulted in significantly lower precipitation over the northern ice sheets.

The Lateglacial: Europe

Terrestrial proxy records, involving a range of geological and biological evidence, indicate that around, or shortly after, 15 ka BP, a significant rise in temperature occurred throughout north-west Europe (Walker, 1995), and in other areas of the North Atlantic also (Kapsner *et al.*, 1995). Oceanographical records also indicate warmer conditions, but reductions in SSTs caused by pulses of meltwater discharge from the decaying northern ice sheets point to accelerated deglaciation throughout the period 15–13 ka BP (Koç *et al.*, 1996). This increase in the rate of glacier wastage around the maritime fringes of the Northern Hemisphere ice sheets and in meltwater flux into the North Atlantic culminated in the short-lived climatic deterioration of the European Younger Dryas Stadial (see above). During that period (~13–11.5 ka BP), readvances occurred around the western margins of the Fennoscandian ice sheet (Boulton *et al.*, 2001), and there was renewed glacier activity in many mountain regions of western Europe (Sibrava *et al.*, 1986). In Britain, a large mountain icefield developed in the western Highlands of Scotland (Figure 4.2), while smaller glaciers and glacier complexes were to be found in the eastern and northern Scottish Highlands and on a number of the Hebridean islands. Further south, cirque glaciers formed in the Southern Uplands, the Lake District, the uplands of North and South Wales (Figure 4.4) and the hills of eastern and south-western Ireland

Figure 4.4 Moraine formed by a Loch Lomond readvance glacier in the Brecon Beacons, South Wales (photo Mike Walker)

(Gray and Coxon, 1991). This episode (the **Loch Lomond Readvance**) was the last occasion when glaciers existed in the British Isles.

The Lateglacial: North America

In North America, the equivalent of the European Younger Dryas climatic oscillation appears to have been most pronounced in the eastern parts of the USA and Canada (see above). Inland, the period from *c*.16 to 11 ka BP saw the gradual contraction of the Laurentide ice sheet, albeit with frequent local readvances particularly in the Great Lakes region. In the Western Cordillera of North America, glaciers were in retreat by ~16 ka BP (Porter and Swanson, 1998), although in both the Canadian Rocky Mountains and in the Front Ranges of Colorado there are indications of renewed glacier activity which may be coeval with the Younger Dryas climatic oscillation (Reasoner *et al*., 1994; Menounos and Reasoner, 1997).

Holocene glacier activity

The global climatic warming which marked the beginning of the present interglacial was reflected in rapid deglaciation on a hemispherical scale. By 11.5 ka BP, or perhaps even before, glacier ice had vanished from the British Isles. In Scandinavia, the wasting ice sheet broke into three domes after *c*.10.6 ka BP: one to the east of the modern ice cap of Svartisen along the Norwegian–Swedish border; a dome centred over the Jotunheimen area in south-west Norway; with a third small dome separating later in Jämtland in central Norway/Sweden (Boulton *et al*., 2001). These residual ice masses appear to have wasted relatively rapidly, with the Jotunheim area deglaciated by ~7900 BP (Matthews *et al*., 2000) and the Jostedalsbreen ice mass (currently the largest ice cap on mainland Europe) disappearing by ~7350 BP (Nesje *et al*., 2000). In North America, by ~7 ka BP the much larger Laurentide ice sheet had been reduced to residual ice masses in Labrador-Ungava, Keewatin and Baffin Island (Andrews, 1989; Dyke and Dredge, 1989), with the Barnes and Penny ice caps on Baffin Island being the only surviving present-day remnants.

There is evidence in the mountain regions of both Europe and North America for glacier activity throughout the Holocene. In many areas, glacier readvances during the early Holocene were followed between ~9 and 6 ka BP by a period of widespread ice recession, albeit interrupted by short-lived glacier advances (Nesje and Dahl, 2000). A similar pattern is evident in the mountains of western North America. There, many mountain glaciers had retreated to near present limits by *c*.11.5 ka BP, and while there are some indications of early Holocene advances, glacier activity during the Altithermal Hypsithermal interval (*c*.9–6 ka BP) was minimal (Davis, 1988; Osborn and Luckman, 1988). Evidence for renewed glacier advances during the **Neoglacial** period (post 6 ka BP), however, is found throughout the Western Cordillera, especially after *c*.9–4 ka BP (Leonard and Reasoner, 1999). In northern Iceland, there is also evidence for glacier readvances from mid-Holocene time (~5 ka BP) onwards (Stötter *et al*., 1999), while glacial advances post *c*.4 ka BP are recorded in the European Alps, in the Pyrenees, in the Tatra Mountains of the Czech Republic and in the Caucasus (Grove, 1997). In Scandinavia, there is also widespread evidence for renewed glacier activity from ~6 ka BP onwards, and particularly from ~3 ka BP. Around the Jostedalsbreen ice cap, for example, three phases of renewed glacier activity date to the periods 6.1–4.4 ka BP, 4.5–2.7 ka BP, and post 2.7 ka BP (Nesje *et al*., 2000). In both Europe and North America, and indeed in other areas of the world also, there is evidence for major glacier readvances during the Little Ice Age (Figure 4.5). This renewed phase of glacier activity appears to have been a global phenomenon, with maximum ice-frontal positions in many mountain areas being achieved between the sixteenth and nineteenth centuries (Nesje and Dahl, 2000).

Periglacial activity

The present periglacial domain covers three of the great vegetational zones of the world: the sub-arctic and northern forests, the arctic tundra and ice-free polar desert zones, and the birch–pine forests and mountain scrublands. Together

Figure 4.5 The Little Ice Age maximum moraine dating from AD 1750 at Boverbreen, Jotunheimen, Norway (photo John Matthews)

these constitute around 25 per cent of the earth's land surface, but during the cold stages of the Quaternary, periglacial conditions characterised some 40–50 per cent (French, 1996).

Europe

The distribution of relict periglacial phenomena shows that at the height of the last cold stage regions beyond the ice margins experienced a range of periglacial conditions. In Europe, the limit of permafrost is believed to have extended from southern France, northwards around the foothills of the Alps and then eastwards approximately along the course of the River Danube (Maarleveld, 1976). The boundary between continuous and discontinuous permafrost lay across northern France (Figure 4.6). A combination of periglacial and coleopteran evidence suggests that mean annual temperatures in the discontinuous permafrost zone were of the order of –4 °C (~14–15 °C below those of the present day). Further to the north and east, mean annual temperatures as low as –8 °C have been inferred, with the mean temperature of the coldest month in the range –20 to –25 °C. The annual temperature amplitude, which was in the range –28 to –33 °C, implies a high degree of continentality at that time (Huijzer and Vandenberghe, 1998).

Following the establishment of warmer North Atlantic waters around the coasts of Europe from around 15 ka BP onwards, more equable conditions replaced the periglacial climatic regime that had prevailed for much of the previous 100 ka. Thawing of the permafrost occurred over large

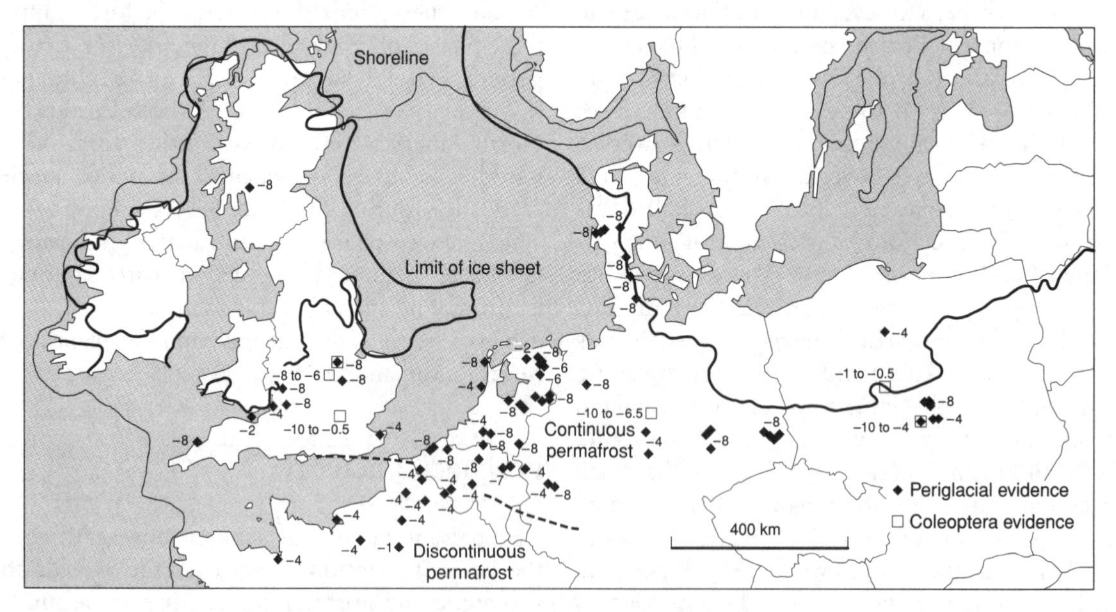

Figure 4.6 Estimates of the mean annual temperature in western and central Europe during the 29–22 ka interval, based on periglacial and coleopteran data (after Huijzer and Vandenberghe, 1998, Figure 15)

areas, although some permafrost either remained throughout the Lateglacial or re-formed during the short-lived cold episode between 13 and 11.5 ka BP. In western Europe, continuous permafrost may have existed in northern Denmark and parts of southern Sweden, while further south discontinuous permafrost occurred throughout the Netherlands, Belgium and northern Germany (Isarin, 1997). In Britain, the distribution of relict periglacial phenomena of Loch Lomond Stadial age suggests continuous permafrost down to sea level in the more northerly parts of the country, with discontinuous permafrost possibly extending to the south of London. Climatic reconstructions indicate mean annual air temperatures of ~–5 °C in central England during the coldest part of the stadial, with winter temperatures as low as –20 °C. In north-east Scotland, temperatures would have been of the order of 2 °C lower (Ballantyne and Harris, 1994). These temperature reconstructions are broadly in line with those from the European mainland, which suggest that the mean annual –8 °C and winter –20 °C isotherms were located around 54–55° N (Renssen and Isarin, 1998).

Following the climatic amelioration at around 11.5 ka BP, permafrost disappeared almost completely from western Europe, and at the present day discontinuous permafrost is to be found only in the mountains of northern Fennoscandia (Seppala, 1987), on the Kola Peninsula, and in central and southern Iceland (Washburn, 1979). However, periglacial conditions (albeit in the absence of permanently frozen ground) exist in all mountain regions of central and northern Europe, and even in the British Isles where the climate of the uplands is characterised by strong winds and extreme wetness rather than severe cold or deep ground freezing, active periglacial geomorphological processes and landforms occur (Ballantyne and Harris, 1994).

North America

In eastern and central North America, relict periglacial phenomena are found in the Appalachian Mountains as far south as South Carolina. In the Western Cordillera, alpine permafrost bounded the mountain glacier complexes, with the permafrost zone perhaps continuing up to 100 km south of the main mountain ice sheet. To the east, the permafrost zone adjacent to the Laurentide ice sheet (Figure 4.3) extended south for some 700 km in the high plains region of Colorado and Kansas but narrowed eastwards towards the northern Appalachians (Péwé, 1984). Relict periglacial phenomena in Iowa suggest mean annual air temperatures at least 14 °C below present during the coldest parts of Late Wisconsinan time (Walters, 1994), values which are broadly in accord with temperature decrease estimates of at least 16 °C for Nebraska (Wayne, 1991) and ~17 °C for Illinois (Johnson, 1990). Fossil coleopteran data from the permafrost zone around the southern margin of the Laurentide ice sheet suggest slightly lower temperature reductions (11–12 °C below present), although mean January temperatures may have been up to 20 °C colder (Elias *et al.*, 1996). Relict periglacial evidence points to a reduction in mean annual temperature of ~9–11 °C in the mountain regions of the Midwest (Péwé, 1983), although a slightly lower fall in temperature (6–7 °C) has been inferred for the full glacial environment of the Pacific north-west (Thompson *et al.*, 1993). During the retreat of the ice sheet from around 16 ka BP onwards, the continental periglacial region migrated northwards. This zone, which ranged from 80 to 250 km wide over much of its length, was narrower than the belt in Europe, reflecting the more southerly position of the ice margin in the USA, and also the influence of extensive proglacial water bodies during the retreat phase of the ice sheet that reduced the permafrost area and may have helped to ameliorate climate (French, 1996).

Large areas of North America have remained under the influence of a periglacial climatic regime at the present day, a pattern that became established following the eventual disappearance of the Laurentide ice sheet during the early/mid-Holocene. Continuous or discontinuous permafrost characterises 80–85 per cent of Alaska and 50 per cent of Canada, while over 100 000 km^2 of alpine permafrost has been mapped in the United States, mostly in the Western Cordillera extending as far south as the Mexican border (Péwé, 1983).

Sea-level change

Components of sea-level change

Few areas of the world have experienced any degree of coastal stability during the course of the Quaternary, but in those areas of the mid-latitudes that lay around the fringes of the great ice sheets, the effects on landscape of changing levels of land and sea have been particularly pronounced. This section examines the components of sea-level change in the Atlantic basin region and considers the effects of fluctuating sea levels on the coastlines of Europe and North America at the close of the last cold stage and during the present interglacial. Changing levels of land and sea reflect the interplay of two major elements: global (**eustatic**) changes in sea level, and localised tectonic activity resulting in vertical displacement of the land. The latter are known as **isostatic** movements, the term **isostasy** referring to the state of balance that exists in the earth's crust so that depression in one locality will be compensated for by a rise in the crust elsewhere (Fairbridge, 1983). Over successive glacial/interglacial cycles, the dominant control on both components is the expansion and contraction of the continental ice sheets. Hence, global sea-level changes that result from the repeated abstraction of water from, and subsequent return to, the ocean basins are referred to as **glacio-eustatic changes**; similarly crustal deformation caused by loading of glacier ice is termed **glacio-isostasy**. The weight of water released into the ocean basins as ice sheets melt also leads to crustal downwarping, a process referred to as **hydro-isostasy**. Changes in sea level that take place through the interplay of these and other factors are known as *relative sea-level changes*, i.e. a change in the position of sea level relative to the land; such changes are essentially local in effect.

There has been considerable discussion as to whether such local sea-level histories can be used to reconstruct global eustatic sea-level changes, in other words to establish a sequence of **absolute** sea levels. Attention has focused on 'far-field sites', i.e. locations that are distant from major centres of glacial activity (so-called 'near-field regions'), and especially areas that are considered to be tectonically stable. One such locality is the island of Barbados, where a long record of sea-level change is preserved in fossil coral sequences, and where the local sea-level history for the last 20 ka closely approximates the changes in eustatic sea level predicted by modelling exercises (Peltier, 2002). However, in view of the complexity of land and sea-level movements (involving glacio-eustasy, glacio-isostasy, hydro-isostasy), and when further complications involving gravitational effects on sea-surface levels relating to earth rotation are taken into account (Milne *et al.*, 2002), it remains questionable whether estimates of sea-level change that are in any way meaningful at the global scale can be obtained from shoreline data alone.

In areas where glacio-isostatic uplift has occurred, the interplay between land uplift and glacio-eustatic rise can produce a complicated history of relative sea-level changes. These can be reconstructed using both geomorphological (e.g. **raised shorelines**: see below) and sedimentary evidence (e.g. lithostratigraphic change from marine to terrestrial sediments, and vice versa). Particularly valuable in this respect are *isolation basins*, which are basins in coastal environments which have been affected by both glacio-isostatic and glacio-eustatic changes, and which have therefore been periodically occupied by the sea (Figure 4.7).

Figure 4.7 An isolation basin in north-west Scotland at Rumach lochdar. The isolation of this basin from marine influence occurred at 12.7-13.1 ka BP. Details in Shennan *et al.* (2000a) (photo Ian Shennan)

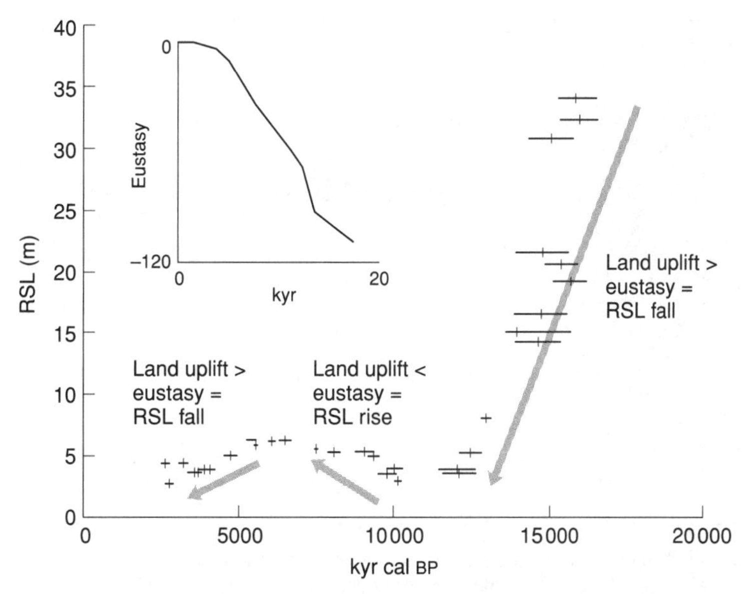

Figure 4.8 Schematic diagram of relative sea-level change showing the influence of glacio-isostatic and glacio-eustatic components, based on data from Arisaig, western Scotland. Deglaciation from *c.*18 ka BP results in rapid land uplift which outstrips eustatic sea-level rise, and hence there is a *relative fall* in local sea level. Decelerating uplift and progressive sea-level rise from 10 ka to *c.*7 ka BP leads to a *relative rise* in local sea level, while the reverse obtains after *c.*6 ka BP. Dated sea-level index points (with error bars) are shown, as is the curve for global eustatic sea-level rise (after Shennan *et al.*, 2000a, Figure 6)

A combination of lithostratigraphic and biostratigraphic (pollen, diatom, etc.) evidence enables **sea-level index points** to be established, which reflect relative fall (isolation) or rise (connection) of sea level at the altitude of the sill of the basin (Shennan *et al.*, 1994, 1996). Where these changes can be dated, a history of local sea-level change, showing relative sea-level rise (*positive sea-level tendency*) and fall (*negative sea-level tendency*), can be reconstructed (Figure 4.8). Sea-level index points can also be established for sites other than isolation basins, however, and together these form the basis for the reconstruction of sea-level histories at the regional and national scales (Figure 4.9).

Glacio-isostatic changes

The lowest sea levels at any time during the last glacial cycle occurred from ~30 to ~19 ka BP when land-ice volumes were ~55×10^6 km^3 greater than present (Lambeck *et al.*, 2002). During that period sufficient volume of water had been removed from the ocean basins and stored in the continental ice sheets and glaciers to reduce global sea level by around 135 m (Yokoyama *et al.*, 2000). This glacio-eustatic lowering was accompanied by the isostatic depression of Fennoscandia, northern Britain and Canada through glacial loading. Ice wastage began around 19 ka BP (Mix *et al.*, 2001), after which sea levels rose rapidly, while thinning and eventual disappearance of these ice masses resulted in progressive glacio-isostatic recovery. Shorelines that formed around the margins of the melting ice sheets were progressively raised above sea level as isostatic uplift outstripped eustatic sea-level rise, these **raised shorelines** which include both depositional features (beaches) and erosional forms (platforms) being progressively tilted away from the centre of isostatic depression. The analysis of raised shorelines and associated features provides evidence of the extent of isostatic uplift since deglaciation. Where prominent shorelines or deposits of the same age are found in different areas, points of equal altitude along these shorelines

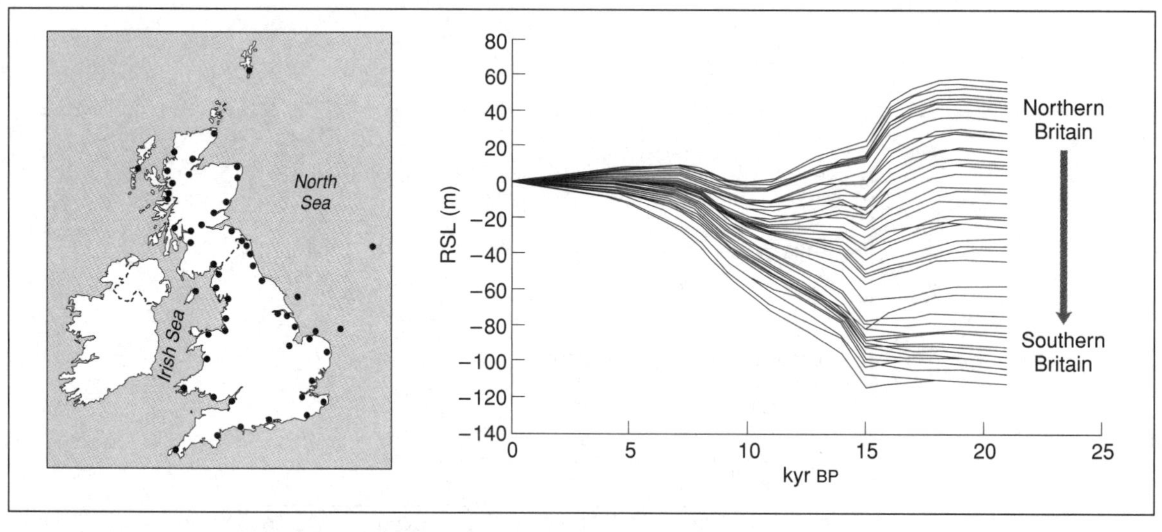

Figure 4.9 Relative sea-level changes for over the past 20 ka for 55 localities around the coast of the British Isles. Areas of northern Britain that were affected by glacio-isostatic recovery show an initial fall and subsequent rise in relative sea level (see Figure 4.7), whereas localities in southern Britain which lie beyond the area of isostatic uplift record a unidirectional rise in relative sea level (after Shennan *et al.*, 2002, Figures 1 and 2)

can be joined to form **isobases** (lines showing equal uplift). Isobase maps then provide a three-dimensional image of the deformation of the land surface by the weight of glacier ice, and provide a basis for estimating the amount of uplift that has taken place since particular shorelines were formed. In Scandinavia, for example, isobase maps (maps showing lines of equal uplift) indicate that in the Baltic region, over 700 m of uplift has taken place during the Holocene (Figure 4.10). In Britain, a combination of shoreline and other data show that the western and northern regions of the British Isles have been glacio-isostatically depressed by more than 250 m (Figure 4.11), while in North America the total amount of depression near the centre of the Laurentide sheet inferred from glaciological modelling may have exceeded 900 m (Denton and Hughes, 1981).

In all of these areas, the process of land emergence has continued throughout the Holocene. In Scotland, isobases on shorelines that developed during the Lateglacial period (Firth *et al.*, 1993) compared to data on global eustatic sea-level rise (Figure 4.12) suggest that glacio-isostatic recovery since the Loch Lomond Stadial (12.6–11.5 ka BP) is in excess of 60 m, with the main uplift phase

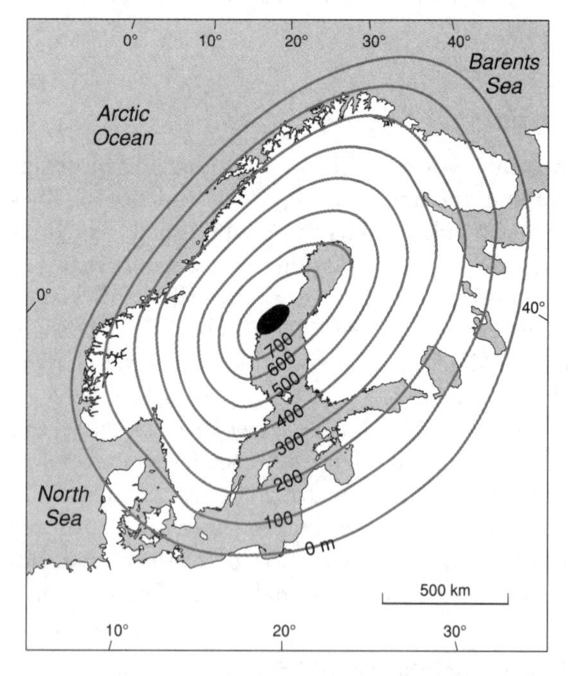

Figure 4.10 Isobases (in metres) showing absolute uplift of Scandinavia during the Holocene (after Mörner, 1980a, Figure 6)

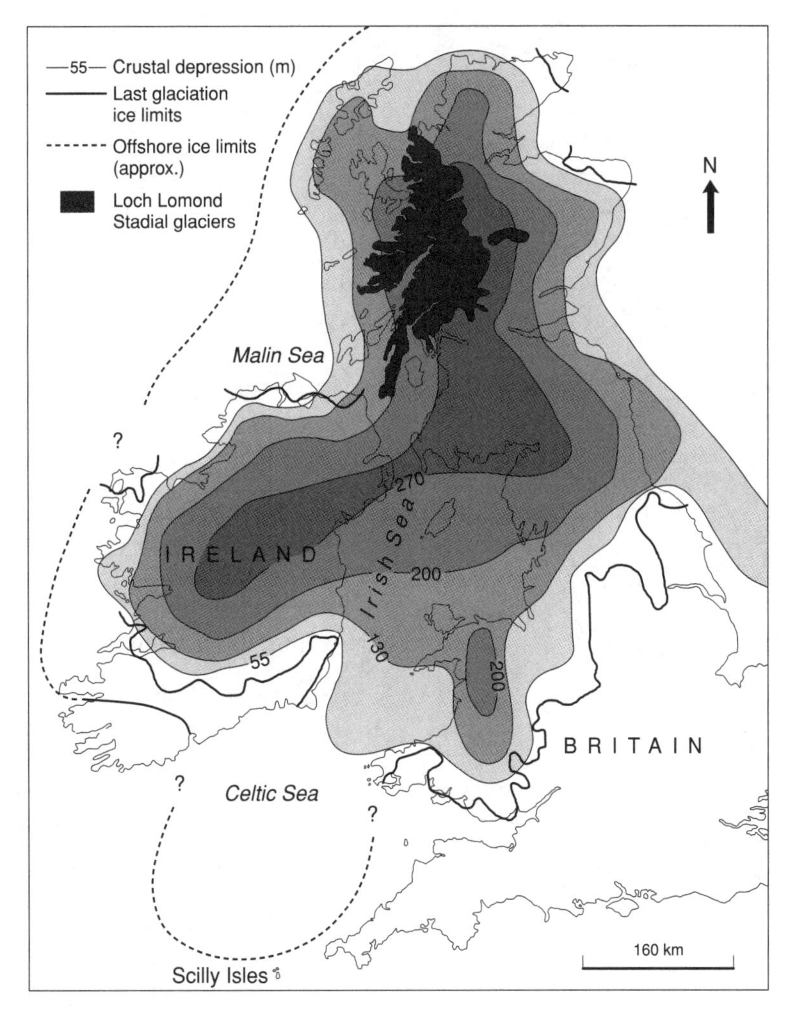

Figure 4.11 The modelled pattern of crustal depression in (metres) in the British Isles at the end of the last glaciation (after Devoy, 1995, in *Island Britain: a Quaternary Perspective* edited by R.C. Preece, Geological Society of London, Special Publication No. 96, London, pp. 181–208, Figure 4. Reprinted by permission of The Geological Society.)

occurring during the early Holocene. Indeed, raised shoreline data imply that even near the centre of isostatic recovery, not much more than 10 m of uplift has occurred in the last 5.8–6.8 ka (Smith *et al.*, 2000). Estimated uplift rates range from 14.4–31.8 mm yr^{-1} during the later part of the Lateglacial, to 4.0–7.3 mm yr^{-1} in the early Holocene to 0.4–4.8 mm yr^{-1} in the mid-Holocene. Uplift is continuing at the present day, at minimum rates of between 0.2 and 1.0 mm yr^{-1} near the former ice centre, and 0.2 ± 0.1 mm yr^{-1} near the margin (Firth and Stewart, 2000). In Scandinavia, data from the Gulf of Bothnia area in the northern Baltic indicate that over 150 m of uplift has occurred over the last 8 ka, with estimates of present rates of uplift ranging from close to zero along the Norwegian coast to 8 mm yr^{-1} in central parts of the Baltic Sea (Fjeldskaar *et al.*, 2000) and 11 mm yr^{-1} near Umeå in northern Sweden (Milne *et al.*, 2001). In North America, over 200 m of uplift has taken place in the area to the east of Hudson Bay in the last ~8 ka, where rates of isostatic recovery have declined from *c.*6 cm yr^{-1} at 8 ka BP to *c.*1.1 cm yr^{-1} at the present day (Hillaire-Marcel, 1980).

In the southern part of the North Sea basin, isostatic depression has continued throughout the

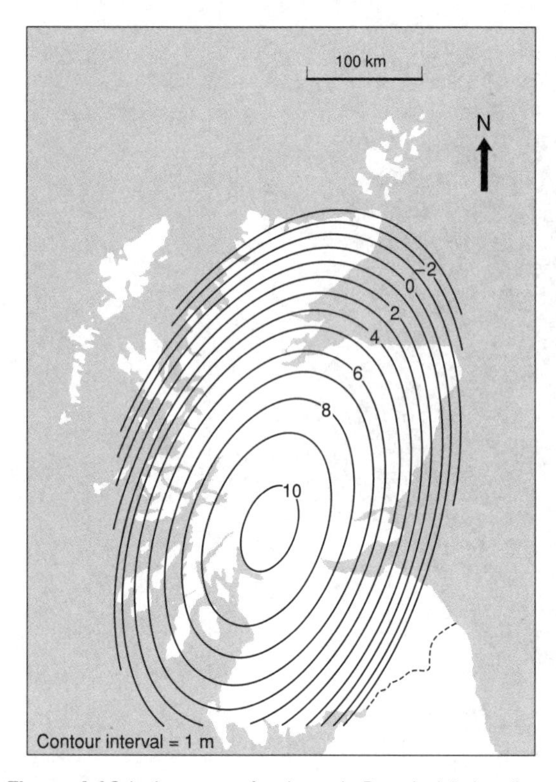

Figure 4.12 Isobase map for the main Postglacial shoreline, a prominent shoreline that has been mapped in a number of localities around the coast of Scotland, and dated to ~5.6–6.8 ka BP (Smith *et al.*, 2000, Figure 8)

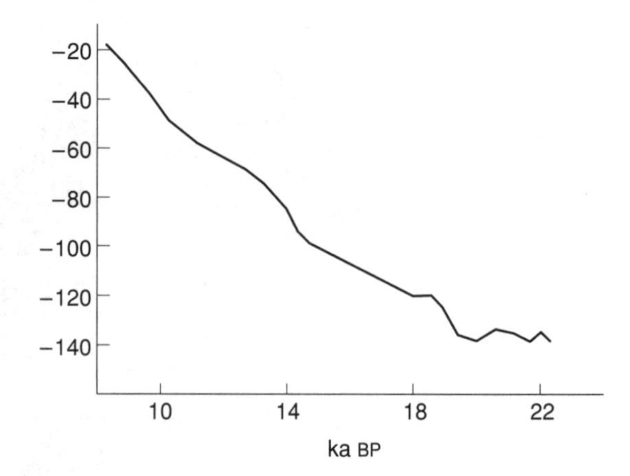

Figure 4.13 Equivalent sea-level estimates based on coral data. Note the rapid rise in eustatic sea level at *c.*19 ka BP (after Mix *et al.*, 2001, Figure 4a)

Holocene. There are indications that the North Sea has long been a focus for crustal downwarping, a trend which has been particularly pronounced following deglaciation, and the collapse of the glacio-isostatic forebulge (the marginal displacement of the crust involving upbulging beyond the ice margin) associated with the Fennoscandian ice sheet (Mörner, 1991). In addition, continued sediment loading from inflowing rivers such as the Thames and Maas–Rhine systems also contributed to the 170 m of subsidence in the southern North Sea since regional deglaciation (Mörner, 1980b). Similar downwarping trends over the past 10 ka are apparent along the east coast of the USA, with maximal subsidence being recorded in the vicinity of Delaware where over 40 m of depression has occurred since ~10 ka BP (Newman *et al.*, 1980).

Glacio-eustatic changes

The pattern of absolute change in eustatic sea level at the close of the last cold stage has hitherto proved difficult to establish, principally because of the limitations of shoreline data outlined above. In recent years, however, new approaches involving sedimentological and biological evidence from deep-ocean sequences, uranium-series dating of submerged coral reefs in low-latitude regions, and oxygen isotope data on marine Foraminifera, have provided new insights into global sea-level trends during the period of ice wastage (Yokoyama *et al.*, 2000; Mix *et al.*, 2001). Recent data (Figure 4.13) suggest a rapid rise in sea level from *c.*–135 m to *c.*–120 m at the end of the Last Glacial Maximum (*c.*19 ka BP), and a further abrupt rise from *c.*–100 m around 14.5 ka BP to *c.*–60 m at the beginning of the Holocene (11.5 ka BP). By that time, global ice volumes may have been reduced by more than 50 per cent. During the Lateglacial period, however, rapid glacio-isostatic recovery in north-west Europe and North America outstripped eustatic sea-level rise and hence shorelines formed during that period now stand well above those of the present day. In Maine isostatic uplift has raised marine deposits by up to 150 m (Bloom, 1983); in Scotland the highest Late Devensian shorelines now stand at over 50 m OD (Ordnance datum) on the west coast and 40 m OD on the east coast

(D.E. Smith, 1997), while in the Oslofjord area of Norway, shorelines that formed during the Late Weichselian/early Holocene have been raised by over 220 m (Hafsten, 1983). During the early Holocene, however, eustatic sea-level rise (at rates of around 10 mm yr^{-1}) began to exceed the rate of isostatic recovery in many areas. This resulted in a major marine transgression (the 'Main Postglacial Transgression') around the coastline of northern Britain between *c.*10 ka BP and 7 ka BP (D.E. Smith, 1997), in the flooding by the sea after *c.*9 ka BP of the extensive freshwater lake that had occupied the Baltic following deglaciation (Björck, 1995b), and in the penetration of marine waters into Hudson Bay, also around *c.*9 ka BP, which led to the rapid disintegration of the remaining mass of the Laurentide ice sheet. Isobases on the elevation of the shoreline that formed in the south-eastern Hudson Bay around 7.8 ka BP show land uplift since that time of more than 200 m (Veillette *et al.*, 1999).

In the North Sea area (Plate 4.1), the coastline at the beginning of the Holocene comprised the area of the Norwegian Trough and western embayment extending midway down the coast of eastern England. Shortly after 10 ka BP a shallow estuary developed to the south of the Dogger Bank, and by ~9 ka BP this land area was cut off from the European mainland, while further south the North Sea was connected to the English Channel via a narrow channel which may have formed some time between 9.7 and 9.4 ka BP. By *c.*7.8 ka BP, the Dogger Bank was only exposed at low tide, complete submergence of this former island occurring around 7 ka BP (Shennan *et al.*, 2000b). In the Irish Sea, there is no indication on Plate 4.1 of a connection between mainland Britain and Ireland, and although intermittent land bridges have been suggested (e.g. around 16 ka BP: Lambeck and Purcell, 2001) firm evidence for such links remains elusive (Devoy, 1995). The separation of Britain from Europe therefore probably occurred at least 6 ka after the severing by eustatic sea-level rise of the last land bridge between mainland Britain and Ireland. Estimates of Holocene sea-level rise in southern Britain (Figure 4.14a) suggest that in the Bristol Channel area, sea level rose from around −35 m OD at *c.*10.5 ka BP, with the rate of sea-level rise between ~10.5 and 7.5 ka BP exceeding 1 cm yr^{-1}. A similar rapid rate of sea-level rise during the early/mid-Holocene was recorded for the Thames estuary (Devoy, 1979). Data from the coast of eastern England point to a rate of sea-level rise of more than 0.5 cm yr^{-1} for the period 8–6 ka BP (Figure 4.14b).

After around 7 ka BP, marine incursions into coastal areas of north-west Europe took place more slowly. The configuration of the British coastline was comparable, in terms of its principal elements, with that of the present day (Plate 4.1), although considerably more indented due to the drowning of wetlands and estuaries which have subsequently silted up (see Figure 5.16a). Isostatic recovery in northern England and Scotland had once again outstripped eustatic sea-level rise, and the same was true of Scandinavia. Indeed, the rate of isostatic uplift in some areas of Fennoscandia was so rapid that on parts of the Norwegian coast, for example, isostatic emergence kept pace with or even exceeded eustatic sea-level rise throughout the Holocene (Hafsten, 1983). In the southern North Sea, continuing crustal subsidence meant that the maximum of the Holocene marine transgression was delayed until after 4.5 ka BP (Jelgersma, 1979). Current estimates of relative land- and sea-level change in the British Isles (Shennan and Horton, 2002) indicate maximum relative land uplift in west-central Scotland (+1.6 mm yr^{-1}) and maximum subsidence in south-west England (−1.2 mm yr^{-1}). The spatial pattern of Late Holocene relative land/sea-level changes in Britain is shown in Figure 4.15.

In eastern North America, relative sea level around 11 ka BP stood at −25 to −32 m, after which rapid early Holocene submergence led to rapid inundation of much of the coastal zone (Bloom, 1984). Deceleration in the rate of sea-level rise is apparent after *c.*8 ka BP (Figure 4.16), although since *c.*7 ka BP, the process of submergence has continued with a relative sea-level rise (reflecting both eustatic rise and crustal subsidence) of between 3 and 17 m along the coastline between Maine and South Carolina (Newman *et al.*, 1980; Gehrels, 1999). Further north in Nova Scotia, where isostatic readjustment continued into the Holocene, a relative sea-level rise of more than 20 m has occurred in the past 6.5 ka (Scott *et al.*, 1995).

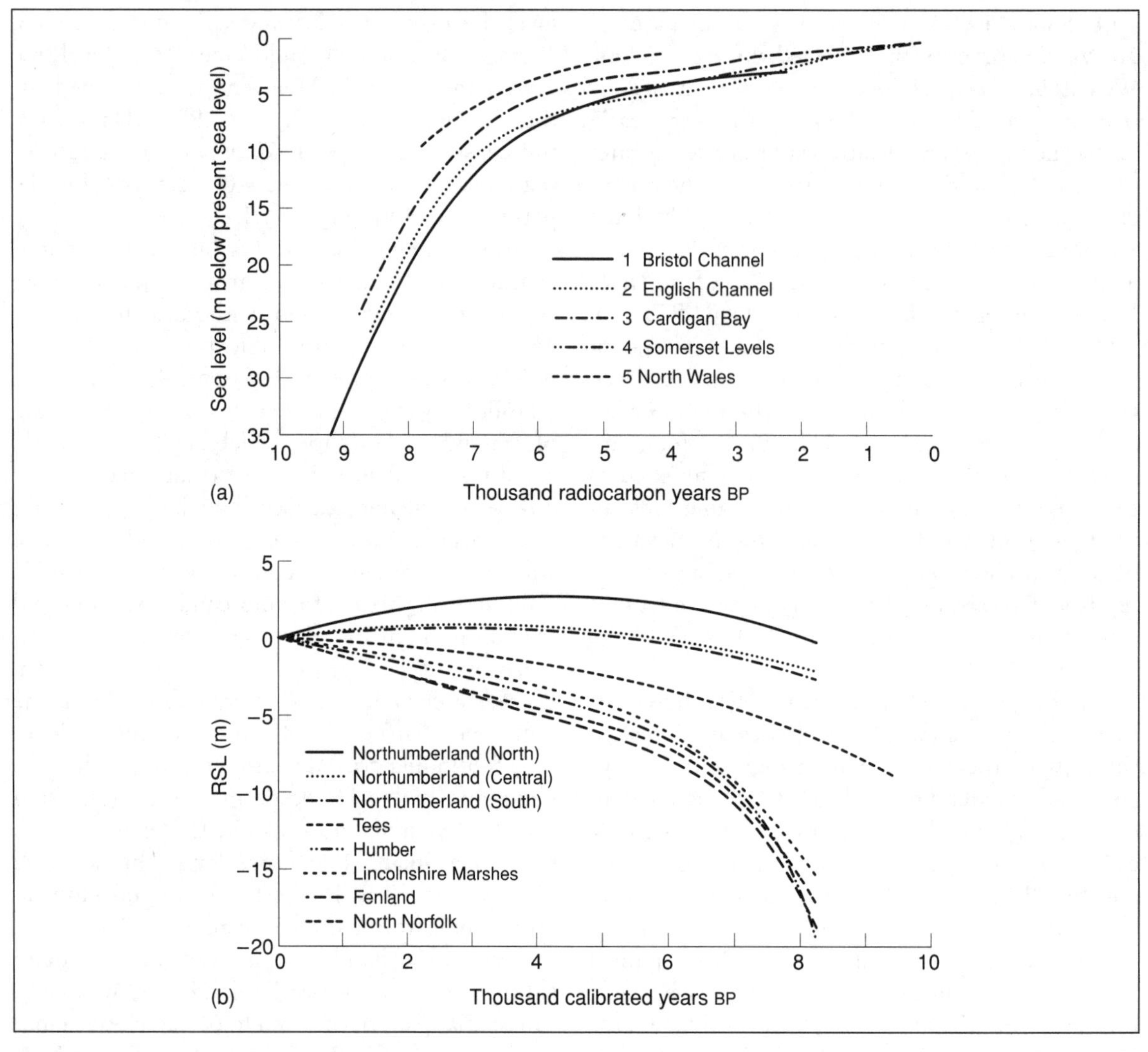

Figure 4.14 Holocene sea-level curves for (a) southern England, south-west England and Wales (after Shennan, 1983); (b) the east coast of England (after Shennan *et al.*, 2000c, in *Holocene Land-Ocean Interaction and Environmental Change around the North Sea* edited by I. Shennan and J.E. Andrews, Geological Society of London, Special Publication No. 96, London, pp. 181–208, Figure 4. Reprinted by permission of The Geological Society). Note that in (b), the three curves for Northumberland have been affected by glacio-isostatic recovery throughout the Holocene, a process that is still ongoing. Note also that the timescale for (a) is in radiocarbon yrs BP (see Table 1.1) and (b) is in calibrated yrs BP

In conclusion it should be noted that eustatic sea level is still rising. Tide-gauge data from numerous localities around the world suggest that over the course of the last century, global sea levels have risen by about 18 cm (Warrick *et al.*, 1996), with estimates of current rate of sea-level rise varying between 1 and 3 mm yr^{-1} (Gornitz, 1995). In Fennoscandia, for example, the present rise in sea level has been estimated at of 2.1 ± 0.3 mm yr^{-1} (Milne *et al.*, 2001), while along the eastern coast of the United States, rates of recent sea-level rise of 2.7 and 3.0 mm yr^{-1} have been recorded for the New York and Connecticut areas respectively (Nydick *et al.*, 1995). In Britain, tide-gauge data from a range of coastal stations show twentieth-century sea-level rise of up to *c.*2.6 mm yr^{-1}

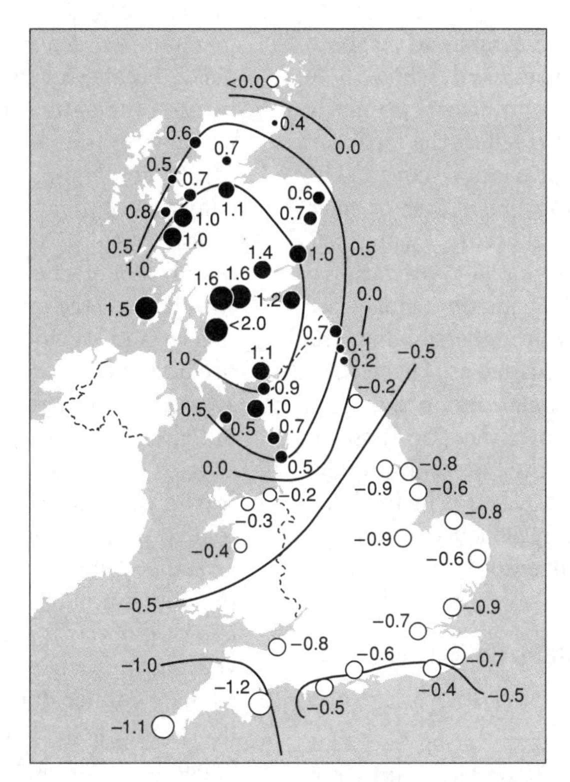

Figure 4.15 Late Holocene relative land/sea-level changes (mm yr^{-1}) in Britain. Positive values indicate relative land uplift or sea-level fall, negative values are relative land subsidence or sea-level rise (after Shennan and Horton, 2002, Figure 6)

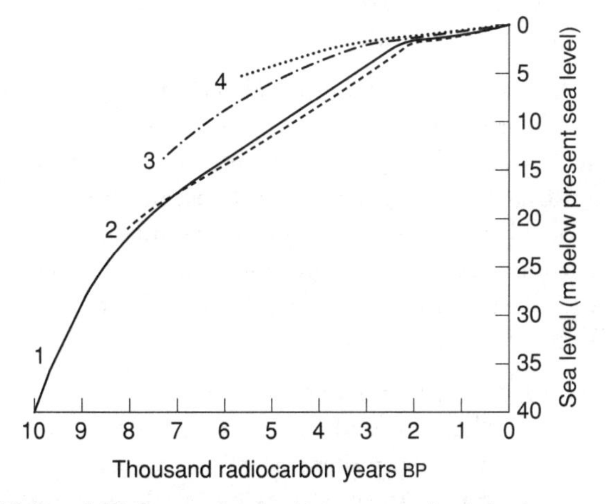

Figure 4.16 Curves showing Holocene sea-level rise in eastern North America: (1) S.E. Massachusetts (Oldale and O'Hara, 1980); (2) Delaware coast (Belknap and Kraft, 1977); (3) Bermuda; (4) Florida (Newman *et al.*, 1980). Note that the timescale is in radiocarbon yrs BP (see Table 1.1 for calibration)

(Woodworth *et al.*, 1999). Comparison between tide-gauge and geological records suggests that ~1 mm yr^{-1} of this figure is an additional rate of sea-level change during the twentieth century (Shennan and Horton, 2002). As this almost certainly reflects the anthropogenically induced global warming of recent decades, present and future sea-level trends are considered further in Chapter 9.

Vegetational and pedological changes

Late Quaternary climatic change had dramatic effects on the terrestrial biosphere. In the temperate mid-latitude regions of the Northern Hemisphere, vegetational belts migrated over several thousands of kilometres in response to the fluctuating rhythms of successive glacial/interglacial cycles. These macroscale vegetational changes were accompanied by (and in some instances were partly in response to) changes in soil properties, and while these pedogenic variations were governed primarily by climatic changes, they also reflect the feedback effects from the regional vegetation cover. The ways in which soil–vegetation relationships have evolved over the course of the Late Quaternary and, in particular, the response of vegetational and pedogenic processes to climatic change, are of fundamental importance in understanding the dynamics of contemporary ecosystems. This section outlines the major vegetational and pedological changes that have occurred over the closing stages of the last glacial and during the present interglacial cycle. The database is provided principally by pollen analysis, augmented by evidence from plant macrofossils and buried palaeosols.

Models of vegetational and pedological change

The pattern of biotic changes during the Late Quaternary can most easily be understood by reference to a simple model of ecological processes, the so-called interglacial cycle. Such a model was outlined by Iversen (1958) to explain the sequence of vegetational changes that occurred over successive interglacials in north-west Europe, and this scheme has subsequently been further elaborated by Birks (1986). The model envisages vegetation and soils changing in response to the influence of climate over four successive stages:

1. *Cryocratic phase:* This is the glacial stage when climate is cold, dry and continental in the extra-glacial areas. Vegetation consists of sparse, species-poor, arctic-alpine or steppe herbaceous communities growing on skeletal mineral soils, often underlain by permafrost, which are disturbed by cryoturbation.
2. *Protocratic phase:* This is the first stage of an interglacial and, in response to rising temperatures, the steppe and tundra communities are replaced successively by species-rich grassland, scrub and open woodland. Depending on parent material, soils are often base-rich and fertile with low humus content. In areas of moderate precipitation, leaching is at a minimum, partly because of the time factor, and partly because the natural buffer capacity of the soil has not yet been overcome.
3. *Mesocratic phase:* This corresponds with the climatic optimum of an interglacial, and is characterised by the expansion and establishment of closed temperate deciduous woodland. Tree immigration varies with species and so woodland diversification occurs throughout this episode. In lowland areas, brown earths are common and are initially at least of high base-status.
4. *Oligocratic phase:* Soil retrogression from brown earths to podzols and, in some areas, to acid peats leads to progressive elimination of some of the nutrient-demanding plants of the mesocratic phase. Climate begins to deteriorate towards the end of this stage, with vegetation changing from mixed woodlands to open conifer-dominated woods, ericaceous heaths and bogs. Some protocratic plants expand, however, with the opening up of the woodland. Major climatic deterioration heralds the arrival of the succeeding cryocratic phase as the woodlands disappear, soils are destroyed by gelifluction and cryoturbation, and herbaceous communities become established once again on the skeletal, mineral soils.

In Britain, pollen analytical evidence has shown that for the cold stages, it is possible to differentiate between full glacial and interstadial spectra, and that for interglacials the data can be broadly resolved into a series of pollen assemblage zones reflecting sequential vegetation changes characteristic of successive interglacial cycles (Turner and West, 1968; West, 1970):

Zone I *Pre-temperate zone:* This zone is characterised by rising values for arboreal pollen, particularly of the boreal trees such as birch (*Betula*) and pine (*Pinus*). Light-demanding shrubs and herbs are initially dominant, but rapidly decline.

Zone II *Early temperate zone:* A zone dominated by trees associated with mixed-oak woodland such as oak (*Quercus*), elm (*Ulmus*) and hazel (*Corylus*). Non-arboreal pollen is minimal.

Zone III *Late temperate zone:* This zone reflects the expansion of trees not present in zone II, including hornbeam (*Carpinus*), alder (*Alnus*) and spruce (*Picea*), and the gradual disappearance of some of the previously established mixed-oak forest genera.

Zone IV *Post-temperate stage:* Boreal trees are dominant once more, but there is a significant increase in non-arboreal pollen, reflecting opening of the woodland and the expansion, in particular, of heathland plants (Ericaceae).

In practice, however, matters are much more complicated, for close examination of the palaeoecological records from successive interglacials shows that this broad framework masks detailed differences in species composition from interglacial to interglacial. Variations in geographical distribution of refugia, historical 'accidents' of migration, or differences in the climatic character of temperate and cold stages have generally been invoked to account for these detailed variations (Tzedakis and Bennett, 1995; Bennett, 1997). Moreover, although vegetation succession tends to be regarded as an essentially deterministic process, with regional-scale vegetational developments being driven by changes in climate, there may also be a stochastic (random) element involved in the spread of tree populations during a particular interglacial because of the complex inter-species and species–climate relationships that often develop (Bennett *et al.*, 1991).

The same line of thinking applies to pedological changes. Traditional soil-forming theory sees soil developing progressively until it is in equilibrium with prevailing environmental conditions, and this is implicit in the models outlined above. However, soils are effectively open systems and rarely, if ever, achieve equilibrium with their environment. Soil evolution can be either developmental or regressive, depending on the influence of both dynamic (e.g. water-table dynamics, frequencies of wetting and drying, effects of organisms) and passive (e.g. parent material, stability of geomorphic surfaces, soil chemical environment) factors (Johnson *et al.*, 1990). Moreover, inconstancy in environmental conditions, and notions of multidirectional changes and multiple steady states (as predicted by non-linear dynamics: see Chapter 2), mean that in the process of evolution, soils can progress, stay the same or retrogress, depending on environmental circumstances (Huggett, 1998). The result is that some old soils show only minimal development or even regression, while younger soils often display evidence of rapid and strong development (Johnson *et al.*, 1990). Overall, therefore, while models of soil and vegetational development during the course of an interglacial provide a useful framework within which to consider vegetational and pedological change during the Late Quaternary, it must be remembered that they are no more than 'models', in other words, general approximations of reality.

The 'cryocratic' phase

At the end of the last cold stage the periglacial zone (see above) was characterised by a vegetation which was frequently sparse, with large areas of bare ground and disturbed or moving soils. The flora of this 'crycratic phase' consisted largely of herbaceous taxa of mixed ecological and geographical affinities, including arctic, alpine, steppe, southern, ruderal, marsh and halophytic taxa (West, 2000). Such diverse floristic elements reflect the polar desert environment that existed beyond the ice margins, where a markedly continental climate prevailed, but within which local variations in edaphic conditions, exposure, and slope stability provided microhabitats for a range of plant communities. Arctic soils, often with loessic materials incorporated into the upper horizons and with evidence of frost disruption, were widespread (Rose *et al.*, 2000), these soils often being recognisable today by the presence of such relict periglacial features as ice wedge casts, sand wedges and cryoturbation structures (see Figure 2.15 and Plates 2.2 and 2.3).

South of the European periglacial zone, steppe characterised by mugwort (*Artemisia*) and associated dryland herbs was to be found across much of southern and south-eastern Europe, although there are indications in the pollen records from the Iberian Peninsula of pine and oak woodland in the south of Spain (Huntley and Birks, 1983; Huntley, 1990). Short-lived interstadials are reflected in many areas of southern and central Europe by the expansion of pine and spruce (Reille *et al.*, 2000). In the eastern and central United States, pollen and plant macrofossil data suggest that boreal woodland was widespread throughout the last cold stage, with open *Picea* woodland grading westwards into tundra along the margins of the Laurentide ice sheet, *Pinus*-dominated woodland throughout the east, and *Picea*-dominated woodland in the continental interior, with temperate hardwoods (including *Quercus*, *Fagus* and *Ulmus*) growing locally near the Mississippi Valley (Jackson *et al.*, 2000). *Artemisia*-steppe and tundra covered large areas of the Western Cordillera, although parkland communities of spruce, pine and hemlock (*Tsuga*) were to be found particularly in the Pacific north-west (Thompson *et al.*, 1993).

The 'protocratic' phase

The 'protocratic phase' of the Iversen model began in north-west Europe around 16 ka BP and ended shortly before 10 ka BP. As such it includes the Lateglacial climatic oscillation and the first millennium of the Holocene. The thermal improvement that began around 15 ka BP saw the gradual replacement of steppe/tundra by open boreal woodland across large areas of Europe (Figure 4.17). By 12.5 ka BP, spruce forest was extensive in the east, while a belt of birch–conifer woodland extended from Poland south-westwards into central France. Steppe vegetation still characterised much of southern Europe, although such areas were less extensive than during the Weichselian, while mixed deciduous forests were widespread in the south of the Iberian Peninsula. In western Europe, the open tundra was invaded by shrubs, and in lowland areas extensive stands of tree birch became established. The uplands of western Britain supported open grassland and acidophilous crowberry (*Empetrum*) heath (Walker, 1995). Only limited pedological data are available for this episode, but buried palaeosols show accumulation of raw humus in wetter areas or shallow incorporation of organic matter in rendzinas (Catt, 1979). At a number of sites in southern Britain, particularly on loessic sediments and in areas where carbonate-rich parent material is present, a rendzina-type soil of Lateglacial Interstadial age has been identified. Known as the 'Pitstone Soil' (Rose *et al.*, 1985), or 'Allerød Soil' (e.g. Catt and Staines, 1998), this soil may be comparable in terms of both age and pedological characteristics with palaeosols that have been described from the Netherlands, Belgium and northern France (Preece, 1998b). In many parts of north-west Europe, Lateglacial pedogenesis is characterised by processes of decalcification, iron pan development and eluviation/illuviation (p. 212; van Vliet-Lonoë, 1990).

The abrupt climatic deterioration of the Loch Lomond/Younger Dryas Stadial was reflected in significant changes in the vegetation pattern throughout the extra-glacial areas of Europe, but especially along the Atlantic seaboard (Lowe, 1994; Walker, 1995). Tundra extended from southern Sweden through the British Isles and across much of France, while in north-east Europe, xeric variants of tundra, mixed forest and birch forest were widespread. Steppe dominated southern Europe, with deciduous woodland restricted to north-west Iberia. Birch–coniferous forests were confined to a wedge trending north-eastwards from Iberia across central Europe (Figure 4.17). In many parts of north-west Europe, the process of soil maturation that began during the preceding interstadial was disrupted by cryoturbation activity associated with discontinuous permafrost as climate became increasingly severe (Rose *et al.*, 2000).

Rapid climatic amelioration around 11.5 ka BP led to dramatic changes in the vegetation of Europe (Figure 4.18). By 10 ka BP, large areas of northern and western Europe had witnessed the replacement of tundra first by open grassland, and then successively by dwarf shrub heath with juniper (*Juniperus*), and willow (*Salix*), *Betula* forest and finally *Betula–Corylus* woodland. Birch forest also covered much of newly deglaciated Fennoscandia, with birch–conifer forests well established around the mountain ice sheet and

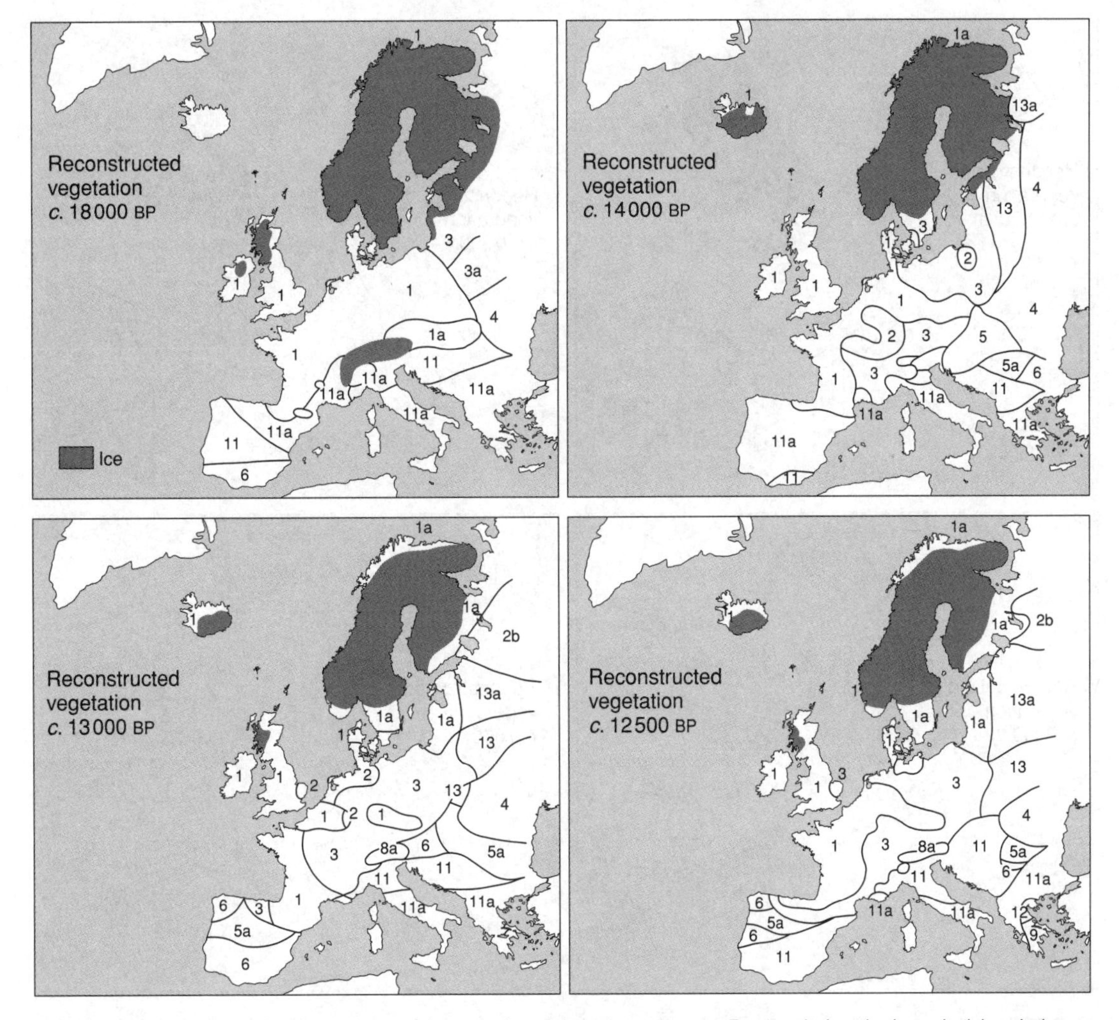

Figure 4.17 Isopollen maps showing patterns of vegetation change in north-west Europe during the Lateglacial period (adapted from Huntley, B. and Birks, H.S.B., 1983, *An Atlas of Past and Present Pollen Maps for Europe: 0–13000 Years Ago*, © Cambridge University Press, reprinted with permission of the publisher and authors)
1 Tundra; 1a Tundra 'xeric' variant; 2 Birch forest; 2b Birch forest 'xeric' variant; 3 Birch–conifer forest; 4 Spruce-dominated forest; 5 Northern mixed conifer–deciduous forest; 5a Northern mixed conifer–deciduous forest: *Pinus* variant; 6 Mixed-deciduous forest; 8a Montane mixed-conifer forest: *Pinus* (Haploxylon) '*Larix*' variant; 9 Mediterranean forest; 10 Xerix Mediterranean vegetation; 11 Steppe; 11a Steppe: treeless variant; 12 Grassland; 13 Mixed forest; 13a Mixed forest: 'xeric' variant

glaciers in central Norway and Sweden (Huntley and Birks, 1983; Huntley, 1990). *Pinus*-dominated woodland covered much of the lowlands of northern Europe, while in the south mixed woodland extended from the Iberian Peninsula to Greece. By this time, steppe and tundra communities had been virtually eliminated from mainland Europe. This phase of vegetational development, the end of the protocratic stage of Iversen or the pre-temperate zone of Turner and West, was also accompanied by the widespread development of base-rich soils (Catt, 1979).

In North America during the Lateglacial, the tundra belt along the ice sheet margin was progressively invaded by boreal forest trees, with *Picea*, fir (*Abies*) and larch (*Larix*) moving northwards onto formerly glaciated terrain. *Betula* became widespread south of the ice margin by 14 ka BP,

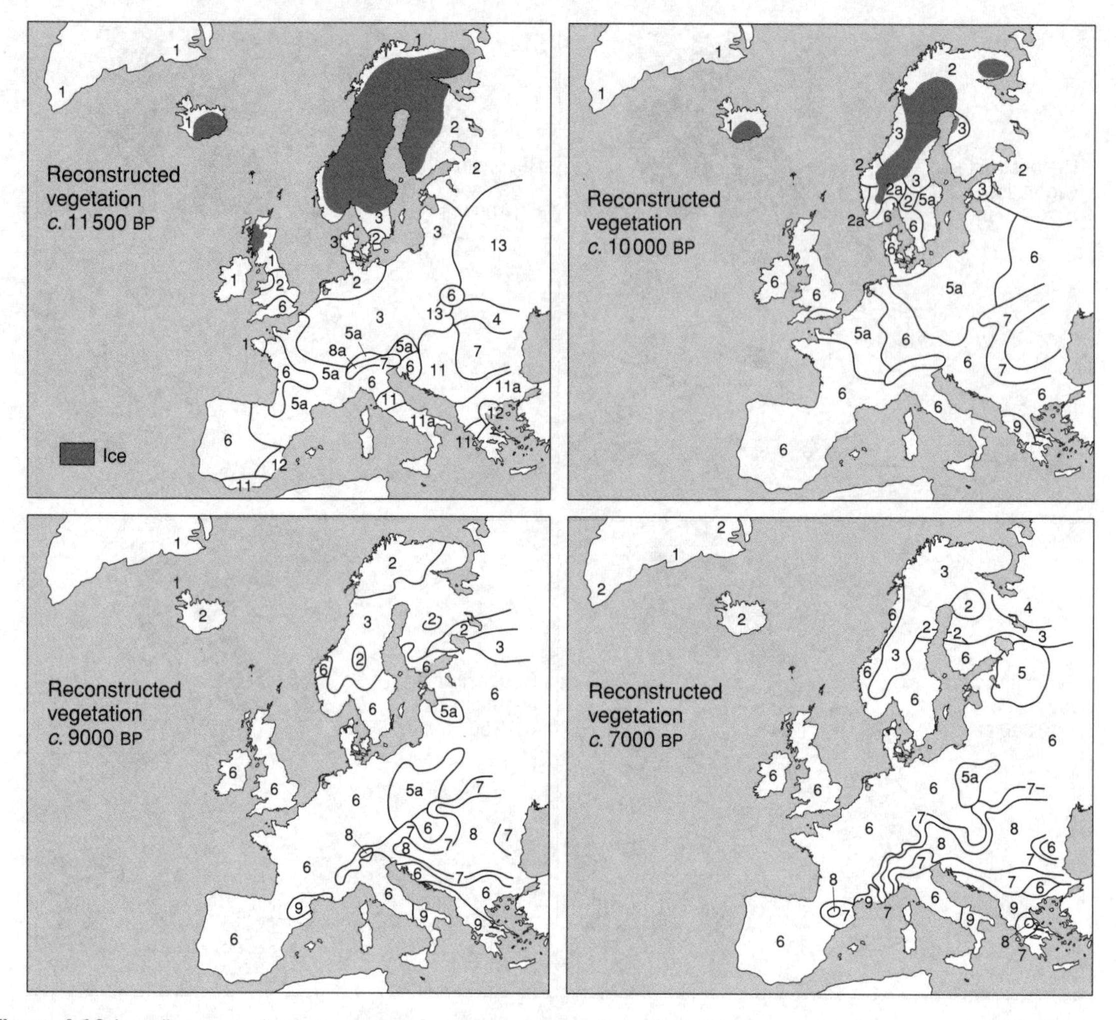

Figure 4.18 Isopollen maps showing patterns of vegetation change in north-west Europe during the early and middle Holocene periods (adapted from Huntley, B. and Birks, H.S.B., 1983, *An Atlas of Past and Present Pollen Maps for Europe: 0–13000 Years Ago*, © Cambridge University Press, reprinted with permission of the publisher and authors)
1 Tundra; 2 Birch forest; 2a Birch forest *Populus* variant; 3 Birch–conifer forest; 4 Spruce-dominated forest; 5 Northern mixed conifer–deciduous forest; 5a northern mixed conifer–deciduous forest: *Pinus* variant; 6 Mixed-deciduous forest; 7 Montane mixed conifer–deciduous forest; 8 Montane mixed-conifer forest: 8a Montane mixed conifer forest: *Pinus* (Haploxylon) *Larix* variant; 9 Mediterranean forest; 11 Steppe; 11a Steppe: treeless variant; 12 Grassland; 13 Mixed forest

while *Pinus* also expanded, all of these changes reflecting an increase in the growing season and also, perhaps, increasing precipitation (Jackson *et al.*, 1997). However, the pollen and plant macrofossil data for this time period suggest a regional vegetational composition unlike any regionally extensive vegetation in eastern North America today, suggesting a unique combination of climatic variables at that time (Overpeck *et al.*, 1992). In mid-continental areas, the evidence points to *Picea*-dominated woodland, with ash (*Fraxinus*), *Ulmus* and hornbeam (*Ostrya/Carpinus*), while further east, the forests were dominated by *Picea*, *Betula* and poplar (*Populus*). Along the east coast in New England, open *Betula* woodlands were succeeded by boreal forest (*Picea*, *Larix*, *Abies*) between *c.*14.5 and 13 ka BP, and by thermophilous deciduous (*Quercus*) woodland and conifers up

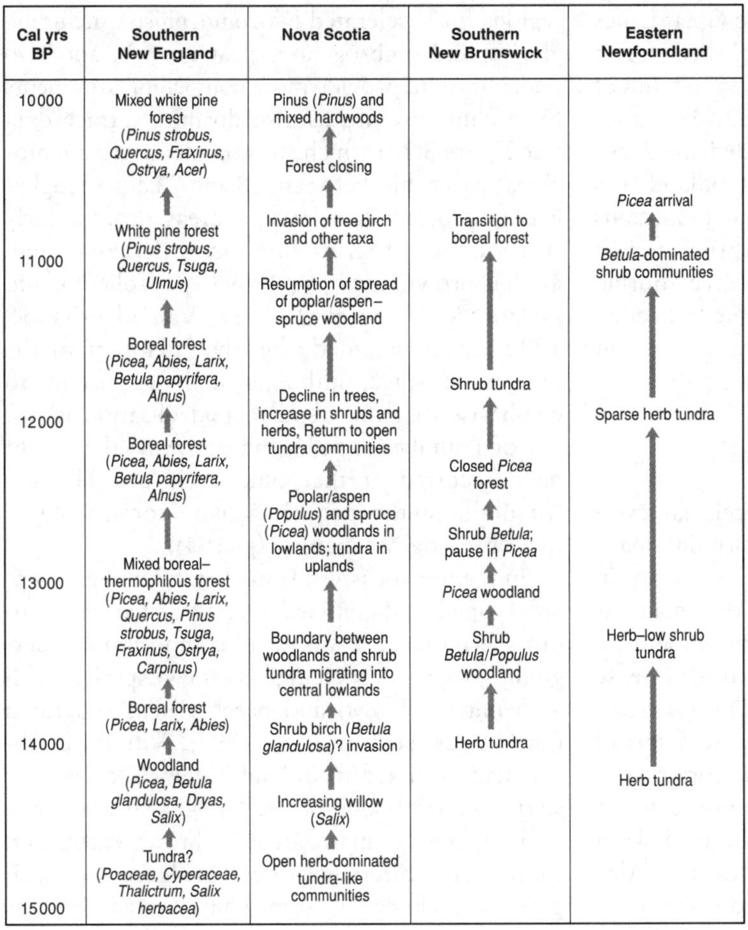

Cal yrs BP	Southern New England	Nova Scotia	Southern New Brunswick	Eastern Newfoundland
10000	Mixed white pine forest (*Pinus strobus, Quercus, Fraxinus, Ostrya, Acer*)	Pinus (*Pinus*) and mixed hardwoods		
		Forest closing		
				Picea arrival
	White pine forest (*Pinus strobus, Quercus, Tsuga, Ulmus*)	Invasion of tree birch and other taxa	Transition to boreal forest	
11000				*Betula*-dominated shrub communities
		Resumption of spread of poplar/aspen–spruce woodland		
	Boreal forest (*Picea, Abies, Larix, Betula papyrifera, Alnus*)		Shrub tundra	
12000		Decline in trees, increase in shrubs and herbs, Return to open tundra communities		Sparse herb tundra
	Boreal forest (*Picea, Abies, Larix, Betula papyrifera, Alnus*)		Closed *Picea* forest	
		Poplar/aspen (*Populus*) and spruce (*Picea*) woodlands in lowlands; tundra in uplands	Shrub *Betula*; pause in *Picea*	
13000	Mixed boreal–thermophilous forest (*Picea, Abies, Larix, Quercus, Pinus strobus, Tsuga, Fraxinus, Ostrya, Carpinus*)		*Picea* woodland	
		Boundary between woodlands and shrub tundra migrating into central lowlands	Shrub *Betula*/*Populus* woodland	Herb–low shrub tundra
	Boreal forest (*Picea, Larix, Abies*)			
14000		Shrub birch (*Betula glandulosa*)? invasion	Herb tundra	
	Woodland (*Picea, Betula glandulosa, Dryas, Salix*)	Increasing willow (*Salix*)		Herb tundra
15000	Tundra? (*Poaceae, Cyperaceae, Thalictrum, Salix herbacea*)	Open herb-dominated tundra-like communities		

Figure 4.19 Lateglacial and early Holocene vegetational history of eastern North America: southern New England (Peteet *et al.*, 1994); Nova Scotia (Mott *et al.*, 1994); southern New Brunswick (Cwynar *et al.*, 1994); eastern Newfoundland (Anderson and Macpherson, 1994)

to 12.5 ka BP (Figure 4.19). The Younger Dryas climatic reversal is reflected in this area by a return to boreal-dominated woodland (Peteet *et al.*, 1994). Further north in the Maritime Provinces of Canada, shrub tundra was suceeded by open boreal woodland with willow (*Salix*), *Betula*, *Picea* and *Populus* during the thermal maximum of the Lateglacial between *c.*13 and 12.5 ka BP, followed by a reversion to herbaceous communities during the cooler phase of the Younger Dryas (Anderson and Macpherson, 1994; Mott, 1994). Elsewhere in the United States, evidence for vegetational retrogression during the Younger Dryas cooling is limited, although the increase in *Picea* in pollen records from the Till Plains areas of Ohio, Indiana, Michigan and Illinois (Shane and Anderson, 1993), and the increase in British Columbia in mountain hemlock (*Tsuga mertensiana*) and a complementary expansion of herbaceous taxa *c.*12.9 and 11.5 ka (Mathews, 1993), may be vegetational expressions of this climatic downturn. In general, pedological data for the Lateglacial are limited, but basiphilous, nitrogen-deficient skeletal soils seem to have been widespread on newly deglaciated sites and throughout the tundra zone where geliflution and frost heaving continued until the establishment of closed forest.

The modern spatial patterns of vegetation in North America first appeared very early in the Holocene (Webb *et al.*, 1993). The northward expansion of boreal forest trees during the early Holocene was accompanied by a northward shift of their southern margins in response to increasing summer temperatures and decreasing available moisture (Jackson *et al.*, 1997). In the Great Lakes region, an invasion by pine established forest similar to that of the southern boreal woodland of the present day, while in the Midwest, deciduous forest of diverse composition succeded the *Picea* phase. Further to the west, *Picea* woodland was succeeded directly by prairie. In the mountains of the western United States, pine, spruce and fir parkland are characteristic of many areas during the early Holocene, with higher treelines reflecting summer temperatures warmer than at present (Thompson *et al.*, 1993).

The 'mesocratic' phase

The 'mesocratic phase' of the interglacial cycle (Zone II – early temperate zone in Britain) spans the period from around 10.5 ka to 6 ka BP. In north-west Europe (Figure 4.18), this stage is characterised by the immigration and expansion of temperate deciduous trees which formed dense mixed woodland dominated by *Corylus*, *Quercus*, *Ulmus* and lime (*Tilia*), and by the widespread development of fertile brown earth forest soils with mull humus (Birks, 1986). The mixed forest zone extended from southern Spain to Poland and around the southern fringes of the Alps where montane deciduous and mixed deciduous–coniferous woodland occurred. To the south, Mediterranean forest developed in southern Italy and in Greece, while in Fennoscandia birch–coniferous forests blanketed the mountains, flanked by mixed coniferous–deciduous woodland on the lower slopes (Huntley, 1990). In terms of immigration and expansion, *Corylus* was the first mesocratic tree to attain dominance and was widely established by 10.5 ka BP. By 9 ka BP, *Quercus* and *Pinus* had expanded across much of north-west Europe, followed between 9 and 8 ka BP by *Alnus* and between 7 and 6 ka BP by ash (*Fraxinus*) (Birks, 1989). The expansion of alder has often been considered to reflect a shift to wetter climatic conditions, but the variable timing of the alder rise in pollen records from the British Isles in particular suggests that local site factors, rather than regional climate, may have been the principal determinants in *Alnus* expansion (Bennett, 1990). However, pedological evidence from upland regions for accelerated hydromorphism, including gleying and ombrogenous peat growth, and also podzolisation, provide more compelling arguments for a climatic shift, as also do mire surface data (see Figure 3.20) which show evidence for a major climatic wet shift between 7.8 and 7 ka BP (Hughes *et al.*, 2000). In some lowland areas also, the early stages of soil deterioration along the continuum argillic brown earth, to brown podzolic soil, to podzol (Anderson *et al.*, 1982; Van Vliet-Lanoë, 1990) are apparent during the later part of the mesocratic stage, although it is important to emphasise that the trend to podzolisation is just one of a number of pedogenic pathways that might have occurred at that time. It is possible that Mesolithic anthropogenic activity contributed to these pedological changes (p. 194).

In the north-eastern United States and adjacent areas of Canada, mixed forest existed throughout the mesocratic stage, its composition undergoing progressive change as new species such as hemlock (*Tsuga*) and beech (*Fagus*) migrated northwards (Jackson *et al.*, 1997). With the establishment of mixed woodland in the Holocene, the sterile, skeletal soils of the Lateglacial may have been replaced in many areas by brown earths and forest soils, although direct evidence for such pedogenic change is somewhat limited. Further west, the spruce and pine woodland of the Lateglacial was succeeded by mixed woodland with *Quercus*, *Ulmus* and *Pinus* in the region of the Great Lakes, and southwards towards the valley of the Mississippi. As in the east, diversification of the woodland occurred with the subsequent immigration or expansion during the early/mid-Holocene of other tree species including beech (*Fagus*), hemlock (*Tsuga*) and maple (*Acer*). To the west of the mixed forest zone in the Great Plains region, *Picea* woodland was succeeded by prairie. During the Holocene, the prairie–forest border moved eastwards in the northern Midwest from 11.5 to 7 ka BP and then retreated westwards (Webb *et al.*, 1993).

The 'oligocratic phase'

In western Europe, the vegetation and pedological changes of the late Holocene ('oligocratic phase';

Late Temperate zone) become increasingly difficult to interpret, because of the widespread evidence for anthropogenic impact on landscape (Chapter 6). From *c.*6 ka BP onwards, the European vegetation patterns were disrupted and a retrogressive sequence of vegetational changes resulted in the progressive replacement of deciduous woodland by an increasingly sporadically forested landscape with heaths, extensive grasslands and, in upland regions in particular, blanket mires (Huntley and Birks, 1983). The climatic shift to wetter conditions, which became especially marked after *c.*3 ka BP, led to accelerated leaching of soil nutrients, the change in the brown soils from mull to mor humus and the widespread development of a range of acid podzolic soils, including gley-podzols, stagnopodzols and podzols *sensu stricto* (Catt, 1979). Hydromorphic activity, which had previously been associated primarily with upland areas, now became characteristic of lowland regions throughout western Europe, the process of soil degradation being further accelerated by anthropogenic clearance of woodland (Macphail, 1986). The results of soil deterioration and the progressive climatic shift to cooler and moister conditions led to a decline in those tree species requiring deep, fertile soils. Hence *Ulmus*, *Tilia* and *Corylus* were replaced progressively by *Quercus* and *Betula* from around 6 ka BP onwards, while in the mountains extensive blanket mire development was accompanied by a progressive fall in treeline altitudes (Birks, 1986). By 4–5 ka BP, the upland regions of the maritime fringes of north-west Europe were becoming increasingly treeless, and covered largely by acidophilous heathland, grassland and ombrogenous blanket mire.

The effect of human impact on the vegetation cover of North America, while well attested in the ethnohistorical literature (p. 193), is less marked in the palaeoecological records than in western Europe, and hence the sequence of natural vegetational retrogression during the oligocratic stage of the present interglacial can be more readily discerned. The climatic cooling of the late Holocene is most apparent in the reappearance of spruce in the pollen records from the northern United States from around 5 ka BP onwards (Webb *et al.*, 1984), and by the contraction southwards and towards lower elevations of *Pinus strobus* (white pine), *Tsuga canadensis* (western hemlock) and *Betula alleghaniensis* (birch). All of these species expanded westward into the western Great Lakes region, however, in response to increasing effective moisture levels during the late Holocene (Jackson *et al.*, 1997). Over the past two millennia, the reappearance of spruce in parts of New England has been accompanied by an increase in fir and alder and a decline in beech and hemlock, trends which have been attributed to a decrease in temperature or a rise in moisture or both (Davis, 1984). Increasing climatic wetness in the Midwest region resulted in the development of the extensive patterned peatlands of Minnesota which date from around 4 ka BP (Griffin, 1975), the time at which *Picea* began to reappear in the pollen records, and in the westward retreat of the prairie/forest border (Webb *et al.*, 1993).

Palaeohydrological changes

River systems are extremely sensitive to climatic change and relict fluvial landforms throughout the temperate zone bear witness to the major climatic and environmental changes that have occurred over the course of the past 25 ka. The principal factor governing hydrological changes is precipitation, but the impact of rainfall variations on fluvial systems is modulated by other landscape components, notably soil and vegetation, which are themselves controlled, *inter alia*, by climate. Hence, variations in river incision, river terrace development, channel change, sediment loading, etc. are determined either directly or indirectly by the course of climatic change. During the late Holocene, however, it is now apparent that people exerted an important influence on hydrological processes (Chapter 7). This section is concerned with those areas of palaeohydrology that reflect climatic as opposed to anthropogenic control.

The fluvial record

Although relict fluvial landforms and deposits provide clear indications of the climatic and landscape changes that occurred at the end of the last cold stage and during the course of the present interglacial, reconstructing the chain of processes and responses that connects palaeoclimatic and palaeohydrological events to discernible field evidence is far from straightforward. In particular, the problems of equifinality (Chapter 2) loom large in palaeohydrological studies. A good example is provided by river terraces. Terraces represent former floodplains abandoned by the river as it becomes incised, and are major components of a fluvial landscape. Sequences of terrace fragments are found in most river valleys of the temperate zone and careful analysis of the altitude, morphology and stratigraphy of the terraces can form the basis for reconstructions of fluvial histories (Gibbard, 1985, 1994). Moreover, as terraces are adjacent to, but raised above, the contemporary river channel, they provide ideal habitats for large animals or potential sites for settlement. As a consequence, terrace gravels often contain mammalian remains (Figure 4.20) or artefacts, and are therefore important in biostratigraphical or archaeological contexts (Bridgland, 1994, 2000; Wymer, 1999). Terrace formation clearly results from changes in river level, but this may be brought about by a range of factors, some of which are external such as fluctuations in sea level, tectonic processes (e.g. isostatic uplift), changes in precipitation, and periodic inputs of glacial meltwaters, whereas others (such as variations in nature and quantity of the sediment load) are internal to the fluvial system (Maddy and Bridgland, 2000). Inferring the major controlling variable (process) from the terrace fragments themselves (form) requires a multiple-working hypothesis approach in which all possibilities are eliminated save one (Chapter 2), and this may not always be possible on available field evidence. Moreover, deciphering temporal variations in fluvial processes, and relating these to resulting landforms and sedimentary sequences, is invariably complicated by the spatial variation in fluvial activity that characterises most river systems, with local changes in fluvial regime often resulting from the transgression of key geomorphological thresholds. These problems notwithstanding, however, significant advances have been made in the field of palaeohydrology over the past two decades and in the palaeo-

(a)

(b)

Figure 4.20 (a) A woolly mammoth tusk exposed in Late Devensian fluvial gravels at Lackford, Essex, England; (b) the mammoth tusk after excavation (photos Mike Walker)

environmental inferences that can be made from palaeohydrological sequences (e.g. Branson *et al.*, 1996; Benito *et al.*, 1998). Consequently, it is now possible to begin to reconstruct the pattern of fluvial response to climatic change at the end of the last cold stage and during the present interglacial over large areas of the northern temperate zone.

Glaciofluvial deposits

It has been estimated that during the deglacial phase of a glacial cycle, rates of sediment yield may increase to about ten times the geological norm (i.e. the natural long-term rate) and then rapidly subside (Church and Ryder, 1972). Hence enormous quantities of outwash material in the form of sandur (outwash plains) and valley trains (spreads of gravel material along valley floors) are characteristic fluvial features of newly deglaciated regions. In upland regions, the pattern of decreasing discharge with deglaciation is matched by concomitant changes in channel pattern, from multi-thread, low-sinuosity channel systems to single-thread, higher-sinuosity channels (Maizels and Aitken, 1991). A change is also apparent from an aggradational to an erosional regime, with rivers becoming deeply incised into the unconsolidated fluvial sediments forming suites of outwash terraces leading away from the former glacier margin. Relict outwash terrace sequences dating from the end of the last cold stage characterise river systems throughout the temperate zone. Meltwater channels that formed either beneath the ice or at the ice margin are also distinctive elements of former glacierised landscapes.

Glacial lakes

In favourable topographic situations around the decaying ice margins proglacial lakes developed, some of which were of spectacular dimensions and hence major palaeohydrological phenomena. In north-west Europe, for example, the Baltic Ice Lake which was impounded to the north by the Fennoscandian ice sheet (Figure 4.21a), occupied a large area of the southern Baltic basin at an early stage in deglaciation (Björck, 1995b) and was over 1000 km in length at its maximum development. A later freshwater episode, the Ancylus Lake, covered almost the entire area of the modern Baltic basin, along with coastal areas of eastern Sweden and much of western Finland (Figure 4.21c).

During deglaciation in North America, enormous lakes developed along the margins of the wasting Laurentide ice sheet (Teller and Kehaw, 1994), including Glacial Lake McConnell in the upper reaches of the present-day Mackenzie drainage system, Glacial Lake Agassiz in the Manitoba area of the eastern prairies, and Glacial Lake Ojibway to the south of Hudson Bay. These lakes varied in size, but at times their dimensions were enormous and often exceeded 100 000 km^2. Indeed, around 8.4 ka BP, Lake Agassiz merged with Lake Ojibway to form a single lake with a surface area of 841 000 km^2 (Figure 4.22), which is more than twice the size of the largest lake in the modern world, the Caspian Sea (Teller *et al.*, 2002). To the south and east of Lake Agassiz, the proto-Great Lakes developed over a protracted period of deglaciation between *c.*21 and 7.5 ka BP. The complex history of evolution and drainage of these vast lakes reflects interactions between ice-marginal positions, crustal rebound and regional topography (Teller, 2001). The lakes drained by various routes to the Gulf of Mexico, the Atlantic and Arctic oceans (Figure 4.22), and received overflows from impounded water bodies along the ice margins to the north and west.

Outbursts from ice-dammed lakes (jokulhlaups) such as these exerted a major influence on the geomorphology of the immediate proglacial area (Maizels, 1997), and their effects may have extended even further afield. For example, a catastrophic release of meltwater from Glacial Lake Agassiz via the Mackenzie River into the Arctic Ocean at *c.*11.3 ka BP led to the discharge of ~21 000 km^3 of meltwater into the Arctic basin over a period of 1.5–3 years (Fisher *et al.*, 2002). This is equivalent to a 6 m rise in the Arctic Ocean (or a 0.062 m rise in global sea level). An even larger discharge event at *c.*8.4 ka BP resulted in the release of 163 000 km^3 into the Arctic basin, a process that may have occurred within a single year (Teller *et al.*, 2002). Because the influx of

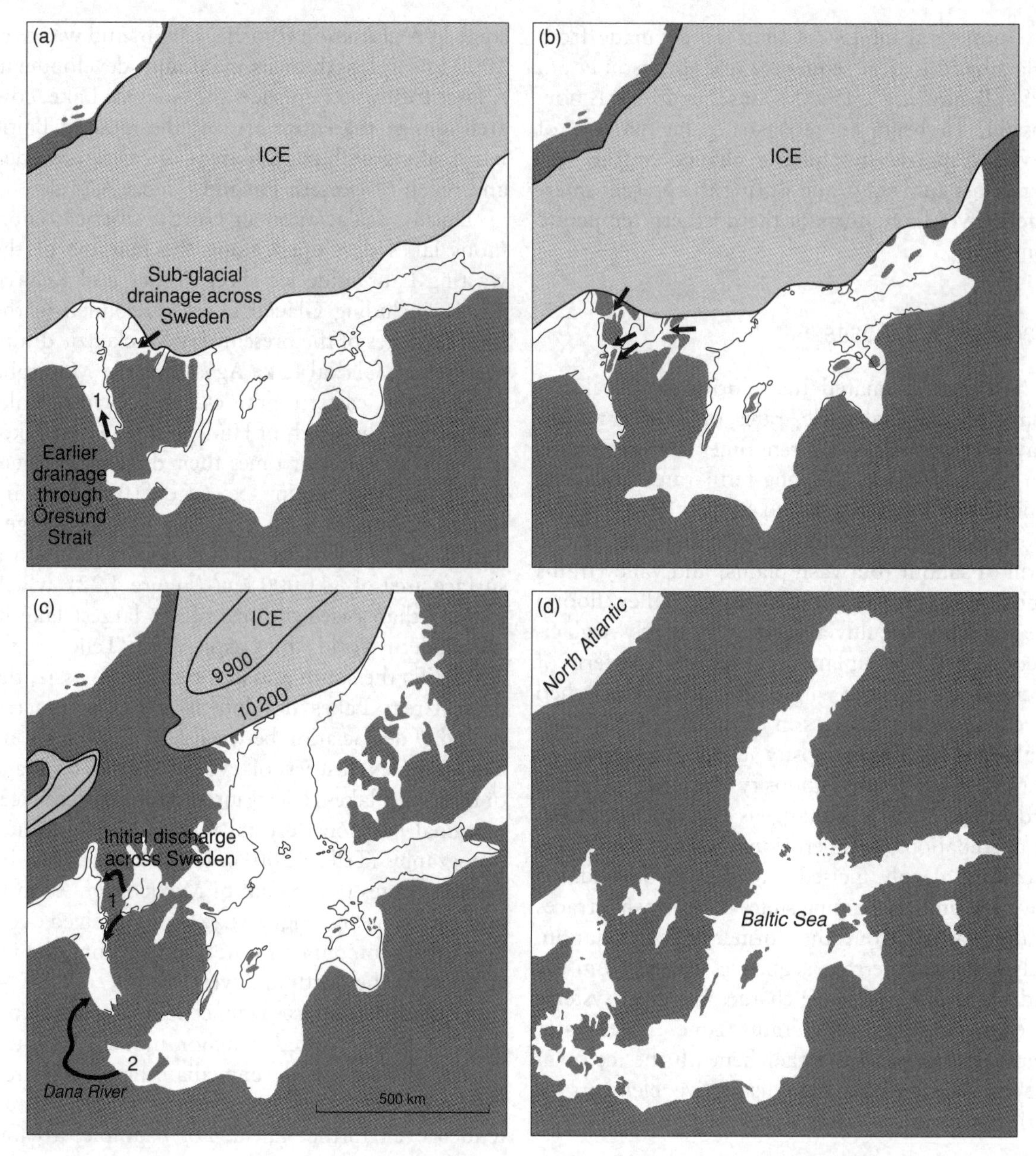

Figure 4.21 The evolution of the Baltic: (a) Baltic ice lake 12.9 ka BP. Bold arrows on this and subsequent maps show outlet (1 = outlet in 14 ka BP; 2 = outlet in 12.9 ka BP); (b) Yoldia Sea *c.*11.3–11 ka BP; (c) Ancylus Lake 10.2–10 ka BP; (d) the Baltic today (after Björk, 1995 a and b)

fresh water reaching the North Atlantic Ocean can inhibit thermohaline circulation (p. 104), such outbursts could have been the triggers for short-term climate change during the course of the last deglaciation (Clark *et al.*, 2001).

Some of the most spectacular phenomena of the entire Quaternary period, however, were the series of cataclysmic floods that occurred in the Columbia River system of the north-west USA between *c.*18 and 14 ka BP (Baker and

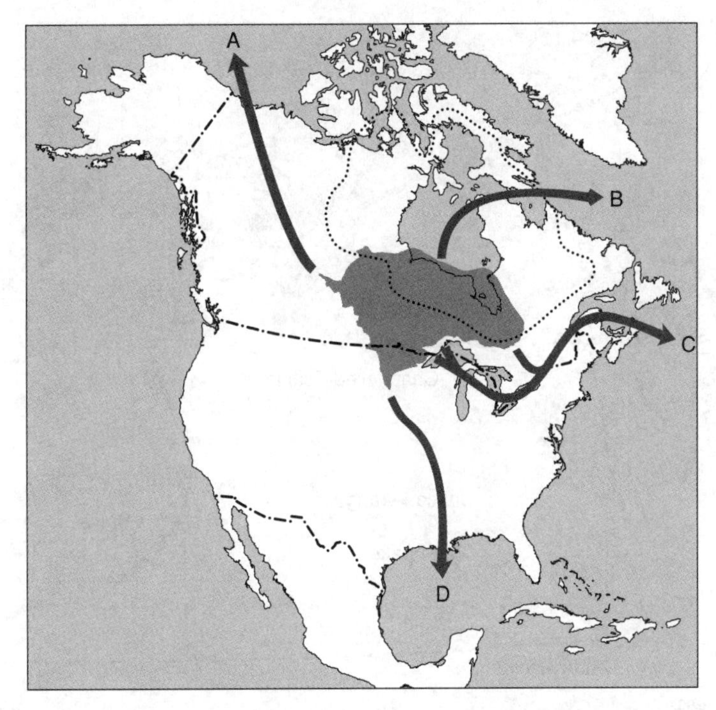

Figure 4.22 The combined extent of Glacial Lakes Agassiz and Ojibway (shaded), and the routing of meltwater outbursts and overflows to the oceans: (A) Mackenzie Valley to Arctic Ocean; (B) Hudson Bay to North Atlantic Ocean; (C) St Lawrence to North Atlantic Ocean; (D) Mississippi Valley to Gulf of Mexico. The extent of the Laurentide ice sheet at around 10 ka BP is shown by the dotted line (after Teller *et al.*, 2002, Figure 1)

Bunker, 1985; Baker and Komar, 1987). These may have been successive jokulhlaups from glacial Lake Missoula (Figure 4.23), and their effects can be seen in the network of deeply dissected river valleys that are incised into the loess-covered lava plateaux of eastern Washington State, an area of over 40 000 km known as the Channelled Scabland. The magnitude of the flood events is reflected in the fact that their effects are also apparent on the abyssal sea floor off the mouth of the Columbia River some 700 km to the west. Palaeodischarge models for Lake Missoula suggest peak discharge hydrographs ranging from 2.7×10^6 to 14×10^6 m^3 s^{-1} over 8–20 day periods (Clarke *et al.*, 1984). Debate over the concept of cataclysmic flooding in the Columbia River region has continued throughout the century, not least because it provides a classic example of the way in which cataclysmic events can be reconciled within a framework of methodological uniformitarianism in the explanation of landscape evolution (p. 22).

Periglacial palaeohydrology

Throughout western Europe and North America, there is also abundant evidence for former periglacial fluvial activity. In many periglacial river valleys, the combined effect of frost weathering and rapid mass movement on valley sides leads to an abundant supply of coarse-grained sediment to river channels (Ballantyne and Harris, 1994). Periglacial rivers, therefore, tend to be bedload-dominated and many contemporary periglacial valleys are characterised by considerable thicknesses of fluvial gravels. Such aggradational sequences relating to Late Weichselian/Late Wisconsinan periglacial fluvial regimes can be found in river valleys throughout Europe and North America (Starkel *et al.*, 1991; Baker, 1984).

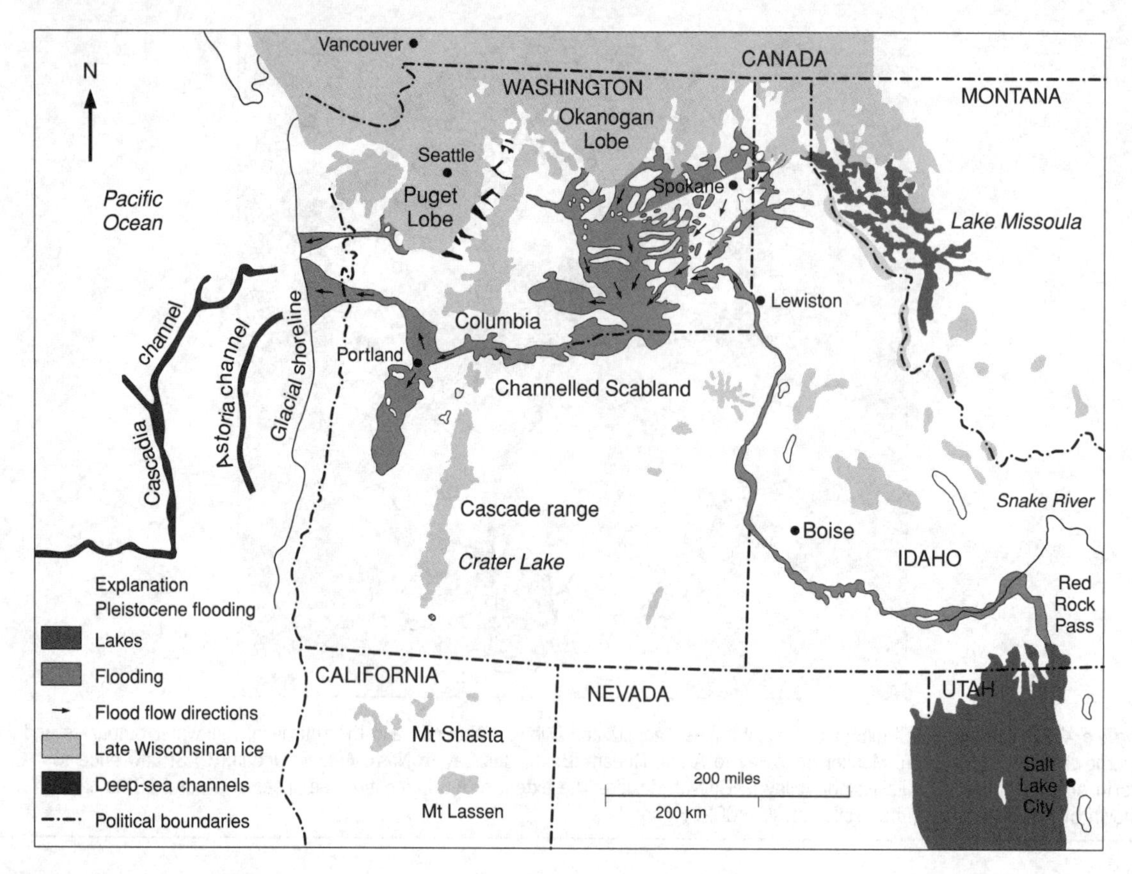

Figure 4.23 Regions of the north-west United States affected by cataclysmic flooding during the late Wisconsinan (after Baker and Bunker, 1985, Figure 1)

A distinctive feature of contemporary periglacial rivers is a marked seasonality in annual discharge regimes, with peak flows in some rivers being concentrated in the spring following snowmelt (Arctic 'nival' rivers), while in others, where there is a glacially derived component of discharge (Arctic 'proglacial' rivers), flows remain high throughout the summer (Bryant, 1983). These processes, along with the presence of permafrost which affects infiltration capacity and soil cohesion, and marked variations in vegetation cover, result in a wide variety of periglacial river types (Vandenberghe, 2001). Under periglacial conditions in Britain, unglacierised catchments in lowland Britain (and elsewhere in Europe) probably experienced runoff patterns broadly similar in character to subarctic (associated with discontinuous permafrost) and arctic (continuous permafrost) nival flow regimes (Ballantyne and Harris, 1994). In addition, the characteristic forms of river valleys within present-day periglacial regions, notably the markedly asymmetric cross-profile which relates, *inter alia*, to slope aspect, have been noted as relict periglacial features of river valleys within the present-day temperate zone (French, 1996).

Lateglacial palaeohydrology

Some of the most dramatic palaeohydrological changes that have occurred in western Europe over the past 15 ka years accompanied the climatic fluctuations and associated vegetational changes of the Lateglacial (Bohncke and Vandenberghe, 1991; Andres *et al.*, 2001). Data from many European river systems suggest that the geomorphological effect of river activity was most

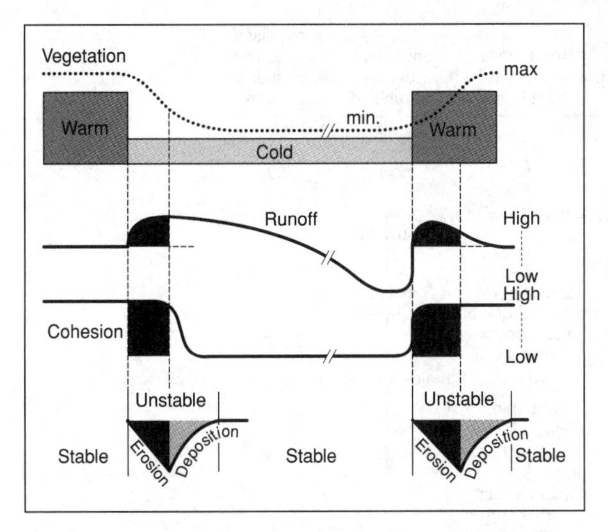

Figure 4.24 Model of fluvial stability–instability changes in relation to climate and climate-derived parameters (after Vandenberghe, 1995, Figure 5)

pronounced at times of climatic transitions, and this has led to the development of a general conceptual model of fluvial response to climate change (Vandenberghe, 1993). Each cycle begins with a temperature drop (at the onset of the Younger Dryas around 12.8 ka BP, for example) which results in lower evapotranspiration, and thus increased surface runoff (Figure 4.24). As the vegetation cover persists for some time after the climatic downturn, the soils remain relatively stable, and so although there is an increase in river discharge, there is no concomitant increase in sediment yield. The result, therefore, is increased fluvial incision. Gradual breakdown of the vegetation cover, however, leads to an increase in surface runoff and sediment supply and to a shift from an erosional to an aggradational fluvial regime, with sedimentation occurring largely in braided river systems. The later part of the cold episode shows a relative balance between erosion and aggradation. When the climate ameliorates, for example at the beginning of the Lateglacial Interstadial (*c*.15 ka BP) or the Holocene (*c*.11.5 ka BP), vegetation development is initially retarded so that evapotranspiration remains low and discharges increase. In time, however, soils become stabilised, inwash declines, the sediment load in the rivers is reduced, and incision resumes. Further increases in evapotranspiration result in reduced discharges, thus bringing about channel infilling and a return to the previous landscape (Figure 4.24). Valley bottom stability during the following interglacial (e.g. during the Holocene) is reflected in the lateral migration of the river across its alluvial plain, and by soil formation (Vandenberghe, 1995). This sequence of events can be recognised in many Lateglacial river valley sequences in western Europe (e.g. van Huissteden and Vandenberghe, 1988; Rose, 1994).

Holocene palaeohydrology

The early millennia of the present interglacial saw a marked reduction in discharge in many rivers of the northern temperate zone. In north-west Europe, there is evidence from lake basins of lower water levels around ~9.5 ka BP reflecting a significant reduction in precipitation levels (Barber and Coope, 1987; Harrison & Digerfeldt, 1993), although wetter conditions appear to have obtained in southern Europe at this time. Reduced precipitation in northern and western Europe, therefore, accompanied by the establishment of mixed woodland across much of the landscape, led to an abrupt decline in stream loading, a gradual cessation of alluviation and the development of single-thread meandering river channels which progressively incised into the Lateglacial valley fills (Macklin and Lewin, 1986). Characteristic features of the river valleys of the European lowlands during the early Holocene were therefore stable river channels with small discharge variation and limited bedload transport. Gradual overbank sedimentation and increasing organic sediment accumulation on the floodplains were also characteristic features of these well-vegetated low-relief environments (Rose, 1994). Wetter conditions after *c*.8 ka BP, however, prompted an increase in fluvial activity most notably in flood frequency, the effects of which were increased stream loading and episodic phases of aggradation and incision (Starkel, 1991). From around 4–5 ka BP onwards, however, with accelerating woodland clearance and the establishment of a settled agricultural economy, human activity is an increasingly important element in the fluvial

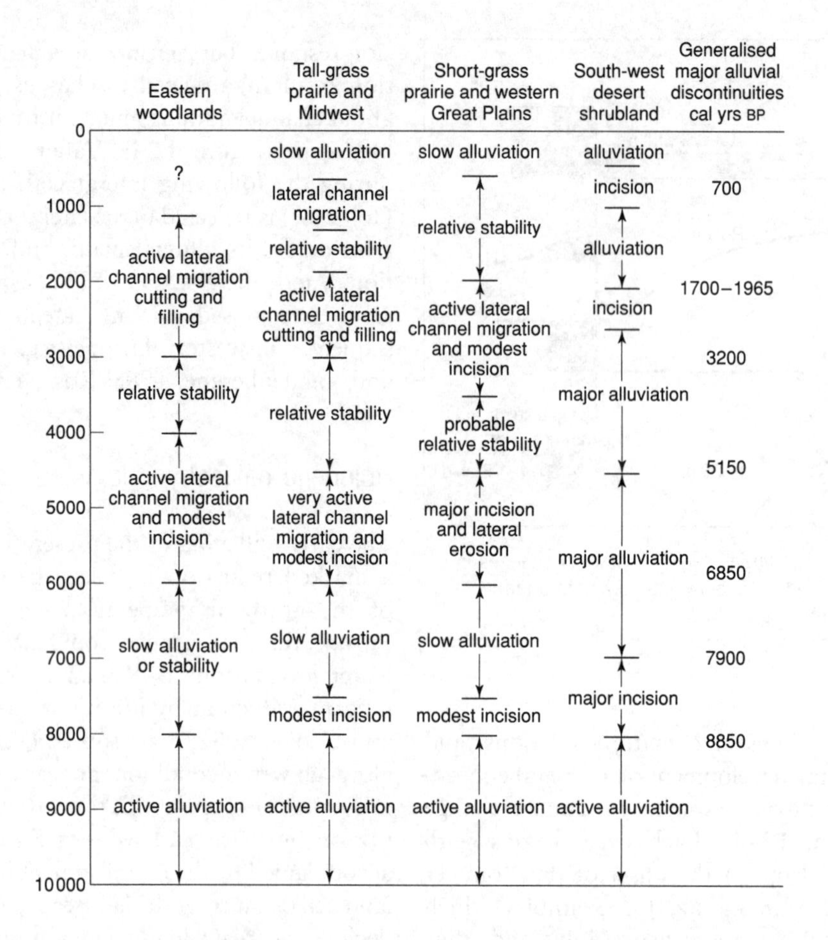

Figure 4.25 Generalised alluvial chronologies in different areas of the United States during the Holocene (after Knox, J.C., 1983, Responses of river systems to the Holocene climates in *Late Quaternary Environments of the United States, Volume 2: The Holocene*, edited by H.E. Wright, Jr. and Stephen C. Porter, Copyright © 1983 by the University of Minnesota, reprinted by permission of the University of Minnesota Press, Figure 3.9). Note that the left-hand axis shows the timescale in radiocarbon yrs BP (see Table 1.1 for calibration), whereas alluvial discontinuities (right-hand column) are in calibrated yrs BP

histories of the northern temperate zone (Macklin and Lewin, 1993; Macklin, 1999), and the climatic signal becomes more difficult to isolate. Human impact on, and interaction with, Holocene river systems are discussed in Chapter 7.

In North America, the response of river systems to Holocene climatic changes has been discussed by Knox (1984) and is summarised in Figure 4.25. The data indicate that drier conditions in many areas between *c.*11.5 and 9 ka BP prompted widespread valley alluviation, especially in the western regions of the USA. An erosional episode between 8.5 and 7.9 ka is apparent in the Midwest and the Great Plains, and especially in the south-west, reflecting, it appears, warmer and wetter conditions in those regions from around 8.5 ka BP onwards. There was a general decline in alluviation up to ~7 ka BP as large-scale vegetation communities became established, but this relatively quiescent phase was succeeded by widespread erosion of early Holocene fills. According to Knox, this is partly a reflection of decreased stream loading due to the more extensive vegetation cover, and partly of an increase in flooding as a consequence of higher precipitation levels and frequency of storms. In more arid areas, however, alluviation continued as the dominant geomorphological fluvial process throughout most of the Holocene. Both erosional and depositional intensity appears to have subsided between *c*.4.8 and 3.0 ka BP, but there is evidence in most regions of renewed fluvial activity after that time. In the mid-continental region in particular, there was a significant increase in flood frequency after *c*.3.3 ka BP under a cooler and wetter climatic regime, with even larger floods occurring between AD 1250 and 1450 during the transition from the Medieval Warm Interval to the Little Ice Age (Knox, 1993).

One feature of many fluvial records is the apparent relationship between high flood frequency and periods of rapid climate change (Knox, 2000). In the south-west USA, late Holocene fluvial records contain evidence of major floods during periods of climate change (Ely, 1997), while Holocene fluvial records from Britain also show a strong relationship between flood frequency and climate change (Macklin and Lewin, 2003). Many European river systems contain evidence of increased flooding at the end of both the Medieval Warm Period and of the succeeding Little Ice Age (Benito *et al*., 1996; Rumsby and Macklin, 1996). If rapid climate changes can bring about an increase in flood frequency, then it is possible that the anomalous concentration of large floods that have been observed in many regions of the world since the early 1950s could be due to rapid global warming during the twentieth century (Knox, 2000; Milly *et al*., 2002). This warming trend, now widely attributable to anthropogenic activity, is discussed more fully in Chapter 9.

5 People in a world of constant change

Introduction

The case for developing an integrated social and environmental perspective on environmental changes of the Late Quaternary has been outlined in Chapter 1 where the need to avoid the often simplistic opposition of people and nature was highlighted. Evidence for environmental change on a wide range of spatial and temporal scales has been outlined in Chapters 2–4; we turn now to a consideration of the interactive relationship between people and their changing environments.

How people cope

A remarkable feature of humanity is the capacity to adapt to diverse environmental conditions, which has made it possible to occupy almost every terrestrial environment on earth. Observation of any culture demonstrates diverse ways in which people and their ways of life are affected by environmental factors. These range from physical anthropological evidence for the cold adaptation of the Inuit, to the mobile savanna lifestyle of the African Masai which has produced the greatest runners on earth.

In addition to physical adaptation there is the diversity of individual, social and cultural strategies which enable human communities to adjust to environmental change. These provide the flexibility to cope with rapid changes of a range of types and timescales, from the regular and predictable to the rare and unexpected. Examples of coping strategies, and the types of environmental changes to which they may be most applicable, are outlined in Figure 5.1. The list of strategies is only a partial one and in practice the strategies frequently overlap and interrelate. The literature has often shown a rather deterministic tendency to focus on a narrow range of responses, particularly migration and economic change, so social factors have often not received the consideration they deserve. What is advocated is a multiple working hypothesis (p. 20) approach which evaluates the diversity of responses that human communities may have to environmental change. Diversity is inevitable given the enormous variability in the timescale and nature of environmental changes. Some are part of longer-term, perhaps perceptible trends of climate, sea level and vegetation succession. Some are of a rhythmical (e.g. seasonal or tidal) nature. Others are of stochastic occurrence such as drought and flood; events sometimes, but not always, of sufficiently frequent recurrence interval that communities may anticipate future occurrence and plan accordingly. More problematic for any community are rare, frequently catastrophic changes, such as volcanism, tsunamis and disease. Such events are, to varying degrees, precisely definable in time and space and are thus of particular interest as pressure points in people/environment relationships that are amenable to detailed study. There are also those events of extreme rarity, such as comet and asteroid impact, which would have been totally unforeseen.

Events which are less frequent, and thus of less perceptible recurrence, are likely to be associated with a narrower and less adequate range of coping strategies. Figure 5.1 hypothesises that social

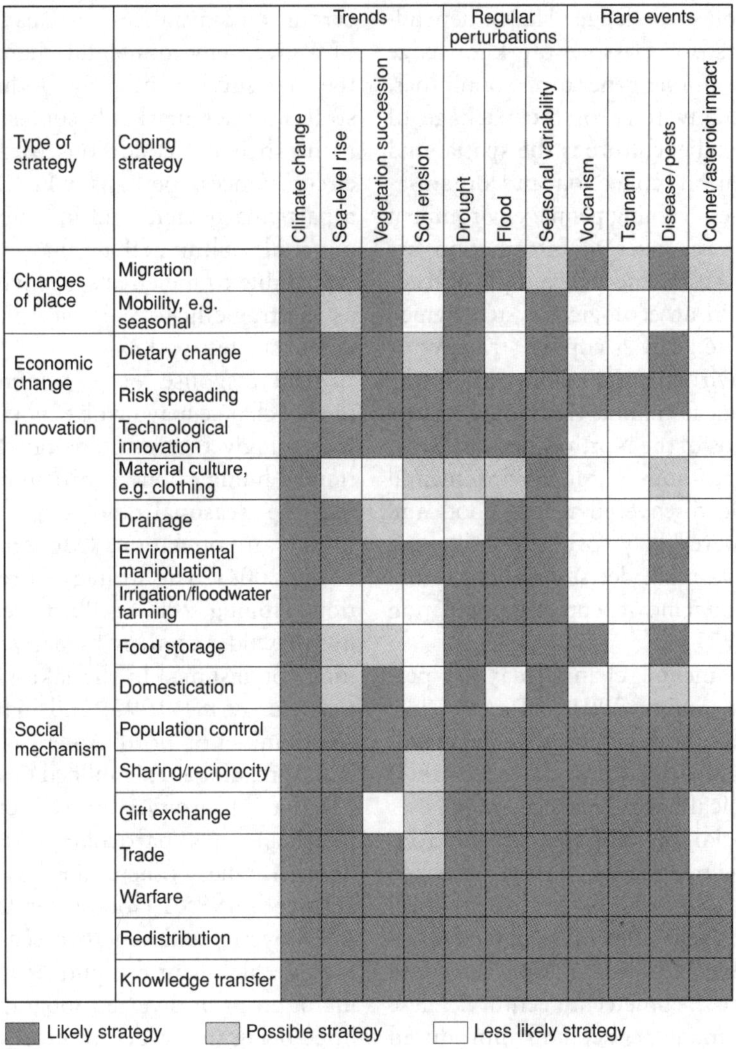

Type of strategy	Coping strategy	Trends				Regular perturbations			Rare events			
		Climate change	Sea-level rise	Vegetation succession	Soil erosion	Drought	Flood	Seasonal variability	Volcanism	Tsunami	Disease/pests	Comet/asteroid impact
Changes of place	Migration	Likely	Likely	Likely	Possible	Likely	Likely	Possible	Likely	Possible	Likely	Possible
	Mobility, e.g. seasonal	Likely	Possible	Likely	Less likely	Likely	Less likely	Likely	Less likely	Less likely	Less likely	Less likely
Economic change	Dietary change	Likely	Likely	Likely	Possible	Likely	Less likely	Likely	Less likely	Less likely	Likely	Less likely
	Risk spreading	Likely	Possible	Likely	Likely	Likely	Likely	Likely	Likely	Likely	Likely	Less likely
Innovation	Technological innovation	Likely	Likely	Likely	Likely	Likely	Likely	Likely	Less likely	Less likely	Less likely	Less likely
	Material culture, e.g. clothing	Likely	Less likely	Less likely	Possible	Possible	Possible	Likely	Less likely	Less likely	Possible	Less likely
	Drainage	Likely	Likely	Possible	Less likely	Less likely	Likely	Likely	Less likely	Less likely	Less likely	Less likely
	Environmental manipulation	Likely	Likely	Likely	Likely	Likely	Likely	Likely	Less likely	Less likely	Likely	Less likely
	Irrigation/floodwater farming	Likely	Less likely	Less likely	Less likely	Likely	Likely	Likely	Less likely	Less likely	Less likely	Less likely
	Food storage	Likely	Possible	Possible	Less likely	Likely	Likely	Likely	Likely	Less likely	Likely	Less likely
	Domestication	Likely	Possible	Likely	Less likely	Likely	Less likely	Likely	Less likely	Less likely	Possible	Less likely
Social mechanism	Population control	Likely	Likely	Likely	Likely	Likely	Likely	Likely	Likely	Likely	Likely	Likely
	Sharing/reciprocity	Likely	Likely	Likely	Possible	Likely	Likely	Likely	Likely	Likely	Likely	Likely
	Gift exchange	Likely	Likely	Less likely	Possible	Likely	Likely	Likely	Likely	Likely	Likely	Less likely
	Trade	Likely	Likely	Likely	Possible	Likely	Likely	Likely	Likely	Likely	Likely	Less likely
	Warfare	Likely	Likely	Likely	Likely	Likely	Likely	Likely	Likely	Likely	Likely	Likely
	Redistribution	Likely	Likely	Likely	Likely	Likely	Likely	Likely	Likely	Likely	Likely	Likely
	Knowledge transfer	Likely	Likely	Likely	Likely	Likely	Likely	Likely	Likely	Likely	Likely	Likely

Likely strategy — Possible strategy — Less likely strategy

Figure 5.1 Examples of the strategies that human communities may adopt to cope with selected environmental changes

mechanisms are likely to enable communities to cope with a wider range of hazards, than, for instance, changes of place or economy. Aspects of many social strategies may be applicable to a range of hazards and will therefore be of special value to those communities which are subject to environmental hazards of varied types and timescales, for example in tectonically highly active areas.

Environmental perception is at the heart of understanding how human communities cope. The basic building block of environmental knowledge is the perception of the individual person, an approach known as **methodological individualism** (Smith, 2001). However, the mechanisms for gathering, processing and acting on information are of a largely social and collective nature. The knowledge which communities have about their environment will be determined by a range of factors. In spatial terms there is the extent of the area from which a community obtains environmental information, which is dependent on its networks of social interaction. In temporal terms there is the timescale of available knowledge. This is dependent on mechanisms for passing on knowledge through time, as from one generation to another. Thus one of the many roles of the package of activities which we label culture is the spatial and temporal dissemination, collection and organisation of information concerning people/environment interactions. Such exchanges of information are the basis for future action and an example of feedback effects (p. 8) in an ecological system which may act to reduce (e.g. by a coping strategy) or enhance (e.g. by environmental manipulation) the effects of an ecological change. Tribal gatherings such as the fandangos of the North American Great Basin were clearing houses for environmental information. People discovered where good and bad nut harvests were likely to be the following year and on this basis made decisions about future scheduling (i.e. seasonal movement) and population levels (Hardesty, 1977).

Extreme environmental change may be perceived as a hazard (Smith, 2001). Hazards (like environments) do not, however, have an independent existence but are defined in relation to their social context. Radical hazard science emphasises the underlying social, as opposed to environmental, causes (Hewitt, 1983). In drought-stricken areas of northern Nigeria the anthropological record demonstrates that there were coping strategies which had evolved over a long time and were effective before the nineteenth century. These collapsed under the monetary economy introduced by colonial administrations, when even small-scale rainfall changes had social repercussions much greater than in previous centuries (Watts, 1983). In many cases the significance of social factors is undeniable, as for instance in the role of political factors and warfare in exacerbating the impact of drought conditions in the Horn of Africa during the 1990s. Clearly the existence and magnitude of a hazard depend on the effectiveness of social mechanisms, the buffering and coping strategies, which a community has in place.

Such strategies may be of very diverse kinds. Material culture (artefacts) and technology form part of adaptation/coping mechanisms. They may create something analogous to a filter between environmental effects and communities. Under certain circumstances this may lessen the effects of a given environmental change. An example is the rare survival of early clothing which demonstrates some remarkably sophisticated adaptations to harsh living conditions (p. 174). Under other circumstances, perhaps when a community has a particularly rigid and inflexible way of life and material culture, that may make them more vulnerable to the effects of environmental change, as the tragic history of the Vikings in Greenland demonstrates (p. 177).

The response of a community to environmental change may involve mobility and relocation. Increasingly the isotopic and elemental composition of human bone is providing a way of assessing the seasonal and longer-term movements of human populations (Montgomery *et al.*, 2000; Sealy, 2001). The strategy of resource diversification (Tipping, 2002) is illustrated by the continued use of wild resources by early farming communities, for instance in the lake villages of the Alps (Schibler *et al.*, 1997) and many of the farming communities of North America (Reitz *et al.*, 1996). Risk spreading is exemplified by the Mediterannean practice of growing mixed crops in one field (maslins); in a particular year one species may flourish when others do less well (Jones and Halstead, 1995). Floodwater farming systems of the Libyan pre-desert involved elaborate systems for the catchment of limited runoff and the growing of crops in diverse topographic situations, the hope being that a decent crop would be obtained somewhere in that area of very low, and highly unpredictable, rainfall (Gilbertson, 1996). The abundant storage pits of Iron Age lowland Britain (Figure 5.2) may have provided a buffering strategy against the uncertainties of harvest and possibly endemic social conflict in an age of many defended hillforts. Even non-agricultural societies in California had in place complex social systems of storage and redistribution (Bean and Lawton, 1993).

Decisions about population structure may include seasonal dispersal of parts of a population to temporary activity camps as in the Fort Ancient communities of mid-continental USA (Wagner, 1996). Decisions concerning population level may, in extreme circumstances, involve birth control, infanticide or the killing of the elderly.

Figure 5.2 Iron Age storage pits at Danebury Hillfort, Hampshire (photo Danebury Trust, courtesy Barry Cunliffe)

Strategies involving relationships with other communities included sharing, reciprocity, gift exchange, trade and warfare (Halstead and O'Shea, 1989). The potlatch ceremonies of the American north-west involved the giving of great feasts and gifts in exchange for prestige. Such practices meant that communities maintained a sufficient surplus in the economy for exchange with others during times of stress. These occurred periodically as a result of the effect of ocean current changes on salmon runs (Hardesty, 1977) and the episodic effects of earthquakes and tsunamis (p. 172). Strategies involving reciprocity, gift exchange and trade (Dodgson *et al.*, 2000) may be hinted at by material cultural evidence, for instance geoarchaeological provenancing studies (Herz and Garrison, 1998), indicating links between communities, particularly those living in diverse ecological zones. The causewayed enclosures of Neolithic southern Britain have particularly abundant artefact evidence for widespread networks of social communication (Mercer and Healy, forthcoming).

In emphasising terms such as environmental hazard and coping strategy it is important that we do not slip into the way of thinking that environmental changes always created problems for communities. Frequently they would have created new opportunities or **affordances** (Ingold, 2000) for different and more productive ways of life. Seasonal flooding of riverine environments might be seen from our twenty-first-century perspective as a hazard, but as Gillings (1998) demonstrates, these floods created affordancies for seasonal exploitation in the Neolithic of Hungary; wetland environments offer many comparable cases (p. 159).

Deliberate human alteration of the environment is another way in which conditions more productive, or desirable, for human activities were created (Chapter 6). Sometimes the intention was to create something new and suitable for particular economic activities. Often, however, environmental manipulation must have been intended not so much to create something new, but to retain aspects of the environment which were valued but were affected by change, i.e. to counter the effects of naturally occurring environmental changes such as vegetation succession. This might, for example, take the form of burning to maintain open areas; drainage, or coastal protection of areas affected by rising water-tables; irrigation of areas subject to increasing aridity; or the construction of animal shelters in times of deteriorating climatic conditions. In altering the environment human communities may also make themselves susceptible to new types of environmental hazard. Thus coping strategies are also important in alleviating the effects, both short and long term, of human communities themselves (Butzer, 1996).

Discussion of the relationship between environmental change and people has often shown a tendency to become polarised around one of the concepts or strategies introduced above, in opposition to the others. In reality individual situations need to be weighed up in terms of a spectrum of influences between extremes: determinism–hazard buffering–coping–affordance–environmental manipulation. A multiple working hypothesis approach seeks to identify the broadest possible spectrum of coping/affordance strategies as a basis for critical appraisal in the light of the evidence. It may often be the case that the archaeological record proves inadequate to identify the subtleties of particular coping strategies. However, examples have been identified of storage, artefact exchange, clothing, environmental manipulation, etc. which leave recoverable traces. It is certainly the case that the archaeological record is inadequate to support the crude determinism of much conventional reasoning, and it is surely right that a diversity of possibilities and human adaptive strategies be identified and considered. A balanced approach is necessary and important because the extraordinary evolutionary success of humanity is after all in large measure due to the ability to adapt and cope with a remarkable range of environmental circumstances as the following case studies show.

Environmental change and human evolution

Since the development of evolutionary theory by Darwin (1859), environmental change has been seen as an important driver in the evolution of hominids. This raises fundamental questions about the relationship between people and nature and forms a background to the shorter timescales and largely Holocene themes considered in the following chapters. Two distinct processes operate on very different timescales. One is the basic Darwinian principle of the survival of the fittest adapted to particular environmental conditions. On this depends which traits in an existing pool of genetic diversity are perpetuated. Darwin's original model emphasised gradualism over long timescales. More recent research has highlighted the role of punctuated equilibrium, stochastic factors and contingency (p. 6) in the perpetuation of genes (Bilsborough, 1999; Foley, 1999). The second key set of processes are the cultural and social mechanisms which people have developed to enable them to cope with environmental circumstances. These mechanisms are situated in an environmental context but are not solely driven by environmental factors; they include, for instance, social practices which enable people to cope by means of interactions with other individuals and groups.

Africa provides the context for human evolution. The earliest hominid is the recent find of *Sahelanthropus tchadensis* from Chad in the Sahara tentatively dated 6–7 ma and associated with a fauna of mixed savannah woodland (Brunet *et al.*, 2002; Vignaud *et al.*, 2002). By 5.6 ma *Australopithecus afarensis* is present at Lothagam, Kenya, and by 4.5 ma at Tabarin, Kenya, with other early Australopithecine species appearing at Awash, Ethiopia (4.5 ma) and Kanapoi, Kenya (4.2–3.9 ma), both apparently associated with woodland environments. That the early Australopithecines were bipedal is evident from the discovery of early human footprints below volcanic ash at Laetoli, Kenya, dated *c*.3.6 ma and the 'Lucy' skeleton from Hadar, Ethiopia (2.9 ma). The development of bipedalism has, since the work of Darwin, been seen as an adaptation to a grassy savannah environment. As Foley (1987) put it, 'we are what we are because an arboreal animal has taken up the challenge of terrestrial life'. Bipedalism is of such paramount importance because it frees hands for tool use and making, thereby creating a multiplicity of fresh opportunities for adaptive strategies and social communication.

During the later Miocene and Pliocene Africa was subject to progressive drying, reduction in equatorial rainforest and increasing savannah environments. Climate change at this time is also marked by the expansion of the Northern Hemisphere ice sheets from *c*.2.4 ma and the onset of clearly defined warm and cold cycles. Interglacials were represented by pluvial episodes in which tropical rainforest was flanked by savannah and glacials by drier episodes with an expanded savannah and limited rainforest refugia. A consequence of this was habitat fragmentation creating isolated populations providing a likely context for speciation (Potts, 1996), which is consistent with evidence for the period of greatest hominid taxonomic diversity that occurs *c*.2 ma.

Although it is widely assumed that a key factor in human evolution was climatic and related environmental changes, it must be acknowledged that hominid fossils are insufficiently numerous and insufficiently precisely dated for it to be possible to relate specific evolutionary changes to particular palaeoenvironmental events (Foley, 1994). The significance of environmental factors has, however, been highlighted by work on long-term faunal change in species for which the fossil record is far richer. Particularly rapid turnover of African mammal species as a result of speciation, immigration and extinction has been demonstrated by Vrba (1995) between 2.9 and 2.7 ma and this is marked at Omo, Ethiopia, at a time when there is widespread evidence for drier conditions in Africa (Bobe *et al.*, 2002).

At about the same time there is the earliest evidence of tool use at Hadar (2.5 ma) and Kada Gona (2.7 ma) in Ethiopia and Kokalalei, Kenya (2.36 ma). The earliest type of hominid which can be associated with tool use is *Homo habilis* at Koobi Fora (2 ma) and Olduvai (1.8 ma), the Oldowan flaked pebble tool industry. By 1.8 ma *Homo erectus* is represented at Koobi Fora. Following this, a rapid geographical expansion (Figure 5.3) is often regarded as being driven, or facilitated by, the increasingly pronounced environmental changes of the early and middle Pleistocene leading to the appearance of *H. erectus* in east Asia by 1.4 ma and Java by 1.2 ma. *Homo heidelbergensis* is present in Europe at Atapuerca, Spain (0.78 ma), Mauer, Germany and Boxgrove, England (0.5 ma). The earliest Acheulean handaxes were produced by *H. erectus* from *c.*1.4 ma. It is striking that the production of these distinctive artefacts continued for more than 1 ma until *c.*200 ka without pronounced change, despite the frequent environmental changes over this period and the contrasting environmental conditions in which hominids were, by this time, living. Even so,

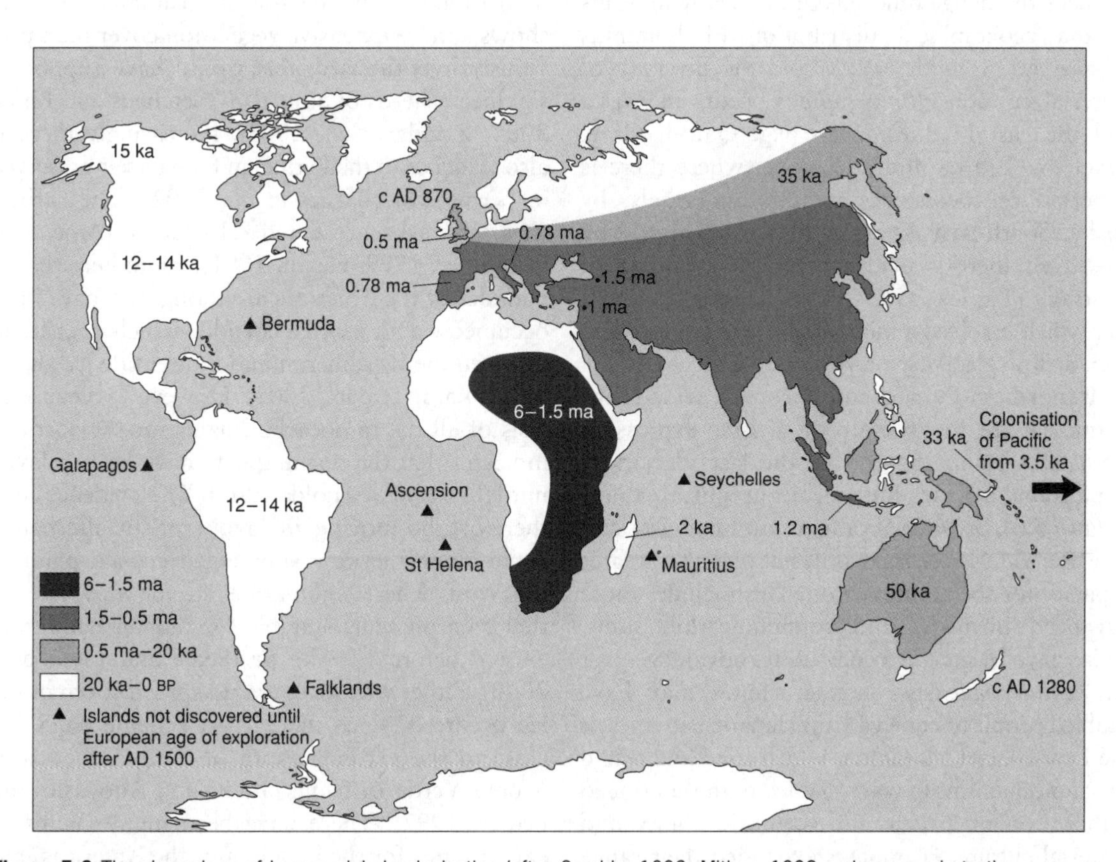

Figure 5.3 The chronology of human global colonisation (after Gamble, 1993; Mithen, 1998 and sources in text)

these handaxes are not ubiquitous, being absent for instance in the South-East Asian hominid sites.

Neanderthals, *Homo neanderthalensis*, appear in south-west Asia and Europe around 120 ka and are characterised by a markedly heavy build, prominent brow ridges, prognathous (protruding) jaw and disproportionately large limbs. These traits are widely regarded as adaptations among a geographically isolated population to the cold conditions of a glaciation (Stringer, 1995). The last Neanderthals occur around 30 ka and, although some would see them as contributing to the ancestry of modern humans, a more widely held view is that they represent an evolutionary dead end, adapted to extreme conditions and less capable of a diversity of adaptive responses than *Homo sapiens*, anatomically modern humans. Genetic evidence again points to an African origin for this species, since it is most genetically diverse in that continent. With the development of *Homo sapiens* the pace of change quickens dramatically in terms of the geographical distribution of hominids (Figure 5.3; Gamble, 1993) and the diversity of material culture. *Homo sapiens* occurs in Africa and the east Mediterranean (e.g. Qafzeh Cave, Israel) by 100 ka and in Europe, where there is a period of coexistence with Neanderthals, by 40 ka. South-East Asia was also reached and by 45–50 ka there is evidence for the peopling of Australia (Turney *et al.*, 2001; Gillespie, 2002), a feat which involved a minimum 60 km sea crossing even at low glacial sea level.

In northern Eurasia and Europe, a wide range of media are the subject of artistic expression from *c.*38 ka to the end of the last glaciation. Expansion of the human geographical range (Figure 5.3), art, burial practice and material culture are widely seen as manifestations of a greater capacity for social interaction. This includes more advanced linguistic and communication skills, which gave *H. sapiens* a competitive advantage over *H. neanderthalensis*. Such attributes may have enabled people to cope with the climatic extremes of the Last Glacial Maximum which tried and tested social mechanisms in ways that led to the continued elaboration of artistic and linguistic ability and material culture. *H. sapiens* expanded their range in south-west Europe from 35 ka BP during a time of frequent climatic oscillation and by about 24 ka the last *H. neanderthalensis* in southern Iberia seem to have died out (d'Errico and Goñi, 2003).

Homo sapiens were able to adapt to the extreme conditions of the northern Eurasian steppe as demonstrated by settlements of mammoth hunters at Mal'ta, near Lake Baikal, *c.*25–13 ka BP and by settlements at the extreme eastern limit of Asia at Ushki Lake, close to the Bering Sea, by 15 ka BP. Adaptation to these harsh environments made possible the crossing of the Straits of Beringia to Alaska at a time of low sea level during and following the Last Glacial Maximum between 25 and 11 ka BP. However, at the glacial maximum the route south was blocked by the vast Laurentide ice sheet (see Figure 4.3). There has been much debate concerning the suitability of the Beringian and Alaskan landscapes for human communities at this time. However, the recent discovery of extensive areas of vegetation preserved by permafrost and buried by volcanic ash fall *c.*17 ka BP shows a more extensive vegetation cover than previously hypothesised; this would have supported significant herds of animals (Goetcheus and Birks, 2001; Zazula *et al.*, 2003). Increasingly there is also evidence of the Beringian fauna itself preserved in permafrost (Barnes *et al.*, 2002). The earliest sites in Alaska are dated *c.*15 ka BP (Ames and Maschner, 1999; Fagan, 1991). From here the remainder of the Americas are thought to have been occupied via an ice-free corridor which opened up between the Cordilleran and Laurentide ice sheets *c.*14.3 ka BP (Fiedel, 2002). However, recent analysis of all the radiocarbon dates for the corridor indicates that the opening may have been delayed until 13 ka BP (Arnold, 2002). The evidence may therefore be moving in favour of an alternative route along the now submerged coastal plain to the west where submarine survey now shows that environments suitable for human habitation existed before 12.2 ka BP (Fedje and Josenhans, 2000). Either way a very rapid spread south of the ice occurred, since the earliest reliable dates are close to the extreme south of South America at Monte Verde in Chile (14 ka BP; Adovasio and Pedler, 1997). There have been many claims of earlier dates for the peopling of the Americas, but none before *c.*12–14 ka BP appears at present to be absolutely convincing and widely accepted. Around 13 ka BP there emerged in North America Clovis hunters, whose distinctive lithic projectile points are on several sites associated with the remains of mammoths and other megafauna and whose way of life may therefore be seen as an adaptation to the terminal Pleistocene (p. 187; Fiedel, 2002).

People and the Lateglacial/Holocene transition

The Last Glacial Maximum occurred around 22 ka BP, when the polar front was as far south as Portugal (see Figure 3.15a) and much of northern Europe was abandoned. Refugia existed centred on south-west France but extending into the Paris basin as far as the Rhine (Street and Terberger, 1999). Other refugia were in Spain, parts of the Mediterranean, and parts of the Ukraine and Russian Plain (Soffer and Gamble, 1990). Notwithstanding the harsh environment, this period saw remarkable developments in terms of art, jewellery, musical instruments and ritual objects which Wobst (1990) described as a 'periglacial behaviour package'. Gamble (1993) likewise refers to these developments as a product of 'arctic hysteria' whereby complex social mechanisms were evolved to ensure survival. The environmental and cultural sequence of north-west Europe from the Last Glacial Maximum to the mid-Holocene is outlined in Figure 5.4.

Palaeolithic art occurs in Europe and the Russian Plain before the Last Glacial Maximum and continues until the end of the last cold stage. The earliest dated art is generally mobile in the form of carved bone or stone slabs. An example

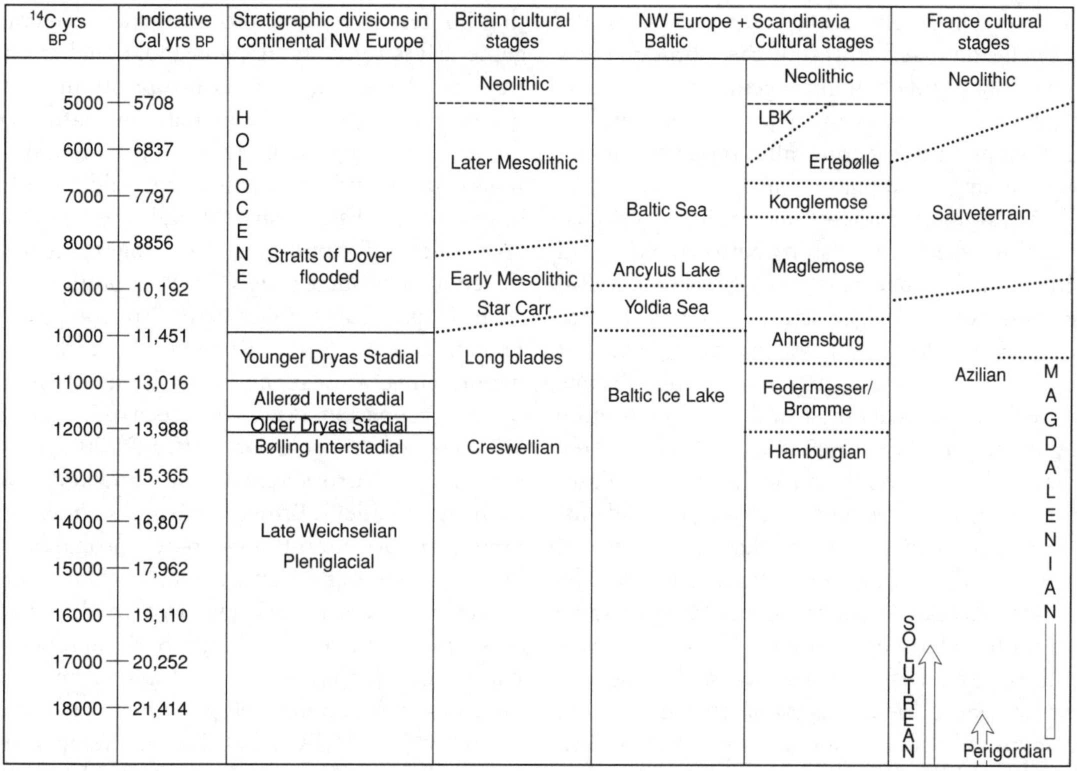

^{14}C yrs BP	Indicative Cal yrs BP	Stratigraphic divisions in continental NW Europe	Britain cultural stages	NW Europe + Scandinavia Baltic	NW Europe + Scandinavia Cultural stages	France cultural stages
		HOLOCENE	Neolithic		Neolithic	Neolithic
5000	5708				LBK	
6000	6837		Later Mesolithic		Ertebølle	
7000	7797			Baltic Sea	Konglemose	Sauveterrain
8000	8856	Straits of Dover flooded				
9000	10,192		Early Mesolithic Star Carr	Ancylus Lake Yoldia Sea	Maglemose	
10000	11,451	Younger Dryas Stadial	Long blades		Ahrensburg	
11000	13,016	Allerød Interstadial		Baltic Ice Lake	Federmesser/ Bromme	Azilian
12000	13,988	Older Dryas Stadial Bølling Interstadial	Creswellian		Hamburgian	MAGDALENIAN
13000	15,365					
14000	16,807	Late Weichselian				
15000	17,962	Pleniglacial				
16000	19,110					
17000	20,252					SOLUTREAN
18000	21,414					Perigordian

Figure 5.4 The relationship between environmental change and selected cultural sequences in western Europe between the Last Glacial Maximum and the mid-Holocene. The indicative calibrated ages are obtained following the procedure described in Table 1.1

is an ivory human figure with a feline head from Hohlenstein-Stadel cave, Germany, dated to 30 ka BP (Bahn and Vertut, 1997). Cave art in France and Spain can now be dated directly using the AMS radiocarbon technique (p. 53) on charcoal which was used to outline sketches. Spanish and French caves have produced dates between 30 and 14.5 ka BP (Valladas *et al.*, 2001). Animals depicted in the art emphasise a steppe and periglacial fauna: horse, bison and wild cattle predominate but many other animals including mammoth and woolly rhinoceros also occur (Bahn and Vertut, 1997).

The caves and rock shelters of the Dordogne (Plate 5.1) are characterised by diverse lithic assemblages which frequently occur in successive occupation horizons within the same cave sediment sequence. The implication is that the area was occupied by a diversity of cultural groups, or in differing interpretations, by groups following contrasting patterns of activity, for example in different seasons. In either case it is difficult to escape the conclusion that at this time the Dordogne was occupied by people coming together from diverse backgrounds and ways of life, the coexistence of Neanderthal and anatomically modern human groups serving to emphasise this point.

Contrasting hypotheses have also developed concerning the relationship between art and environmental change. Some such as Jochim (1983) emphasise the Dordogne as a refugia, a sort of Garden of Eden, rich in resources at a time of otherwise harsh conditions. Here people became concentrated, settled down and developed more complex societies in which social relations were mediated by art. As increasingly precise dates and environmental sequences become available for the occupied sites it is clear that at the glacial maximum full tundra conditions existed on the Dordogne plateaux with forest surviving in some more protected valley locations. There are fewer sites at the Last Glacial Maximum and those that occur are not as rich, suggesting that during the maximum at least the population was not swelled by those seeking a refuge (Rigaud and Simek, 1990). An alternative hypothesis founded on cognitive principles has been advanced by Mithen (1990, 1991). This model might be described as a

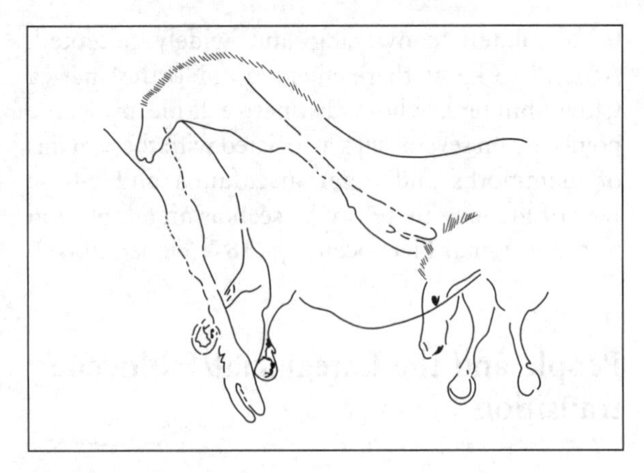

Figure 5.5 Lascaux, France: engraved cow and horse with hooves shown in twisted perspective (after Mithen, 1990, Figure 8.1a)

coping strategy, a response to declining hunting yields at the Last Glacial Maximum. It is suggested that hunters were forced to turn from cooperative game drives to the tracking of individual animals. Aspects of the art, such as footprints in twisted perspective (Figure 5.5), animals in mating condition, etc., represent knowledge of animals, tracks, trails and signs which it would have been important to disseminate through the medium of art to develop appropriate hunting strategies in the context of increasingly harsh conditions.

It is particularly instructive to compare the French case with the Russian Plain, the other major refuge area at the Last Glacial Maximum, where occupation occurs between 250 and 800 km south of the ice front (Soffer, 1990). Activity was in a permafrost steppe with woodland refugia in the river valleys. Bison, horse and reindeer were hunted and mammoth bone, partly probably from hunted animals and partly from natural mammoth graveyards, was extensively used in the making of shelters and pit dwellings. Both summer and winter occupation sites are known; modelling indicates that summer temperatures were similar to today (*c.* +14 °C), but January temperatures were −30 °C at the glacial maximum. These communities had clearly evolved coping strategies by which they could occupy the plain in all seasons, living at individual sites for periods of up to

9 months and sometimes in communities of up to 50 people. Survival in this productive yet highly unpredictable steppe environment involved the storage of meat in pits cut in the permafrost. What is particularly remarkable is that this area was likewise characterised by art from as early as 21 ka BP, with female figurines and horses carved in mammoth ivory, mammoths carved in bone and a range of decorated bones, some used as personal ornaments (Gvozdover, 1995). In this context the relationship between a society coping with harsh environmental conditions and the social basis of coping strategies including art and personal ornamentation is especially striking.

Following the Last Glacial Maximum conditions gradually became less severe (pp. 69–75). Magdalenian communities expanded into western and central Europe *c.*16 ka BP. At Stellmore in Germany, in the treeless tundra of a recently deglaciated landscape, former subglacial meltwater tunnel valleys flooded to become lakes, creating topographic bottlenecks in which reindeer could be intercepted by hunters armed with weapons carrying shouldered points of Hamburgian type (Bokelmann, 1991). During the Lateglacial Interstadial occupation extended to the British Isles centred on the period between 14.8 and 14.2 ka BP. Creswellian communities hunted reindeer, bovids and horse (Housley *et al.*, 1997; Barton *et al.*, 2003). The first discovery of Palaeolithic cave art in Britain at Creswell Crags is also likely to date to this episode (Bahn *et al.*, 2003). In south-west Europe the Lateglacial amelioration led to a reduction in reindeer numbers and increasing human dietary diversification and material culture change which, from *c.*11 ka BP, gave rise to the Azilian culture (Straus, 1996).

In northern Europe and Scandinavia the Allerød period (p. 86) is seen by Fischer (1991) as offering very particular opportunities to the hunting communities, conditions he describes as 'a paradise on earth': a recently deglaciated landscape with a low human population density and a high carrying capacity. During the Younger Dryas (Figure 3.16), a period of much harsher climatic conditions, there is evidence for continuing human activity in previously occupied areas of north-west Europe. Sites of this phase are represented by the so-called tanged (or Ahrensburg) point complex. An example is a second stage of activity at Stellmore, this time the hunting of reindeer with an innovation, bow and arrow, the wood shafts of the arrows being preserved in the lake sediments (Bokelmann, 1991).

The very rapid climate change at the beginning of the Holocene 11.5 ka BP, much of it in less than a human generation (p. 69), poses the question: how did human communities cope? It would have been very marked in terms of temperatures, sea level, and the response of plants and animals to this change involved much less time lag than once believed (Ammann *et al.*, 2000). Disharmonious ecosystems were created for which no modern analogues exist. Open grassland communities would gradually have been invaded by birch and pine woodland. The first few hundred years of the Holocene in Britain are characterised by continued use of long blade flint industries of Lateglacial type which were used by communities which hunted reindeer and horse on the initial Holocene grass steppe (Barton, 1991). It is only with the development of woodland that the material culture changes; microliths (small lithic elements used in the production of composite artefacts) become important and flint axes are used. These are seen for instance in the Maglemose culture of Denmark (Fischer, 1991) and some of the earliest Mesolithic sites in Britain at Star Carr and Thatcham which were established around 10.9 ka BP (Dark, 2000b).

Although it may be anticipated that the dramatic environmental changes at the beginning of the Holocene would be marked by equally significant cultural changes, matters are not quite that simple. We have seen that many of the most interesting social developments in terms of art, burial practice and personal ornamentation can be traced back into the hostile environments of the last glacial and each in a European context is poorly attested for millennia after the start of the Holocene. Microliths, the bow and arrow and even the exploitation of diverse faunal resources are also prefigured by the activities of certain earlier communities in the closing stages of the last glaciation. Communities of the Holocene may therefore be seen as selecting from an available repertoire of adaptive and coping strategies, many of which seem actively to have evolved during the exceptionally rapid environmental changes towards the end of the last cold stage.

The origins of agriculture

Introduction

The domestication of plants and animals and the emergence of agricultural communities was of major significance in human terms. Sedentary (less mobile) communities developed because larger populations could be fed in one place, leading ultimately to the development of complex societies in which individuals could specialise in activities other than food production and the emergence of civilisation was fostered. Through domestication people have transformed vast tracts of the earth by taking species to new areas and creating the conditions in which they could flourish. Domestication involves control of the breeding processes of plants and animals (B.D. Smith, 1995a). Consciously and unconsciously people have selected certain traits favoured by humans, or the agricultural regimes that they have created. Genetic change leads to morphological change through which wild and domestic forms may be distinguished. In the wheats of south-west Asia, for instance, mutant cereal ears with a tough rachis (connecting axis of the ear) were preferentially, and largely unconsciously, collected in the harvesting process because others with a brittle rachis tended to shatter at a touch. People in the Americas also selected larger forms of maize cob for instance, and those with better food value. So too with domestic animals; humans bred from those with particular traits of size, docility, coat or for the retention of particular juvenile characteristics which made animals easier to control. This eventually led to recognisable changes in bone morphology. Many early domesticates, such as sheep, cattle and pigs, are smaller than their wild ancestors (Davis, 1987).

Domestication was not a one-off event, for it occurred independently at many times and places as the examples shown in Figure 5.6 demonstrate, but here we focus specifically on south-west Asia and the Americas. Other areas were undoubtedly of great importance but have generally not been investigated to the same extent, so the dates and environmental contexts of initial domestication are less securely known for rice and millet in China (Glover and Higham, 1996) and the crops of Africa such as sorghum, forms of millet and rice (Harlan, 1992). It has often been argued that tropical 'vegiculture', crops such as yams, taro and manioc, were early domesticates in Africa, south-east Asia and Oceania, but the evidence for such crops less frequently survives (Hather, 1994).

The study of domestication has in recent years been transformed by important technical advances. The direct radiocarbon dating by AMS of domesticates, especially individual seeds, provides high-quality dates. Previously we had to rely on larger samples of biological material such as wood charcoal from the same context, which were *assumed* to be associated with the domesticate. It is now clear that some of the supposed associations were not correct, for instance because small numbers of seeds were intrusive in earlier layers. As a result the dates of some domestications have had to be revised downwards by millennia (B.D. Smith, 1995a).

Equally revolutionary is accumulating genetic evidence for the evolutionary history of domestic species. DNA studies of living species provides information on those wild populations which are most similar to the domesticate, and thus the most probable geographical area of domestication. The survival of ancient biomolecular evidence is also beginning to contribute to our understanding of the history of domesticates, despite significant challenges created by preservation, contamination and taphonomy (T. Brown, 1998; Jones, 2001). Isotopic evidence (p. 46) of past diet from human bone is also providing records of dietary change and the balance between dietary components which is independent of the taphonomically distorted record of plant remains and animal bones (Sealy, 2001).

Conceptual aspects of the domestication issue are informed by a growing appreciation from anthropology of the diversity of relationships which exist between people, plants and animals. A distinction needs to be made between domestication and other forms of environmental manipulation, such as cultivation, which can be carried out to

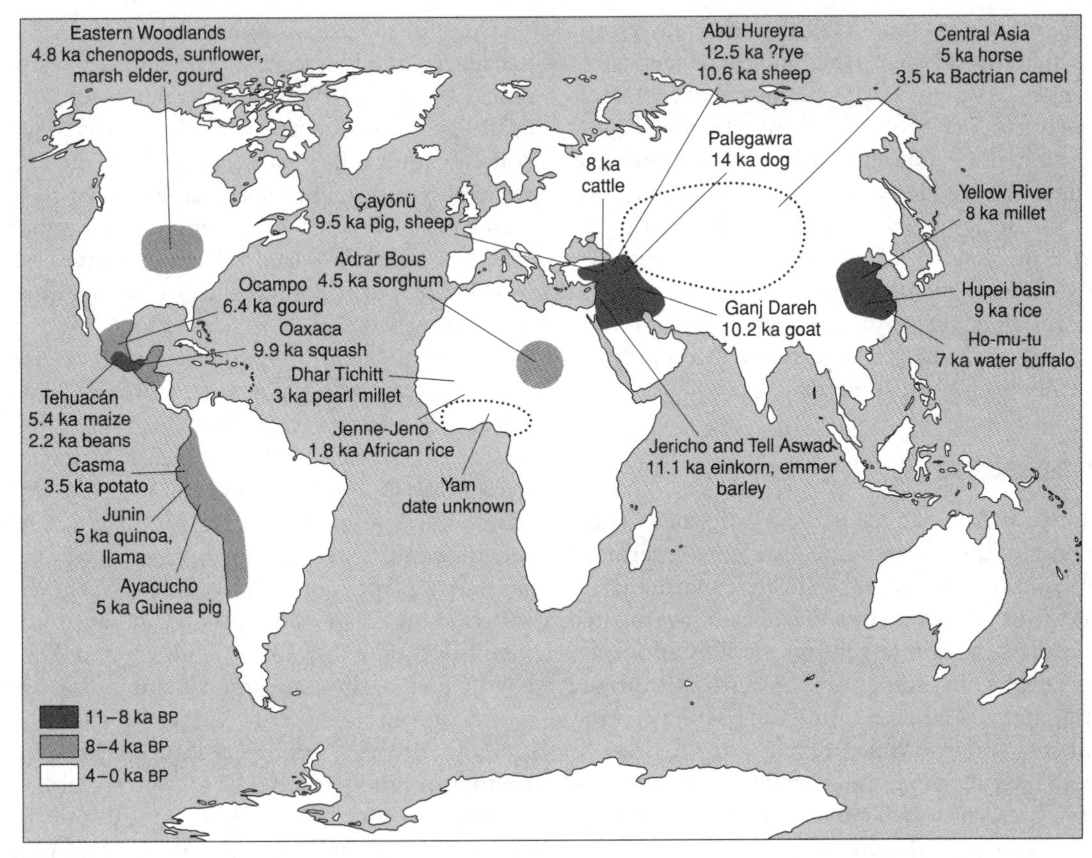

Figure 5.6 The regions, sites and dates at which some of the most important domestications took place (after Simmons, 1989, Figure 4.1, with modifications after B.D. Smith, 1995a and sources in text)

encourage undomesticated species. Australian Aborigines lacked domesticated plants but manipulated their environment by the practice of burning (p. 193) and the replacement planting of yams. Domestication is increasingly seen in terms of a gradient of gradually intensifying relationships between people and biota. The 'dump heap' theory holds that domestication came about because waste plant material was discarded and multiplied in the artificial, disturbed and nitrogen-rich areas around settlements. Hunter-gatherer camps may have around them 'gardens' of favoured species growing naturally from seeds germinated from dumped food and faeces (Blackburn and Anderson, 1993).

Domestication would seldom have been a passport to an easier lifestyle. It is well documented that many foragers spend less time getting food and have more leisure than agriculturalists (Lee and De Vore, 1968). Vegetable resources, particularly cereals, are regarded by hunter-gatherers as low-preference foods. As Flannery (1973) observed, 'since farming represents a decision to work harder and eat more third choice food, I suspect that people did it because they felt they had to, not because they wanted to'.

In considering the origins of domestication some writers put particular emphasis on environmental, influences others on social factors or population levels. Such distinctions are seldom helpful because it is increasingly evident that various factors contributed in intimately linked ways. More productive may be concepts of coevolution: considering the interactive relationships between climate, human agency, plant and animal communities. Each may be responsible for modifying the other, giving rise to conditions of increasing interdependency. In a simple case, both natural disturbance factors (Chapter 6) and people may have modified plant and animal communities thereby creating new conditions which led to changes of human behaviour and social interaction, in turn leading to the increasing control of domestication. There are hints of such relationships in what happens at the woodland edge in south-west Asia and in the plant communities associated with floodplain settlements of North America.

South-west Asia

South-west Asia has the earliest well-documented evidence of domestication, close to the Lateglacial Holocene transition, and here the environmental context of domestication has been investigated in detail. Remarkable cultural manifestations are associated with some of the earliest farming communities. The area in question is the Levant, southern Turkey, Mesopotamia and the Zagros Mountains, an area traditionally known as the Fertile Crescent which curves round three sides of the Syrian Desert (Figure 5.7).

Despite the environmental changes of the Lateglacial and Holocene, the present-day distributions of many of the wild progenitors of domesticates provide evidence of the likely areas in which domestication took place (Figure 5.8). Such modern analogue distributions must, however, be used with caution because the dramatic environmental changes of the Lateglacial will at times have created very different distributions and associations of plant communities as Figure 5.9 shows.

The three main domestic cereals were emmer wheat (*Triticum turgidum* ssp. *dicoccum*), einkorn wheat (*Triticum monococcum*) and barley (*Hordeum vulgare*). Morphological changes to cereals, such as the replacement of the brittle rachis adapted to wild seed dispersal by a tougher rachis, which would have retained more seed in the harvested plant, could have taken place within a period of 200–300 years given appropriate conditions of cultivation (Hillman and Davies, 1990). Five early companion plant domesticates also occur in the Fertile Crescent: pea (*Pisum sativum*), lentil (*Lens culinovis*), chickpea (*Cier arietium*), bitter vetch (*Vicia ervilia*) and flax (*Linum vsitatissimum*). Wheat, pea and lentil cultivars have only a fraction of the genetic

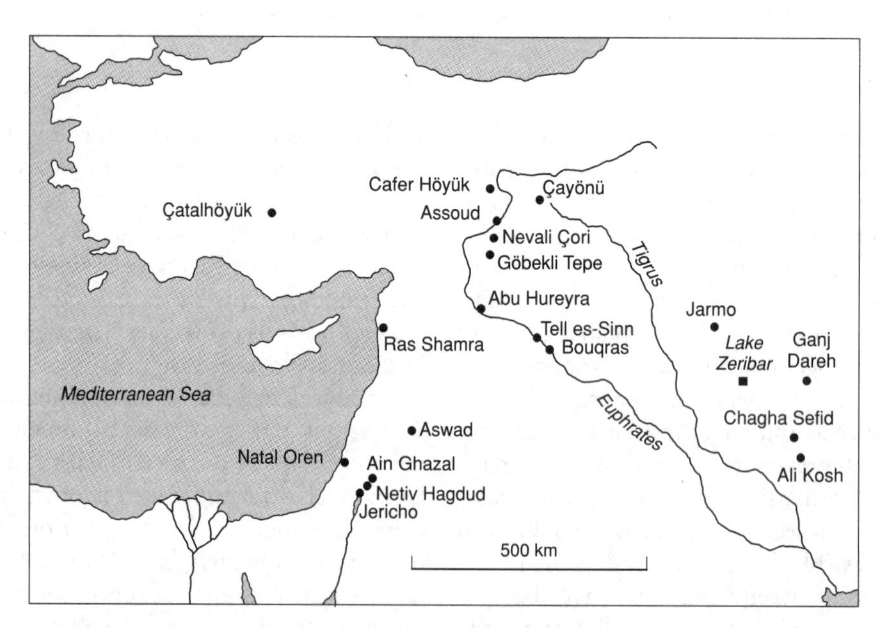

Figure 5.7 Main south-west Asian sites associated with the early history of domestication

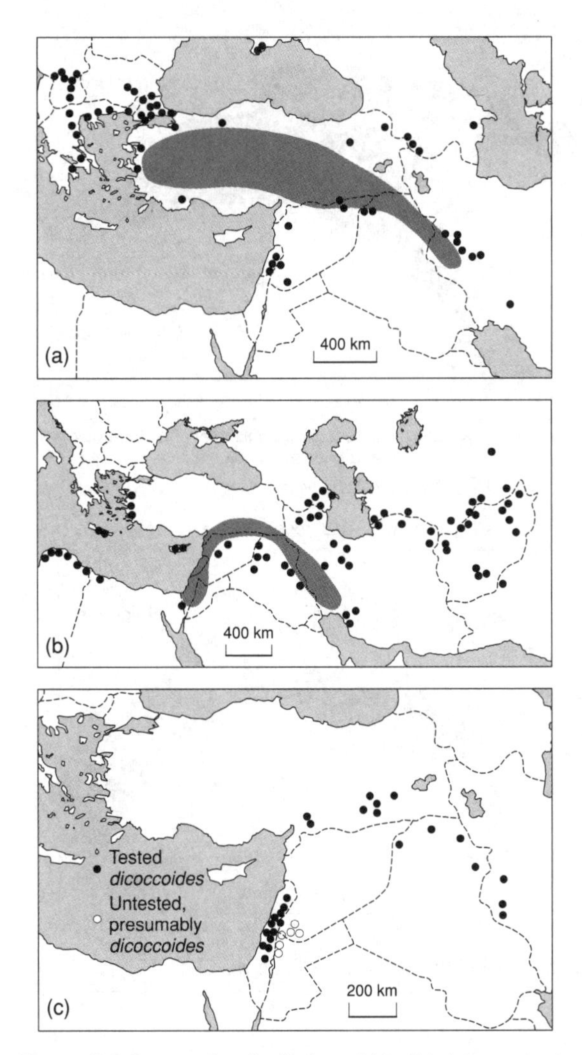

Figure 5.8 Present day distribution of (a) wild einkorn wheat (*Triticum boeoticum*); (b) wild barley (*Hordeum spontaneum*); and (c) wild emmer wheat (*Triticum dicoccoides*); shaded area shows main distribution; dots indicate isolated population (after Zohary, 1989, Figures 22.1, 22.2 and 22.3)

polymorphism of the wild populations, indicating that these plants may have been taken into cultivation only once, while the presence of two non-shattering genes in barley implies it has been domesticated more than once (Zohary, 1996).

As climate ameliorated during the Lateglacial, so environmental conditions and ways of life were established which provided the context for domestication. During the last glaciation arid conditions had prevailed in much of south-west Asia; as rainfall increased, at first grassland, then open woodland spread east along the Fertile Crescent (Roberts and Wright, 1993; Wilkinson, 2003). Already from the middle of the last glaciation a broad spectrum of dietary resources was being exploited (P.C. Edwards, 1989). Close to the Last Glacial Maximum *c.*22 ka BP there is the earliest exploitation of wild cereals as part of an economy which involved fishing, hunting and gathering at Ohalo by the Sea of Galilee (Kislev *et al.*, 1992).

The extent of woodland, grass steppe and areas rich in wild cereals during the Lateglacial Interstadial is shown in Figure 5.9b. These favourable conditions were marked by Natufian settlements in the Levant and Epipalaeolithic communities in the central Fertile Crescent. By 14.5 ka BP some of these settlements were characterised by well-built houses, evidence of sedentism and three traits of special significance because they emphasise the contribution of wild cereals to the economy of these sites: sickle blades, grinding stones and storage structures. This is evidence of a strategy to cope with wild cereal resources which were episodically superabundant (Bar-Yosef, 1995). Areas of wild grasses would have been as productive as later fields of domestic cereals and it has been calculated that a small family could gather enough wild cereals in three weeks to feed themselves for a year (B.D. Smith, 1995a).

The process and context of domestication have been investigated in particular detail at Abu Hureyra in the Euphrates Valley, Syria (Moore *et al.*, 2000). Epipalaeolithic settlement began *c.*13.2 ka BP with a sedentary settlement of hunters and gatherers, who collected a wide range of 157 wild plant taxa including wild cereals. At this stage gazelles accounted for 80 per cent of the animals and wild ovicaprids (sheep/goat – many anatomical elements of the two species are difficult to distinguish) 20 per cent. The first morphological changes to cereals associated with domestication are claimed during the Epipalaeolithic *c.*13–12.5 ka BP; this comes from a small number of AMS radiocarbon-dated rye (*Secale cereale*) grains (Hillman *et al.*, 2001). Rye is not a widely recorded early domesticate. At the same time there is an increase in seeds

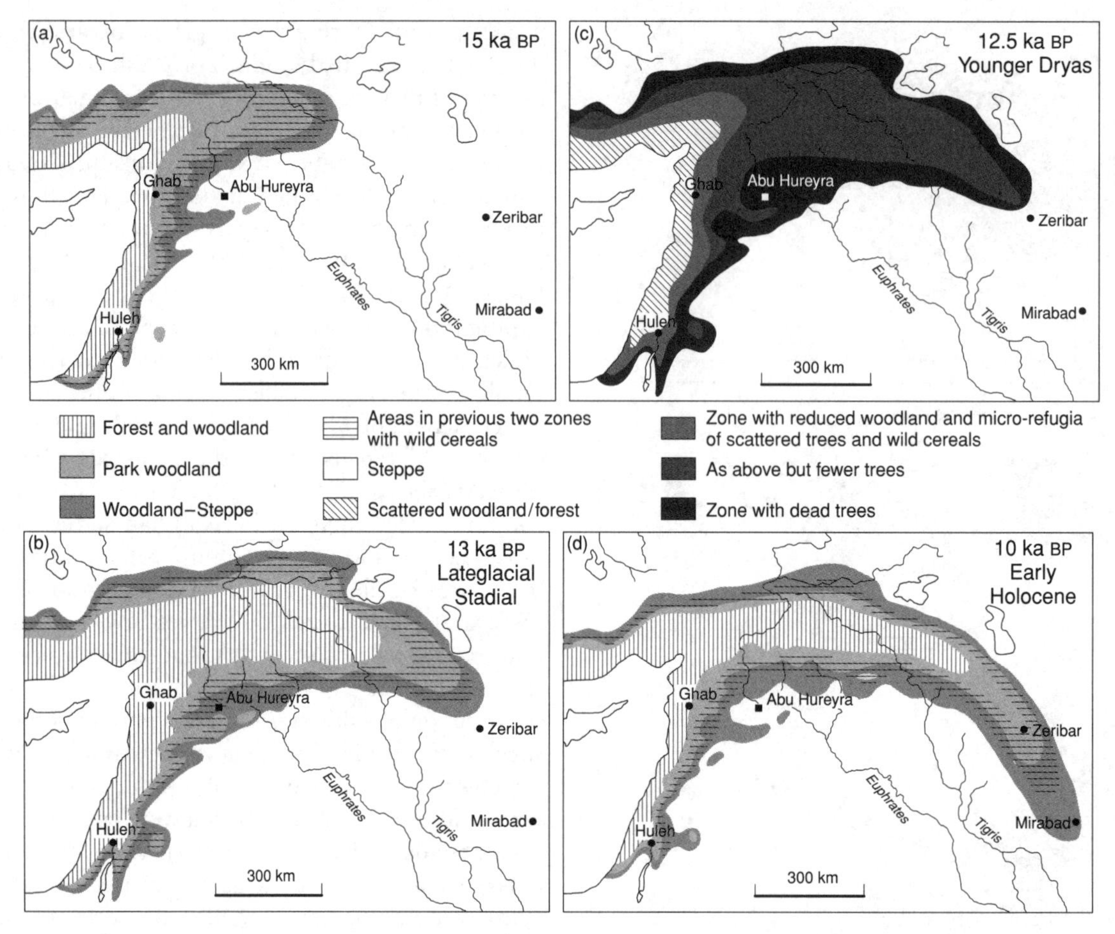

Figure 5.9 Models of south-west Asian vegetation change 15–10 ka BP (after Moore *et al.*, 2000, Figure 3.18)

of weeds which are characteristic of cultivated land in the arid zone. Plant macrofossil evidence shows that these changes occurred during an episode of increased aridity, and there was a reduction in moisture-demanding plants including wild cereals. As the regional environmental reconstruction (Figure 5.9c) shows, the open woodland and grass steppe plant communities, which had spread eastwards in the Lateglacial Interstadial between 15 and 13 ka BP and are associated with abundant with wild cereals, underwent a dramatic reduction in area, with trees being reduced to scattered patches. Hillman *et al.* (2001) argue that these changes are a direct result of the Younger Dryas climatic downturn (p. 87; Bar-Yosef, 2000). Plant domestication, they argue, was largely prompted by the need to increase the abundance and predictability of plant resources. This can be seen as a coping strategy to help maintain existing population levels and settlement patterns under less favourable climatic conditions.

It must be acknowledged that no other site has produced such early evidence of domestication as Abu Hureyra; other sites are about 1400 years later (B.D. Smith, 1995a). Domesticated einkorn, emmer and barley occur at Jericho and Aswad from *c.*11.1 ka BP. Barley was domesticated at Netiv Hagdud between 11.3 and 11.1 ka BP, and domesticated einkorn and emmer are present between 10.7 and10.2 ka BP at Abu Hureyra, Ain Ghazel, Natal Oran, Cayönü and Gang Darah. All but the last of these sites are in the western area of the

Fertile Crescent where cereal domestication seems to have taken place. However, one proviso is that for political reasons fieldwork has been very limited in the eastern Fertile Crescent in recent decades.

At Ali Kosh in the eastern Fertile Crescent occupation occurred 9.5–9 ka BP and at first domesticates contributed just 10 per cent of plant resources, though later that increased. This is an area marginal for agriculture where domesticates never dominated the economy but provided a measure of security and reduced risk. Of the companion plants, lentils were domesticated by 9.9 ka BP, pea, chickpea, bitter vetch and flax by 9 ka BP (Zohary and Hopf, 1994). Assemblages dominated by domestic cereals subsequently spread much more widely; between 9 and 8 ka BP they formed the basis of arable agriculture in Greece, the Nile Valley, Turkmenistan and the Indus Valley of Pakistan (Zohary and Hopf, 1994).

The earliest animal domestication is the dog with evidence for morphological changes by *c.*13 ka BP in both the Old and New Worlds, thus conceivably a migrant across Beringia with people (p. 146). This was a model coevolutionary partnership born of mutual advantage which both gained from complementary hunting skills. The next animal domestication in south-west Asia is significantly later than plant domestication, 800 years (possibly, given the Abu Hureyra evidence, 2000 years) later. However, contrasting rates of morphological change between plants and animals could contribute to this contrast. The main domestic animals were goats (*Capra hircus*), sheep (*Ovis aries*), pig (*Sus scrofa*) and cattle (*Bos taurus*). Sheep and goats have a mainly Fertile Crescent distribution extending to the east; pig and cattle have much wider natural distributions. At Ganj Darah, Iran, the age and sex profile of goat herds *c.*10.2 ka BP suggests domestication, as is the case a little later at Jarmo. Goats were a key resource in this eastern arm of the Fertile Crescent, both before and after domestication, and the belief is that their domestication took place in this area (B.D. Smith, 1995a). From 9.5 to 9 ka BP domestic sheep occur at Bouqras, Cayönü, Cafer Höyük, Tell-es-Sinn, Assoud, Ras Shamra, Jarmo and Ali Kosh. Sheep were a more exploited resource in the central part of the Fertile Crescent and it is likely that their domestication occurred here (B.D. Smith, 1995a). At Abu Hureyra domestic sheep and goats occur in levels between 10.6 and 9.3 ka BP, reversing previous proportions; sheep and goats now make up 80 per cent of the bone assemblage with gazelles at 20 per cent (Moore *et al.*, 2000). Interestingly, a similar change occurs at this date at the opposite end of the gazelle migration route at Azraq in Jordan. In the Jordanian Desert stone-shaped kite structures are associated with the intensive hunting of gazelle. It seems probable that as human populations expanded in the Holocene, the gazelle population was decimated and this may have been an important stimulus to the domestication of sheep. Pigs are thought to have been domesticated in the wooded landscape of the western central Fertile Crescent, the earliest evidence occurring at Cayönü, Turkey, *c.*9.5 ka BP. Cattle seem to have been domesticated rather later; there is possible evidence of domestication during the life of the Çatalhöyük settlement between 9 and 8.5 ka BP. Both cattle and pigs occur at Abu Hureyra *c.*9.4 ka BP.

A case for the first domestication being driven by the climatic downturn of the Younger Dryas has been put by several authors (e.g. Sherratt, 1997; Bar-Yosef, 2000). Firm supporting evidence seems to rest largely on the small number of dated domestic seeds in Epipalaeolithic contexts at Abu Hureyra (Hillman *et al.*, 2001). Elsewhere, however, Epipalaeolithic and Natufian sites seem to have exploited wild resources, and early evidence of morphological changes to wheat and barley occurs *c.*11.1–8.9 ka BP in pre-pottery Neolithic, early Holocene contexts. Thus the initial experiments with cultivation and manipulation of the reproductive processes of plants may have taken place during the Younger Dryas. Domesticates came to play a key role in the ecology of communities in the early Holocene by which time open woodland, grass and steppe communities had again spread east along the Fertile Crescent (Figure 5.9d). Given the very rapid nature of early Holocene environmental change (p. 69), this is likely to have been a time of transient and disharmonious ecological communities, including episodes of plenty interspersed with reduced numbers of particular plant and animal taxa, as populations interacted with climate and stochastic factors to adjust to new conditions. People may have contributed to the ecological instability of this period by the use of fire to create favourable conditions for some taxa, and in this way, have retarded the process of woodland colonisation (Roberts, 2002) as hunter-gatherer communities did at the same time in Britain (p. 194).

Arid episodes and unpredictable rainfall will have presented particular challenges. It is striking how many of the key agricultural sites are in geographical situations, i.e. close to springs, on alluvial fans and at the edges of lakes, where communities would have been most able to cope with arid episodes. Attempts by communities to cope with, and control, the instability and unpredictability of early Holocene environments may, given currently available dates from most sites, be as important in understanding the origins of agriculture as the more extreme conditions of the Younger Dryas.

Population levels must also have represented a key factor. The development of sedentism from 14.5 ka BP would have led to a relaxation of natural population controls; together with the expansion of favourable environments for settlement in the Lateglacial Interstadial and early Holocene, this would have led to marked population expansion. However, settled populations would also have had a greater susceptibility to disease (Groube, 1996). As settlements became larger and more numerous it was no longer an option to cope with environmental hazards by moving to new settlements, or budding off part of the population elsewhere. The most favoured sites are likely already to have been settled, hence the increasing importance of exerting greater control over the predictability and abundance of resources through domestication.

The social context of domestication is also increasingly given emphasis by remarkable cultural manifestations that occur at this time. In the upper Euphrates in south-east Turkey, sites with cult structures decorated with reliefs and carvings of wild animals and humans are found both at Gobeki Tepe (9.8 ka BP) and Nevali Cori (10.2 ka BP; Hauptmann, 1999). At the first site there is no evidence of domestication and at the second there are domesticated sheep/goats and plants. Einkorn may have been domesticated at or near this site; it is near a modern einkorn population which has been claimed as genetically closest to domestic varieties and thus a possible site of domestication (Heun *et al*., 1997). The association of cult centre, art and domestication has led Mithen (2003, Chapter 8) to argue for the primacy of social factors as a driver of domestication. Clay female figurines occur on several sites: 13 lime plaster figures from Ain Ghazel are dated *c*.9.6 ka BP and human skulls plastered and decorated as heads at Jericho. These figurines, skulls, and burials within buildings, which occur from the time of Natufian pre-agricultural communities, suggest an ancestor cult by which settled populations legimated claims to landscape and resources. At Jericho a substantial tower and stone walls, perhaps Neolithic flood defences, are evidence of community action and coping strategies. Most striking perhaps is the evidence from Çatalhöyük (7.4–6.2 ka BP) where there are terracotta female figurines, wall paintings, plaster altars and figures of bulls (Hodder, 1996; Cessford, 2001). This evidence highlights the special importance of women in the new agricultural economy and the ritual and symbolic significance of cattle at a time when they are thought to have been undergoing domestication. Wider contacts and social relationships are attested by the exchange of obsidian and beads. The availability of new domestic resources would have played an important role in mediating social relationships, a novel coping strategy by which resources became more predictable and abundant. This is often seen as people exerting greater control over the environment. It was also, however, a two-way, or coevolutionary process, because by adopting these strategies human communities also effectively limited the range of options available to them in coping with future environmental perturbations.

America

Mesoamerica

The most important Mesoamerican crop is maize (*Zea mays*) which genetic research demonstrates was domesticated from the closely related group

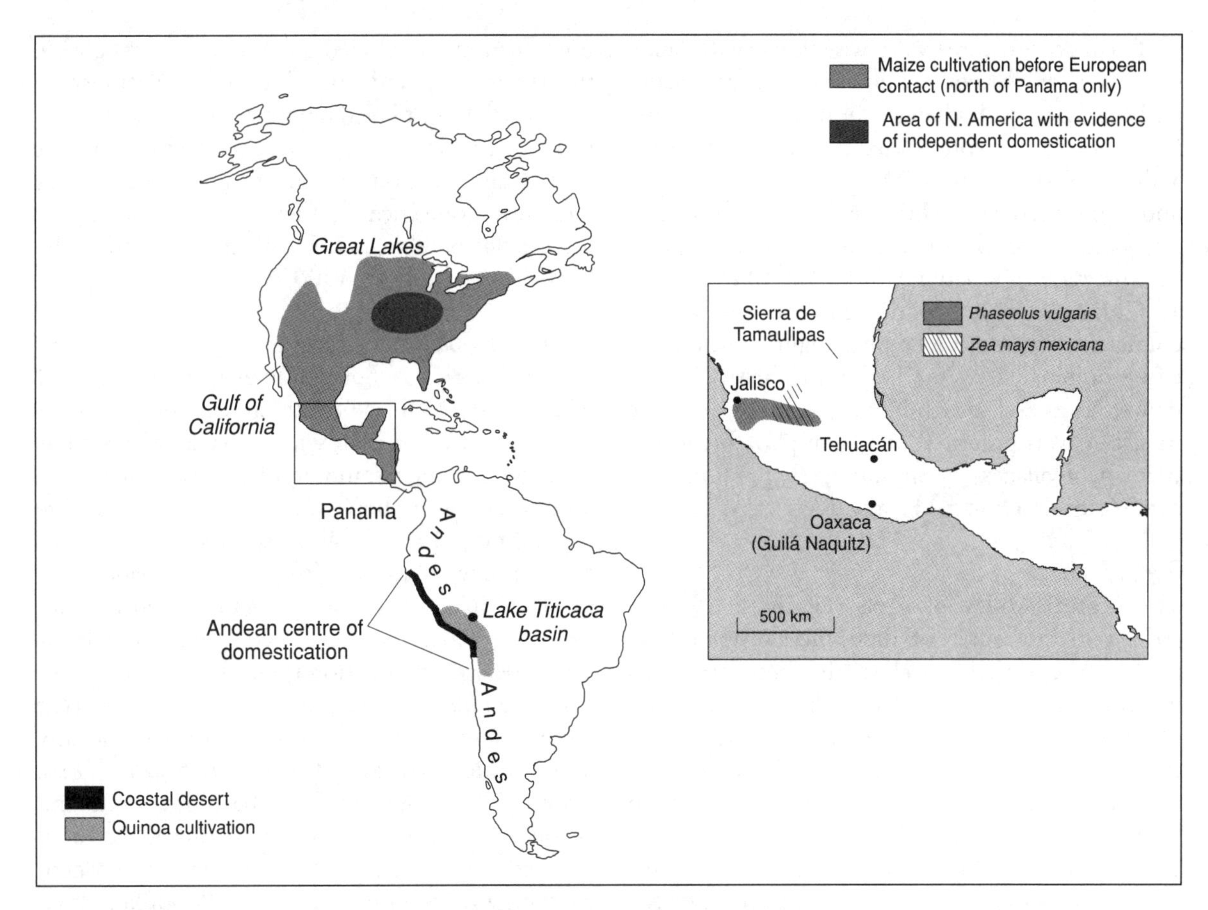

Figure 5.10 Locations of early domesticates in the Americas and the area of North America where native Americans grew maize before European contact (adapted from McAndrews, 1988, Figure 1, Smith, 1995a)

of wild grasses called teosinte. Proteins of the domestic maize are shown to be biochemically closest to wild progenitors in central Mexico (B.D. Smith, 1995a); a probable area of domestication in the Jalisco area overlaps with the wild progenitor of the common bean (*Phaseolus vulgaris*) (Figure 5.10). The earliest macrofossils of domesticates are mostly found in cave sites in drier interior areas: the Tehuacán Valley, the Sierra de Tamaulipas and Oaxaca (Figure 5.10). Direct AMS radiocarbon dating of maize macrofossils in the Tehuacán Valley shows the earliest examples are 5.4 ka BP, several millennia later than originally claimed on the basis of less reliable dates (B.D. Smith, 1995a). The earliest beans reached the Tehuacán Valley 2.3 ka BP. Other crops domesticated in middle America include the squash (*Cucurbita pepo*); manioc (*Manihot esculenta*), found in coprolites from Tehuacán *c.*2 ka BP; chilli peppers (*Capsicum* sp.); and a wide range of other plants (McClung deTapia, 1992). Direct AMS radiocarbon dating of plant macrofossils from the Ocampo caves, Tamaulipas, demonstrates domestic bottle gourd (*Laganaria siceraria*) and squash from 6.4 ka BP and maize from 4.3 ka BP (B.D. Smith, 1997a). Once again direct dating has had the effect of considerably reducing the timing of the earliest domestications. There is always the problem, however, of the patchy distribution of the record, and Piperno and Pearsall (1998) argue, particularly on the basis of phytolith and pollen evidence, for the earlier occurrence of maize and other crop plants in the lowland neotropics, for example Panama.

Early domesticates were associated with small seasonally mobile groups who adopted domesticates with little overall change to their way of life, as the evidence from Guila Naquitz in Oaxaca Valley, Mexico, shows (Flannery, 1986; Marcus and Flannery, 1996). Here AMS radiocarbon dating demonstrates the early domestication of squash (*Cucurbita pepo*) from 9.9 ka BP (B.D. Smith, 1997b). An economy with a significant agricultural emphasis appears *c*.3.5 ka BP and agriculture came to be dominated by the highly productive crops, maize, beans and squash by 2.3 ka BP. By this time irrigation was practised and semi-permanent and permanent villages appear, ultimately providing the context for the Olmec, Maya and Aztec civilisations.

South America

The central Andes of Peru (Figure 5.10) is a second major centre of independent domestication in the Americas. Here the best preserved macrofossil evidence is from the dry coastal deserts of Peru, whereas many of the domesticates derive from higher Andean locations. Again the earliest dates about which we can be confident have been shown by AMS radiocarbon dating to be significantly younger which indicates that the earliest domesticates are *c*.5.7–4.5 ka BP; these include the potato (*Solanum*), which genetic evidence indicates was domesticated in the Lake Titicaca basin, and quinoa (*Chenopodium quinoa*). A locally domesticated bean occurs from 2.4 ka BP and squash was also domesticated. Maize was introduced by contact with Mesoamerica *c*.3.4 ka BP (B.D. Smith, 1995a). Other writers argue for dates up to 3 millennia earlier than the earliest so far available from the direct AMS radiocarbon dating of the domesticates themselves (Pearsall, 1992; Hastorf, 1999). Some domestications such as manioc (*Manihot esculenta*) and sweet potato (*Ipomoea batatas*) will have taken place in tropical lowlands, which have been little investigated.

Domesticated animals in the Andes were the camellids, llamas (*Lama glama*) and alpacas (*Lama pacos*) and also guinea pigs (*Cavia porcellus*). Domestication of these species occurred between *c*.5.7 ka BP and 4.5 ka BP, by which time camellid herding was widely established. Interestingly, quinoa is a favourite food of llamas and it is likely to have flourished in abandoned animal pens, highlighting the recriprocal and coevolutionary relationships which will have played a part in the domestication process. From 4.5 ka BP civic architecture appears in coastal Peru, but among fishing communities with few domesticates, then the development of increasingly sophisticated Andean states led to the Inca empire from AD 1400.

North America

North America is not a domestication centre of what today would be regarded as major crops. It is, however, an area in which recent research has transformed understanding of the transition from a hunting and gathering to a farming way of life (B.D. Smith, 1995b). What is striking about this area is new evidence for local domestications of plants in eastern North America (Figure 5.10), many of which have now gone out of cultivation. Local domestication took place of goosefoot (*Chenopodium berlandieri*), marsh elder (*Iva annua*), sunflower (*Helianthus annuus*) and locally domesticated gourds (*Cucurbita pepo*). These plants are thought to have been favoured and encouraged by the disturbed conditions created by a combination of human settlement and geomorphic processes in river valley floodplain environments (B.D. Smith, 1995b). Macrofossils which are morphologically distinct (e.g. thinner seed coats, larger seeds) have been AMS radiocarbon dated, and appear from 4.8 to 3.2 ka BP. Later, around 3.2 ka BP, maize, beans and squash spread from Mesoamerica into the south-west (Minnis, 1992) and eastern North America. Initial adoption was by hunter-gatherers, both sedentary and mobile, who did not at this stage undergo a major change of lifestyle but used the domesticates to supplement their diet (Wills, 1995). Even some of the most complex societies, such as the Hopewell of Illinois (2.1–1.5 ka BP) who created huge geometric earthworks, had diets to which maize made only a modest contribution according to isotopic evidence (van der Merwe, 1992). Later around 1 ka BP the nutritionally balanced triad of maize, beans and squash came to dominate the economy of selected areas such as some of the main river valleys and the arid zone marginal communities of the American south-west (p. 230). Domestic animals were few: the turkey, duck, dog and bee. Agriculture was hardly practised north of a line from the Great Lakes to the Gulf of California (Figure 5.10); even so, beyond that line there is abundant evidence for the manipulation of plant communities which at times amounts to cultivation, but not domestication (p. 193).

Coastal wetlands

At the Last Glacial Maximum sea level stood *c.*130 m lower than today (Figure 4.13). Glacially lowered sea level is a significant factor in our understanding of the pattern of prehistoric global colonisation, for example of Australasia *c.*45–50 ka BP and the Americas *c.*14 ka BP (Figure 5.3; Gamble, 1993). As sea level rose at the end of the last cold stage and in the early Holocene vast new tracts of coastal wetland were created and progressively inundated by the sea. These offered a diversity of highly attractive resources especially for communities of fishers, hunters and gatherers (Louwe Kooijmans, 1993). They were subject to marine influences the extent of which fluctuated through time. Dynamism is increased by a range of disturbance factors: storms, floods, concentrations of animals, etc. (Chapter 6), which particularly in areas of low coastal relief, create mosaics of subclimax plant communities. The affordances which these environments offer have to be balanced by an ability to cope with environmental changes, both those which are of regular and predictable nature (tides and seasons) and stochastic events (major storms, tsunamis and the effects of earthquakes).

Eustatic sea-level rise following the wastage of the ice sheets (Chapter 4) was rapid in the early Holocene; after *c.*6.8 ka BP a gradually diminishing rate of rise obtains (Figure 4.16). In the mid to high northern latitudes tracts of coastal woodland were submerged and are exposed today at low tide as 'submerged forests'. In coastal sequences transgressive phases are marked by minerogenic marine sediments (clays, silts and sometimes sands). Their deposition is interrupted by regression phases when peat-forming communities extended into former marine environments at times when the rate of sea-level rise slowed, or there were short-term falls within the generally rising trend of Holocene sea level. Reduced marine influence can also be created by local topographic factors, such as the development of coastal barriers of shingle, or sand bars or by isostatic uplift in those areas once covered by ice (p. 117). Holocene sea-level rise has submerged sites (Figure 5.11), transformed the original setting of many others (see Figure 1.3) and given rise to extensive wetlands which offer exceptional conditions for the preservation of environmental evidence and artefacts (Coles and Coles, 1989, 1996).

Figure 5.11 Kernic, Brittany, France: a Neolithic tomb exposed at low tide on the beach (photo Martin Bell)

The Baltic

The Late Pleistocene and Holocene of the Baltic has a particularly complex history reflecting the interplay between retreating ice, eustatic sea-level rise and isostatic uplift (Björck, 1995b). Stages in the development of the Baltic saw very different sets of environmental conditions and salinities (see Figure 4.21). This was the dynamic context for the colonisation of Scandinavia (Larsson, 2003). During the Allerød, environments in the early stages of vegetation succession would have carried substantial herds of herbivores and highly favourable conditions for colonisation. Holocene transgression of low-lying areas created many islands and inlets, forming an extensive ecotonal coastal fringe. Innumerable islands forming archipelagos in the northern Baltic were in some cases only suitable for specialised seasonal exploitation, but as isostatic readjustment proceeded, larger land masses were created which were able to support permanent settlement (Robertsson *et al.*, 1996). Isostatic readjustment gives rise to a stepped sequence of former shorelines in which the settlement patterns and economies of particular emergent stages can be identified.

A marked transgressive stage between *c.*9 and 8.1 ka BP probably results from the collapse of the Laurentide ice sheet (p. 121); this involved a rate of rise in sea level of up to 2.3 m per century which must have been highly perceptible to contemporary communities (Christensen *et al.*, 1997). This was also a time of marked cultural change from the Maglemosian to the Kongemose *c.*8.3 ka BP, and from *c.*7.5 ka BP there are coastal shell middens, some very substantial, especially those of the late Mesolithic Ertebølle culture (Andersen, 2000). Although shells are highly conspicuous, fishing, other marine resources and terrestrial resources made a greater contribution to the diet. Evidence for the coastal exploitation strategies of the later Mesolithic has been greatly enhanced by recent discoveries on the route of the Storebaelt bridge/tunnel, between Fyn and Zealand, Denmark, where beautifully made woven Mesolithic fish baskets and dugout canoes have been found at sites such as Halsskov Fjord and Margrethes Naes (Pedersen *et al.*, 1997).

In southern Denmark, Holocene eustatic sea-level rise exceeded isostatic uplift (Figure 5.12), with the result that Mesolithic coastal sites are found below present sea level which gives rise to exceptional preservation of decorated wood paddles and other organic artefacts at Tybrind Vig (Andersen, 1985). North of the 'tilt line' shown on Figure 5.12 isostatic uplift has exceeded eustatic sea-level rise, creating raised Mesolithic shorelines. The shell midden site of Ertebølle is on the raised shore of the Limfjord (Andersen, 1995; Figure 5.13). In this area a number of middens show compositional change from a predominantly oyster midden in the Ertebølle period to mainly cockles in overlying Neolithic midden layers (Andersen, 2000), clear evidence of fluctuating oyster populations on offshore banks to which these coastal communities had to adapt. This reflects episodic fluctuation in oyster populations in offshore banks

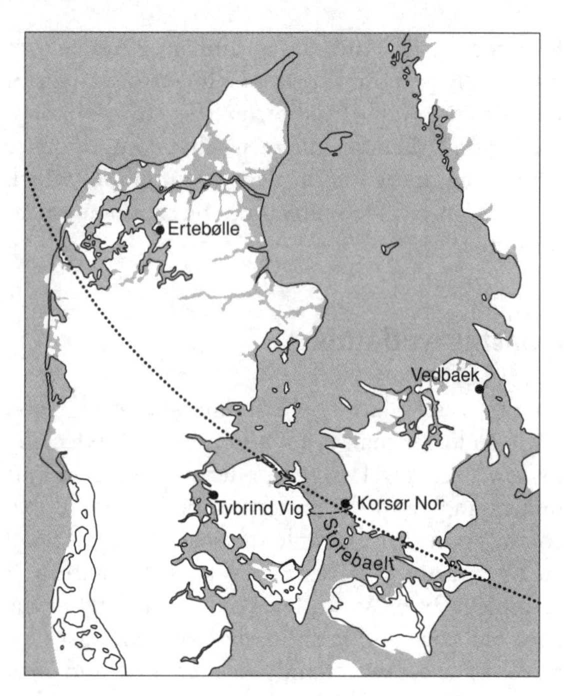

Figure 5.12 Denmark showing the maximum extent of the Holocene transgression in the later Mesolithic and the 'tilt line' separating submerged coasts to the south-west and isostatically raised shores to the north-east (after Andersen, 1995, Figure 1, with additions)

Figure 5.13 Excavation of the shell midden at Ertebølle, Denmark (photo courtesy of Søren Andersen)

as a result of changing marine conditions. Notwithstanding these changes, sites continued to be occupied, sometimes for 1–1.5 ka. The richness and diversity of artefact assemblages and biological resources exploited on some late Mesolithic sites indicate sedentary settlements as does the presence of many burials at sites such as Vedbaek and Korsør Nor (Denmark) and Skateholm (Sweden). Analysis of human bone isotopic composition (p. 46) demonstrates a high proportion of marine food, reinforcing the view that these communities lived year-round on the coast (Richards *et al.*, 2003b).

'Doggerland', the Netherlands and north Germany

The Holocene evolution of the North Sea is increasingly well understood as a result of carefully selected sea-level index points, the precise tidal relationships of which have been established using a range of sources, pollen, Foraminifera, diatoms, etc. We have already seen how Shennan *et al.* (2000b) have modelled the palaeogeography of the North Sea in steps of 1 or 0.5 ka between 11 and 6.8 ka BP (see Plate 4.1). In the early Holocene what is now the North Sea was a terrestrial landscape which has been called 'Doggerland' with extensive tracts of sandy landscapes and wetlands joining Britain to continental Europe (Coles, 1998). In the early Holocene human communities would have had to constantly migrate and adapt in the face of progressive inundations. The Straits of Dover were breached *c.*9.3 ka BP. This separated the British and continental populations and must account for some of the differences in the material culture and ways of life between these two areas in the later Mesolithic. By 6.5 ka BP the last vestiges of dry land in Dogger Bank were submerged.

The Holocene transgression crossed the present coast of the Netherlands *c.*7.9 ka BP and marine influence spread over much of this low-lying country, about half of which is at, or below, present sea level and has been subject to Holocene sedimentation. The sedimentary sequence is known in considerable detail and can be closely correlated with archaeological evidence (Louwe Kooijmans, 1974; Fokkens, 1998). Figure 5.14a is a reconstruction of the Netherlands landscape *c.*4.9 ka BP illustrating the main Holocene environments represented by distinct sedimentary units. To seaward is a dune sand barrier which is first attested from *c.*6.1 ka BP, behind this are tidal flats, extending inland to salt marsh and inland of this peat and in the river valleys minerogenic fluvial sediments (Berendsen and Stouthamer, 2001). The boundaries between these sedimentary units varied spatially through time as a result, particularly, of transgressive and regressive marine tendencies. That gave rise to the interleaving of sediments which is shown in the section through the Holocene stratigraphy of the Netherlands in Figure 5.14b.

Within this vast wetland many of the earliest settlements were on small areas of slightly higher ground provided by Pleistocene river dunes (donken) or levees along river edges. Such sites are generally small and wetland-focused, as exemplified for instance by major excavations of Mesolithic and Neolithic sites at Hardinxveld or the Neolithic sites at Hazendonk and Swifterbant (Louwe Kooijmans, 1987; 2003). Throughout the Neolithic, this wetland economy remained only semi-agrarian with some domestic animals and cereals, since hunting, fishing and fowling were still an important part of the economy and some sites were only seasonally occupied.

Peat growth became more extensive and from the middle Bronze Age to the early medieval period settlements are absent from the peatlands. In the Bronze Age there was a reduction in the range of wetland resource exploitation and a narrowing of the subsistence strategy. This was a period of settlement on higher clay areas and sandy creek ridges during which communities adopted a rather rigid mixed farming system (Louwe Kooijmans, 1993). Perhaps as a consequence they became more vulnerable to the effects of environmental change than was the case with Neolithic or Iron Age communities which exploited a more diverse range of settlement locations and wetland resources.

A distinctive series of regressive (peat) and transgressive (clays) can be traced up the coast of continental Europe, and the regressive phases are particularly associated with greater concentrations of settlement on the wetland (Louwe Kooijmans, 1980). However, this is not invariably the case;

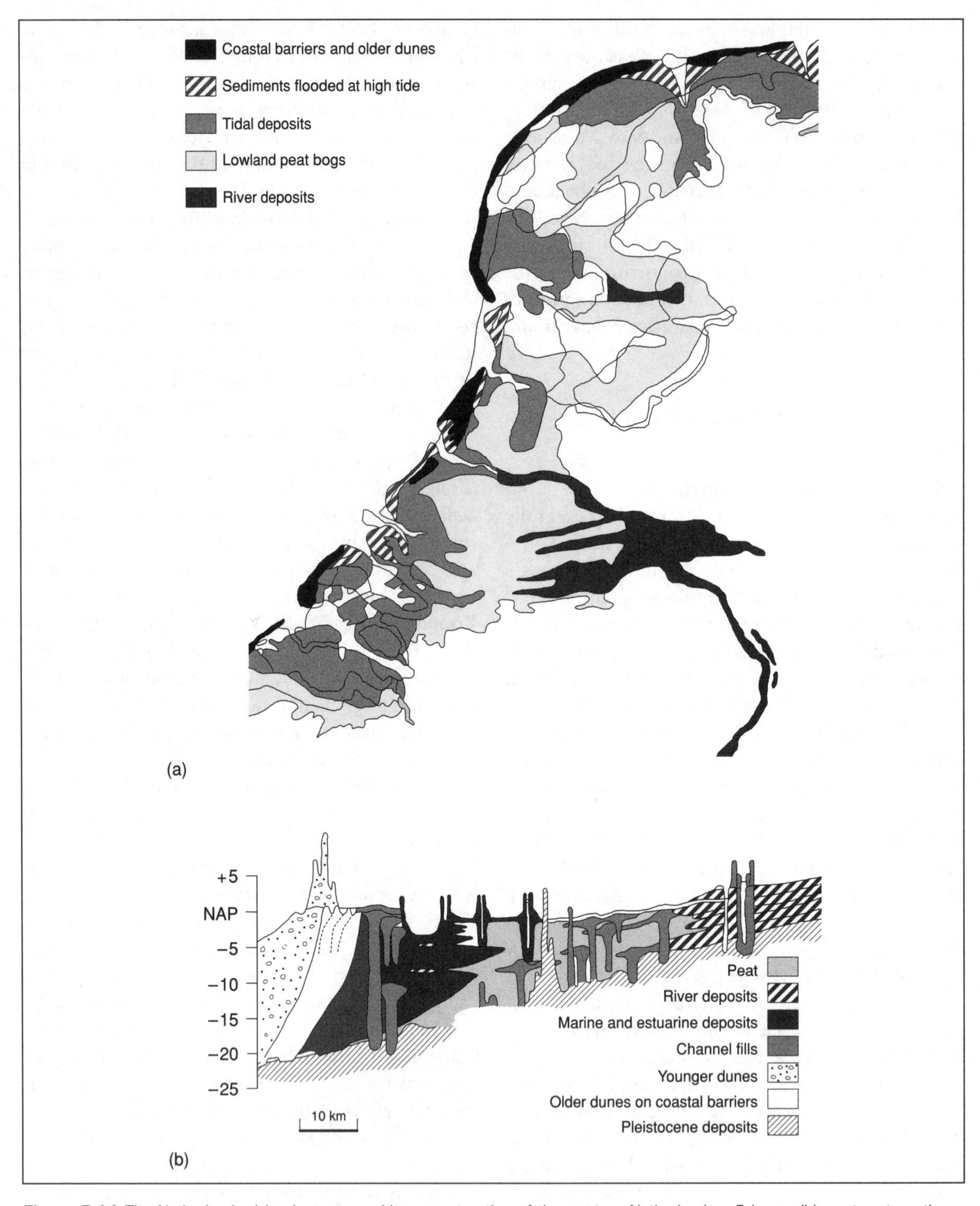

Figure 5.14 The Netherlands: (a) palaeogeographic reconstruction of the western Netherlands *c*.5 ka BP; (b) west-east section through the Rhine/Meuse delta (after Louwe Kooijmans, L.P., 1987, Neolithic settlement and subsidence in the wetlands of the Rhine Meuse Delta of the Netherlands in *European Westlands in Prehistory* edited by J.M. Coles and A.J. Lawson, Figure 139 and 140. By permission of Oxford University Press)

sometimes transgressions widened inlets and extended the system of creeks giving rise to better drained surfaces and conditions suitable for settlement, as during the Iron Age in the Assendelver Polder (Brandt *et al.*, 1987; Louwe Kooijmans, 1993). During the Iron Age some communities also began to improve drainage by digging ditches and by the late Iron Age the technological innovation of hollow logs with flap valves allowed ditches to drain through dams into creeks at low tide (Rippon, 2000, Figure 31).

On the exposed coast of the northern Netherlands and Germany the earliest settlements are on the raised levees along river banks, then as salt marshes accumulate there is evidence of short-term occupation by seasonal pastoralists. By 2.5 ka BP communities in the Netherlands learnt to cope with permanent occupation of this environment by the creation of artificial settlement mounds (terps) which raised settlement above the highest spring tide inundation (Fokkens, 1998). In north Germany (Figure 5.15) settlement extends onto the salt marsh *c.*2 ka BP, and mounds appear *c.*1.7 ka BP (Petzelberger, 2000). The waterlogged wooden buildings at Feddersen Wierde, Germany, are an example of this type of settlement mound set within marshes subject to flooding. In situations so vulnerable to environmental change it is not surprising that the history of settlements is often interrupted. In the third century AD there is often a break in settlement in the Netherlands and from the fifth century AD in Germany when a transgression occurred. This is also the period of the great post-Roman migrations and the economic changes consequent upon the collapse of the Roman world. It is thus difficult to disentangle the effects of transgression from wider social influences. Landscape abandonment at this time occurs very much more widely than just the coastal wetlands, emphasising the contribution of social factors (Rippon, 2000).

From about the ninth century AD onwards a collective and ambitious strategy developed which enabled communities to cope with the constantly changing nature of these coastal environments. Banks were created to keep the sea out, initially low banks which were effective only during summer

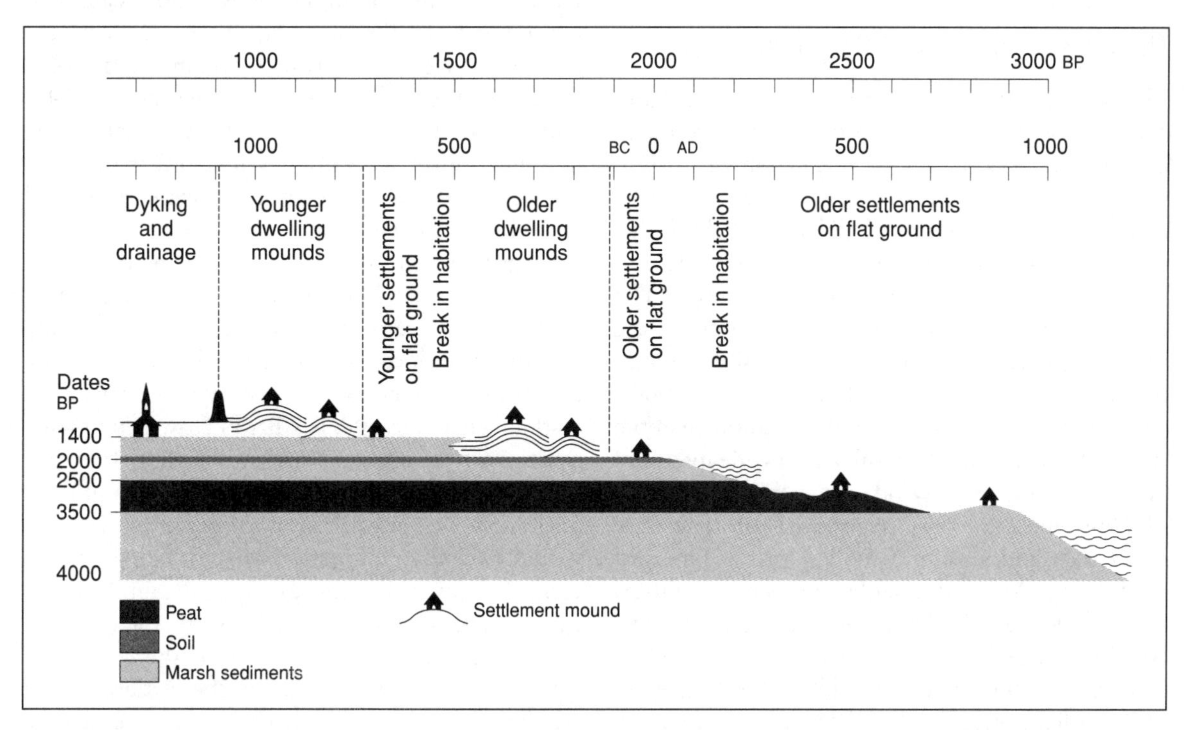

Figure 5.15 North German coastal marsh area: a schematic diagram showing the sedimentary development and occupation history including dwelling mounds (Wurts) (after Behre, 1995; Petzelberger, 2000)

periods of reduced tidal range, but later in twelfth and thirteenth century AD Germany (Figure 5.15) winter banks which allowed these wetlands to be used for a greater range of activities year-round (Rippon, 2000).

Despite an increasing sea defensive capability these environments remained vulnerable to dramatic change as a result both of extreme coastal events and the effects of human activities. The Zuyderzee was formed by a major marine incursion during a storm surge in AD 1170. Along the coast of the Netherlands low older dunes had formed in prehistory and within these occupation sites, fields and ploughmarks are often found, as on the west coast of Jutland in Denmark (Figure 2.18). In the early medieval a marked change took place in the Dutch dune belt with the development of tall parabolic so-called Younger Dunes (Figure 5.14b). Deposition of these began in a period of warm and dry conditions, the early medieval climatic optimum *c.* AD 800 (p. 93; Zagwijn, 1984), and there was extensive dune development during the succeeding Little Ice Age (p. 94).

The Severn Estuary and Somerset Levels

The Severn Estuary in south-west Britain (Figure 5.16a) is a funnel-shaped estuary with expanses of now reclaimed Holocene salt-marsh silts round its margin. To its south are the Somerset Levels, comprising a coastal dune barrier and behind this estuarine silts and then peats, a smaller-scale version of the sequence seen in the Netherlands. The estuary and levels together comprise one of the most intensively studied areas of coastal wetland in Britain (Coles and Coles, 1986; Bell *et al.*, 2000). The Severn has the second highest tidal range in the world (14.8 m), with the result that a wide intertidal area is exposed at low tide, revealing a sequence of Holocene sediments (Figure 5.16b). The base is an old land surface with a forest submerged *c.*7.8 ka BP. Mesolithic settlement occurs on drier islands such as Goldcliff and the wetland edge. An abundance of estuarine resources is attested by footprints in laminated silts (Aldhouse-Green *et al.*, 1992; J.R.L. Allen, 1997). These include deer, aurochsen, birds, canids and humans, adults as well as children, showing that coastal exploitation involved all sections of the Mesolithic population (Figure 5.17). Mesolithic activity on the wetland seems to have been seasonal, associated with hunting, fishing and obtaining raw materials.

From *c.*6.8 ka BP the silts alternate with peats, representing transgressive and regressive phases. In areas more remote from marine influence, such as the Somerset Levels, successive peat-forming plant communities developed: reed peat, fen woodland and raised bog and, within these, many wooden trackways of Neolithic and Bronze Age date. The Sweet Track is the most sophisticated and earliest of these, dated dendrochronologically to 3807/06 BC (Hillam *et al.*, 1990). It was a raised walkway linking areas of dry ground at a time when plant macrofossil and insect evidence shows that flooding was occurring. Inundation of two dendrochronologically dated coastal submerged forests at Stolford and Woolaston occurred at about the same time (Morgan *et al.*, 1987). Construction of this, or other, trackways might be perceived in various ways: an act determined by an environmental change, such as flooding, or a coping strategy whereby social communications were created, or maintained, despite flooding. In the earlier Neolithic there were many trackways, including some hurdle walkways (Figure 6.14); then between 4.9 and 3.4 ka BP few tracks and after that in the middle Bronze Age numbers increase (Coles and Coles, 1998). Some, but not all, were associated with flooding events.

There is an absence of settlement in the Somerset Levels wetland until the later Iron Age. Settlements were on the dryland ridges and sandy islands, their locations occasionally hinted at by the convergence points of trackways. A Bronze Age wetland edge settlement at Brean Down comprised a series of occupations on stabilisation horizons within coastal sand dunes (Figures 5.16b and Plate 5.2). This site lies at the northern edge of what would have been a large area of salt marsh and its economy combined the grazing of animals with the extraction of salt from sea water (Bell, 1990).

In the Severn Estuary itself there are no trackways and few sites within the wetland in the Neolithic and early Bronze Age. In the middle Bronze Age *c.*3.5 ka BP there is a marked change, with the establishment of seasonal settlements on

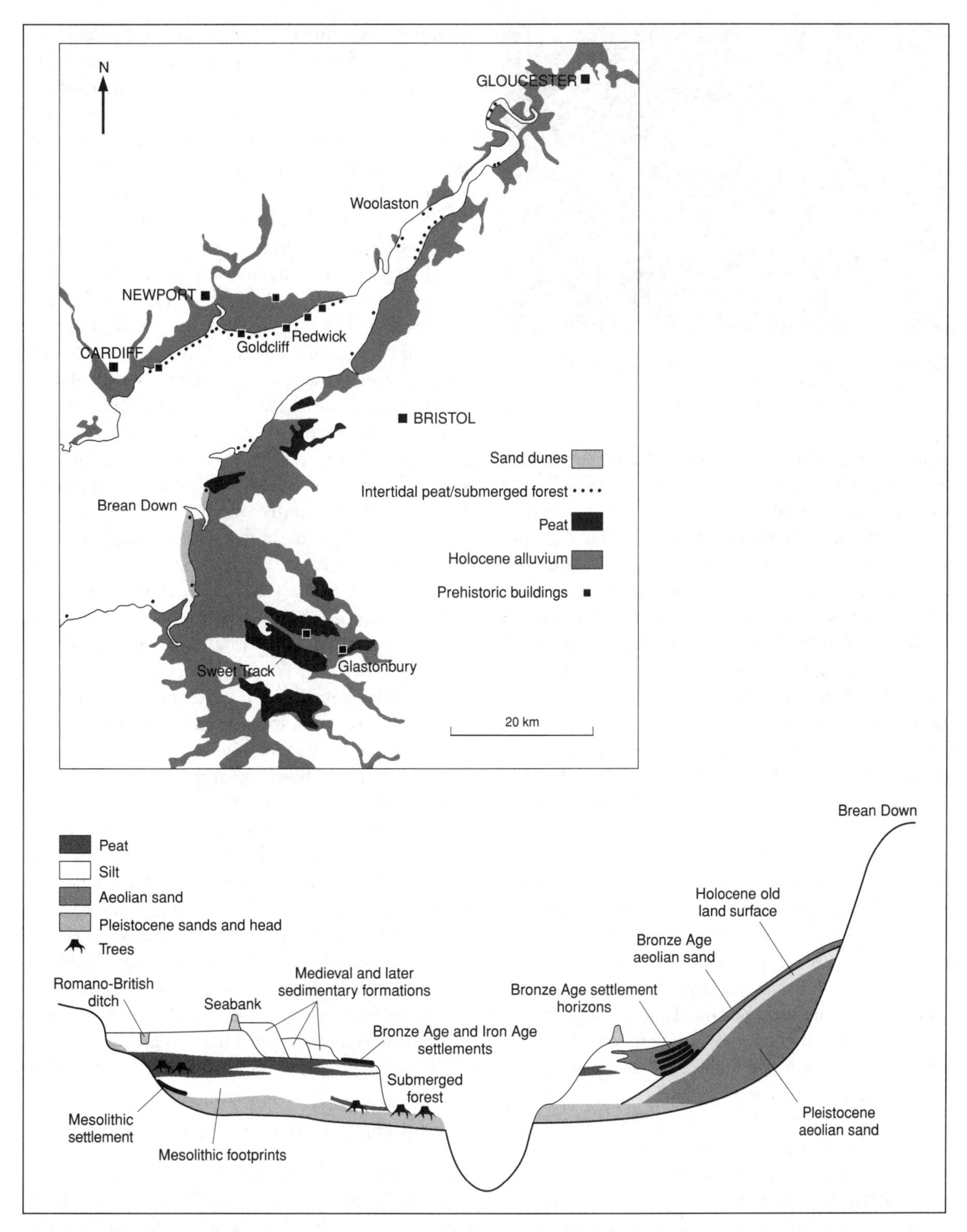

Figure 5.16 Holocene sediment sequence in the Severn Estuary showing the main sediment types and types of archaeological context in (a) plan and (b) schematic section; ((b) adapted from J.R.L. Allen, 1997, Figure 17a)

Figure 5.17 Mesolithic human footprint at Goldcliff East, Wales (photo Edward Sacre)

peat surfaces often at the edge of salt marsh (Bell *et al.*, 2000). Settlements of rectangular buildings at Redwick and Goldcliff were seasonally occupied in spring and summer during the Bronze Age and Iron Age. Innumerable hoofprints indicate they were associated with cattle pastoralism. Dendrochronology shows the wood for one building at Goldcliff (Figure 5.18) was cut in April/May 273 BC, supporting the idea of spring occupation and providing a precise date for marine transgression onto a peat surface at that point, since artefacts and footprints are here interleaved at the peat–silt transition. However, dated structures from the surroundings show that the transgression in the wider area occurred over a century, and probably more. It was, therefore, a process rather than an event. There is increased flooding and marine influence in the late Bronze Age and early Iron Age in the Somerset Levels. These coastal changes may have marked the cessation of dune deposition at Brean Down and marine influence increased more widely in the estuaries of southern Britain (Long *et al.*, 2000).

The first settlement in the Somerset Levels is over a millennium later, *c.*2.2 ka BP, at Meare on the margins of a raised mire where clay mounds were made. Some probably carried tents and represent seasonal activity because the site is likely to have been too wet in winter (J. Coles, 1987). Here scattered communities came together on the wetland for animal grazing, craft production and the exchange of diverse material culture, suggesting that their purpose was both socially and environmentally determined. Glastonbury 'lake village' has a comparably rich material culture but is a site of contrasting character (Coles and Minnitt, 1995). It was occupied from *c.*2.2 to 2 ka BP and at its height was a settlement of 13 round houses on clay mounds, occupied year-round by about 200 people living within a palisaded enclosure. This lay on a slightly drier patch of fen woodland with open water and reedswamp around, located in wetland perhaps partly for defensive reasons. In its final phase the settlement became smaller, probably seasonal, and was abandoned as a result of flooding.

Thus in prehistory this area of south-west England and Wales exhibits diverse kinds of wetland utilisation: some activities were based at dryland sites but involved trackway construction for access to wetland resources; some settlement was seasonal and Glastonbury was settled year round. Activity was essentially opportunistic, attuned to the natural rhythms and cycles of the estuarine environment. From the Romano-British period in the second century AD people/landscape relationships in this area took on a very different form with drainage ditches and sea defence banks (Rippon, 1997, 2000). Initial reclamation seems to have been Roman military-inspired, but it later became more widespread enabling year-round settlement and a wider range of activities in landscapes previously flooded for part of the year.

In medieval and later times there were extensive sea defence embankments and peats no longer formed. Continuing episodic coastal instability is manifest as erosion phases alternating with the deposition of salt-marsh sedimentary units which form a sequence of step-like salt-marsh surfaces, many of them fossilised from continued marine increments by successive sea defences. For this reason surfaces to seaward, which are still receiving marine sediment, are higher than those previously reclaimed (Figure 5.16b; J.R.L. Allen, 2000).

Coastal change: causes and consequences

Rapid Lateglacial and early Holocene sea-level rise created a constantly changing context for human activity. In some areas such as Doggerland this was mainly a story of progressive marine advance. In Scandinavia the situation was made complex

Figure 5.18 Iron Age building at Goldcliff, Wales constructed 273 BC with a reconstruction showing its environmental settling in a raised mire undergoing marine transgression (photo Lesley Boulton; reconstruction drawing Steven Allen)

by the interplay of isostasy and eustasy creating dramatic changes in Baltic geography. The coastal changes are of both human and biological significance. The isolation of Britain after *c.*9.5 ka BP contributes to its floral and faunal impoverishment by comparison with continental Europe and may help to explain the differences between the later Mesolithic in the two areas. Ireland has been separated from the rest of the UK since the Lateglacial (p. 121). Many thermophilous taxa are thought to have reached Ireland not by land but across the water of St George's Channel. This helps to explain the impoverished nature of Irish biota, with 30 per cent fewer plant species than Britain (Preece, 1995). Red deer and aurochsen did not reach Ireland until after 4 ka BP (Woodman *et al.*, 1998) and this is widely considered to account for the contrasting strongly riverine and coastal focus of the Irish Mesolithic by comparison with the deer-centred economy of England in the Mesolithic.

In the later Holocene sea level continued to rise, but more gradually, and within this overall rising trend there are transgressive and regressive phases. Some affected widely separated areas at approximately the same time (Tooley, 1985; van Geel *et al.*, 1996; Long *et al.*, 2000), and *may* reflect climatic changes, particularly over shorter timescales which could have given rise to short-term sea-level falls or reductions in the rate of rise. However, local factors also contribute; the breaching and reformation of coastal bars and the initiation of erosion cycles may have set in train long-term coastal change.

From the perspective of humanly relevant timescales, events will often have been more important than gradual processes. Historical sources highlight the human significance of events, for example catastrophic flooding in the Severn Estuary and Bristol Channel documented in AD 1606–7 (Boon, 1980). Extreme events of this kind have a return period of hundreds of years. However, such events are likely to have been more frequent during periods of unstable and transitional climate such as the beginning of the Little Ice Age (p. 94).

As our understanding of the subtleties of environmental change and chronology improves, so earlier assumptions that regressions equal human activity and transgressions equal abandonment are seen as simplistic. Transgressions sometimes provided suitable contexts for settlement, as at Assendelver, Netherlands, and often for a range of other activities, such as grazing and fishing in the Severn Estuary. It is likewise evident that different communities responded to the processes and affordances in different ways, creating regionally contrasting ways of life. Environmental factors are one of a series of contributory factors. In parts of the Netherlands where vast areas of wetland were distant from dryland, exploitation demanded settlement of slightly higher levees and dunes within the wetland itself. In the Somerset Levels, by contrast, nowhere is far from ridges, or islands, of dry ground; consequently the archaeology is dominated by trackways until the later Iron Age when the wetland was settled, presumably from deliberate cultural preference, allied perhaps to defensive considerations in the case of Glastonbury. Cultural circumstances will often have been as important as environment in deciding what happened. Rich coastal environments existed in Britain and Ireland but in the Mesolithic these areas lacked the large coastal middens with evidence of sedentism which are found in Scandinavia. Markedly seasonal patterns of coastal exploitation remained prevalent in Britain into the Iron Age, perhaps providing a direct link with seasonal upland/lowland, wetland/dryland patterns of exploitation which are attested by the ethnohistorical record in western Britain and Ireland during medieval times. In Ireland the role which fishing played in medieval and later coastal areas has been particularly well documented by the discovery of many fish traps in the Shannon Estuary (O'Sullivan, 2001) and Strangford Lough (McErlean *et al.*, 2002).

Contrasts have been noted in the dates at which communities moved from opportunistic exploitation of coastal environments, when conditions were suitable, to their large-scale transformation by draining and the making of sea banks. The earliest such evidence is in the Severn Estuary in the second century AD, around 1000 years before these practices were adopted in the Netherlands and Germany north of the Roman world. As communities insulated themselves against the effects of regular (tidal and seasonal) coastal change and against events on a predictable scale so it became possible for them to settle permanently on wetland and to carry out a wider range of activities. In so doing they became more vulnerable to high-magnitude, low-frequency major flood events. Human activity also occurs in the context of longer-term secular trends (p. 8). Land-use strategies in one period based on drainage and reclamation may not be sustainable in the long term.

Geohazards

In the following sections a range of geohazards is considered: volcanoes, earthquakes, tsunamis and extraterrestrial bodies. Each is of special interest for research on the relationship between environmental change and human communities; they are precisely definable in terms of time, spatial extent and magnitude. In specific cases, however, each of these parameters has often proved controversial, providing an object lesson in the difficulties of achieving precision in the palaeorecord and interpreting the effects of known events on human communities.

Volcanism

Close to the volcano the effects are lava flows and mud flows, the latter resulting from the mobilisation of ash by groundwater liberated by the eruption and by rainfall. Affecting a wider area up to at least tens of kilometres are pyroclastic flows, very fast moving, and extremely hot masses of gas, ash and liquid which are destructive of all in their advance. Affecting a far greater area is the deposition of ash (tephra). The ash from an individual eruption is identified on the basis of its distinctive mineralogy and geochemistry. Tephras thus form isochrones: time parallel marker horizons which are the basis of tephrochronology and facilitate the correlation and dating of palaeoenvironmental sequences and archaeological occupations (Haflidason *et al.*, 2000). In areas much affected by volcanism this technique is of great value, as in Iceland and parts of the 'Pacific ring of fire' – New Zealand, New Guinea, Japan, the west coast of North America and Mesoamerica. Beyond the areas of visible ash layers minute tephra shards can identify eruption horizons in peats and lake sequences (Hall and Pilcher, 2002). We have already seen how the occurrence of tephra shards and acidity peaks in ice cores enables eruptions to be identified within these sequences (Figure 3.27). Dendrochronological records also produce evidence of periods of anomalous tree growth, some of which correspond to known eruptions. These are manifest in frost-damaged rings in the bristlecone pine sequence of California (LaMarche and Hirschboeck, 1984) and periods of little or no growth in bog oaks in Ireland (Baillie and Munro, 1988; Baillie, 1995). There are, however, tree growth anomalies which do not correspond to evidence for contemporary eruptions (Buckland *et al.*, 1997).

The effect of volcanic eruptions on present-day complex and literate societies has been documented in detail (e.g. Blong, 1984). However, the instrumental record and recent historical sources only cover a limited time range and a limited number and magnitude of events affecting only a sample of types of social context. If we are to expand our understanding of the effects of volcanism then we must draw on the palaeorecord from archaeology and on the limited but highly informative ethnohistorical record.

For the eruption of Vesuvius in AD 79 there is direct information regarding its effects, including an eyewitness account by Pliny the Younger. Pompeii and villas in its vicinity were buried by ash and Herculaneum by a mud flow (Jashemski, 1979). The towns were never rebuilt, although by the third century AD there was some resumption of activity around the affected area. Most of the population escaped and had time to remove many of their valuables. Around 2000 people died, many of fume asphyxiation; some have recently been discovered crowded in the harbour at Herculaneum in their attempts to escape. Some of the dead are represented by evocative casts of their dying bodies in the solidified ash.

The most intensively studied and controversial case of the effects of volcanism on past civilisation is Thera (Santorini) in the Aegean (Manning, 1999). An eruption of this island volcano during the Bronze Age is estimated to have produced 30–40 km^3 of ejecta. Tephra from the eruption is found in Crete, mainland Turkey and in cores in the Aegean, east Mediterranean and Black Sea (Figure 5.19). The date of the eruption remains highly controversial. It buried horizons containing artefacts of mature Late Minoan IA style which traditionally, on the basis of typological links to Egyptian chronology, are dated about 3450 BP (1500 BC). However, the ice core and tree-ring records contain no evidence of a major eruption around this time, whereas there is a very clear signal more than a century earlier variously dated in ice cores: Dye 3, 3594 BP (1644 BC); GRIP, 3586 BP (1636 BC); and GISP 2, 3573 BP (1623 BC). The Californian and Irish tree-ring records both display a major anomaly in 3576–8 BP (1626–1628 BC; LaMarche and Hirschboeck, 1984; Baillie and Munro, 1988).

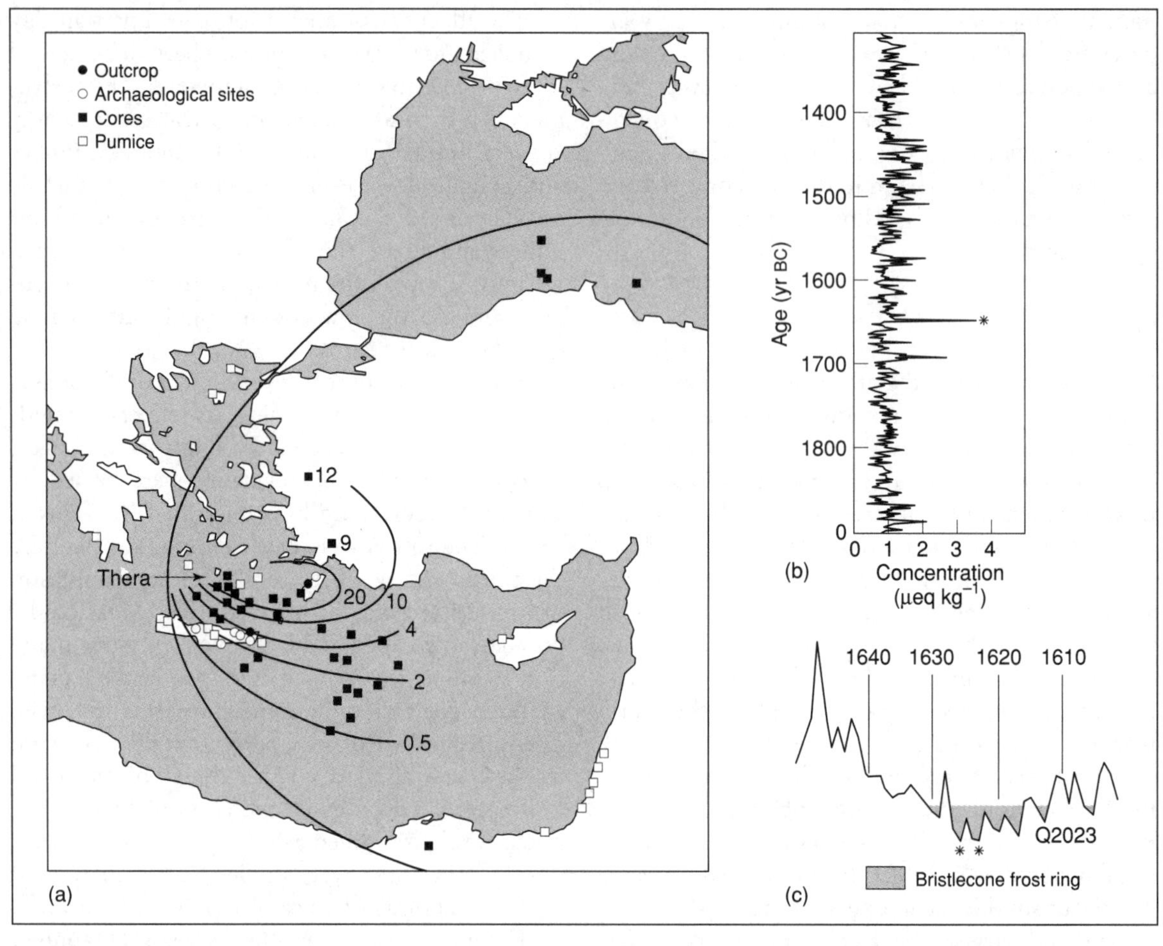

Figure 5.19 The Thera eruption: (a) the distribution of tephra (after McCoy and Heiken, 2000a, Figure 13); (b) acidity curves in Dye 3 core, Greenland; asterisk marks the possible eruption (reprinted with permission from *Nature*, after Hammer, C.U. *et al.* 1987, The Minoan eruption of Santorini in Greece dated to 1645BC, *Nature*, **328**, pp. 517–19, Figure 2. Copyright 1987 Macmillan Magazines Ltd.); (c) tree-ring width in northern Ireland, shaded areas are particularly narrow rings, the asterisk the narrowest ring in the tree's life. The shaded box marks the Californian bristlecone pine frost rings which are tentatively correlated with the Thera eruption (reprinted with permission from *Nature*, after Baillie, M.G.L. and Munro, M.A.R., 1988, Irish tree rings, Santorini and volcanic dust veils, *Nature*, **332**, pp. 344–46, Figure 1. Copyright 1988 Macmillon Magazines Ltd.)

There is also a pronounced growth anomaly in a floating juniper tree-ring sequence which is dated by AMS radiocarbon wiggle-matching (p. 54) to about this time (Kuniholm *et al.*, 1996; Manning *et al.*, 2001). Radiocarbon evidence from short-lived samples buried by the eruption also indicate a most likely date range in the seventeenth century BC. There is, therefore, evidence for a major environmental perturbation variously dated between 1645 and 1623 BC which is consistent with Manning's (1999) revised archaeological chronology for the east Mediterranean and may represent the eruption of Thera (Hammer *et al.*, 2003).

Thera itself was devastated by the eruption. The Bronze Age site at Akrotiri was buried by ash, its buildings surviving up to a height of 6 m, their walls decorated with magnificent Minoan frescos. The population escaped from the excavated area with some portable valuables but may have succumbed to the huge tsunami which accompanied collapse of the Thera caldera (McCoy and Heiken, 2000a). Many writers have followed Marinatos

(1939) in arguing that this eruption impacted on Crete, 120 km away, causing the demise of Minoan palatial civilisation. However, a study on Crete found Thera pumice but without evidence of tsunami impact and no sign of a significant impact on Cretan vegetation (Bottema and Sarpaki, 2003). Furthermore Cretan abandonment layers are characterised by Late Minoan IB pottery which is considered to be 50–100 years later than the Late Minoan 1A pottery buried by ash on Thera. The Minoan collapse is now thought more likely to have been due to the social and economic tensions within this early state society.

Iceland is a highly volcanically active area with many known eruptions and tephra layers which are of great importance as isochrones in environmental and archaeological research (Haflidason *et al.*, 2000). Following Norse colonisation in the AD 870s (p. 179) the effects of volcanism on human communities have been well documented. An example is the eruption of Hekla in AD 1104 which caused the desertion of farms, mostly within 25 km of the eruption, but one was 70 km away; none was ever resettled. Synthesising the effects of six major eruptions since AD 1104 Thorarinsson (1971) has shown that, in all but the most severely affected areas, impact has been short term: 10 cm of tephra led to abandonments of up to 1 year, 15 cm of 1–5 years and 20 cm periods of some decades.

Tephra of Icelandic origin reached the British Isles and minute shards from at least 20 separate eruptions are recorded in peat bogs and lake sequences in Ireland (Hall and Pilcher, 2002), Scotland and England (Dugmore *et al.*, 1995), where they are proving to be extremely valuable in the dating and correlation of environmental sequences. Blackford *et al.*, (1992) showed that Hekla 4 tephra was coeval with the mid-Holocene pine decline in Scotland. However, in Ireland the pine decline is not associated with this tephra, nor do the earlier Lairg tephras (*c.*6.6 ka BP) produce any detectable vegetation response in pollen sequences (Hall *et al.* 1994; Hall, 2003). The eruption of Laki in AD 1783 is documented as having led to breathing problems for people, death of fish and damage to plants in Britain (Dodgson *et al.*, 2000).

A major acidity peak in the Greenland ice core *c.*1100 BC may be a Hekla eruption which, it has been suggested, is represented by a particularly narrow group of Irish tree rings from 1159 BC (Baillie and Munro, 1988). It has been argued that this eruption led to upland abandonment in Scotland (Burgess, 1989). But the settlement pattern changes are not well dated and the effects claimed are much greater than would be anticipated from what we know of the effects of historic period eruptions (Buckland *et al.*, 1997). Dated palynological sequences from marginal upland sites in Scotland show considerable chronological and spatial variation in the pattern of cereal growing, pastoral activity and reductions in land-use intensity (Tipping, 2002), which does not support the emphasis that some writers have given to volcanism as a cause of abandonment. If this factor did have an effect it is more likely in terms of tephra deposition reinforcing an existing tendency of upland soils towards acidification (p. 215; Grattan and Gilbertson, 1994).

An ethnohistoric perspective is particularly valuable in helping to understand how societies very different from our own may respond to hazard. In parts of Papua New Guinea oral legend tells of a 'time of darkness' lasting several days in which plants and animals were severely affected and in about 30 per cent of communities people died, mostly in collapsing houses, or of starvation. One group even turned briefly to cannibalism as a consequence (Blong, 1984). Blong's (1982) study represents a fascinating integration of evidence for human perception from ethnohistory and the earth science record. Distributions of legends correspond closely with those of tephra and constitute a reliable, albeit oral, historical record of a previously unknown eruption on Long Island in the Bismarck Sea probably between AD 1630 and 1670. The eruption produced 30 km^3 of ejecta which covers an area of 80 km^2 at depths of between 16 and 1.5 cm. The legends record differing attitudes to the eruption; some communities regarded it as very harmful and felt they had never been the same since. Surprisingly, by contrast, about 20 per cent of legends record beneficial effects, a time of great plenty brought about by the fertilising properties of the ash. Some saw the eruption as presaging great cultural achievements, one group even invoked earth magic to secure a repetition! Positive and negative perceptions of the eruption do not correlate with the evidence of its purely physical effects and may relate to local social and cultural context. An archaeological study of the effects of volcanic eruptions in parts of Papua New Guinea during the last 6000 years showed that two severe eruptions resulted in abandonments of a millennium or more, but there was no simple correlation between eruption severity and the duration and nature of human response (Torrence *et al.*, 2000). Abandonment duration reduced through time as landscapes became more intensively utilised.

The recent and past effects of eruptions may also be compared around the Aleutian Peninsula, Alaska. Mount Katmai erupted in 1912 producing 28 km^3 of ejecta, creating a devastated lunar-like landscape today used for training space crews. Tephra deposition produced serious short-term effects but where there was less than 10 cm of tephra ecological recovery was rapid. A return to previous ways of life was possible over most of the area within 20 years (Dumond, 1979; Riehle *et al.*, 2000). This was because the affected area had not been densely settled by Aleut communities, while kin networks and exploitation territories extended beyond the devastated area. People moved away as refugees, later to return. The ethnohistorically recorded effects are consistent with archaeological evidence from 15 prehistoric eruptions over the last 4500 years (Dumond, 1979).

Some thin ash layers occur within cultural layers and may have caused little disturbance. Others separate cultural phases, suggesting periods of abandonment sufficiently long for cultural change to occur. Interestingly, however, the cultural changes extend well beyond the affected area and the conclusion is that, despite the severity and frequency of volcanic eruptions around the Aleutians, this was not a major driver of cultural change.

Tsunamis and earthquakes

Tsunamis, seismic sea waves, are created when water is displaced by large-scale movement of the seafloor by earthquakes, submarine slides, volcanic eruptions, or, theoretically at least, by water displacement as a result of explosions or impacts of extraterrestrial origin (p. 173). Volcanic eruption of Krakatau in 1883 generated a 36 m high tsunami which killed 36 800 people in the coastal communities of Java and Sumatra (Carey *et al.*, 2000). The effects registered on harbour tide gauges worldwide. The earthquake-prone areas of the Pacific rim are at particular risk and in Japan there are written records of more than 150 tsunamis.

The Holocene sedimentary record of the American north-west coast (Oregon, Washington and British Columbia) has a well-documented tsunami and earthquake record related to the archaeology of the area. Offshore coring reveals marine turbidite evidence for 13 great earthquakes since the time when Mount Mazama ash (Plate 2.8) covered this area *c.*7.8 ka BP (Jacoby *et al.*, 1992). Lesser earthquakes occur on this coast about once every 400 years. In this area there are submerged forests buried by salt-marsh sediments as a result of sudden earthquake submergence, in some cases associated with sandy tsunami sediments (Atwater, 1987). One event which originated in this area on 27 January 1700 is precisely dated because of its historically recorded impact in coastal communities on the opposite side of the Pacific in Japan. Dendrochronological dating of drowned forests in the north-west also points to a date close to AD 1700 (Atwater and Yamaguchi, 1991). A number of archaeological sites on the north-west coast were buried by sediments from tsunamis and as a result of earthquake-related land displacement (Hutchinson and McMillan, 1997; Minor and Grant, 1996). Tsunamis and earthquakes both figure prominently in the stories and oral histories of first nation (native American) communities. The AD 1700 event is thought to be recorded in the 264-year-old oral history of the Nuu-chah-nulth people of Vancouver Island which describes shaking land and high waves that destroyed the community of Pachema Bay (Reksten, 2001). Even so, many of these communities had their most important settlements of great cedar wood houses in highly vulnerable positions on low terraces just above the beach; an example is Ozette, Washington State (Figure 5.20; p. 243; Ames and Maschner, 1999).

On the opposite side of the world an early tsunami deposit is especially well documented at Storegga off the Norwegian coast (Bondevik *et al.*,

Figure 5.20 Ozette, USA: site of former settlement on terrace above the beach. The settlement was buried and preserved by a mudslide in c.AD 1750 (photo Martin Bell)

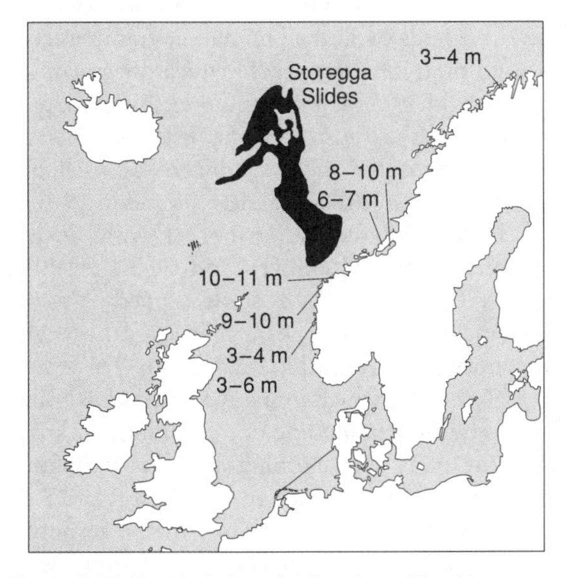

Figure 5.21 The North Sea showing sites with evidence of the Storegga Tsunami (reproduced from Bondevik, S., Svendsan, J.I., Johnsen, G., Mangerud, J. and Kaland, P.E., 1997, The Storegga tsunami along the Norwegian coast, its age and run up, from *Boreas*, **26**, pp. 29–53, www.tandf.no/boreas, by permission of Taylor & Francis AS)

1997; Dawson and Smith, 1997). This was a great submarine sediment slide which involved an area of 20 000 km^2 at *c*.8 ka BP; this impacted on the Norwegian coast and the east coast of Scotland, where marine sand layers were deposited within otherwise fine estuarine sediments or peats on 11 dated sites (Figure 5.21). On some sites there is evidence that the event marks the end of Mesolithic occupation phases (Dawson *et al.*, 1990; D.E. Smith, 2002). Coles (2000b) has suggested that this event contributed to the inundation of Doggerland.

Earthquakes alone can have a hugely devastating effect on property and communities. Many of the great classical cities and sites of the Mediterranean show evidence of earthquake damage and many were rebuilt after successive quakes as testimony to the resilience and determination of communities (McGuire *et al.*, 2000). Akrotiri on Thera suffered a major Bronze Age earthquake but the city was repaired and rebuilt on the rubble, only to be totally destroyed later by volcanism (p. 169; McCoy and Heiken, 2000a). The city of Lisbon was rebuilt after the catastrophic earthquake and tsunami which destroyed it on All Saints' Day AD 1755, an event variously rationalised by John Wesley as a punishment for sin and by Kant and Rousseau as an act of nature (Duff, 1998).

Asteroids, comets and meteorites

The environmental and social effects of extraterrestrial impacts are both hotly debated and perhaps the finest illustration of the difficulties we face in evaluating the effects of high-magnitude, low-frequency events. Asteroids are minor planetary rock bodies, meteorites are smaller rock bodies which impact on earth and comets are bodies of dusty ice. Events at the Cretaceous/Tertiary boundary (65 ma), when there was mass extinction of dinosaurs and many other taxa leading to the ascendancy of the mammals, emphasise the importance of this factor. A worldwide boundary layer of clay, soot and glassy microtectites is rich in iridium and other platinum group metals which are found in extraterrestrial bodies (Alvarez *et al.*, 1980). This catastrophic impact is thought to have produced a crater 200 km in diameter in the Yucatan Peninsula, Mexico. A lesser magnitude event occurred much more recently. On 30 June 1908 at Tunguska, Siberia, a 50 m diameter meteor exploded 8.5 km above the earth with energy equivalent to 50 megatons of TNT (Lewis, 1996). This airburst created no crater but all trees within

20 km were blown down outwards from the centre and charred by thermal radiation from the fireball. The explosion was heard over 1 million km^2 and atmospheric effects were seen as far as England. The area was sparsely populated, but all 20 people within 50 km were injured and 2 died, along with around 1000 reindeer.

A recently identified small crater at Abruzzo, Italy, is radiocarbon-dated to the fifth century AD and is thought to have inspired a local oral legend, which was passed down for 1.6 ka, of a star falling to earth at the time of this area's conversion from paganism to Christianity (Santilli *et al.*, 2003).

Baillie (1999) suggests that some of the narrow ring events recorded in the tree-ring record may reflect extraterrestrial impacts. Some happened at times when there is no evidence of major volcanic eruptions in the ice-core record (Figure 3.27). Baillie identified several events which he thinks may have been caused in this way. He points to tantalising early historic records which mention bright lights or, in the case of eastern sources, dragons in the sky. His assumption is that these observations are environmentally descriptive rather than metaphorical. Where such an event is described as occurring at, or around, the time of a cultural change such as the beginning and end of the Shang dynasty, or the Justinian plague, his assumption is that the environmental change in part caused the cultural change but that relationship remains to be established.

Historical records assembled by Lewis (1996) refer to 148 impacts that resulted in property damage, injury or death. The earliest is a biblical reference in *c.*1420 BC (Joshua 10: 11), 8 of the events predate AD 1000 and 63 per cent occurred since AD 1900. This pattern strengthens Lewis's premise that because of the patchy distribution of scientific observers, and the fact that many objects will fall in the ocean, the recorded incidence of these highly stochastic phenomena will be much less than the actual incidence. Modelling is beginning to provide more information; impacts up to 100 megatons are likely to take place around once in 2 ka, those of 1000 megatons can be expected every 15 ka (Lewis, 1996). It takes little imagination to see how even a modest event, of Tunguska proportions, might have had devastating consequences if the asteroid fell in a particularly populous area or where it generated giant tsunamis (Kerr, 2003). At such times even the most sophisticated and technologically advanced coping strategies would be as nothing compared to the devastating impact of such events.

Coping with the cold

Sometime between 20 and 40 ka BP, certainly by the Upper Palaeolithic, *Homo sapiens* had developed forms of clothing, shelter and economy which meant they could cope with conditions of extreme cold (Gamble, 1993), thus making possible the colonisation of Siberia, Beringia and the Americas (Figure 5.3). In the Holocene there are some remarkable finds of clothed human bodies preserved by conditions of extreme cold which document the ways in which human communities have adapted to some of the most extreme environments on earth. Perhaps the most significant of such discoveries is the so-called 'ice man' found in 1991 at Hauslabjoch on the Italian border with Austria (Spindler, 1993; Fleckinger and Steiner, 1999). Discovered in the high snow-capped Alps at 3210 m, it is the body of a man aged *c.*46 who died in the later Neolithic *c.*5.3 ka BP. This was a stochastic event which in many respects has transformed our understanding of the Alpine Neolithic. With the body was clothing and a toolkit for a self-sufficient existence in the high Alps (Plate 5.3). These remains provide a record of the resources he used, what he did and where he came from (Borlenschlager and Oeggl, 2000). His clothing was beautifully adapted to the extreme cold. Shoes had outers of bearskin and deerskin and an inner shoe net stuffed with grass for insulation. He wore leggings of domestic goat hide, a leather loincloth and a belt of calf leather. Over these was a coat of goat hide and then a cape of plaited grass and for his head a bearskin cap. So well adapted is the clothing, that the grass cape is of a type worn by shepherds in the high Alps into recent times.

Associated artefacts included a bow, quiver and arrows, a hafted copper flat axe, a rucksack and two sewn birch bark boxes; one had contained live embers wrapped in maple leaves. A pouch contained some flint implements, a bone awl, fungus used as tinder and a small dagger. The charcoals, wood, pollen and other plant macrofossils found on the clothing reflect the successive altitudinal zones of woodland, mixed oak, then mature spruce forest, through which he had passed. His last meal was of cereals (einkorn and barley), vegetables and meat. Spikelets from cereal processing were found in his cloak and ember container (Dickson *et al.*, 2000).

His clothing and artefacts are reflective of a transitory way of life in which hunting is combined with access to domestic plants and animals. The combination is seen also in the economy of the Neolithic Alpine lake villages (p. 205). He appears to have come from, or had recent contact with, agricultural communities in the valleys south of the Alps. Mosses (e.g. *Neckera* spp.) with the body and hop hornbeam (*Ostrya carpenifolia*) from the intestines, grow to the south (Dickson *et al.*, 1996). The state of digestion of pollen in his intestine indicates that he had been in these valleys as little as 12 hours before his death. The most likely settlement suggested by this botanical evidence is Juval in the lower Senales Valley (Figure 5.22). His final journey seems, from a combination of botanical evidence, to have taken place in late spring or early summer, probably June.

Almost a millennium earlier, by about 6.1 ka BP, high Alpine pastures near by on the north side

Figure 5.22 The Senelles Valley in north Italy identified as the likely home settlement from which the iceman began his journey across the Alps (photo Martin Bell)

of the Alps show pollen evidence for changes in plant communities which are interpreted as pointing to the onset of seasonal transhumance to high pastures. These pastures have traditionally been exploited by communities from south of the Alps (Borlenschlager and Oeggl, 2000). One favoured hypothesis is that the ice man was involved in transhumance. In his shoulder was a flint arrowhead. This wounded and exhausted man may have frozen to death in a storm and have quickly become encapsulated in ice for 5250 years until an exceptional ice melt, a manifestation perhaps of recent global warming (Chapter 9), led to his discovery (see cover photograph).

Evidence of adaptation to extreme cold also comes from North America. In the south Yukon there are patches of ice, which in 1997 melted for the first time in 8 ka, again perhaps because of global warming (Greer, 2002). Melting snow patches have produced masses of hunting equipment: atlatles (spear throwers), darts, bows and arrows dating between 7.3 ka and 90 BP. These were hunting sites for caribou, which at times congregate on the snow patches. Their well-preserved dung is abundant in the melting snow and itself represents a source of long-term palaeobotanical evidence. A melting glacier in northern British Columbia has also produced the body of a male, late teens or early twenties, who died *c.* AD 1415–45; he is known as Kwäday Dän Ts'ı̄nchi, in the local first nation language 'long ago dead person' (Beattie, 2000; Pringle, 2002). Associated artefacts included a hat of woven plant fibres, a robe of squirrel fur and other artefacts, some of more recent date suggesting perhaps a pattern of artefact loss on a frequented route. Pollen from the body indicates a widely travelled individual journeying from the coast to high mountains.

From Qilakitsoq, near Disko on the west coast of Greenland (Figure 5.23), comes another remarkable find of human bodies and clothing (Hart Hansen *et al.*, 1991). Two graves contained six women and two children who were interred around AD 1475 and were preserved by a form of freeze drying in a rock tomb below an overhang. Their clothing was wonderfully adapted to extreme cold, two layers everywhere, but also allowing freedom of movement vital to survival. Outer parkas

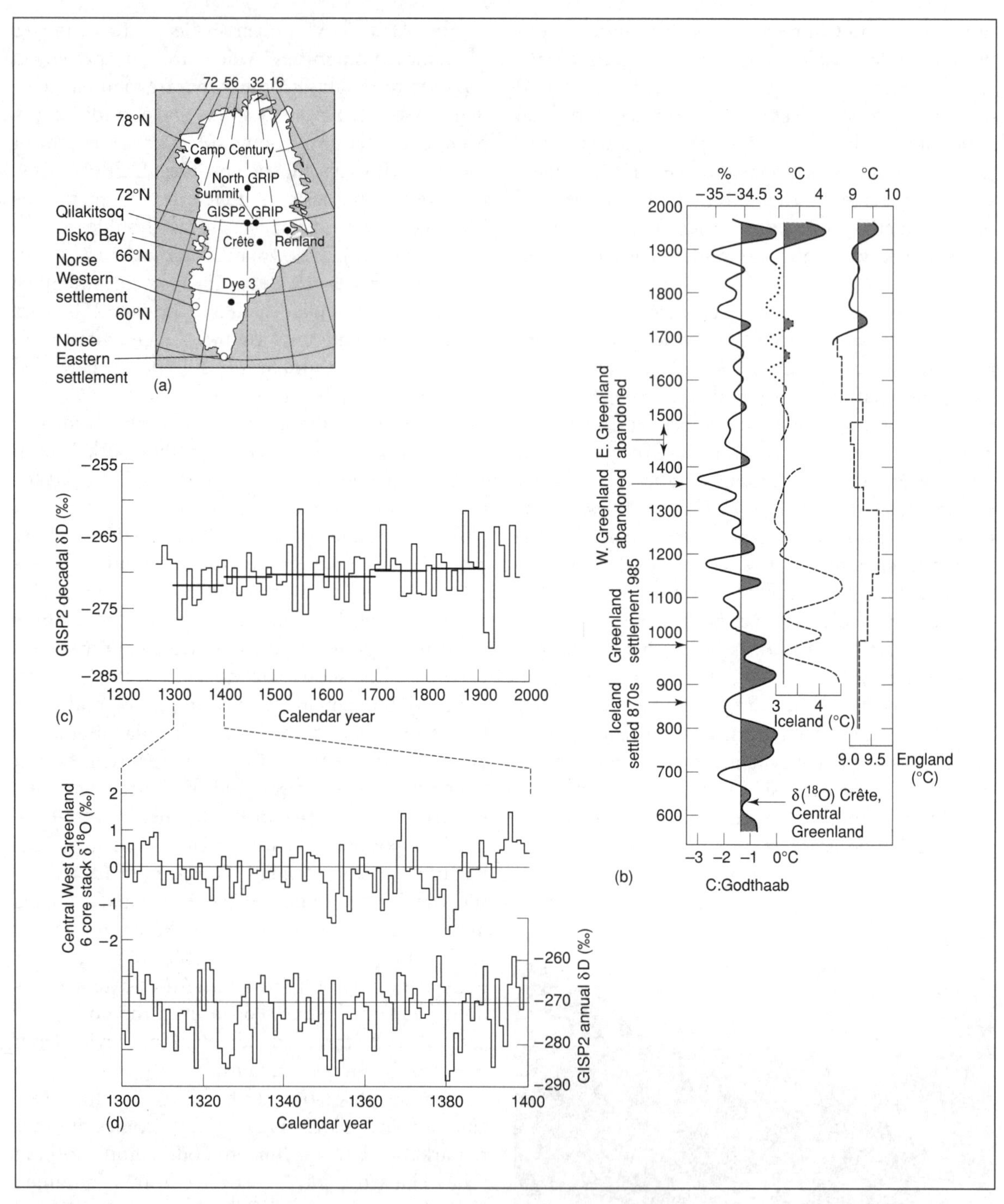

Figure 5.23 Greenland: (a) the locations of ice cores and key Norse and Inuit settlements (adapted from Hart Hansen *et al.*, 1991) (b) Oxygen isotope curve (^{18}O from Crête, Central Greenland compared to temperatures in Iceland and England. Dashed curves based on indirect evidence, the broken Icelandic curve on sea-ice data and the solid Icelandic and English curve on instrumental records (reprinted with permission from *Nature*, after Dansgaard, W. *et al.*, 1975, climatic changes, Norseman and modern man, *Nature*, **255**, pp. 24–28, Figure 2. Copyright 1975 Macmillan Magazines Ltd.); (c) GISP 2 decadal isotopic signal AD 1270–1986; (d) detail of isotopic record for the fourteenth century AD (c and d after Barlow *et al.*, 1997, Figure 4)

were of sealskin, inners of birdskin, trousers and socks of caribou and sealskin. Boots of waterproof skin were lined with lyme grass. The clothing is that worn by the Inuit into the twentieth century, demonstrating that such sophisticated adaptation occurred early. Isotopic evidence for diet shows that 75 per cent was from marine sources; mineral grains and the raw materials used provide clues to the pattern of movement in their nomadic lifestyle. The detail provided by these eight individuals is all the more valuable because it can be compared with a rich ethnohistoric and pictographic record of Inuit life, clothing and belief systems. Hart Hansen *et al.* (1991) make the point that the adaptation of the Inuit is not primarily genetic and physiographic, but rather the result of establishing a society and a way of life which can provide protection from extreme cold. The Inuit would have had to cope with cold conditions for much of the year, to do so offered the opportunity to exploit locally rich, but seasonally highly variable resources, fish, seals, whales, caribou, etc. Fluctuations in climate and population level are likely to have functioned as a pump[2], encouraging occupation of more extreme environments and better forms of adaptation. The Italian 'ice man' and Kwäday Dän Ts'ī̄nchi suggest, however, that this is not the entire story: the maintenance of social communications was also of great importance, especially perhaps when it involved the crossing of hostile environments to acquire exotic objects and arcane knowledge of high status. Powerful stimuli perhaps to take to the limits that remarkable human capacity to adapt to environments of extreme hostility which on occasions proved fatal.

The Little Optimum and Little Ice Age

We have already seen that the warm period of the Little Optimum between *c.* AD 800–1200, followed by the Little Ice Age *c.* AD 1350–1850, has been identified from a wide range of proxy sources (Chapter 3; Grove, 2002; Fagan, 2000). The latter part of the Little Ice Age and the subsequent warming since *c.*1850 are also reflected in instrumental records. These provide the opportunity for comparison of proxy, historical and instrumental sources of evidence for environmental changes and their effects on human communities. As the chronological precision of the ice-core record has increased, so it has become apparent that the century-scale secular climatic episodes are actually an oversimplification of a more complex pattern with significant spatial and temporal variability (Barlow, 2001). On a decadal scale there are periods with relatively low temperatures but these alternate with warmer phases, both highly relevant to the decision making of human communities.

The Little Optimum was a time of audacious Norse seafaring and colonisation in the North Atlantic which took people to the climatically hostile northern lands of Iceland, Greenland and west to the first European discovery of America. In Greenland and Iceland there is an exceptionally rich palaeoenvironmental record preserved by extreme cold and waterlogging in middens and occupation horizons within buildings, providing insights into the economies and ways of life of both Norse and Inuit communities (Arneborg and Gulløv, 1998; Appelt *et al.*, 2000). Norse colonisation in both areas gave rise to marked changes in vegetation communities, a 'landnam' phase[2] with charcoal and a substantial increase of grazing indicator species. The biogeographical effects of Norse communities were especially marked with plants and particularly synanthropic invertebrates introduced from more southerly areas with transported goods and fodder (Buckland *et al.*, 1998).

Vinland and Greenland

Norse sagas describe the discovery of Vinland and a settlement has been excavated at L'Anse aux Meadows in Newfoundland, an area within the subarctic climatic zone. It comprised three domestic units with characteristic Norse artefacts and evidence of ironworking, unknown to native Americans (Ingstad, 1977). No certain evidence of agriculture was found but plant macrofossils of butternuts (*Juglans cinerea*) and vines, from which they named their discovered land, shows they had contact with areas at least 800 km to the south, where the northern limits of these taxa occur (Clausen, 1993). The settlement is thought to have been established

Figure 5.24 Excavation of the Gården under Sandet, Norse site in Greenland, note the preservation of wood and organics under permafrost conditions. Excavations by the National Museums of Denmark and Greenland (photo Paul Buckland)

c. AD 1000 and occupied for only a few years. A factor in the abandonment of this precocious colony is likely to have been the extremely attenuated communication lines to Scandinavia.

Greenland had been settled by the Norse earlier in *c.* AD 985. There were two settlements, an eastern settlement of 200 farms and a western one of 100 farms (Figures 5.23 and 5.24). Cereal production was not possible in this harsh climate and the economy was based largely upon cattle, sheep and goats, kept for dairy products rather than meat. Most animals, particularly the farms' horses and breeding stock, would have had to be kept in byres through the long winter with the result that communities were highly dependent on a good hay harvest using both twigs and seaweed to supplement their grass fodder. Agriculture was supported by the hunting of seals and caribou. The maintenance of trade links with Norway was based particularly on walrus ivory and polar bear skins, many obtained from hunting grounds at Disko Bay 800 km north of the western settlement. At first the Norse colonies flourished. They had their own bishop from AD 1125 to 1378 and there was a major programme of church building from AD 1125 to 1300. Then a marked decline set in resulting in the abandonment of the western settlement *c.* AD 1356–60. The circumstances are mysterious; a party from the eastern settlement found the other settlement deserted, with domestic animals wandering around. The last official contact with the eastern settlement was in AD 1408 and communication finally ceased in the second half of the fifteenth century.

The climatic downturn of the Little Ice Age certainly contributed to the demise of these settlements. Oxygen isotope records from the Crête ice core and particularly the GISP 2 ice core shows that Greenland temperatures in the fourteenth century were the lowest in 700 years (Figure 5.23). A period of particularly low temperatures occurred between AD 1349 and 1356 when the western settlement was deserted (Figure 5.23c and d; Barlow *et al.*, 1997). There was no such clear climatic downturn at the time the eastern settlement was abandoned between AD 1450 and 1500. Modelling shows that these communities were particularly vulnerable to the effects on fodder production of a run of 3–5 bad years out of 10. The timing and spacing of climatic perturbation are seen to be more important than the magnitude. Animal bone assemblages were highly fragmented, every scrap of meat and marrow was consumed by these communities under stress. The final days of the western settlement farm at Nipaitsoq were in late winter; hares and ptarmigan had been caught but desperation led to the slaughter and butchery of domestic cattle and finally the hunting dogs (Buckland *et al.*, 1996). Changing patterns of trade may also have contributed to the demise of these communities; the Norse world increasingly obtained its ivory from Africa and Russia, and trade with Greenland ceased in AD 1368. Their plight was further exacerbated because Norse communities were competing for some food resources with Inuit who had become established on the outer fringes of the western settlement area by *c.* AD 1300. Conflict between Norse and Inuit communities is historically documented at the time of the western settlement abandonment. Yet there is no evidence of human remains in abandonment horizons; the last occupants, no longer able to cope, may simply have departed.

Contacts between the two cultures there certainly were; there are Inuit carvings of Norsemen, Norse artefacts in Inuit settlements and occasional Inuit artefacts in Norse contexts. Yet the Norse seem to have learnt nothing from the 'Skraellings' as they called the native people. They never adopted Inuit styles of harpoons and hunting technology which would have enabled them to cope with periods of greater sea-ice cover in the fourteenth century. Fishing made a limited contribution to the economy of the Greenland settlements (McGovern, 1992). Evidence from bone isotopic chemistry does, however, show increased use of marine resources through the period of Norse settlement (Arneborg *et al.*, 1999). Especially striking is the evidence of Norse clothing, wonderfully preserved by permafrost in the cemetery at Hergolfsnaes, eastern settlement (Hovgaard, 1925). The permafrost, and thus favourable preservation conditions, became more extensive as the Little Ice Age progressed. Costumes of wool, not fur, followed the European fashions of the fifteenth century, including fashions which apparently never reached Norway and demonstrate some continuing contact with more southerly areas of Europe which may have continued after the last known contact with Norway (Nordtorp-Madson, 2000). This clothing has no hint of special adaptation to the Greenland climate, nor of any borrowings from Inuit dress. The contrast between this and the remarkably sophisticated and adapted clothing of the Inuit mummies from Qilakitsoq (p. 175) is striking. It may have been the very conservatism of the Norse communities, bolstered by the strong influence of the church, which sealed the fate of these communities as the only established outpost of European society (so far!) to be extinguished. Hostile circumstances led these people not to adapt but to reinforce their own identity as Europeans.

Iceland

Iceland was first settled by the Norse towards the end of the ninth century AD. Pollen evidence for a landnam occurs below tephra of a composition found in Greenland ice cores in AD 871. Substantial areas became rapidly colonised by farms and in Norse times barley was grown, although this had declined by the thirteenth century AD and ceased by the sixteenth century. During the Little Ice Age glaciers were advancing and several Norse farms were engulfed (Grove, 2002). The eighteenth century was a time of particular hardship with failed harvests, declining fish catches, and an impoverished, sometimes starving population with an increasingly dispossessed underclass retreating from the most severely affected areas in the north. Several additional factors exacerbated the problems of the Little Ice Age. There were a succession of major volcanic eruptions, the most severe that of Laki (Lakagiga) in 1783 (Thorarinsson, 1979) which poisoned pastures and led to the death of many animals. A fifth of the human population died of famine and at this time the abandonment of Iceland was seriously considered. The effects of volcanism and climate are not easy to separate in Iceland. Many glaciers are on volcanoes and eruptions cause glacier surges and catastrophic floods (*jokulhlaups*) from ice-dammed lakes (p. 133). Evidence that short-term climatic deteriorations of 1–3 years follow volcanic eruptions has also been noted (p. 100), and acidity profiles from the Greenland ice cores (Figure 3.27) show that much of the period of the Little Ice Age is marked by higher than normal levels of volcanic activity (Haflidason *et al.*, 2000). As if the effects of severe weather and volcanism were not enough, the weakened population suffered a smallpox outbreak in AD 1709 which wiped out a third of the population.

A proxy climatic record is provided by data on the extent of sea ice (Bergthórsson, 1969) which was calibrated by Lamb's (1972) early instrumental records from 1846 (Figure 5.23b). The sea-ice curve is in general agreement with the Crête isotopic curve, indicating a period of low temperatures from AD 1600 to 1900. At times sea ice was so extensive that polar bears came ashore. Such extensive ice had a major effect on trade and communications. Biological evidence from excavated settlements such as Reykholt (Buckland *et al.*, 1992) points to somewhat squalid domestic conditions with an abundance of synanthropic insects, human head lice and lice from wool processing. The Svalbard settlement shows how communities coped with hard times; in the Little Ice Age there is evidence for the use of a wider range of resources, a shift from cattle to caprines, the abundant use of fish and the exploitation of polar bear and seals, both associated with extensive pack ice.

Britain and European mainland

Upland and moorland areas of Britain show extensive evidence for medieval and later agriculture in the form of deserted farms and fields sometimes with cultivation ridges, well beyond the present limits of economic crop growing. Clearly there is an altitudinal limit as to how far the long-term boundary of crop growing may be pushed. Given the natural variability of climate some holdings will be **marginal**, i.e. viable for crop growth for a period but not over the more variable conditions of an extended timescale. The palaeoenvironmental record helps to establish where these boundaries lie, but also highlights the rather problematic nature of the marginality concept. This is now seen not so much as a rigid boundary, but one which is relative to particular sets of environmental conditions and economic and social structures which influence the ability of communities to cope (Coles and Mills, 1998).

Investigations of upland desertion in the Lammermuir Hills of south-east Scotland (Parry, 1978, 1981; Parry and Carter, 1985) involved the development of models based on present-day crop–climate relationships and on the early instrumental record from AD 1659, assembled by Manley (1974) from which the changing frequency of crop failure was modelled. Recurrent crop failure was perceived as the most likely cause of desertion. Farms above 450 m were likely to have suffered failures in 11 successive years between 1688 and 1698. The modelled changing altitudinal limits of agriculture correlate with the areas of abandoned arable, supporting the climatic hypothesis. The chronology of abandonment has been subject to further critical evaluation 65 km away on the Cheviot Hills where evidence of land-use history is provided by well-dated pollen sequences (Tipping, 1998, 2002). Here agriculture did not cease at the height of the Little Ice Age and on some sites continued right through to the nineteenth century. Eighteenth-century population reduction in this case is seen as relating more to agricultural and social change. The site at Houndtor, Dartmoor, encapsulates these contrasting hypotheses. Beresford (1981) saw climate change as the cause of desertion, whereas Austin (1985) notes deserted and surviving settlements at the same altitude, arguing that contingent factors, in this case perhaps legal constraints, the nature of landholding and individual manorial policy, were more important.

During the Little Optimum vineyards extended 3–5° of latitude further north and 100–200 m higher up hillsides, and the Domesday Book (AD 1086) records 38 vineyards in southern England. The dates of vine harvests, carefully recorded from the late fifteenth century, provide a valuable proxy climate source (see Figure 2.26) and correlate with glacier fluctuations (Chapter 2, p. 48). Even so, Le Roy Ladurie (1972) has shown that a decline in vine growing in the fourteenth century north of Paris is not explicable purely in climatic terms but is reflective of more general economic decline during which vine cultivation ceased to be profitable. Episodes of very severe climate are well documented by historical sources during the seventeenth and eighteenth centuries AD. In France, under the highly conservative agriculture of the *ancien régime*, there was a massive peasant population whose diet was largely bread. Bad harvests were a significant contributor to the social turmoil which led to the revolution of 1789 (Fagan, 2000). In Britain, by contrast, these climatic conditions and poor harvests had less effect, partly because by this time Britain was benefiting from a diversified range of crops and increased productivity, reflecting the early stages of the agricultural and industrial revolutions.

On a local scale the marked periods of storm conditions which are documented during the Little Ice Age had a dramatic impact on the lives of some coastal communities. Settlements, churches and field systems were inundated by dunes along the Atlantic seaboard of Europe from Brittany (Figure 5.25) to Denmark, at various dates between the thirteenth and eighteenth centuries AD (Lamb, 1995).

In the medieval and post-medieval world, even up to the later nineteenth century, hostile weather conditions were often rationalised as evidence of God's wrath with human communities. During the sixteenth century this became particularly associated with accusations of witchcraft which in England and France peaked in the bad winter years of 1587–88. Germany in the 1560s actively debated God's authority over the weather and witches were

Figure 5.25 The church, cemetery and presbytery of Iliz Koz, Brittany, France which together with its village and agricultural landscape was buried by sand dunes in AD 1720–22 (photo Martin Bell)

held responsible for climatic misfortune. Between AD 1580 and 1620, a thousand witches were burnt in the Berne region of Switzerland alone (Behringer, 1999). Earlier, the last recorded Norse contact with Greenland in AD 1408 (p. 178) described a witch burning, evidence perhaps of the social tensions created by the self-destructive course of the Greenlanders' way of life and the part which religious conservatism played.

Notes

1. Population pump. The idea that episodically fluctuating climate, population levels and spatially variable resources could have acted like a pump: periodically drawing in and expelling populations. Such processes may particularly have driven the colonisation of remote areas with markedly patchy and changing resources. They may have contributed to the colonisation of Eurasia and Beringia in the latter part of the Last Glaciation (p. 146) and the Inuit colonisation of the Arctic north and Greenland (p. 177) in the later Holocene.
2. Landnam. The vegetational and other ecological effects which accompanied the initial clearance of land for agriculture. Landnam events were first identified in pollen studies in early Neolithic Denmark (p. 203). Later ecological changes also called landnam have been described at the time of the first colonisation of Iceland and Greenland by Norse communities (p. 177).

6 Cultural landscapes: human agency and environmental change

Introduction

Humans cope with their environment and its constantly changing nature by adaptation and the diverse strategies outlined in Chapter 5. An important part of these strategies is the purposeful alteration of the environment itself. People play an active role in the creation of their own environment to the extent that many parts of the world have been totally transformed by long histories of human activity. Whereas early impact was once thought to be mainly in Europe, it is now evident that before European contact in the Americas there were destructive environmental practices (Butzer, 1996) and similarly in the Polynesian islands (Kirch and Hunt, 1997) and parts of the Amazonian rainforest once thought to be pristine (Heckenberger *et al*., 2003). The concept of people changing the environment has been challenged by Ingold (1993b) because it presents people as external agents, in the words of McGlade (1995) a sort of 'pathology' on nature. In reality people live not on or off the environment but within it (Ingold, 1993b). As argued in Chapter 1, people are themselves key ecological factors.

The tendency to look on human relations with the environment as external or 'pathological' is a natural enough reaction in a world increasingly alarmed by what people are doing. It is, however, a school of thought with problematic consequences. It has sometimes meant that the full range of environmental disturbance factors and the interactions between them are not evaluated. Setting people apart from nature may lead to a tendency to explain environmental changes in *either* natural *or* anthropogenic terms. At times this seems to depend as much on the disciplinary background, or study area, of the researcher as on the specifics of the evidence itself. The historical scientist faces particular problems in trying to infer cause from observed effect: equifinality (p. 50). Trying to distinguish environmental changes due to people from those due to climate is a particular problem in north-west Europe where there is a long history of humans influencing their environment. In North America the problem is less acute because human impact before the last few centuries was less.

In Chapter 1 it was noted that earlier notions of steady and predictable ecological succession towards a stable, and equally predictable, climax have now been eclipsed by a greater emphasis on the dynamism of ecosystems. Contingent (chance) perturbations are seen as important influences on the structure and function of ecosystems (Simmons, 1999). The interplay of a range of disturbance factors gives rise to ecosystems that are patchy and of mosaic character, a concept called **patch dynamics** (Pickett and White, 1985). Another important concept, particularly in the development of later Holocene environments, is **plagioclimax**: an ecological succession which is prevented from progressing to a theoretical climax by environmental factors, of which two of the most important are grazing and human pressure. People and their animals are not, however, the only factors which interrupt succession or disturb environments such as woodland, as Huggett (1995), Simmons (1996) and T. Brown (1997) have emphasised. When considering environmental changes it is therefore important to evaluate a wide range of multiple working hypotheses.

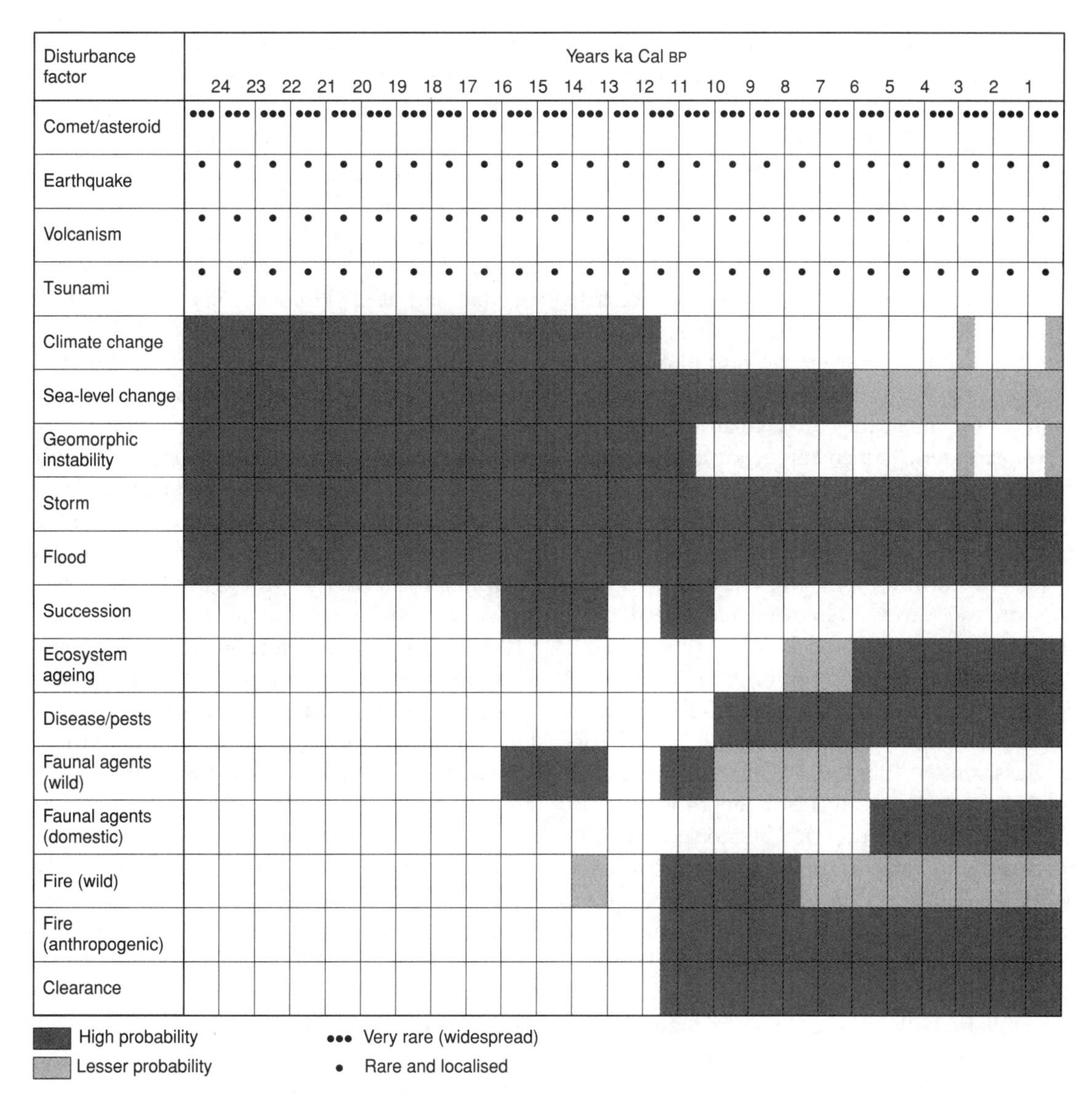

Figure 6.1 List of environmental disturbance factors with suggestions of the likelihood of their occurrence over the last 24 ka BP, based on a British landscape

Figure 6.1 suggests some, but by no means all, of the disturbance factors to consider.

Environmental disturbance factors

The focus of this chapter is on the role of people in transforming the environment. Human capability in this regard has increased through time with a more complex and ecologically demanding economic base, i.e. hunter-gatherer → agricultural → early complex societies → urban → industrial societies. People modify their environment in many different ways: woodland clearance, burning, introduction of domesticates and other taxa and disease, drainage, movement of raw materials for settlement construction and artefact making, creation of monuments, sites and routes, etc. The human modification of landscape is not just about the ecology of plants and animals but also the creation of monuments, paths and areas of distinctive land-use, which have a formative and influencing effect on subsequent patterns of human activity. Later monuments make reference to those of earlier periods and routes once established tend to be perpetuated over time.

While people are undoubtedly a very important cause of environmental disturbance we need to set their activities in the context of other, particularly short-term, disturbance factors. These can be difficult to distinguish from the results of human agency but they are also of interest because, for human communities, they often create **affordances**, i.e. opportunities, settlement or ritual locations, resources.

Of particular importance in the development of patch dynamics are weather-related phenomena, such as storms, floods, etc. The effects of such events on tree throw were well illustrated by a storm in southern England in 1987 (Figure 1.5). The distinctive pits left by uprooting of the root bole are frequently encountered in archaeological excavations. Tree-throw histories in Harvard Forest, Massachusetts and the Appalachian forests of the USA show tree turnover on timescales of 100–300 years (Huggett, 1995; Simmons, 1996: 13), emphasising the importance of storm disturbance in woodland ecology. The openings created would have offered affordances of food and resources for early human communities. Prehistoric artefacts are commonly found in tree-throw pits, presumably used as shelters. Because of the affordances they provide, the openings created by storms are liable to be confused with deliberate human clearance (T. Brown, 1997). Pests and diseases (p. 199) may similarly have been responsible for disturbance, either diseases affecting trees themselves or plagues of defoliating pests.

The significance of faunal agents is most clearly illustrated by the case of beavers (Coles and Orme, 1983). In Canada trees are felled for lodges, dams and food, wetland habitats are greatly extended, creating water bodies, and as these silt up fertile herb-rich, grassy beaver meadows are created on former ponds. Some beaver-dammed lakes have been shown to have existed for millennia. In prehistory beaver (*Castor fiber*) were widely present in England and north-west Europe. Today, with effective conservation, they are again extending their range in France. Here careful documentation of their effects demonstrates that, contrary to earlier opinion, they have a comparable environmental transformatory effect to their Canadian relatives; one stream had 35 dams in 1 km (Coles, 2001).

Vera (1997, 2000) has argued that grazing animals had a significant effect on the woodland of the first half of the Holocene in north-west Europe and eastern North America. He challenges the traditional view of a dense closed woodland and contends that grazers would create a more open woodland more like the wood pasture of recent history. In support of this argument he shows that oak and hazel, which are abundant in many pollen spectra, appear unable to regenerate in closed woodlands today. Using pollen alone it is difficult to evaluate the relative proportion of trees to open areas, as most trees are more abundant pollen producers and the pollen productivity of open taxa was possibly suppressed by grazing. Studies of modern pollen recruitment show that park-like landscapes can produce spectra similar to those which others interpret as closed woodland in earlier Holocene contexts (Groenman-van Waateringe, 1993). However, in areas where grazing fauna was impoverished, for example in Zealand once it became an island after 7 ka BP (p. 211) and Ireland throughout the Holocene, oak and hazel values are also high (Bradshaw *et al.*, 2003). Not just grazing, but the range of disturbance factors considered here would have created a woodland which alongside closed areas would also have had more park-like patches (Svenning, 2002).

An important factor in favour of what might be described as the 'woodland with openings mosaic model' is that all disturbance factors, but especially grazing animals and people, are not evenly distributed across the landscape. They will tend to concentrate in certain places, close to water bodies, at the woodland edge and lineally in those topographic situations that create natural routeways, for instance cols or passes that link the heads of upland valleys, or crossing places of water. Animals will also congregate in particular places in order to escape pests or predators, or enjoy a particular microclimate. In all these places a concentration of grazing is likely to create more open areas and to have prevented regeneration. The effects of grazing will be particularly concentrated in transitional (ecotonal) zones between woodland and open conditions, such as along coasts. This applies to wild animals but particularly domestic animals which were contained in particular areas. A concentration of grazing is likely to prevent any tree regeneration and if continued for the average life of a tree, perhaps 300–400 years, will result in decreasing numbers of scattered trees and then in a totally open landscape. The likelihood is that this unconscious process was one of the most important ways in which open environments were created. The effects of grazing animals are particularly important in preventing regeneration and are likely to account for the longevity of many of the openings represented in pollen diagrams, however they were initially created (Buckland and Edwards, 1984).

Judging by the amount of charcoal that is found in later Quaternary peat and lake sediments, fire was a widespread cause of environmental perturbation. Fire may be wild, naturally caused by lightning strikes. These will particularly affect the boreal forest with highly combustible resin-rich conifers; other particularly susceptible areas will be those with marked seasonal dryness such as the continental interiors and Mediterranean lands. In continental North America the natural incidence of fire has been investigated from annually laminated sediments in lakes and fire scars within tree-ring sequences (Clark, 1988, 1989; Wright and Clark, 1994; Clark and Royall, 1995). Fire incidence is driven by the relationship between episodically drier periods and fuel supply, which is a function of the period since last burn. The result is cyclical patterns, the timescale (called by Clark the renewal function) of which varies according to secular climate change, with, for instance, less frequent but more intense fires recorded during the cooler and moister period of the Little Ice Age (p. 94). Natural fires may be expected in Europe when pine was widespread in the early Holocene boreal woodland, and later in the Neolithic and Bronze Age there is evidence that natural fire was an important part of the ecology of pine woodland in the early stages of mire successions (Figure 8.6; Whitehouse, 2000).

The much more extensive deciduous woodlands of Britain, it is claimed, were almost impossible to burn (Rackham, 2003: 103). Certainly lightning fires affecting more than small patches are rare (Simmons, 1996) and woods failed to ignite under incendiary bombs in the Second World War (T. Brown, 1997). This view is, however, difficult to reconcile with growing evidence of extensive charcoal horizons associated with prehistoric human activity. Perhaps prehistoric communities knew how to read the signs and when to burn under suitable conditions of drought, dryness and wind which may only have obtained every few decades. A particular problem with charcoal occurrence is establishing its source, whether from domestic hearths, natural or anthropogenic fires (Edwards, 1989). These issues may to some extent be resolved by detailed investigation of fire histories including work on various charcoal size fractions (large–local, small–distant), high temporal resolution studies (e.g. annually laminated sediments) as in the work of Clark (1988, 1989), and the specific identification of the material burnt, as in the work of Mellars and Dark (1998).

The natural ageing of ecosystems can also lead to change as illustrated by the glacial–interglacial cycle (p. 124). Soils in the early stages of interglacials have been rejuvenated by weathering, are calcareous and give rise to fertile brown earth soils under deciduous woodland in the mesocratic phase (p. 212). Later with progressive leaching, acidification and podzolisation there is an increase in conifers and heath taxa in the oligocratic phase of late interglacials. That these changes are natural is evident from their occurrence in earlier interglacial phases. However, in the Holocene there is abundant evidence that they have taken place earlier and more extensively as a result of activity by people and domestic animals.

Many of these disturbance factors are clearly closely interlinked. Often one factor increases the sensitivity (p. 8) of an environment to change by another factor. Human activity may increase the susceptibility of animals or plants to disease. Disease outbreaks may be triggered by particular patterns of weather. Grazing animals perpetuate clearings created in other ways. Thus the emergence of extensive areas of cultural landscape in later prehistory is the result of human agency interplaying with a broad spectrum of other factors, natural as well as cultural.

Late Pleistocene extinctions

Geological time has seen a number of marked extinction episodes which, in part, are responsible for the contrasting plant and animal assemblages of different geological epochs. Some species also became extinct at each major Pleistocene climatic transition, but the numbers and geographical spread of extinct species at the end of the last glaciation far exceeded its predecessors. The animals affected were particularly **megafauna,** those over 44 kg in weight. Many were present in the later part of the last glaciation and had disappeared before the earliest Holocene. Paul Martin developed the theory that these extinctions were caused by early hunters. His arguments, together with the counter-arguments of those who emphasise climatic factors, have been rehearsed in four symposia volumes (Martin and Wright, 1967; Martin and Klein, 1984; Mead and Meltzer, 1985; MacPhee, 1999) and reviews by Grayson (2001) and Grayson and Meltzer (2003). Martin (1984) showed that the extinctions were concentrated on land masses where people had settled late, such as the Americas and the Antipodes, and also occurred in the later Holocene on remote islands at approximately the time of initial human settlement (Figure 6.2). Fewer species became extinct in Africa and Eurasia where there has been a long evolutionary development of people/animal relationships.

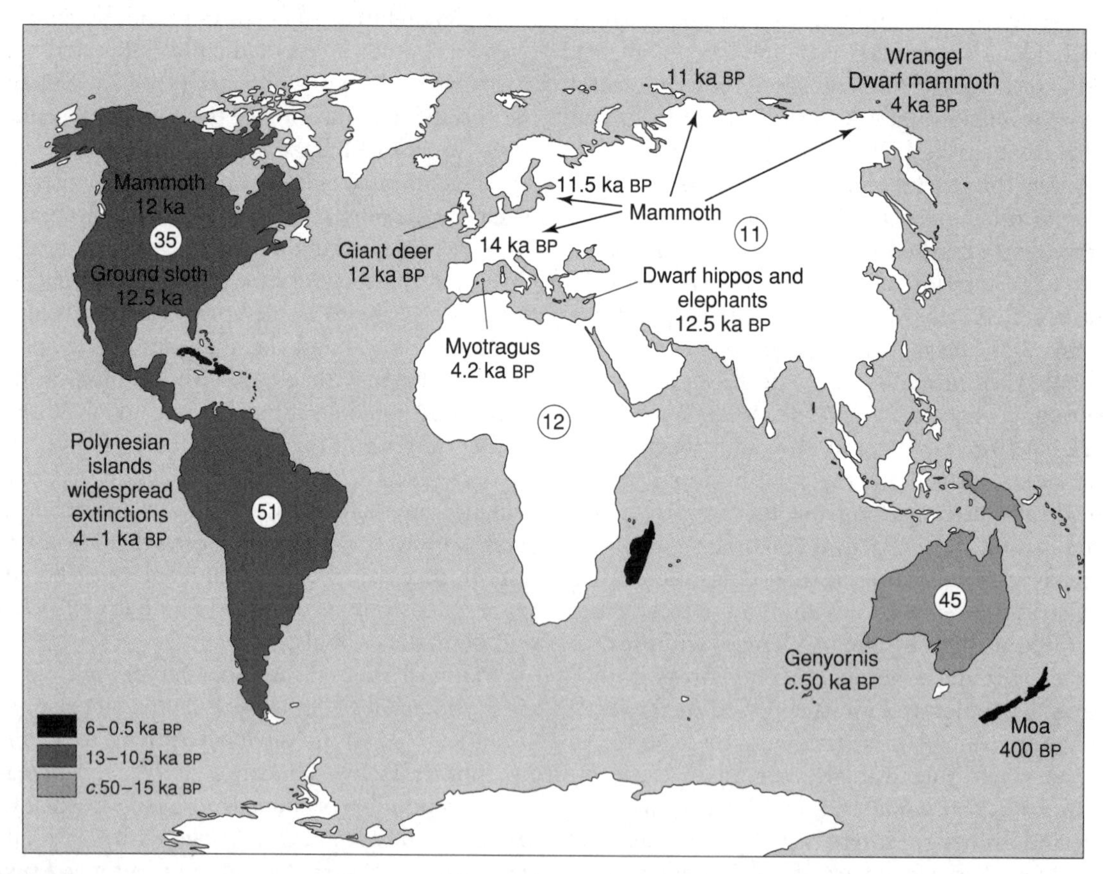

Figure 6.2 Global Late Quaternary extinction periods with dates for selected taxa which are securely dated. Numbers in circles are the approximate number of taxa which became extinct in each area

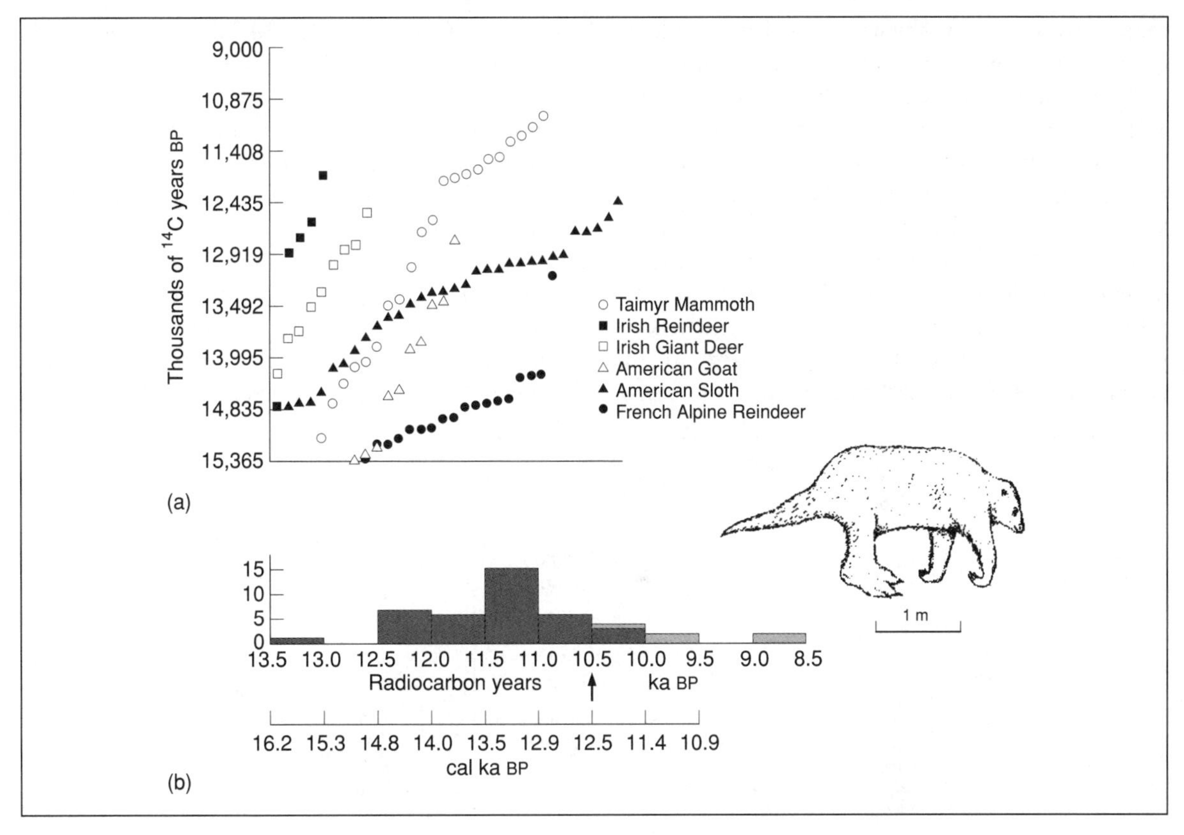

Figure 6.3 Extinction dates: (a) the time distribution of radiocarbon dates for selected north American and Eurasian extinct and regionally extinct species (after Mead and Meltzer (eds) 1985, Figure 1); (b) the youngest radiocarbon dates on dung of the Shasta ground sloth *Nothrotheriops shastensis*, the dates regarded as reliable are in solid black (after Martin *et al.*, 1985, Figure 2).

North America lost 35 genera at the end of the Wisconsinan glaciation. The losses included mammoths, mastodon, giant sloths, horses, bears and cats representing perhaps 70 per cent of the big game. In South America 51 genera were lost. Martin (1967, 1984; Martin and Steadman, 1999) hypothesised a wave of extinctions resulting from overkill by newly arrived human groups, the users of Clovis points, which are distinctively shaped flaked stone projectile points dating between *c.*13.2 ka and 12.8 ka BP. A 'blitzkrieg' or Pleistocene overkill model was proposed whereby animals totally unaccustomed to human predation were very quickly exterminated by a rapidly moving wave of advance. The model still retains currency in the literature, providing an example of continental-scale environmental impact by pre-agricultural communities. It was, however, developed in 1967 and now requires critical re-examination in the light of new knowledge, particularly concerning the rapid timescale and nature of Late Pleistocene environmental change (p. 89; Grayson, 2001).

Dating is at the heart of the extinction debate (Grayson and Meltzer, 2003). Of the North American taxa for which there is dating evidence one or two species may not have survived the Last Glacial Maximum, about 15 survived to 14 ka BP and a smaller number into the Younger Dryas *c.*12.5 ka BP. The extinction of the Shasta ground sloth at about this time is the most well dated (Figure 6.3). The main period of extinction in North America appears to be during the Younger Dryas (Fiedel, 2002). This is the time of Clovis but the evidence for direct predation on megafauna is limited. Associations between prehistoric artefacts and megafauna occur on about 50 sites and there are only 14 on which there is direct evidence of predation and these only concern mammoths and mastodons (Krech, 1999; Grayson, 2001). There is no certain evidence of human predation on any of the other megafauna. Martin explains the absence of kill sites by the speed of the 'blitzkrieg'. Recent modelling has shown that human colonisation *could* have taken place in less than a millennium (Steele *et al.*, 2000). The date of colonisation is much debated (p. 146), but it is striking that decades of search have failed to reveal any convincing evidence of occupation in *North* America much before 12–14 ka BP. Thus, while the chronological evidence is broadly consistent with the overkill hypothesis the paucity of kill sites brings it into question, the more so because there are very similar extinction dates for some taxa in Europe where people had been hunting them for millennia (Figure 6.3).

Extinctions in Eurasia do not appear to occur at one time (Figure 6.4). Some, mostly interglacial adapted species in southern Europe, became extinct before the Last Glacial Maximum, some in the Lateglacial Interstadial and others in the

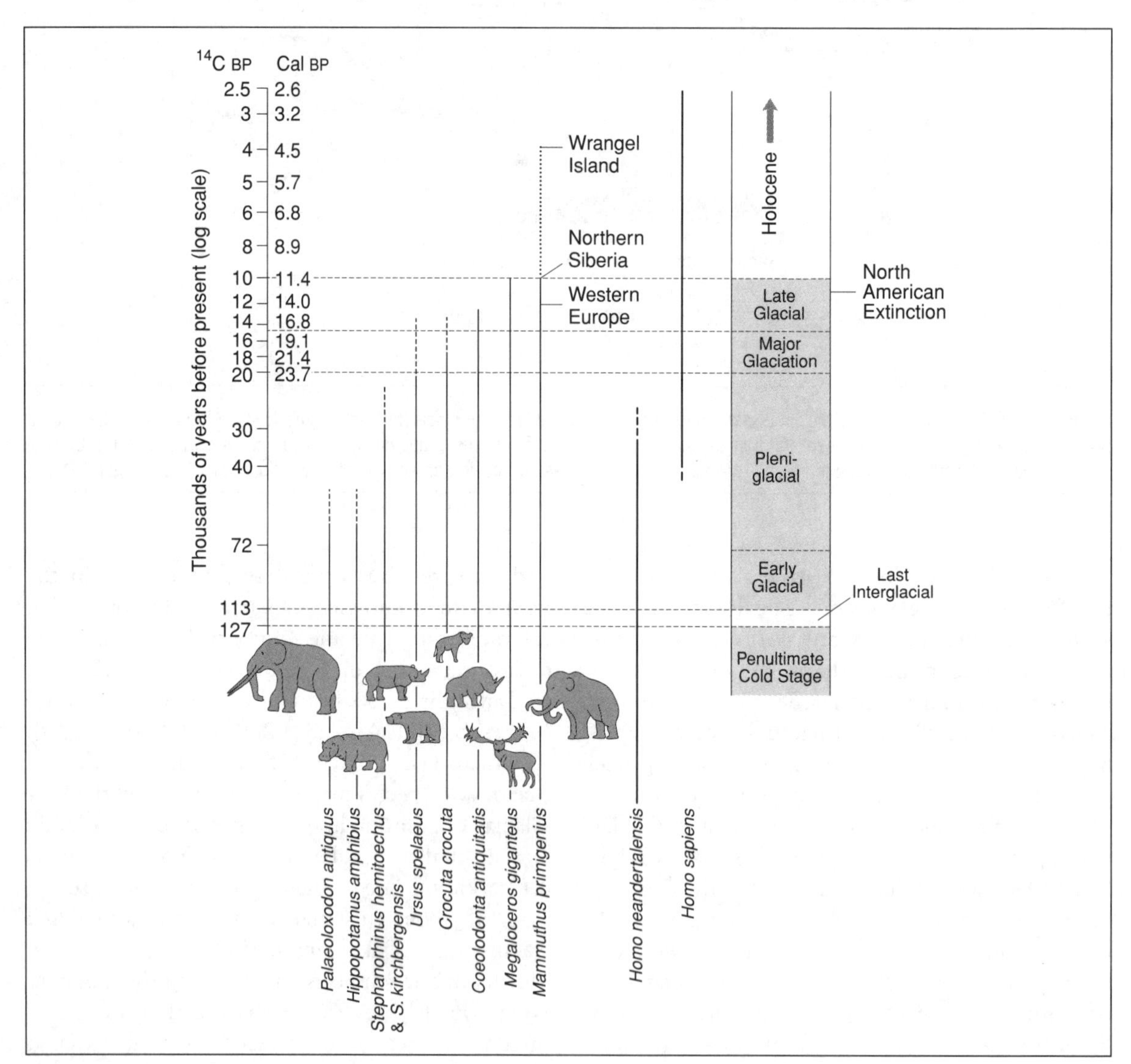

Figure 6.4 The pattern of megafaunal extinctions in northern Eurasia (after Stuart, 1999, Figure 2)

Younger Dryas (Stuart, 1999). Mammoths are lost in both mainland Eurasia and North America between 14 and 12 ka BP, with slightly later survivals east of the Baltic dating as late as 11 ka BP from the Taimyr Peninsula, Siberia (Lister and Bahn, 1994; Lõugas *et al.*, 2002). In Eurasia mammoths and people had long been associated. There are representations of mammoths in cave art at sites such as Rouffignac, France, and as portable art images in Siberia, often carved from mammoth ivory. Settlement of northern Eurasia by anatomically modern hominids around 30–40 ka BP is likely to have impacted on animal populations in Russia and Siberia (p. 145). Such areas may in earlier climatically adverse periods have provided mammoths and other taxa with **refugia**: limited areas of favourable conditions which could support small populations at times of generally adverse conditions.

The vast mammoth steppe of North America, Beringia and Eurasia was a mosaic of vegetation types, in places of high biotic diversity and capable of supporting substantial herds of animals (Willerslev *et al.*, 2003; Zazula *et al.*, 2003). By 11 ka BP this comparatively rich environment was reduced to small relict areas and widely replaced by circumpolar boggy tundra, and to the south of this expanding boreal forest (Lister and Sher, 1995). The biotic changes between the Last Glacial Maximum and Lateglacial were rapid (Ammann *et al.*, 2000), creating disharmonious, or non-analogue, assemblages of animals – those which do not occur in association today (Lundelius, 1989; Stafford *et al.*, 1999). Modelling the effects of climatic and habitat changes on mammoth populations in North America suggests that the distribution of animals became progressively more patchy, leaving isolated populations vulnerable to both human predation and further environmental changes (Mithen, 1996).

A dramatic twist to the mammoth story emerged in 1993 with discoveries on Wrangel Island, once part of the Beringian land mass and now a remote island in the Siberian Sea (Figure 6.2). A series of radiocarbon dates shows that mammoths survived here well into the Holocene, to 4 ka BP (Vartanyan *et al.*, 1993). They had undergone the process of **island dwarfism**, a size reduction which affects isolated animal populations with a finite food supply and a lack of predators. Interestingly, the vegetation of the island is a relict of the Pleistocene steppe communities, helping to explain why such prolonged survival was possible here. People may have been implicated in their final extinction because it occurred at about the time that artefacts appear in the wider region.

Another key species in the extinction debate is the giant deer (*Megaloceros giganteus*; Figure 6.5) which was abundant in Ireland during the Woodgrange Lateglacial Interstadial; most radiocarbon-dated examples are between 14.5 and 12.5 ka BP (Woodman *et al.*, 1998). So numerous was the deer at that time that it is considered to have been a key factor in bringing about vegetation change from juniper scrub to extensive grassland (Bradshaw and Mitchell, 1999). It then rapidly declined to extinction in Ireland during the succeeding Nahanagan (Younger Dryas) Stadial, apparently because this brief cold spell led to a drastic reduction in the quality of its browse (Stuart and van Wijngaarden-Bakker, 1985; Barnosky, 1986). Significantly, it was extinct before the earliest recorded human activity in Ireland around 10 ka BP (Mitchell, 1986). It did, however, survive into the very early Holocene in the Isle of Man (then, due to low sea level, connected to the British mainland, Plate 4.1) where an example is dated *c.*9.9 ka BP (Gill and West, 2001). In Britain as a whole, 75 per cent of the Lateglacial cold stage mammal species became locally extinct at the start of the Holocene, but some taxa show evidence of a time lag with the last examples around *c.*10.3 ka by which time woodland vegetation was colonising open habitats, although a population of reindeer survived in Scotland until 8.8 ka BP (Coard and Chamberlain, 1999).

Australia presents a contrasting picture. Extinction affected 45 species, representing 85 per cent of the terrestrial megafauna. The animals included giant kangaroos, and koalas and a *Diprotodon*. Good dating evidence is rather limited except in the case of *Genyornis*, an ostrich-sized flightless bird (Miller *et al.*, 1999a). Amino acid geochronological analysis (Lowe and Walker, 1997: 285) of a large sample of eggs indicates extinction at *c.*50±5 ka BP (Figure 6.6). The latest

***Figure* 6.5** Giant deer (*Megaloceros giganteus*), (photo National Museum of Wales)

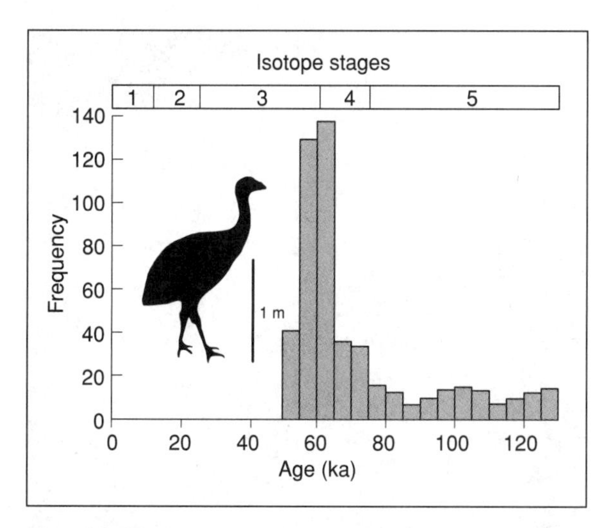

Figure 6.6 Extinction dates for mihirung (*Genyornis*) in Australia. The distribution of dates (total number 505) is in 5 ka classes from 0 to 130 ka BP. Dates are obtained from amino acid racemisation and indicate probable extinction at 50–55 ka BP (reprinted with permission from Miller, G.H. *et al.*, 1999a, Pleistocene extinction of *Genyornis newtoni*: human impact on Australian megafauna, *Science*, **283**, pp. 205–8, Figure 3. Copyright 1999 AAAS.)

reliable dates for megafauna are *c.*46 ka BP so these extinctions seem to occur at about the time Australia was colonised by people (Gillespie, 2002). Within 20 ka all the main ecological regions of the country had been occupied (Jones, 1989; Bowler *et al.*, 2003). No major megafaunal kill sites have been found and there are only 11 archaeological sites containing bones of extinct species (Kohen, 1995). There is a well-documented aridity phase from *c.*40 ka BP, marked, for instance, by progressive drying of Lake Mungo, with important archaeological sites on its former shore (Bowler *et al.*, 2003). Vegetational reconstructions, inferred from pollen and other records, also indicate a major episode of aridity between 30 and 10 ka BP, when coastal rainforests and woodlands were drastically reduced, but scrub and grassland habitats expanded (Dodson, 1989). Horton (1984) concluded that the extinctions were a result of increasing aridity, the megafauna becoming ecologically tethered at a declining number of waterholes which then dried out. Other have argued that the presence of people, and possibly the early practice of aboriginal burning (p. 193) leading to landscape modification, may have tipped the ecological balance towards extinctions (Jones, 1989; Kohen, 1995).

As chronological precision improves so similarities and contrasts emerge in the pattern of Late Pleistocene extinctions. They begin early, perhaps *c.*50 ka BP, in Australia. In Eurasia they appear staggered, some perhaps as early as Australia, most apparently *c.*11 ka BP at the end of the last glacial. In North America the well-dated extinctions also occur *c.*11 ka BP. It is striking that in America, apparently newly colonised, and Europe, where people had coexisted with megafauna for half a million years, the most well-dated extinctions are broadly contemporary. The small number of American and Australian sites with associations between megafauna and archaeological artefacts does not support the focus on megafaunal exploitation which was explicit in the original overkill model. Since that was put forward the rapid timescale of Late Pleistocene climatic change and its effects on vegetation have become far better understood. Changes such as the shrinking mammoth steppe would have depleted megafauna populations and left the survivors concentrated in the contracting refuge areas. Some of these would have been wiped out by continuing climate change; others are likely to have been extinguished by human hunters struggling themselves to cope with an environment undergoing very rapid change. Climate may be seen as the main driver, but hunting was the reason why some reduced refugial populations were finally extinguished. Acting in the same direction towards reduced megafaunal populations, these two factors seem to have had mutually reinforcing effects giving rise to a wave of extinctions of global proportions.

Holocene island extinctions

Islands which have been isolated for extended periods develop bizarre endemic species as a result of a lack of predators, competition and a limited food supply. Large species have a tendency to dwarfism (p. 189), small species to giantism, and birds lacking predators may become flightless (Balouet, 1990). Islands can be a treasure chest of biodiversity as recognised long ago by Darwin on the Galapagos. However, what survives today is often but a remnant of the biodiversity which existed earlier in the Holocene and in most cases there is persuasive evidence that it was the arrival of people on islands which was the decisive factor leading to extinctions.

There is evidence of extinct endemics on many Mediterranean islands including Cyprus, Crete, Sicily, Corsica, Sardinia and Mallorca (Davis, 1987). On Cyprus endemic pygmy elephants and hippos were hunted by pre-agricultural communities *c.*12.5 ka BP and this is thought to have contributed to their extinction before the earliest farming communities arrived (Simmons, 1998). However, initial Holocene vegetation change around the east Mediterranean (see Figure 5.9) will also have been a factor. Mallorcan endemics include a giant doormouse, shrew, owl and a member of the antelope family, *Myotragus balearicus*, which was uniquely adapted to life on the rocky northern slopes of the island. Large numbers of the bones of these creatures and layers of their dung accumulated in rock shelters (Figure 6.7) and cave systems (Waldren, 1982). Their grazing activities would have played a part in maintaining open conditions and thus in the survival of endemic plants (Juniper, 1984). *Myotragus* survived until colonisation by people and domestic animals *c.*4.2 ka BP (Ramis *et al.*, 2002). The effects of grazing domesticates on vegetation may well have been a more important cause of extinction than hunting.

Figure 6.7 Son Matge, Mallorca: (a) the rock shelter where bones and coprolite beds of the extinct endemic *Myotragus balearicus* which is shown in (b) have been excavated (courtesy of Jennifer Foster)

New Zealand has been isolated for the last 80 ma leading, in the absence of predators, to the evolution of a remarkable endemic bird fauna exemplified by the flightless moa. Late Quaternary environmental sequences are well known and there is no evidence of major extinctions here at the Lateglacial–Holocene transition (Martin and Steadman, 1999). Colonisation by Polynesians occurred *c.*AD 1280 (Lowe *et al.*, 2000). Thirty-four species of bird, living at that time, had become extinct before European arrival in AD 1642 (Trotter and McCullock, 1984). Over 300 moa hunting sites are known and date between 900 and 400 BP (Anderson, 1989; Holdaway, 1999). Overkill is suggested because sometimes only the most palatable parts were eaten, suggesting conspicuous consumption by groups competing for social supremacy. Even so, the single most important factor is likely to have been loss of habitat, arising from woodland destruction by the Maori practice of burning which began close to the time of initial colonisation (Lowe *et al.*, 2000; Ogden *et al.*, 2003). By 500 BP this had destroyed most of the lowland forest. The preserved gizzard contents of moa indicate they fed in the forest and forest margin (Anderson, 1988, 1989). Polynesian communities had an equally dramatic impact on the fauna of many Pacific islands which were colonised between 6 and 1 ka BP bringing a wave of bird and other faunal extinctions (Martin and Steadman, 1999), sometimes evidence for dramatic deforestation, as on Easter Island (Flenley and Bahn, 2003), and erosion (Kirch and Hunt, 1997).

Burning by hunter-gatherers

Whatever our conclusions regarding the causes of Late Pleistocene extinctions, there is no doubt that hunter-gatherer communities can have a significant environmental impact. Chroniclers of early expeditions to North America, such as Drake and Hudson, recorded burning by 'Indians', not just for the purpose of growing crops such as maize, but also on a more widespread basis by hunter-gatherers. The ethnohistorical record also documents widespread evidence for burning and the many reasons why it was done (Mellars, 1976; Boyd, 1999). Fire was used as part of hunting practice to drive animals, to improve pasture and human mobility, to make it easier to collect certain types of food, to increase the yield of berries and nuts and to produce suitable basket-making materials. Burning releases nutrients and promotes tender new plant growth attractive to grazing animals. It can increase the carrying capacity for herbivores by between 300 and 700 per cent (Mellars, 1976). Carefully timed serial burns have the effect of creating a mosaic of successional stages, with increased species diversity and an abundance of those woodland edge plants which we know made such an important contribution to the diet of American north-west coast communities (Turner, 1999).

The west coast of America exhibits a diversity of ecological regions and all the way from California to British Columbia burning was practised by non-agricultural communities, albeit for a variety of purposes and at different times of year, attuned to local ecological and climatic conditions and the seasonal round (Boyd, 1999; Blackburn and Anderson, 1993). The practice is particularly well documented by nineteenth-century travellers through the Willamette Valley, Oregon, who encountered an extensive grassland environment with scattered oak groves and much evidence of burning (Boyd, 1999). Palynology shows that some of these areas of oak savannah had existed from 6 ka BP and they are thought to have originated in the drier Hypsithermal, 6–4 ka BP (p. 89), being subsequently maintained by grazing and burning (Leopold and Boyd, 1999). If that is so, it is an interesting example of people coping with ecological change by manipulating the environment itself (p. 143). It is clear that these were fire climax communities because once Europeans arrived and burning ceased the grasslands of the Willamette Valley, and others which covered Whidbey Island, Washington (White, 1999), underwent succession to woodland.

Ethnohistorical sources demonstrate that burning was practised in the woodlands of eastern North America. The most direct evidence of environmental manipulation comes from areas such as the southern coast of New England where agriculture was practised and the greatest population density occurred (Cronon, 1983; Patterson and Backman, 1988). Further north in the boreal forest marked fluctuations in charcoal occurrence within annually varved sediments demonstrate regular fires (Cwynar, 1978; Clark, 1989). In the Midwest, with its seasonally dry climate and greater lightning incidence, it is more difficult to distinguish natural from anthropogenic burns at the prairie/woodland edge. However, burning is well attested ethnohistorically and, in the Ojark Highlands of Missouri for instance, a significant increase in woodland once again coincided with the suppression of native burning practice by Europeans (King and King, 1996).

Australian archaeologists have placed considerable emphasis on the extent of Aboriginal burning, adopting the phrase 'fire stick farming' to highlight the importance of environmental manipulation to the Aboriginal way of life (Jones, 1969; Yen, 1989). Charcoal peaks occur in last interglacial contexts (Singh and Geissler, 1985; Kershaw, 1993), but these must surely have a climatic explanation because there is no clear evidence of human presence before *c.*50 ka BP. Fluctuating charcoal values in the first half of the Holocene are variously interpreted as the result of human activity (Bowdler, 1988; Jones, 1989) or natural climate-related wildfire (Horton, 1982). A marked increase in charcoal occurrence and fire-tolerant vegetation occurs on many sites between 5 and 3 ka BP. Some writers hold that this represents the beginning of the ethnographically attested firing practices at a time (*c.*4 ka BP) when there are also some very marked changes in material culture and increased plant use (Head, 1994; Kohen, 1995). By the time Europeans came on the scene many areas of parkland and open habitat were maintained by regular Aboriginal burning which is well attested ethnohistorically. Tasmania would naturally have been almost totally forested but there were significant areas of sedgefield and heathland which, following the decimation of Tasmanian Aboriginal populations, rapidly underwent succession to woodland.

Comparative ethnohistorical studies of fire use among hunter-gatherers in the highly contrasting environments of Australia and the American northwest show marked similarities in the sophisticated way in which communities controlled the seasonality, frequency, intensity and habitat selectivity of fire, creating cleared patches (yards) and axes of movement (corridors) (Lewis and Ferguson, 1999). In both Tasmania and Vancouver Island, Canada, those open areas were later selected by Europeans for settlement because they found the landscape congenial and suitable for agriculture (Kohen, 1995: 42; Turner, 1999: 196). This is a striking example of how earlier land-use can play a part in the structuring of later settlement even by groups radically different in terms of their culture and economy.

Mesolithic forest clearance in the British Isles and continental Europe

During the early and middle Holocene, nearly all of the British Isles was covered by woodland, successive stages in forest development being shown in Figure 4.18. By 5.7 ka BP a complex woodland mosaic had developed (Figure 6.8), for which

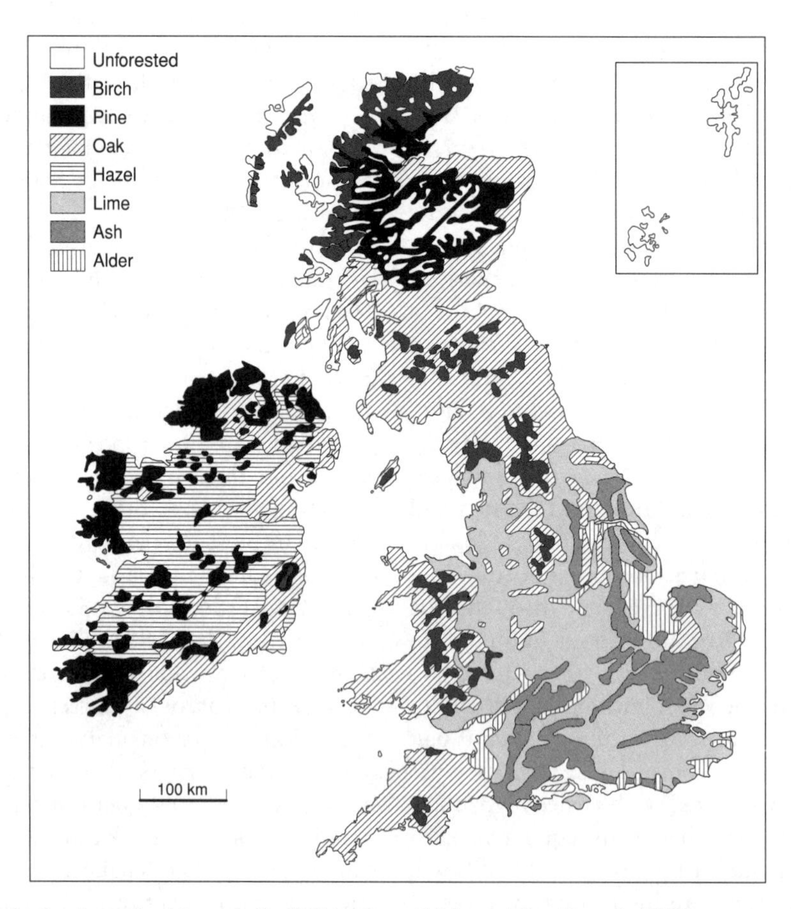

Figure 6.8 Map of dominant woodland types in the British Isles *c*.5.7 ka BP (after Bennett, 1989)

Rackham (2003) coined the name 'wildwood'. Unwooded areas were very limited: the upland summits of Wales, the Pennines, the Lake District and more extensive areas of Scottish upland. Parts of central Ireland were unforested because they were already covered by raised bog. Localised open areas would have existed where disturbance factors were especially concentrated, for example on coasts. Exposed western coasts of the Outer Hebrides, Orkney, Shetland and parts of Caithness were partly open with patchy woodland in sheltered places.

Figure 6.9 Waun-Fignen-Felen, Wales: computer reconstruction of the early Mesolithic lake (from Barton *et al.*, 1995, Figure 17). By Loretta Nikolic

In Britain the episode between the end of the last glaciation 11.5 ka BP and 6 ka BP represents the Mesolithic period of hunter-gatherer activity, which ends with the appearance of the first farmers. Recently, the clear distinction which earlier authors made between Mesolithic and Neolithic ways of life has become more blurred (Zvelebil, 1986). The view of Mesolithic communities as mobile hunter-gatherers constrained by the environment and its seasonal cycles but having minimal effect on it, has been modified in the face of increasing evidence of fire and vegetation disturbance during the Mesolithic. The evidence is most abundant in upland areas which are today moorland, such as the Pennines, North York Moors and Dartmoor (Simmons, 1996). Here there are Mesolithic old land surfaces sealed by blanket peat. The minerogenic soils below peat preserve pollen sequences with evidence of a reduction of woodland, an increase of open taxa including heath species and ultimately the accumulation of blanket peat, which in places occurred within the Mesolithic. On some sites scatters of Mesolithic artefacts and/or charcoal occur in the basal soil, which has been taken to implicate both people and fire in the vegetation changes. At Waun-Fignen-Felen in the Black Mountains of South Wales a small upland lake was surrounded by open birch woodland creating an attractive setting for seasonal (probably summer) visits by Mesolithic communities (Figure 6.9). Charcoal and lithic tools below peat show that people were active in the landscape from the early Mesolithic (Barton *et al.*, 1995). Smith and Cloutman (1988) emphasise the role of burning by these visitors in preventing the development of closed woodland and leading to increased heathland and the extension of blanket peat, but natural factors such as grazing animals may have helped to maintain open conditions. At North Gill on the North York Moors, multiple profiles reveal a charcoal spread which has been resolved into a series of distinct woodland disturbance and charcoal deposition events extending over a period of at least 3.5 ka from 9 to 5.5 ka BP (Figure 6.10; Simmons, 1996). Repeated burning of this kind within a small spatial area is difficult to explain other than by human agency, although in this case it is not a settlement site and artefacts are lacking.

The presence of wood, and less directly pollen evidence, below peat indicates that the altitudinal limit of tree growth in Britain lay somewhere between 500 and 700 m OD. However, woodland edge plant communities are frequently found at lower elevations down to *c.*250–450 m. It has been argued that the effect of burning was artificially to lower the treeline and it is in this area of former, retreating, woodland edge that Mesolithic upland sites are concentrated (Jacobi *et al.*, 1976). Mellars (1976) and Simmons (1996) have interpreted evidence for burning as deliberate management by Mesolithic communities to encourage and attract large herbivores, particularly red deer, the most abundant animals on most Mesolithic sites.

Simmons (1996) sees burning as mainly affecting the upland woodland edge and becoming increasingly prevalent in the later Mesolithic (e.g. Figure 6.10). However, a growing number of sites have evidence of fire early in the Mesolithic and from an increasing range of topographic situations. The most important lowland site is the famous Star Carr, Yorkshire, where at the edge of a

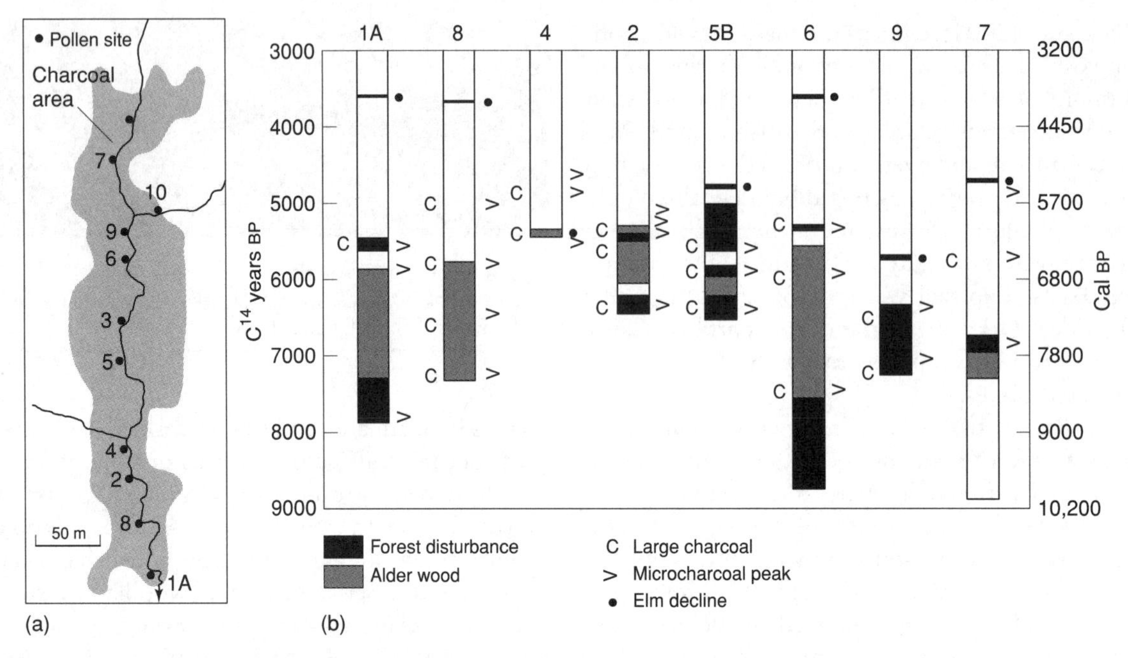

Figure 6.10 North Gill, Yorkshire UK: (a) The spatial distribution of charcoal and pollen sites; (b) the temporal range of vegetation change represented by forest disturbance and charcoal (after Simmons, 1996, Figures 2.4 and 2.5)

Lateglacial relict lake, a settlement was established (Clark, 1954). High-resolution pollen analysis and wiggle-match AMS dating demonstrated that occupation began here within 600 years of the start of the Holocene (Mellars and Dark, 1998; Dark, 2000b). Coeval with occupation are two distinct phases of charcoal deposition: one dated 10.9 ka BP, the other 10.7 ka BP. The burning was of reeds fringing the lake, perhaps to improve visibility and the productivity of lake-edge plant resources both for human use and to attract deer.

Broadly contemporary with Mesolithic activity at Star Carr, there is evidence that fire played a part in the patch dynamic of early Holocene river valley environments, in the lower Thames tributaries at Uxbridge and Thatcham, and a little later *c.*9.2 ka BP there is evidence of burning at Peacock's Farm in the East Anglian Fens (Smith *et al.*, 1989). Similarly in coastal situations in the Severn Estuary and Bristol Channel, extensive charcoal spreads (*c.*7.4 ka BP) now exposed in the intertidal zone are associated with charred submerged forest trees and the burning of reeds which at Goldcliff East is coeval with an excavated Mesolithic settlement (Bell, 2000; Bell *et al.*, 2002). It appears, therefore, that, starting early in the Mesolithic, fire was selectively modifying parts of the woodland edge at both the treeline in the uplands and in the lowland river valleys and coasts.

Evidence from the islands on the west coast of Scotland challenges any neat assumption that charcoal in the Mesolithic is invariably the product of human agency. Both the Inner and Outer Hebrides have a number of sites where pollen sequences show a reduction of woodland and charcoal peaks. Mesolithic sites are present on the Inner Hebrides (Mithen, 2000) but on the Outer Hebrides, where there is similar evidence for fire and vegetation disturbance *c.*9–7 ka BP, there is no evidence of Mesolithic settlement (Edwards, 1996; Edwards and Sugden, 2003). This leads Tipping (1996) to argue that fire frequency in western and northern Scotland relates mainly to phases of climatic dryness, and there is a measure of support for this view in the occurrence of charcoal peaks in Lateglacial sites before there is evidence for human occupation in Scotland (Edwards *et al.*, 2000). However, the geographical and spatial patterning of Holocene

Plate 5.1 Laugerie Haute, Dordogne, France: rock shelter with occupation horizons of the Gravettian, Solutrean and Magdalenian separated by rock falls (photo Martin Bell)

Plate 5.3 A reconstruction of the 'ice man' from the Austrian Italian border illustrating his sophisticated clothing and equipment adapted to a harsh Alpine environment (photo from Photo Archives, South Tyrol Museum of Archaeology–www.iceman.it)

Plate 5.2 Sand dune sequence at Brean Down, Somerset showing Bronze Age occupation layers separated by blown sand and colluvium (photo Anthony Philpott)

Plate 6.1 Hornstaad Hörnle, Germany, landscape reconstruction in 5.8 ka BP showing the Neolithic lakeside settlement and inland fields and secondary woodland (Schlichtherle, 1997. Reproduced by permission of Landesdenkmalamt Baden-Würtemberg, H. Schlichtherle and T. Leonhardt)

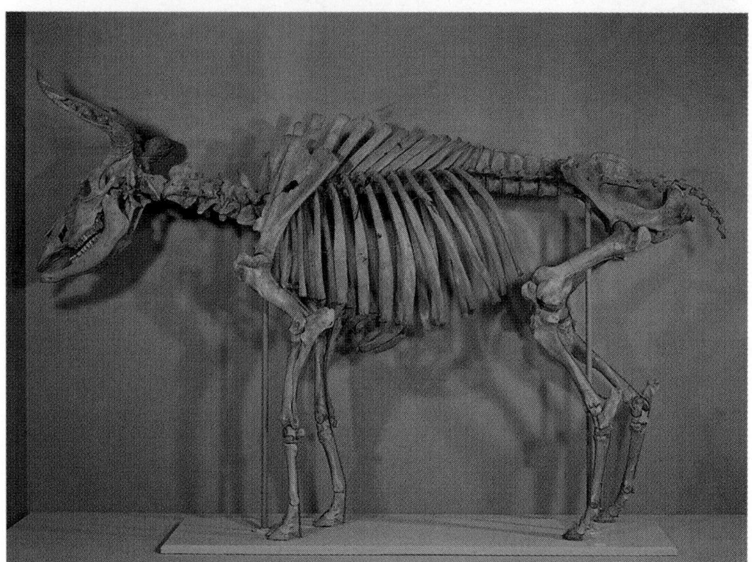

Plate 6.2 Aurochsen (*Bos primigenius*) from Vig, Zealand, Denmark *c.*10 ka BP. This animal was wounded by Mesolithic hunters three times; at least one wound had healed, but a spear thrust through the shoulder blade proved fatal (photo Lennart Larsen. Published by permission of Danish National Museum, Copenhagen)

Plate 6.3 Section of a Bronze Age barrow at Moor Green, Hampshire, England showing old land surface with evidence of podzolisation and podsol turves making up the core of the barrow (photo Paul Ashbee)

(a)

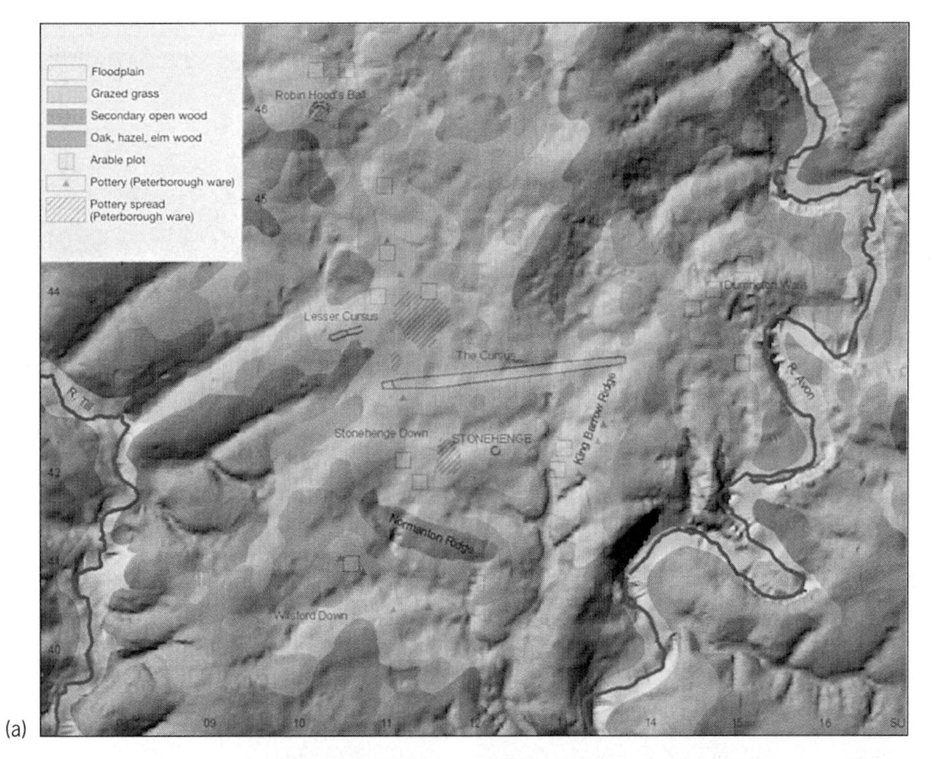

(b)

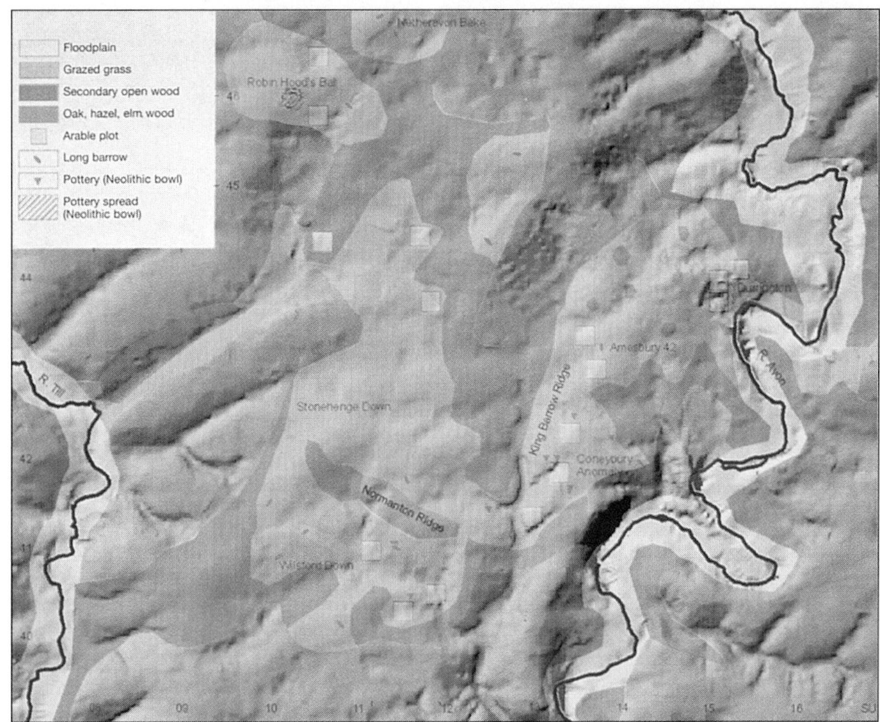

Plate 6.4 Stonehenge, England: (a) the early to middle Neolithic landscape 6-5 ka BP; (b) the middle Neolithic landscape at the time of the first phase of Stonehenge 5-4.8 ka BP (after Allen, 1997 and courtesy of Wessex Archaeology)

Plate 7.1 Colluvial sediment sequence at Kiln Combe, England (photo Brenda Westley)

Plate 7.2 River valley sediments at Woolhampton in the Kennet Valley, England: (a) late Pleistocene gravels, (b) clay with organic bands, (c) calcareous marl, (d) ? palaeosol, (e) peat, (f) tufa (photo Jodi Davison and Shaun Buckley)

Plate 8.1 Federsee, Germany: air photograph of the archaeologically rich wetland which is currently the subject of wetland conservation strategies (photo from Schlichtherle, 1997. Reproduced by permission of Landesdenkmalamt Baden-Württemberg and O. Broasch)

charcoal occurrence is not wholly consistent with climatically driven periodicity (Edwards, 1996), so the possibility remains that there was Mesolithic settlement in the Outer Hebrides but the evidence has not yet been found.

Ireland presents an interesting contrast to England. There are sites such as Newferry, Co. Antrim (Smith, 1984) with evidence of vegetation disturbance and charcoal horizons associated with Mesolithic occupation, but overall the evidence for Mesolithic burning has been described as limited and equivocal (Cooney, 1999: 43). This may be explicable in terms of the impoverished fauna of Ireland which, in this period, lacks the key herbivore species, (elk, probably red and roe deer, and aurochsen) (Woodman *et al.*, 1998), which would have contributed to the maintenance of open areas and would also have been encouraged by the practice of burning. This faunal impoverishment accounts for the greater focus of Irish Mesolithic communities on aquatic resources.

In continental Europe pine charcoal horizons are widespread in layers in the coversand of the Netherlands and north-west Germany and are dated to the Allerød Interstadial, although it is unclear whether they are the result of lightning strikes or anthropogenic fires (Behre, 1988). At Milheeze in the Netherlands, an association with human activity is more direct. There is Upper Palaeolithic settlement during the Younger Dryas associated with four marked reductions in pine–birch forest with pronounced charcoal peaks (Bos and Janssen, 1996). At Zutphen, Netherlands, open grass-land associated with an early Mesolithic site of *c.*10.2 ka BP appears to have been maintained by fire (Groenewoudt *et al.*, 2001). These sites are of similar date to Star Carr and point to similar practices in fire use in areas which, at the time, were joined by Doggerland (p. 161).

Danish sites have been widely subject to palaeo-environmental study and show little evidence of environmental manipulation during the Mesolithic (e.g. Andersen and Rasmussen, 1993). The view is that, as woodland became denser through the middle Holocene, so settlement became increasingly concentrated on rich coastal resources (Iversen, 1973). The Ertebølle settlement site at Ringkloster is an exception with evidence for some reduction in woodland, burning and an increase in open taxa at the time of occupation 6.4 ka BP (Rasmussen, 1995). Further north in Scandinavia there are small numbers of sites in Sweden, Norway and Finland (Welinder, 1990; Regnell *et al.*, 1995; Berglund *et al.*, 1996a: 167), with charcoal peaks associated with a reduction of woodland often of pine and deciduous character.

Lateglacial and very early Holocene evidence for burning in the Netherlands, Germany and Star Carr, England, is apparently associated with human activity. This suggests that the practice of anthropogenic burning may have originated in the attempts of people to cope with the privations created by rapid climate change in the Allerød and Younger Dryas. This may have led them to develop practices which were later employed, perhaps helping to retard the spread of the boreal forest and increasing the productivity of the woodland edge. What is striking is that following the limited early Holocene evidence for the use of fire on the continent it becomes very rare, except perhaps in the boreal forest fringes of Scandinavia. By comparison, in Britain the evidence is abundant, spans the Mesolithic and is present in diverse topographic situations, suggesting that the practice of burning developed as a distinctive insular cultural tradition.

The transition to agriculture in central and north-west Europe

Once they had domestic plants and animals, people had the means and motivation for landscape transformation on an altogether larger scale. This process extended, however, over millennia and it is now apparent that some of the initial changes are significantly complicated by the effects and affordances of natural disturbance factors. Among the main Neolithic domesticates, cereals, sheep and goats were not native to Europe and were introduced into south-east Europe from the south-west Asian centre of domestication (p. 152). Occurrence of generally charred macrofossils of domestic crops charts their arrival in Greece by 8.7 ka BP, over south-east Europe by 7.8 ka BP and across central Europe by 6.8 ka BP (Zohary and Hopf, 1994).

The Neolithic way of life also involved husbandry of cattle and pigs. Neolithic communities also used pottery and ground stone technology. By 7.4 ka BP farmers using distinctive Linearbandkeramik (LBK) pottery and living in substantial settlements of long rectangular houses had occupied central Europe as far as the west of Germany. The LBK settlement pattern was largely restricted to fertile and light loess soils across which farming spread rapidly. The speed of spread has been explained in terms of rapid population growth leading to colonisation by a successful new economy. More recent models emphasise acculturation: the adoption of a new way of life by indigenous hunter-gatherer communities (Whittle, 1996; Price, 2000). Environmental evidence contributes to this debate by demonstrating that the scale of environmental impact by early farmers in south-east Europe is very limited, thus weakening the colonisation argument based on rapid population growth, land shortage and expansion (Willis and Bennett, 1994).

What is particularly striking is that, having spread rapidly across central Europe at *c*.7 ka BP, the spread of farming then halted for a millennium inland of the Baltic and North Sea coastal zone. One suggestion is that further expansion was blocked by the greater density of hunter-gatherer-fisher communities, who had adopted a successful coastal way of life and moved eastward as Doggerland (p. 161) was progressively inundated (Coles, 2000b). The transition to farming can be seen in terms of three key stages which are well illustrated in north-west Europe: an availability phase when it was becoming known through contact with farmers, but was not adopted; a substitution phase when farming was becoming part of the ecology of a frontier zone; and a consolidation phase representing the maturation of the social and economic structures associated with an agricultural community (Zvelebil and Rowley-Conwy, 1986). The dates and lengths of these phases vary considerably within Europe depending on local social and ecological factors. Denmark has a particularly lengthy availability phase, the Ertebølle, between 7 and 6 ka BP when settlement was concentrated on large coastal middens (see Figure 5.13). People here adopted some of the trappings of a Neolithic way of life such as pottery and ground stone axes. Then, after a millennium of availability, agriculture is adopted and the isotopic evidence of human diet (p. 46) shows a rapid shift from marine to terrestrial food sources (Richards *et al.*, 2003b). Some middens continue in use but show evidence for a change in composition from oysters to cockles, on others fishing was abandoned (Andersen, 1991, 2000). Some of these changes may be due to marked coastal environmental changes at this time (Pedersen *et al.*, 1997) and there have been attempts to argue that these were so detrimental to the hunter-gatherer way of life that they led to the adoption of farming (Zvelebil and Rowley-Conwy, 1986). A solely environmental driver appears, however, overly deterministic and most writers currently give greater emphasis to the role of social factors leading to acculturation (Whittle, 1996).

In the rich coastal and estuarine environment of the Netherlands (p. 161) there were attenuated phases of both availability and substitution. Communities of hunter-gatherer-fishers known as Swifterbant, similar to Ertebølle, first adopted the use of pottery *c*.6.8 ka BP at sites such as Hoge Vaart (Hogestijn and Peters, 2001), then *c*.6.6 ka BP domestic animal bones appear at Hardinxveld, although they may not have been kept locally in the wetland and were mostly ritually deposited (Louwe Kooijmans, 2003). Finally, from *c*.6 ka BP the use of cereals appears at sites such as Swifterbant and Hazendonk. On each of these sites the domesticates are not apparently much more than supplements to an economy largely reliant on wild resources. Throughout the Neolithic the ratio of domestic to wild animals in the wetland never rises above 50 per cent (Louwe Kooijmans, 1993).

To the south in the Alps there is little evidence for attenuated availability and substitution phases, but 'lake village' settlements continued to make significant use of wild animals and gathered plant resources (p. 205). Less surprisingly, perhaps, to the north of the Baltic and in European Russia less favourable climatic areas show a very long substitution phase. There is little agriculture east of the Baltic and in European Russia until after 3.4 ka BP (Dolukhanov and Khotinskiy, 1984). Climatically favourable areas in the south of Norway and Sweden adopted agriculture from 5.3 ka BP (Berglund,1991), but its northerly limits fluctuated with climatic change and hunting remained the economy of the north (Berglund, 1985). In Finland some areas only adopted agriculture in the last millennium (Vuorela and Hicks, 1996; Simola, 1994).

In Britain domesticates must have arrived by boat so there may have been an element of migration rather than purely acculturation. The transition to agriculture *c.*5.8 ka BP is close in time to that in both the Netherlands and Denmark despite strikingly different ecological and social contexts. The availability phase is short, few late Mesolithic sites are known and these such as Oronsay, Scotland (Mellars, 1987; Richards and Sheridan, 2000), only just overlap with the earliest Neolithic sites in southern Britain. Even so, the prevailing model is of a gradual transition to an agricultural way of life. This interpretation arises because, although non-domestic Neolithic sites such as causewayed enclosures, tombs, and even flint mines, are well represented from an early date, clearly defined settlement sites have proved elusive. From this has developed the view that early Neolithic communities remained quite mobile, their economy based largely on animal husbandry with rather small-scale crop growing (Thomas, 1999). Many show the continued use of wild resources, particularly hazelnuts, together with cereals forming what might be called a 'muesli' economy (Moffet *et al.*, 1989; Robinson, 2000). On some sites, however, cereals greatly outnumber wild plant macrofossils and these include some sites with substantial timber houses suggestive of more sedentary occupations (Jones, 2000). These latter challenge the gradualistic model, as do the many pollen diagrams which are interpreted as showing evidence of clearance and sometimes cereals close to the beginning of the Neolithic (Smith, 1981). Gradualism is further called into question by accumulating isotopic evidence of human dietary change. Whereas coastal Mesolithic communities had subsisted largely on marine resources, by the beginning of the Neolithic there was a rapid and almost total shift to terrestrial resources (Richards and Hedges, 1999; Richards *et al.*, 2003a).

The elm decline

A decline in elm pollen occurs at about 5.8 ka BP in pollen diagrams from sites throughout northern and north-west Europe. The decline is typically to half its former frequency (Figure 6.11), and occurs at about the time when farming reached much of this area. That apparent contemporaneity has obscured and complicated our understanding of this important environmental change and the Mesolithic/Neolithic transition. The tacit assumption was that people caused the elm decline and that it marked a cultural boundary: pre-elm decline activity being Mesolithic, post-elm decline Neolithic. What caused the decline has been much debated and the various hypotheses reflect evolving paradigms in archaeology and environmental studies (Parker *et al.*, 2002).

Past uncertainties and debates have been substantially resolved in recent years by advances in chronological precision. Most notable is a high-resolution, mostly annual, pollen study of the elm decline in laminated sediments in Diss Mere, eastern England (Figure 6.12; Peglar, 1993b; Peglar and Birks, 1993). Here elm pollen declined from 6 to 2 per cent in just 6 years. There are no radiocarbon dates for the Diss sequence. However, analysis of dated elm declines at 138 sites in the British Isles shows ages in the range of 6347–5281 BP (Parker *et al.*, 2002). There is no significant difference in the mean dates of the decline in the regions of the British Isles, suggesting that, within the precision of radiocarbon dating, the event was synchronous. The same appears to be true across much of north-west Europe, and at the Scandinavian sites of Agorods Mose and Hasing, wiggle-match radiocarbon dating produces similar evidence of a rapid elm decline (Andersen and Rasmussen, 1993; Skog and Regnell, 1995).

Initially a climatic explanation for the elm decline was favoured. Iversen (1941) used the decline, together with a decline in ivy and a rise in ash pollen, to define the boundary between the Atlantic and Sub-Boreal pollen zones. This change was attributed to increasing continentality. There is some evidence for this from the GISP2 ice core at *c.*5.9 ka BP (O'Brien *et al.*, 1995). However,

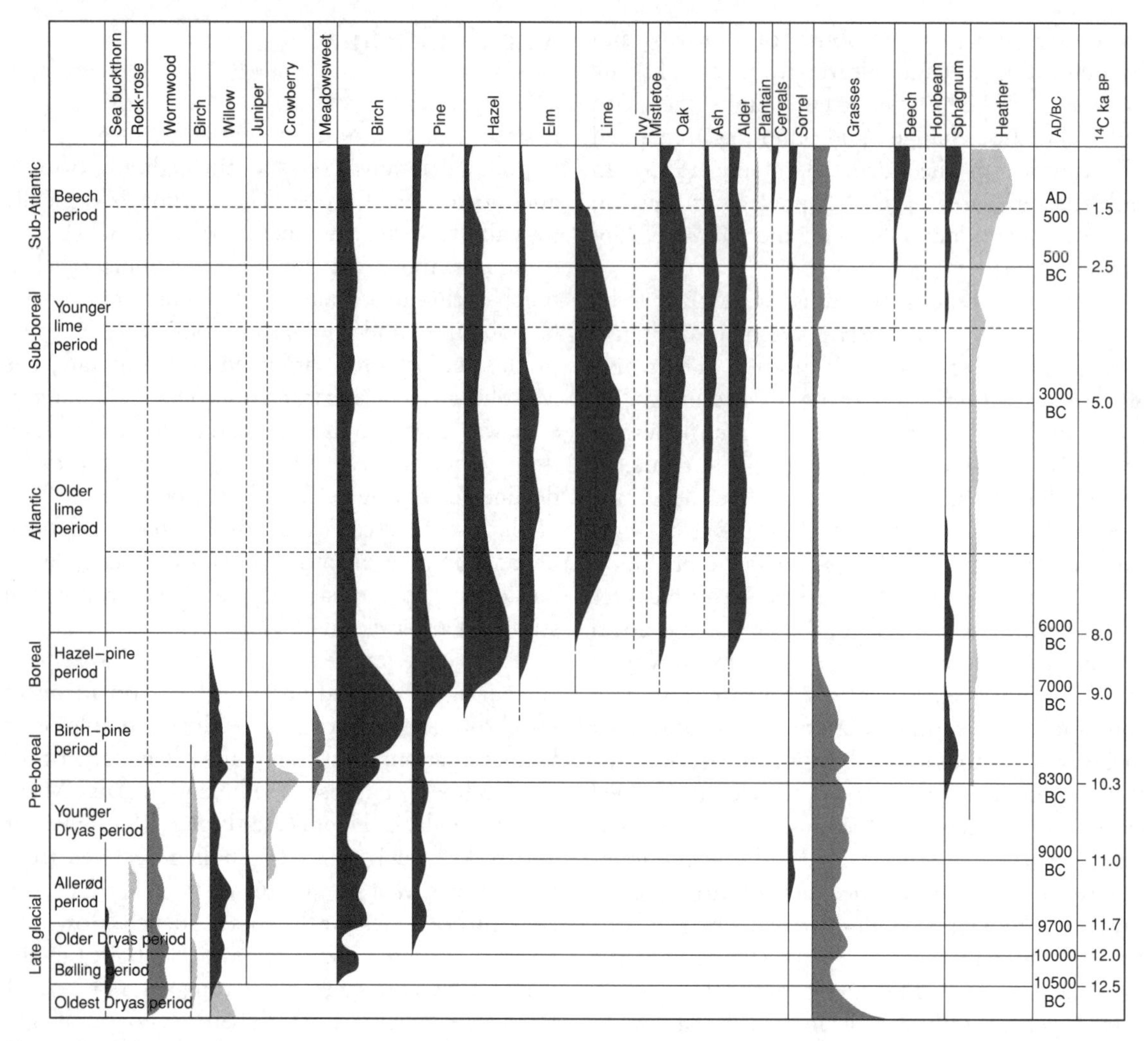

Figure 6.11 Schematic Holocene pollen diagram with curves smoothed from Jutland, Denmark (modified from Iversen, J. 1973, The development of Denmark's nature since the last glacial, *Danmarks Geologiske Undersøgelse, V Raekke*, **7–C**, pp. 1–126, Figre 29. Copyright: Geological Survey of Denmark and Greenland). Note the scale is in radiocarbon yrs BP; for a calibrated timescale see Table 1.1

apart from ivy in Denmark, temperature-sensitive plants show little evidence of reduction at the elm decline, and the overall geographical pattern and chronological spread of the elm decline do not easily fit any climatic hypothesis such as greater coolness, wetness or continentality (Huntley and Birks, 1983). Certainly, there is no evidence for a climatic deterioration sufficiently dramatic to have given rise to the rapid decline seen at Diss.

An anthropogenic explanation for the decline has also been favoured because it coincides *broadly* with the earliest Neolithic sites in Denmark and Britain and is commonly associated with slight increases in open ground plants and a reduction in other woodland taxa. Troels-Smith (1960) came up with the ingenious theory that this reflected the particular circumstances of pioneer animal husbandry in forest where grassland was limited, and leaf fodder, particularly elm, was collected and stall fed to animals. If tree pollarding (p. 205) was practised every two to three years this would have prevented flowering. There is ethnohistorical evidence for the use of leaf fodder and elm foliage is found in cattle stalls of the Swiss Neolithic lake

(a)

Figure 6.12 The elm decline: (a) Diss Mere (photo Martin Bell);

village at Weier, but here it is only a small component of the types of tree foliage used for fodder (Rasmussen, 1989).

Solely anthropogenic explanations for the elm decline suffer from a fatal flaw, in that they fail to take account of the scale of the event, which has become particularly apparent with developing knowledge of its rapid timescale. Elms, according to Rackham (2003: 266), made up about one-eighth of the woodland of Britain, covering perhaps in aggregate some 10 million acres (24 million hectares). The pollarding of this area, to the extent that pollen production was halved, would require a human population of something like half a million. Rowley-Conwy (1982) did similar calculations for Denmark and both show that the numbers of people and animals demanded by an anthropogenic explanation are far larger than is possible. Furthermore, a wholly anthropogenic explanation implies synchronous cultural change over a wide area of north-west Europe which is inconsistent with evidence for the considerable variation in the nature of the Mesolithic/Neolithic transition which was identified in the previous section.

Disease, long considered as a possible cause, received much more serious consideration when scientists could review the effects of a modern analogue, the outbreak of Dutch elm disease which spread to Britain around 1965. Within 13 years, 60 per cent of non-woodland elms had been killed (Rackham, 1986, 2003). The disease is an ascomycete fungus, *Ophiostoma (Ceratocystis) ulmi*, which partly blocks water-conducting vessels and thus causes death of the tree. It is spread by two species of bark beetle, *Scolytus scolytus* and *S. multistriatus*. At first it was argued that this disease was recent and not endemic to Europe. Rackham (2003) was able to demonstrate from literary sources convincing evidence of earlier outbreaks and there is now tree-ring evidence from Norway and Britain for nineteenth-century outbreaks. The beetle vector was around in prehistory; *S. scolytus* has been found 20 cm below the elm decline in a peat sequence at the Mesolithic site at Hampstead Heath, London (Girling, 1988), and elsewhere in early Holocene contexts (Dinnin and Sadler, 1999; Parker *et al.*, 2002). Wood showing the tunnel galleries made by *Scolytus* is present in Neolithic contexts at Amosen, Denmark, and Weier, Switzerland (Kolstrup, 1988; Rasmussen, 1989). Radiocarbon dates suggest that the elm decline spread at *c.*4 km per year, which is comparable to the recent spread of Dutch elm disease. The rapid diffusion of elm disease throughout the north European lowland *c.*5.8 ka BP can probably be equated with the emergence of a more virulent strain (Huntley and Birks, 1983).

Modern analogues strengthen the argument that a tree pathogen was responsible, by demonstrating that these can give rise to equally rapid declines as Figure 6.12b shows. A pollen study of the effects of the twentieth-century Dutch elm disease outbreak at Scords Wood, England, showed a decline over four years (Figure 6.12b; Perry and Moore, 1987). Associated with this was an increase in pollen of ruderal plants now reaching the receiving site from beyond the woodland due to the loss of the tree canopy. This ties in with the ideas of Groenman-van Waateringe (1983, 1988) who has argued that the association between pollen from plants suggestive of agricultural activity and the elm decline may be partly illusory. In a totally wooded landscape small-scale clearances are likely to have been masked in the pollen record, unless, as Edwards (1982) has shown, they happened to be very close (i.e. within *c.*30 m) of the site of pollen deposition. Under the more open conditions following the elm decline even small-scale

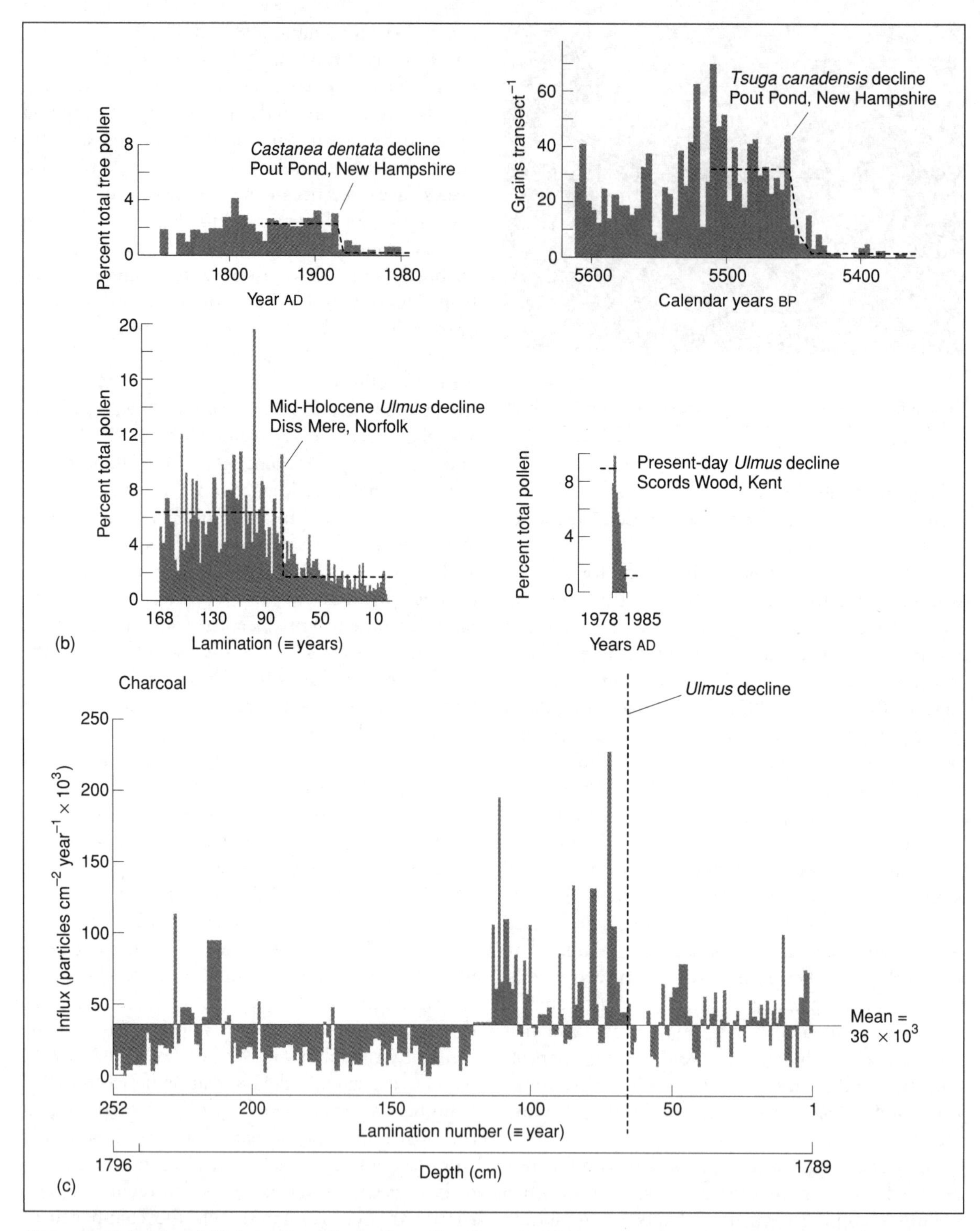

Figure 6.12 *(continued)* (b) timescale of the decline at Diss Mere, Norfolk compared to the decline of chestnut and hemlock in USA and the recent elm decline at Scords Wood, Kent (after Peglar, 1993b, Figure 8); (c) charcoal influx at Diss Mere (plotted as deviation from the mean in a sequence of 250 annual laminae (after Peglar and Birks, 1993, Figure 5)

clearances would be more likely to register in pollen diagrams.

In 1904–50 there was an outbreak of chestnut (*Castanea dentata*) blight caused by the fungus *Endothia parasitica* which virtually exterminated populations of chestnut throughout its range in eastern North America (Davis, 1981). This provides a valuable marker horizon in pollen studies of recent land-use history in the area (Anderson, 1974). About 5.5 ka BP there was a dramatic decline of hemlock (*Tsuga* sp.) which is evident in pollen diagrams from eastern North America and Canada (Davis, 1981; Bennett and Fuller, 2002). It was sudden and synchronous, within the limits of radiocarbon dating, over an area from New Brunswick to upper Michigan; 1.5 million km^2 were affected. In annually laminated sediments at Pout Pond, New Hampshire (Allison *et al.*, 1986) the speed of the hemlock decline is closely comparable to that of the documented chestnut decline. Both involved reductions (around 75 per cent) over timescales as short as 7–8 years (Figure 6.12b). Recovery of the hemlock population took a millennium. The decline is most likely caused by looper moths that eat buds and cause defoliation (Davis, 1981). Minor outbreaks occur today every 11–17 years and tree-ring studies show past episodic biotic stress which could have been caused in this way (Filion and Quinty, 1993). Climatic reconstructions across the period of the hemlock decline suggest that dry periods and/or winter cold may have increased the susceptibility of the trees to insect attack and slowed subsequent recovery of hemlock populations (Calcote, 2003).

Disease is probably the only single factor capable of producing changes of the character and timescale observed at the elm decline. Other factors may, however, have contributed to the conjunction of conditions which allowed a disease outbreak to achieve a catastrophic scale. Climatic conditions, possibly related to increased continentality, could have played a part (Parker *et al.*, 2002) but the evidence for this is limited. The case for a link to human agency is much stronger. In the laminated sediments at Diss there is evidence for human impact and local cultivation beginning 160–120 years before the decline, then 40 years before the decline charcoal increases (Figure 6.12c; Peglar, 1993b; Peglar and Birks, 1993). Significantly, however, at the decline itself there is no evidence of additional human impact. In Ireland 11 out of 34 sites show some evidence of woodland disturbance, such as plantain, preceding the elm decline (O'Connell and Molloy, 2001). A relationship to human activity may also help to explain those sites where there are a number of successive elm declines, as at Hassing Mose, Denmark (Andersen and Rasmussen, 1993) and Bikeberg, Sweden (Regnell *et al.*, 1995). It may also help to explain those sites such as North Gill (Figure 6.10) where there is evidence of elm declines at a range of dates.

The *Scolytus* beetles are particularly associated with the woodland edge and cleared areas and the recent outbreak of Dutch elm disease was particularly virulent when trees were damaged by stock or pollarding (Moe and Rackham, 1992). It is possible, therefore, that the activities of pioneer farmers in the wildwood made elms particularly susceptible to a disease which may have been present at a low level for some time. The appearance of openings within the wildwood as elms died may also have acted in a coevolutionary way to facilitate the rapid take-up of agriculture, which the isotopic evidence of human diet from both Britain and Denmark now suggests (p. 198). As Rackham (2003: 266) so perceptively put it: 'civilisation helping the spread of the disease and the disease helping the spread of civilisation'.

Clearance by early farmers

The traditional model developed by Iversen (1973) of short-term **landnam** or clearance episodes was seen as part of a shifting slash-and-burn (**swidden**) agricultural system associated with frequent settlement mobility. An early experiment in agriculture using prehistoric methods suggested a shifting system was driven by falling crop yields a few years after burning (Steensberg, 1979). Now, however, we have good resolution radiocarbon-dated pollen diagrams which show that many landnam episodes, both in Scandinavia (Rowley-Conwy, 1981; Göransson, 1986) and in the British Isles (Smith, 1981; O'Connell and Molloy,

2001), lasted for hundreds of years, with open conditions being maintained by grazing, and perhaps also by periodic crop growing within managed woodland (Göransson, 1986). What was interpreted as a brief hazel phase in the secondary succession can now be shown to have lasted a considerable period, apparently because hazel was maintained as coppice (Andersen, 1993a). Thus it is becoming apparent that the effects of Neolithic communities were not limited to clearance; they had a significant effect on the character of woodland, and the development during the Neolithic of areas of birch and hazel woodland (Andersen, 1993a, b). Landnams may often represent an amalgam of several clearance events which may only be individually resolvable in local pollen spectra from small bogs (Ammann, 1988). A greater diversity of forms of Neolithic land-use is now becoming apparent in Denmark, with lime woods being cleared for pasture and secondary birch woods burnt prior to cereal cultivation (Andersen, 1993a).

Spatial and temporal contrasts in the extent of clearance and early farming activity are increasingly evident throughout prehistoric Europe as syntheses of well-dated palaeoecological data become available (Berglund *et al.*, 1996a). Scandinavia illustrates this. Deforestation occurs earlier during the early and mid-Neolithic (from *c.*5.8 ka BP) in parts of Jutland, where woodland was naturally less dense, than in Zealand where landscapes dominated by fields and pasture appear from *c.*3 ka BP but significant woodland remained (Andersen *et al.*, 1996). On good agricultural land in south Sweden human impact only became extensive from 3 ka BP and cereals common from 1.5 ka BP, by which time deforestation was largely complete during the Viking period (Berglund, 1991; Berglund *et al.*, 1996b). Further north agriculture is smaller scale and later. In Finland there is some evidence of vegetation disturbance from 5 to 4 ka Cal BP but the appearance of cereals *c.*2 ka BP apparently relates to the development of shifting slash-and-burn agriculture which was only replaced by permanent agriculture from 0.9 ka BP (Vasari *et al.*, 1996).

In the British Isles there are also marked spatial and temporal contrasts in the dates of clearance (Greig, 1996; Edwards and Sadler, 1999). The main concentration of clearance before 4 ka BP was in the chalkland of Wessex, on the East Anglian Breckland, the Cumbrian coastal plain, in County Durham and in parts of Ireland. As a generalisation, clearance occurs earlier and more permanently on lowland sites in the south and east, mostly during the Neolithic and early Bronze Age. During these periods agriculture seems to have been of largely pastoral character and arable was limited and short term. Formal field systems of extensive character appear as early as the middle Neolithic in the west of Ireland (p. 217) and widely within England in the river valleys of the south-east (Yates, 1999), and the chalk and moorland of Dartmoor (p. 218) during the middle Bronze Age *c.*3.5 ka BC (see Figure 6.25). In the north and west of the British Isles, Neolithic and Bronze Age clearance was localised and temporary except in particularly favourable locations. Extensive clearance in the Midlands, Wales and northern England tends to be concentrated during the Iron Age from *c.*2.5 ka BP. Clearance occurs a little later in lowland Scotland and northernmost England *c.*2 ka BP where landscape transects buried by Hadrian's Wall, the Antonine Wall and associated Roman military structures show palynological evidence of landscapes cleared a few decades to a century earlier (Dumayne-Peaty and Barber, 1998; Dark, 2000a).

Within this general picture there are, however, marked contrasts in clearance dates which are explicable, not so much in geographical terms as in terms of long-lived foci of prehistoric activity and centres of social significance (Bradley, 1993). In central Wessex, southern England, the areas of Stonehenge and Avebury were at least partially open grassland by the middle Neolithic, before they developed as the great monument complexes of the later Neolithic (Evans, 1993). Other areas of comparable chalk geology, such as the South Downs, which lack such concentrations of monuments, were extensively cleared later during the Bronze Age (Bell, 1983). Similarly in Ireland the Neolithic and Beaker tombs of the Boyne Valley were constructed in grassland (Groenman-van Wateringe, 1983). The origins of these pre-monument cleared areas are considered further at the end of this chapter.

Early farmers in the circum-Alpine zone

The lake villages of the Alps and Alpine foreland in eastern France, Switzerland, south Germany, Austria and north Italy (Figure 6.13) provide us with some of the highest quality evidence for the nature of Neolithic and Bronze Age farming communities, their way of life and economy (Schlichtherle, 1997; Petrequin and Petrequin, 1988). Preserved wooden structures are in many cases precisely dated by dendrochronology, making it possible to establish when particular buildings were constructed, repaired and, usually with less precision, abandoned.

At Hornstaad Hörnle on Lake Constance in southern Germany dendrochronology provides evidence of the history of tree growth, management and use (Schlichtherle, 1990; Billamboz, 1992). When linked to high-resolution pollen analysis (Rösch, 1996), detailed work on plant macrofossils and soil history (Maier and Vogt, 2001), a picture has been developed of settlement shifting at intervals of about 20–30 years following establishment of the first lake shore settlement in 3917 BC (5867 BP; Plate 6.1). Other lake villages were occupied for similar durations, for example Hauterive-Champréveyres on Lake Neuchatel and Charavines, France. Indeed many Neolithic sites and structures in other parts of Europe had similarly brief lives (Coles and Coles, 1996; Coles, 2000a).

The settlement pattern evidence might support a model of shifting cultivation but there are aspects of the evidence here which do not fit that model. An abundant plant macrofossil record includes annual weed plants and evidence of intensive soil treatment by hoeing or digging; plants of the woodland edge or fallow are lacking (Maier, 1999). Such characteristics do not fit a slash-and-burn model but one large cultivated area on dryland 300–700 m from the settlement in use throughout its life (Figure 6.13c). The soils were rich and abandonment is unlikely to be due to soil exhaustion as the landnam model predicts. Abandonment may reflect a desire to move from a polluted site, social factors or lake-level fluctuations. Some fluctuations are synchronous from lake to lake and apparently climatically driven (Behre *et al.*, 1996). Hornstaad was a settlement raised on piles because the lake level changes seasonally due to snowmelt (Figure 6.13b). Whatever the driver of settlement shift recorded in Figure 6.13d, it created a mosaic of woodland successional stages and resulted in widespread changes in forest composition (Rösch, 1996; Behre *et al.*, 1996).

The economy of this site, and Alpine lake villages generally, also drew on wild plant resources including fruit and nuts from the woodland edge. Wild animals, particularly deer, frequently represent up to half of the meat consumed. Such diversity may have helped communities cope with hard times (Schibler *et al.*, 1997). Transhumance may also have diversified the resource base of some communities. Analysis of the composition of sheep faeces at the site of Horgen Scheller, Switzerland, suggests the animals were only present at the parent settlement for part of the year (Aberet and Jacomet, 1997). High on the Otztaler Alps, well away from the lake villages, at the treeline there is also evidence for vegetation changes associated with the seasonal use of Alpine pastures from 6.1 ka BP, an activity in which the Neolithic 'ice man' was possibly engaged (p. 174; Spindler, 1993).

Neolithic communities clearly had an effect on the vegetation of the woodland edge locally, perhaps on the high Alps and more extensively through a shifting settlement pattern around the lakes. Although the character of the woodland was changed, the lowlands remained an essentially wooded landscape and it is only with the Bronze Age, *c.*4 ka BP, that we have extensive clearance and the emergence of permanently cleared areas and grassland within remaining woodland; some of this was not cleared until medieval or later times (Rösch, 1996; Behre *et al.*, 1996).

Woodland management

Evidence for the extensive survival of ancient woodland in Britain as a result of its deliberate management as a valued resource comes from an eclectic combination of historical studies, fieldwork on wood boundaries and botanical composition (Rackham, 1986, 2003). Plant species have been identified which do not occur in the secondary

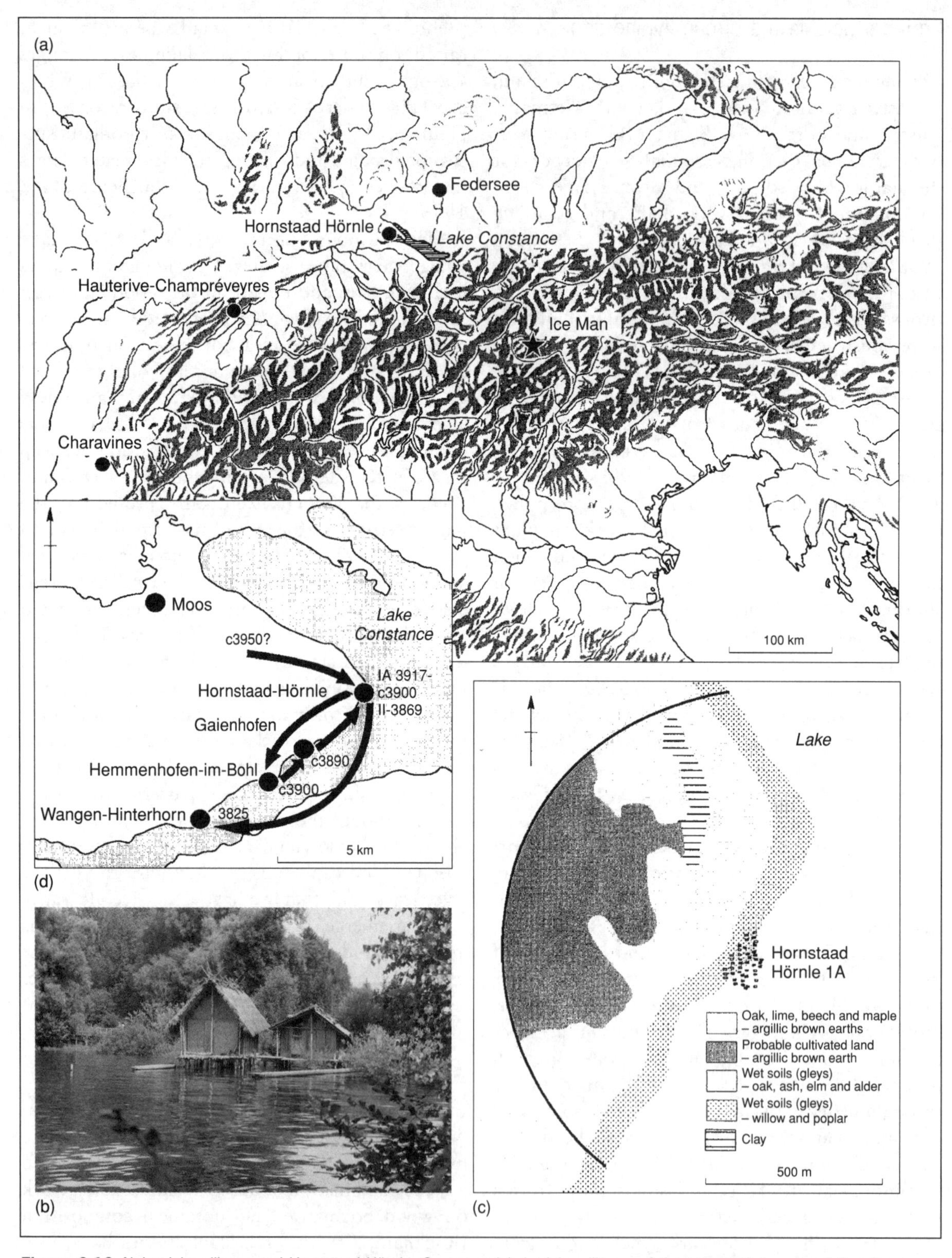

Figure 6.13 Alpine lake villages and Hornstaad Hörnle, Germany: (a) the lake villages of the circum-Alpine zone (after A. Kalkowski in Schlichtherle, 1997. Reproduced by permission of Landesdenkmalamt Baden-Württemberg and A. Kalkowski); (b) reconstruction of Hornstaad Hörnle Neolithic house at Unteruhldingen on the shore of Lake Constance (photo Martin Bell); (c) lakeshore settlement in relation to Neolithic soils and vegetation (after Maier and Vogt 2001, Figure 3); (d) a tentative model of shifting settlement patterns (adapted from Coles and Coles, 1996, Figure 26)

woodland of recent centuries and are restricted to woods which, on historical evidence, are ancient, sometimes medieval. These plants are known as ancient woodland indicators (AWIs) and have been used to identify sites of particular nature conservation significance (Rackham, 2003). One of the most important is small-leaved lime (*Tilia cordata*), which was an important component of the woodland in lowland England (Figure 6.8). In Britain much of the lime wood disappeared during the Bronze Age, perhaps because of grazing pressure; it survived on some sites to Anglo-Saxon times and in places to the present, but it tends not to be found in secondary woodland. AWIs have been argued as indicating continuity of woodland back to the wildwood. However, when this hypothesis was tested by pollen analysis at Sidlings Copse, England, a wood with many AWIs and a documented history back to the thirteenth century AD, it was found to have been cleared from *c.*AD 300 to 1000, demonstrating that AWIs do not necessarily indicate unbroken woodland continuity (Day, 1993).

Ancient woods have survived, not as unconquered wilderness, but because they provided resources which would have been of importance, and were therefore carefully conserved and managed in all periods. Woodland provides timber, for instance by coppicing (cutting close to the ground to produce a crop of long straight poles) and pollarding (cutting above the height of browsing cattle), as well as a wide range of plant and animal resources including wood pasture, pannage for pigs and leaf fodder.

Waterlogged wood on archaeological sites and, less directly, pollen analysis, provide evidence for the early origins of deliberately managed woodland. In Mesolithic Denmark, at sites such as Halsskov (6.7 ka BP) and Tybrind Vig, wooden fish traps were made of regularly sized hazel stakes which had been cut on a coppice cycle of 6–9 years (Christensen, 1997). Production of basketry fish traps would have required careful management of woodland to produce materials of requisite quality, in the way that Native American non-agricultural communities managed woodland to supply basketry and many other raw materials and foods (Blackburn and Anderson, 1993). An abundance of hazel in pollen spectra close to Danish later Mesolithic coastal sites is interpreted as evidence of coppicing, which stimulates flowering and pollen production. This is an alternative view to set beside Vera's (2000) hypothesis that high hazel values reflect open park-like woodland maintained by grazing herbivores. Of course both people and herbivores are environmental disturbance factors which are likely to have produced similar effects on vegetation communities, once again reflecting the problems of equifinality (p. 50).

High hazel values also follow Neolithic landnam in Denmark and Sweden and again an extended period of coppice is envisaged (Göransson, 1987; Andersen, 1993b) although in some cases, as at Ystad, Sweden, this may be more like stump-sprout forest rather than the more strictly and cyclically managed coppices of historic record in Britain and parts of central Europe (Berglund 1991: 168). The later Neolithic hurdle trackways of the Somerset Levels, England, such as the Walton Heath Track (4.3 ka BP) were also made of hazel rods from coppice (Figure 6.14).

Woodland clearance in the Americas

The extensive impact of prehistoric farmers in most areas of Europe is in dramatic contrast to the situation in North America where, in the eastern states, European colonists encountered a landscape of deciduous woodland (Figure 6.15). Accounts by the first European colonists demonstrate, however, that in places the woodland was park-like with grassy areas. These openings have been attributed to the activities of grazing animals, particularly bison, which formerly had a much more extensive distribution (Figure 6.16; Vera 2000). However, there is much evidence for grassy glades and park-like landscapes in areas beyond the main distribution of bison on the east coast in New England and on the west coast in Washington State. In these and other areas the ethnohistorical record leaves little doubt that burning by Native American communities was a key factor in the maintenance of these grassy and park-like landscapes (p. 193). In New England Cronon (1983) records that it was particularly in areas where agriculture was practised, and there were higher densities of population, that burning took place, although by no

Figure 6.14 Late Neolithic hurdle, Walton Heath trackway, 4.8 ka BP, Somerset Levels, England (photo Somerset Levels Project)

means exclusively for agricultural purposes. Here agriculture was by seasonally mobile communities practising slash-and-burn methods and moving on after 8–10 years when fertility declined.

Native American farming practices did not leave their mark on anything like the scale we have seen in prehistoric Europe. According to McAndrews (1988) prehistoric agricultural impacts register in less than 7 per cent of Holocene pollen diagrams in North America. An indication of human vegetation disturbance is a peak of ragweed (*Ambrosia*) pollen in a similar way that plantains (*Plantago lanceolata*) indicate prehistoric activity in Europe. The more limited impact of agriculture in much of America reflects the fact that Native American communities did not keep stock and in some areas the transition to agriculture was gradual, with small-scale agriculture combined with continued use of wild resources (Wills, 1995), which were of particular importance during seasons, or years, of shortage (O'Shea, 1989).

Maize cultivation reached the southern United States from Mexico (p. 158) by 3.2 ka BP, the eastern deciduous woodlands by 2 ka BP and New England by the first millennium BP (Mulholland, 1988; B.D. Smith, 1995b). Palaeoenvironmental sequences from these eastern areas show evidence of limited small-scale agriculture (Davis, 1984; Thorbahn and Cox, 1988). At Crawford Lake, Ontario, an episode of maize cultivation began AD 1360, as dated by annual sediment laminations (McAndrews, 1988). A coeval decline in cedar pollen may relate to the use of its timber and bark in construction of a nearby village which was probably occupied for 10–20 years. A correlation between these changes and increased charcoal led to the view that Iroquois burning may have been responsible for a marked change in forest composition at this time (Clark and Royall, 1995).

The effects of prehistoric communities are more marked along the major river corridors such as the Mississippi, Tennessee and their tributaries. Plant communities in the Little Tennessee Valley were increasingly modified with high *Ambrosia* pollen values from 1500 BP (Delcourt and Delcourt, 1991). Significantly, the disturbed areas spread out

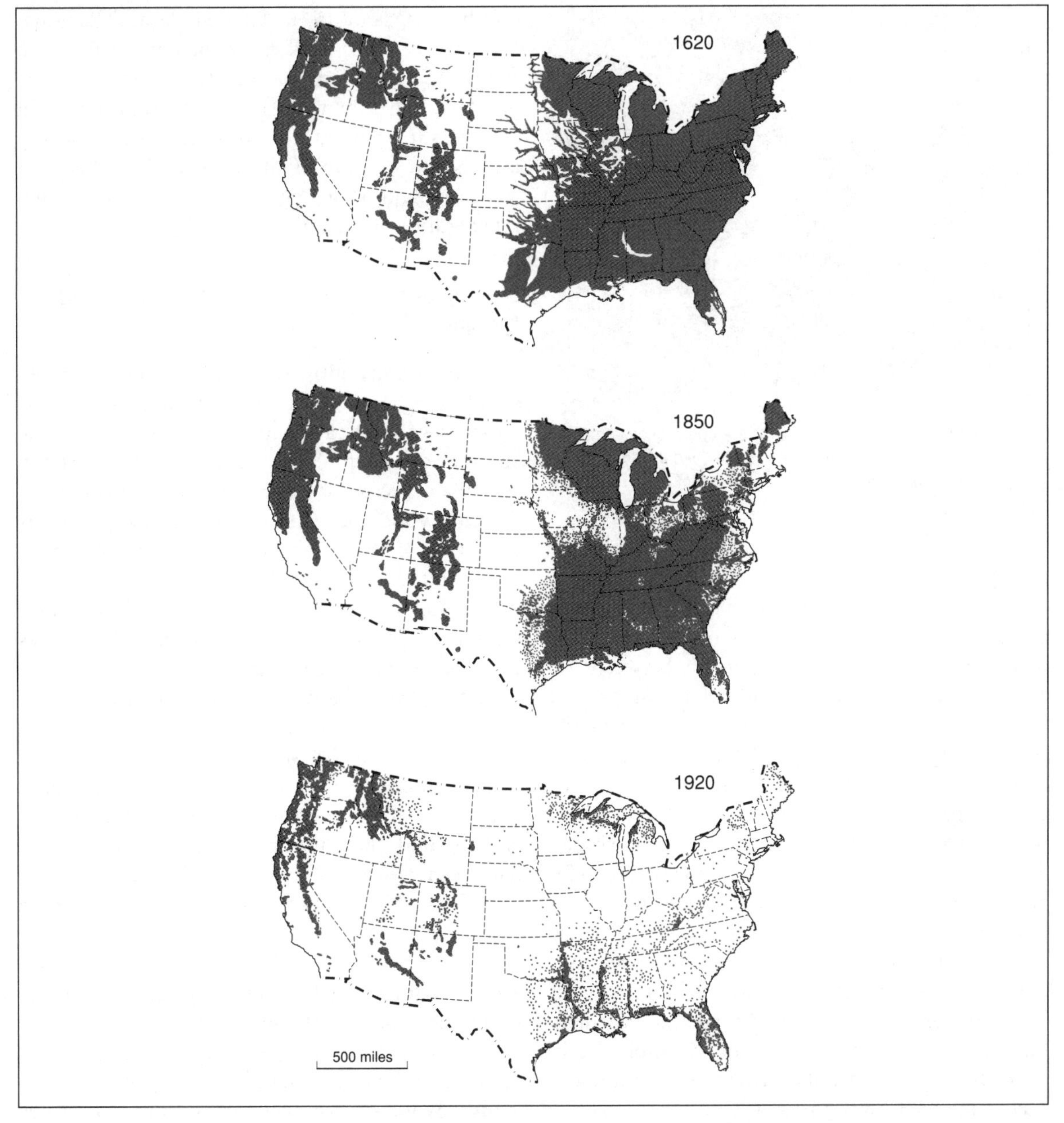

Figure 6.15 The United States showing the estimated extent of virgin forest in 1620, 1850 and 1920. This diagram does not include secondary regenerated woodland and thus underestimates the extent of twentieth-century tree cover (after Greeley, 1925; Williams, 1990. Copyright © 1990 from Clearing of the Forests by M. Williams in *The Making of the American Landscape*, edited by M.P. Conzen. Reproduced by permission of Routledge/Taylor & Francis Books, Inc.)

from areas along the river where natural disturbance would particularly have been concentrated, highlighting the synergistic relationship between natural and anthropogenic disturbance and early agriculture which is found in North America (p. 158; B.D. Smith, 1995a).

When Europeans settled in North America they introduced alien crops and agricultural practices

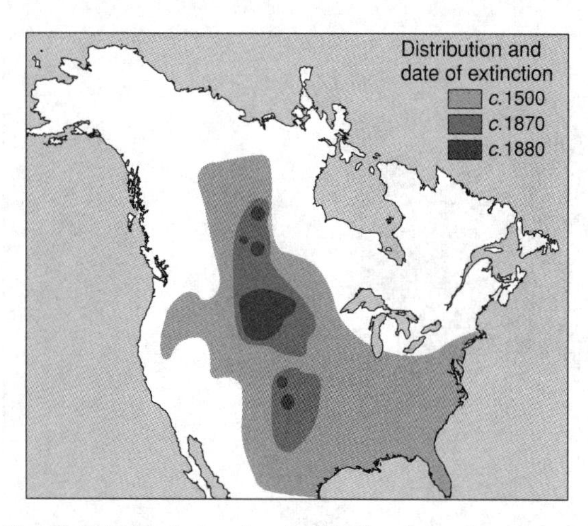

Figure 6.16 North America showing the contracting range and near extermination of the bison (*Bison bison*) from AD 1500 to 1880 (after Vera, 2000, Figure 2.3)

that had evolved over millennia in quite different environmental conditions. Cronon (1983) highlights the cross-cultural comparison between the environmental perception and relationships and ways of life of Native American and European communities in New England. In this area deforestation was under way from the seventeenth century partly to supply raw materials for trade with Europe. The forests of the eastern United States were cleared on a vast scale (Figure 6.15), particularly in the second half of the nineteenth century (Conzen, 1990; Williams, 2003). Hence the most profound impact on the Holocene woodland of North America came not, as in north-west Europe, from the activities of prehistoric farmers but from the depredations of technologically sophisticated immigrants particularly within the timeframe of the last 150 years.

A traditional model in which relatively benign Pre-Columbian land-use is contrasted with the environmentally destructive effects of Europeans is, however, increasingly challenged by palaeoenvironmental records from Middle and South America (Butzer, 1996: Endfield *et al.*, 2000). Intensive Pre-Columbian agriculture led to large-scale erosion in parts of Mexico. Some areas of tropical rainforest once thought to be pristine are revealing previously unsuspected histories of slash-and-burn agriculture in Guatemala and Panama, for instance, with marked periods of regeneration following the Spanish conquest (Cooke *et al.*, 1996). Similarly parts of the Amazonian rainforest can now be shown to have been settled, partly cleared and their vegetation significantly modified between AD 1250 and 1600, again prior to post-colonial regeneration (Heckenberger *et al.*, 2003).

Biological consequences of clearance and farming

Clearance and cultivation, while they remained small-scale and shifting, contributed to the creation of landscape mosaics and thus, in some specific areas, to an increase in biodiversity. North American areas (p. 193) illustrate this as do Scottish woodlands (Tipping *et al.*, 1999). In parts of Denmark maximum biodiversity seems to have obtained in the Iron Age followed by a pronounced decline, as increasingly uniform agricultural landscapes were created (Andersen, 1993a). Key issues concern the scale of human effects and the linear continuity of habitats. Once islands are created, whether in the physical geographic sense (e.g. by sea-level rise) or islands of particular habitat (e.g. woodland), then species become much more vulnerable to the effects of local extinction. It is estimated that about 50 per cent of Britain had been cleared of woodland by the Iron Age (Rackham, 2003), whereas in many parts of Europe extensive areas of forest remained into medieval and even post-medieval times (Berglund *et al.*, 1996a; Williams, 2003). Increasingly through the Holocene, species of closed woodland declined and those preferring open conditions, disturbed habitats and the woodland edge expand dramatically beyond the refugia they occupied in times of extensive woodland cover. Domesticates were introduced from other geographical areas, bringing with them their diseases and parasites and the weed floras of crop plants (di Castri *et al.*, 1990). Near Eastern and Mediterranean weed species colonised north-west Europe with the first farmers (Sykora, 1990). Much later, weeds from Europe were carried with seed corn and other products to colonies all round the world (Crosby, 1986; Grove, 1995). In New Zealand 80 per cent of the weeds are now of European origin (Salisbury, 1964). Common plantains introduced to America by Europeans were called 'whiteman's foot' because they followed the colonists everywhere (Iversen, 1941; McCracken Peck, 1990).

In Europe people were a major cause of biological change after 6 ka BP, making it increasingly difficult to identify changes in plant and animal distributions that are due to climatic and other factors. The dramatic reduction of Scots pine (*Pinus sylvestris*) in Ireland and England *c*.4.5 ka BP was partly due to increasingly wet conditions in the boggy areas it inhabited, but human agency contributed to rising water-tables and the eventual local extinction of Scots pine in Ireland in the first century AD (Pilcher and Hall, 2001). Lime (*Tilia*) may also have been affected by declining temperatures after the Climatic Optimum (p. 91), however, the decline is not synchronous and often associated with evidence in pollen diagrams for grazing. Other trees benefited from the conditions people created; beech (*Fagus*) and hornbeam (*Carpinus*) did well in the secondary woodland of abandoned agricultural land, for instance at the time of post-Roman regeneration in northern Europe (Behre, 1988).

The later Holocene saw reductions in the ranges of many mammals, islands suffering particularly early and severe losses (p. 191). Aurochsen, or wild cattle, were common in the European wildwood but on the Danish island of Zealand became extinct, along with five other mammals, early in the Climatic Optimum (Aaris-Sørensen, 1980). This is partly a product of isolation by rising Holocene sea levels as the present geography of the Baltic took shape (see Figure 4.21). Hunting by Mesolithic communities certainly contributed, however, for there are finds of aurochsen which had suffered multiple injuries with microlith-tipped weapons (Plate 6.2; Aaris-Sørensen, 1998). In most of Europe it was a combination of habitat loss and hunting by later prehistoric communities that decimated the aurochsen. They were locally extinct in mainland Denmark *c*.2 ka BP and in Britain *c*.3.5 ka BP, and became totally extinct with the death of the last animal in Poland in AD 1627. Beaver, bear, wolf and boar all suffered habitat destruction and human predation resulting in their extinction in Britain in early historical times (Rackham, 1986) and in drastic reductions in their mainland European ranges.

The effects on invertebrates, including beetles, of human activity combined with subtle climate change, are particularly well documented (Buckland and Coope, 1991; Robinson, 2001). In Britain 44 species of insect found in mature woodland, wetland and species-rich grassland have become locally extinct as a result of habitat fragmentation, possible temperature reductions since the Bronze Age and other changes such as the loss of decaying wood in managed woodland (Dinnin and Sadler, 1999). Many of the species lost are those of the Urwald, or primary undisturbed forest. On Thorne and Hatfield Moors, England, Urwald species now extinct in Britain are associated with the last remnants of Neolithic and Bronze Age pine woodland (see Figure 8.6; Whitehouse, 2000, 2004). *Rhysodes sulcatus* is an inhabitant of rotting wood known from this and other Neolithic and Bronze Age contexts (Buckland, 1979; Speight, 1991). Today it has a rare and disjunct (i.e. patchy) European distribution which continues to decline; relict populations recorded in the nineteenth century in south Sweden and Germany have since died out (Figure 6.17). Relict populations of a few Urwald species hang on today in Britain on ancient wood pasture sites such as the Royal Forest at Windsor. Significant levels of grazing may not, therefore, be incompatible with the survival of at least some Urwald taxa. Figure 6.18 illustrates the changing beetle faunas of selected British sites between the later Mesolithic and the Iron Age (Dinnin and Sadler, 1999). Mesolithic sites have predominantly woodland taxa with small proportions of open ground and dung beetles, which may be significant given current debates concerning the role of grazing in early Holocene woods (p. 184). Neolithic activity had a modest impact on woodland taxa but the proportion of open country species and synanthropic species (those dependent on human activity) increases. Some, but not all, of the Bronze Age assemblages are significantly less wooded, but the really marked change comes with Iron Age faunas which have predominantly dung and open ground taxa, indicating pastoralism.

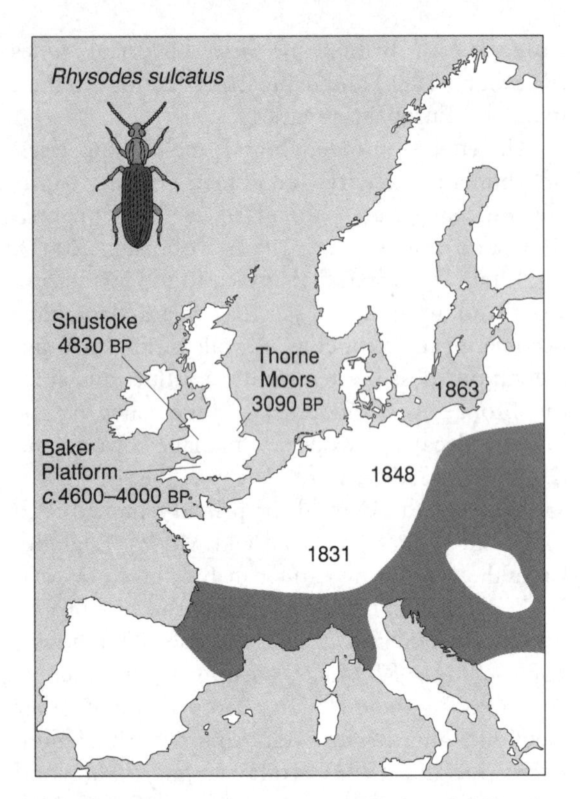

Figure 6.17 The past and present distribution of the beetle *Rhysodes sulcatus* (after Buckland, 1979, Figure 12)

Human activity also transported invertebrates to new areas and created artificial environments in which they were able to flourish. In Roman Britain pests of stored grain appear, sometimes in huge numbers (Buckland, 1991). Norse communities introduced many beetle taxa, particularly synanthropic species, to the North Atlantic islands with imported fodder and produce. Today these make up around half of the beetle fauna of Iceland and Greenland (Buckland *et al.*, 1991; Sadler, 1991; Amorosi *et al.*, 1994). The interplay between people and climate is illustrated by the death-watch beetle (*Xestobium rufovillosum*) which was once present in the wildwood of northern Britain where today there are only disjunct populations in the artificial habitats created by old buildings. The southward reduction of their range has been attributed by Buckland (1975) to the effects of the Little Ice Age (p. 94).

The wildwood was inhabited by a number of mollusc species which, though not totally extinct in Britain, are now reduced to small relict distributions by deforestation and drainage (Kerney, 1999). *Vertigo alpestris* is an example; widely distributed in the first half of the Holocene, it now survives in relict patches in north-west England, Wales and Scotland (Figure 6.19). The loss of woodland occurred widely in calcareous areas during the Neolithic and Bronze Age because these areas of light soils were particularly suitable for prehistoric agriculture. The result was extensive areas of grassland dominated by open country assemblages (Evans, 1993). Some had been characteristic of the open environments of the Lateglacial (p. 125), others, such as some Helicellids, were introduced to Britain by coastal trade from further south in Europe, flourished initially in dry coastal dune environments, and then colonised the seasonally hot and dry habitats created by people in arable land and elsewhere (Preece, 2001).

Dramatic changes in the ecology of agricultural landscapes continue to take place. Increasing use of pesticides and herbicides means that many of the once ubiquitous weeds of arable land are becoming increasingly scarce (Greig, 1996) and some, such as the corncockle, are now threatened with extinction in Britain. Today human activity in many parts of the world has created agricultural landscapes dominated by monocultures of introduced species, be they cereals or genetically identical conifer plantations. Significant future effects may be brought about by the use of genetically modified organisms which, research suggests, may be detrimental to biodiversity.

Holocene pedogenesis

Pedogenic processes and Holocene soil history are reviewed by Limbrey (1975) and French (2003). The broad-scale history of northern temperate latitudes may be outlined in terms of the interglacial cycle which was introduced on p. 124. The base-rich soils of the Lateglacial and early Holocene developed under woodland of the mesocratic phase into brown earths (cambisols) of circum-neutral to slightly acidic character. Trees, being deep

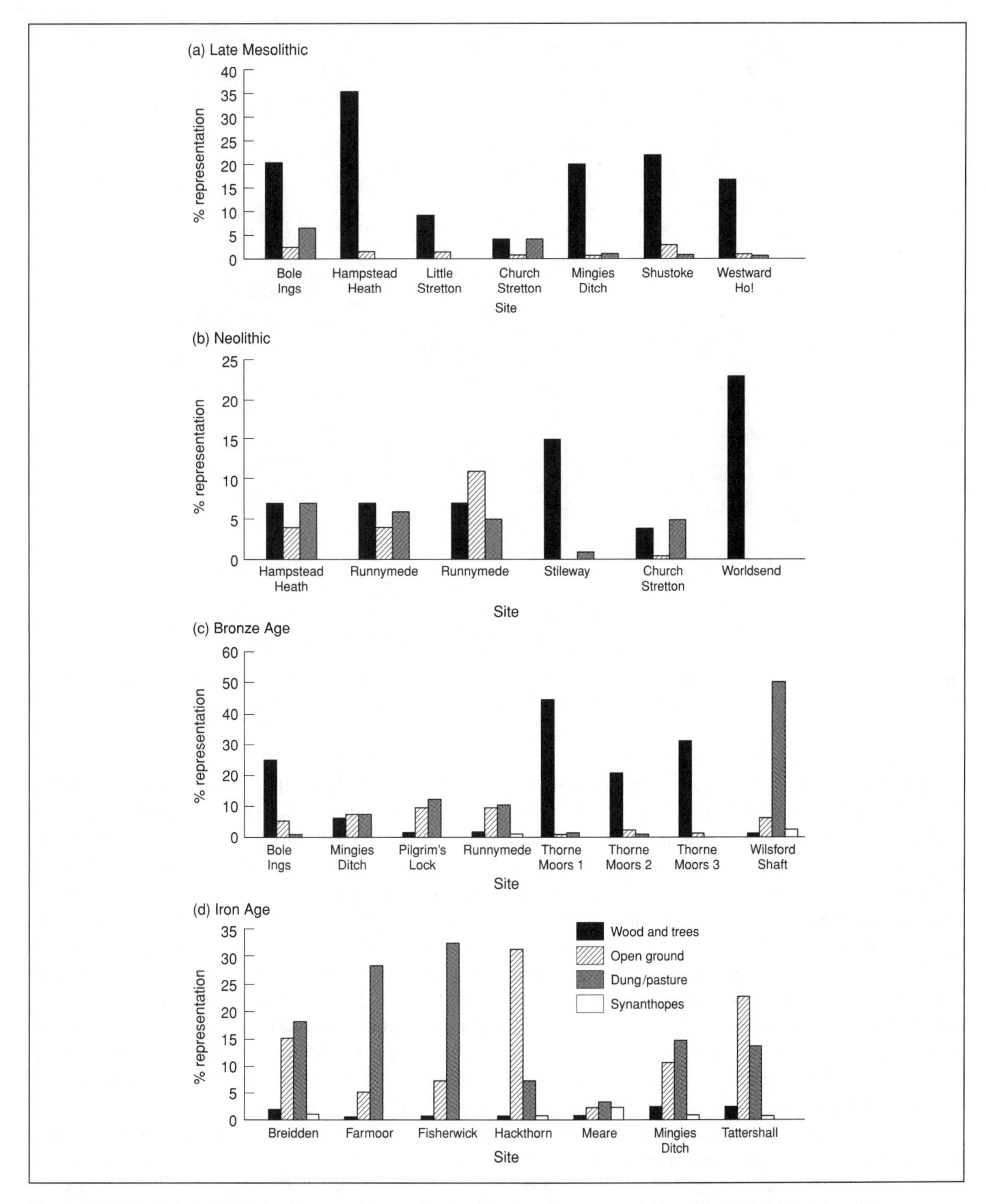

Figure 6.18 Prehistoric British beetle assemblages showing the changing representation of broad habitat categories (after Dinnin and Sadler, 1999, Figure 2)

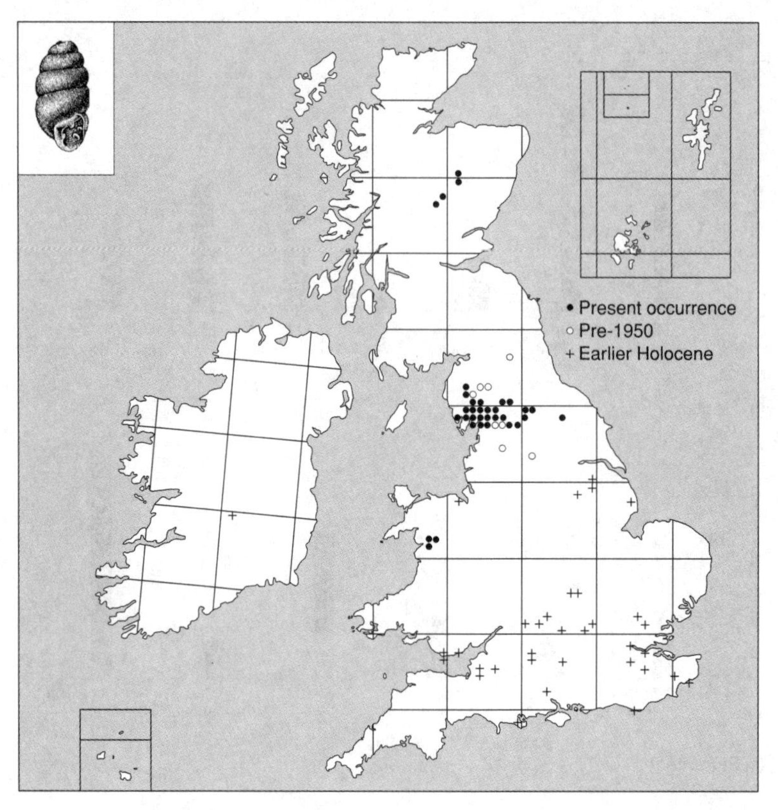

Figure 6.19 The past and present distribution of the mollusc *Vertigo alpestris*. Map produced by the Biological Records Centre, CEH Monks Wood from records of the non-marine Mollusca Recording Scheme (after Kerney, 1999)

rooted, draw nutrients from the lower part of the soil profile but shed leaves, etc., on the soil surface; this recycling counteracts the natural tendency of soils to leach (Figure 6.20). Evapotranspiration by trees draws water from the soil and the tree canopy intercepts rainfall facilitating its gradual infiltration into the soil. Thus trees ameliorate the effects of water runoff and leaching which would otherwise be much greater, particularly in high rainfall areas.

Deciduous woodland soils support an active fauna of earthworms, mites and enchytraeid worms. Under base-rich conditions, faunal activity creates intimate mixtures of organic and mineral soil fractions, for example in earthworm casts and excrements that form the crumb structure of a soil, which is of great importance in retaining nutrients and soil structural stability. A base-rich soil with a high well-mixed organic content is described as mull humus.

To the north of the deciduous forest in the boreal coniferous woodland of northern Europe the vegetation produces a more acid litter and low temperatures lead to limited microbiological activity. Under these conditions bases are leached from the upper (eluvial) horizons which eventually become bleached as clay is lost and iron, aluminium and organic matter are moved down the soil profile and deposited as illuvial horizons. This characteristic horizonation is known as a podzol in European terminology or as spodosols in North America; this represents the natural climax soil in the boreal vegetation zone.

South of the boreal zone, in Denmark and eastern England, natural (i.e. pre-human impact) podzol formation occurred locally on sandy base-poor soils (Iversen, 1969; Valentine and Dalrymple, 1975). However more widespread podzol occurrence follows the loss of woodland. Increased leaching, acidification and associated

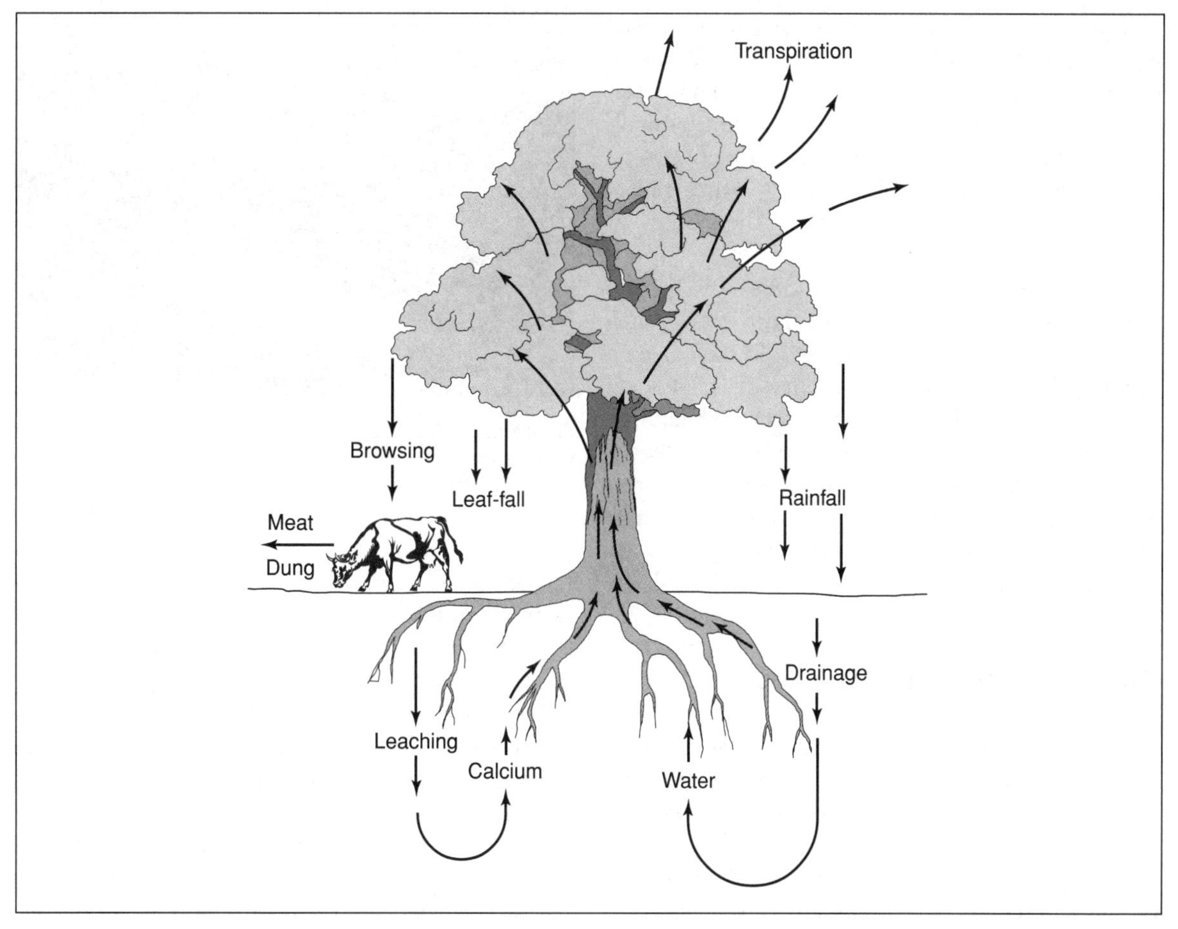

Figure 6.20 Diagram showing the cycling of calcium and water through the soil in a wooded environment (after Evans, 1975, Figure 37)

vegetational change characterise the oligocratic phase of the interglacial cycle. The processes involved can be summarised as follows. Loss of woodland (however caused) meant that nutrients were no longer recycled from depth in the profile, there was a loss of soil structural stability and gradually nutrient-demanding plants were replaced by those tolerant of low nutrient levels. The litter produced was poor in bases and rich in tannins which inhibit decay. Consequently mull gave way to mor humus characterised by low levels of faunal mixing and thus accumulation of organic plant material on the surface. This process was reinforced by hydrological changes. Loss of woodland reduced evapotranspiration and produced more rapid run-off, increasing stream discharge by between 10 and 40 per cent (Moore, 1985, 1986). Increased groundwater resulted in lower microbial activity, soil temperatures and decomposition rates. Drainage was often impeded by the development of a semi-permeable podzol ironpan. The trend towards increased leaching and hydromorphism (impeded drainage) would have been further exacerbated by the effects of later Holocene climatic deterioration (p. 91).

Prehistoric soils were buried below prehistoric monuments, such as the burial mounds which are widespread in Europe (Plate 6.3), or within river valley alluvial sequences which are the main contexts for archaeological work on Holocene soils in North America (Holliday, 1992; Waters, 1996). Information on the structure of the buried soils and the biological evidence they contain is increasingly supplemented by examination of soil micromorphological thin sections (p. 39). These can reveal evidence of leaching, translocation, podzolisation, cultivation, colluviation, alluviation, etc. Sequences of micromorphological features reflect successive developmental stages in the history of that soil (French, 2003).

The general model of Holocene soil history outlined is, of course, a simplification as French (2003: 67) has argued. Reality would have been a mosaic of different soil types reflecting geological background, patches of earlier soils which had escaped erosion and vegetation mosaics created by a range of disturbance factors. Sections of this mosaic offered contrasting opportunities for past communities whose activities often gave rise to significant soil changes as the following sections show.

Blanket bogs and raised mires

Peat formation occurs where surface organic matter does not decompose, particularly because of a high water-table. Two types of peat context are discussed here. Both are essentially ombrogenous, nutrient-poor, rain-fed peats (p. 32). Their growth is, therefore, largely determined by climatic factors and they provide important palaeoclimatic records. Raised mires form in lowland basins where peat, particularly of water-retentive *Sphagnum* species, grows above the level at which it receives water and nutrients from drainage or underlying minerogenic sediments. Blanket bogs, which are considered first, mantle the whole topography with peat sometimes metres thick. They occur in high rainfall areas (>1000 mm p.a.) and are well developed on the west coast of Ireland, Scotland and Norway. Mires are also extensive in more continental areas of the boreal region and at their maximum extent covered one-third of Finland and a tenth of Sweden. Many areas covered by blanket bog once carried woodland, as shown by pollen analysis and by the presence of waterlogged tree stumps (Figure 6.21). A dramatic landscape change has occurred but opinion is divided as to what brought this about. Scientists variously emphasise the role of climate, pedogenesis and anthropogenic factors in what is essentially a classic illustration of the equifinality problem (p. 5).

Figure 6.21 Connemara, Ireland: pine trees within blanket peat (photo Martin Bell)

Initially the prime cause of blanket bog initiation in Britain was seen as climatic (Godwin, 1981). However, as the number of sites with radiocarbon dates for peat initiation increased, it became clear that the process occurred over a protracted time period without clear clustering in wetter phases. Blanket peats cover earlier mineral soil and it is possible that acid brown soils supporting woodland gradually became more base deficient and leached, leading to podzol development and the accumulation of mor humus. However, not all blanket peats exhibit evidence of earlier podzolisation.

Sites in both Scotland and Ireland show blanket peat formation prior to significant human impact. There is a growing number of sites, however, at which there is evidence for charcoal, vegetation disturbance, or agriculture at around the mineral soil/peat transition (Smith, 1981; Moore, 1993). It has been suggested that fine charcoal particles block the soil pore structure and exacerbate the tendency towards waterlogging. Reference has already been made (p. 194) to sites on the Pennines, Black Mountains and other uplands where blanket peat started to form during the Mesolithic, sometimes close to settlements. Elsewhere, it is argued, small-scale Neolithic clearance, or limited grazing within woodland was sufficient to cross critical thresholds and trigger peat formation (Wiltshire and

Moore, 1983). In Wales the most extensive blanket peat formation occurred in the Bronze Age but at some sites as late as 1.4 ka BP (Chambers, 1996).

Below peat on the north coast of County Mayo, Ireland, a complete buried Neolithic landscape of field walls, tombs and settlement sites has been revealed (Figure 6.22). It covers 1000 ha and is the best preserved Neolithic landscape in Europe (Molloy and O'Connell, 1995; Caufield *et al.*, 1998). The field walls are on mineral soil, they are associated with herb-rich grassland and a phase of agriculture from *c.*5.4 ka BP. Farming was largely pastoral with limited arable areas indicated by ploughmarks, a lynchet (see p. 228) and cereal pollen. After perhaps 500 years most of the fields were abandoned and subject to encroachment by

(a)

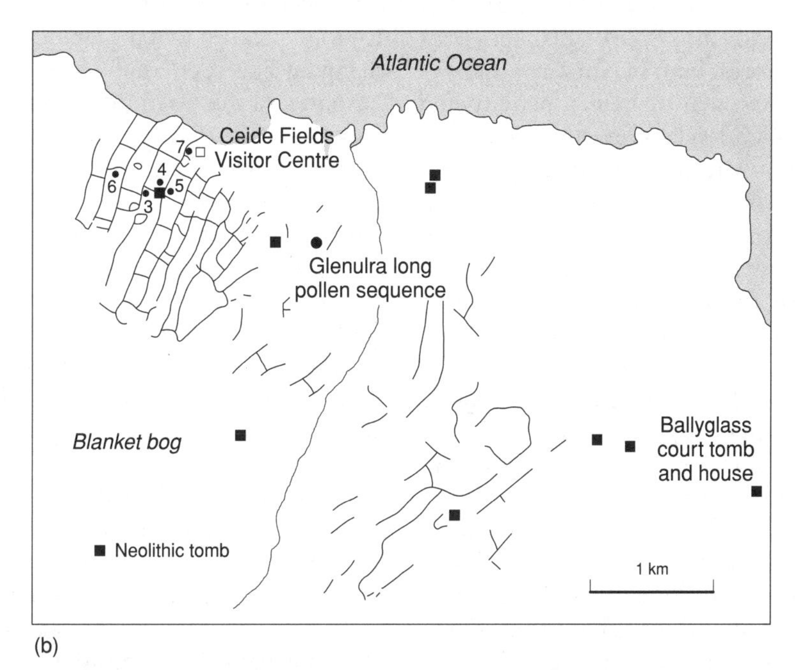

(b)

Figure 6.22 Ceide Fields, Co. Mayo, Ireland. Neolithic field walls under blanket peat: (a) photo of walls excavated from below peat. In the background is the pyramid-like modernist interpretation centre (photo Martin Bell); (b) plan of the field system (after S. Caufield *et al.*, 1998 and Molloy and O'Connell, 1995, Figure 2A)

blanket peat. Its spread was diachronous (time transgressive), and areas not encroached upon until later had some further agricultural use in the late Bronze Age and Iron Age. At both Ceide and in the Connemara National Park (O'Connell, 1994), peat initiation appears to have followed, rather than caused, abandonment of Neolithic agricultural activity. The subsequent spread of blanket peat in parts of Connemara also followed a reduction in the level of agricultural activity in the Iron Age (Molloy and O'Connell, 1993; O'Connell and Molloy, 2001).

Some of the west Norwegian blanket mires also started to form following woodland clearance, well-developed blanket bog being restricted to areas of early clearance (Kaland, 1986, 1988). On the island of Haramsøy blanket mire began to form at 3 ka BP on an unwooded upland plateau following intensification of land-use indicated by charcoal evidence for regular burning (Solem, 1989). The plateau was largely used for grazing and some crop growing. However, there are other sites where human activity was minimal and peat initiation is believed to have been caused by increased waterlogging as a result primarily of climatic factors (Solem, 1986). Blanket bogs around the alpine forest limit in Norway also have histories unrelated to that of human activity and appear to reflect high levels of precipitation.

We have seen how records of changing surface wetness in raised mires constitute an important palaeoclimatic record (Figure 320). Raised mires in Northern Ireland and blanket bogs in Scotland have been shown to demonstrate a comparable response to the climatic downturn of the Little Ice Age (see Figure 2.12; Barber *et al.*, 2000). Mires also contain evidence of a dry phase *c*.4.5 ka BP when pine (*Pinus sylvestris*) colonised bogs including Ceide Fields and Connemara National Park (Figure 6.21; Caufield *et al.*, 1998; O'Connell and Molloy, 2001). The same episode is represented by dendrochronologically-dated phases of tree growth on bogs in Northern Ireland and three sites in northern England including Thorne and Hatfield Moors (see Figure 8.6; Lageard *et al.*, 1999; Boswijk and Whitehouse, 2002).

An additional factor requiring consideration is the effect of people and grazing animals on

Figure 6.23 The Iron Age road at Corlea, Ireland (photo courtesy of Barry Raftery, University College, Dublin)

bog growth and hydrology, given extensive evidence from lowland wetlands that at least the less hazardous parts of these environments were grazed and utilised by people in various ways (p. 166). Use of the Irish bog environments is evident from the many trackways, of which the most dramatic is the great bog road at Corlea dated dendrochronologically to 148 BC (Figure 6.23; Raftery, 1996). Detailed stratigraphic investigation of changing bog topography and hydrology has led Casparie (1986 and 2001) to suggest that in some instances trackway construction led to bog bursts (catastrophic discharges of saturated peat) at Derryville in Ireland and in the Netherlands. Trackway constructions designed to create, or maintain, networks of social communication in wetlands sometimes made matters worse! The Corlea road was so heavy it rapidly sank into the bog and can only have been used for a short time.

The present state of the blanket peat debate may be summarised by noting that in some localised very high rainfall areas, such as the west of Scotland, nuclei in the west of Ireland and upland Norway, peat formation was under way before human activity on any scale and may, therefore, be seen as a result of natural Holocene conditions. It is also increasingly clear that the changing plant macrofossil composition of bogs constitutes an important palaeoclimatic record (p. 32). Many areas of peat have evidence of charcoal at their base, although we should perhaps keep an open mind as to whether the fire is wild or anthropogenic (p. 196). Often anthropogenic influence at the soil/peat interface is evident from other sources, for example pollen, artefacts and walls. Thus, a subtle interplay between fluctuating wetness and the effects of human activity was responsible for the widespread formation of Holocene peats. In some areas what may have happened much later in the present interglacial was greatly accelerated by human activity, while elsewhere the area covered by blanket peat may have been extended by the dramatic ecological changes which people brought about.

The development of moorland

Nearly all moorland areas of Britain and Atlantic Europe were tree covered at the Climatic Optimum. Today, trees are absent or few and the vegetation consists of a restricted range of species tolerant of poor soils. Moorland is best developed in the Highland zone on the west of the British Isles where rainfall is high, and the soils are gley-podzols, often with a peaty top forming blanket peat in the highest and wettest areas. The vegetation consists of such species as heather (*Calluna*), bilberry (*Vaccinium*), grasses and *Sphagnum* moss (Pearsall, 1950). The very clear contrast we see today between moorland and the surrounding agricultural landscape (Figure 6.24) is largely a human artefact. Before the Bronze Age much of what is today moorland carried woodland and had

Figure 6.24 The moorland edge near Widecombe-in-the-Moor, Dartmoor, England (photo Mike Walker)

a resource potential similar to its surroundings. Moorland plant communities began to develop in higher areas with patchy trees and in those places where disturbance factors, including people and grazing animals, were concentrated (Caseldine and Hatton, 1993). Evidence for reduction of the treeline and peat formation in several moorland areas at the time of Mesolithic activity has already been noted (p. 195).

Moorland areas often seem to have been used in less intensive ways during the Neolithic and early Bronze Age and charcoal inputs frequently decrease after the elm decline (Edwards, 1998) as agricultural communities focused more on fertile lowland soils. Bronze Age activity has been studied in particular detail on Dartmoor, England (Balaam *et al.*, 1982; Fleming, 1988). The early Bronze Age (4.4–3.8 ka BP) saw the construction of many burial and ritual monuments such as cairns, stone rows and standing stones, in a landscape which was becoming increasingly open as a result of grazing and regular burning. The really important change occurred in the middle Bronze Age, *c.*3.5 ka BP, with the construction of stone boundary walls or reaves which delimit territory on the periphery of the moor (Figures 6.25 and 8.1), and were associated with enclosures and hut circles. Collectively this represents perhaps the largest area of preserved prehistoric landscape in Europe. The reaves are all the more remarkable because those which have been radiocarbon dated appear to have been constructed over a short period between 3.5 and 3.2 ka BP and the concept governing their layout was so powerful that the orientation sometimes completely ignored major topographic features such as river valleys (e.g. the Dart Valley in Figure 6.25a). This major human impact on the landscape occurred at a time of accelerated clearance which gave rise to scrubby grassland maintained by grazing. Land-use was predominantly pastoral with only small-scale crop growing. Soils below the reaves were acidic; micromorphological analysis shows they were mixed by soil fauna and with only localised podzolisation (Balaam *et al.*,1982; Caseldine, 1999). The reaves fringe the high moorland (Figure 6.25a) and the suggestion has been made that they may encircle the area of Bronze Age podzol/peat development. The intention may

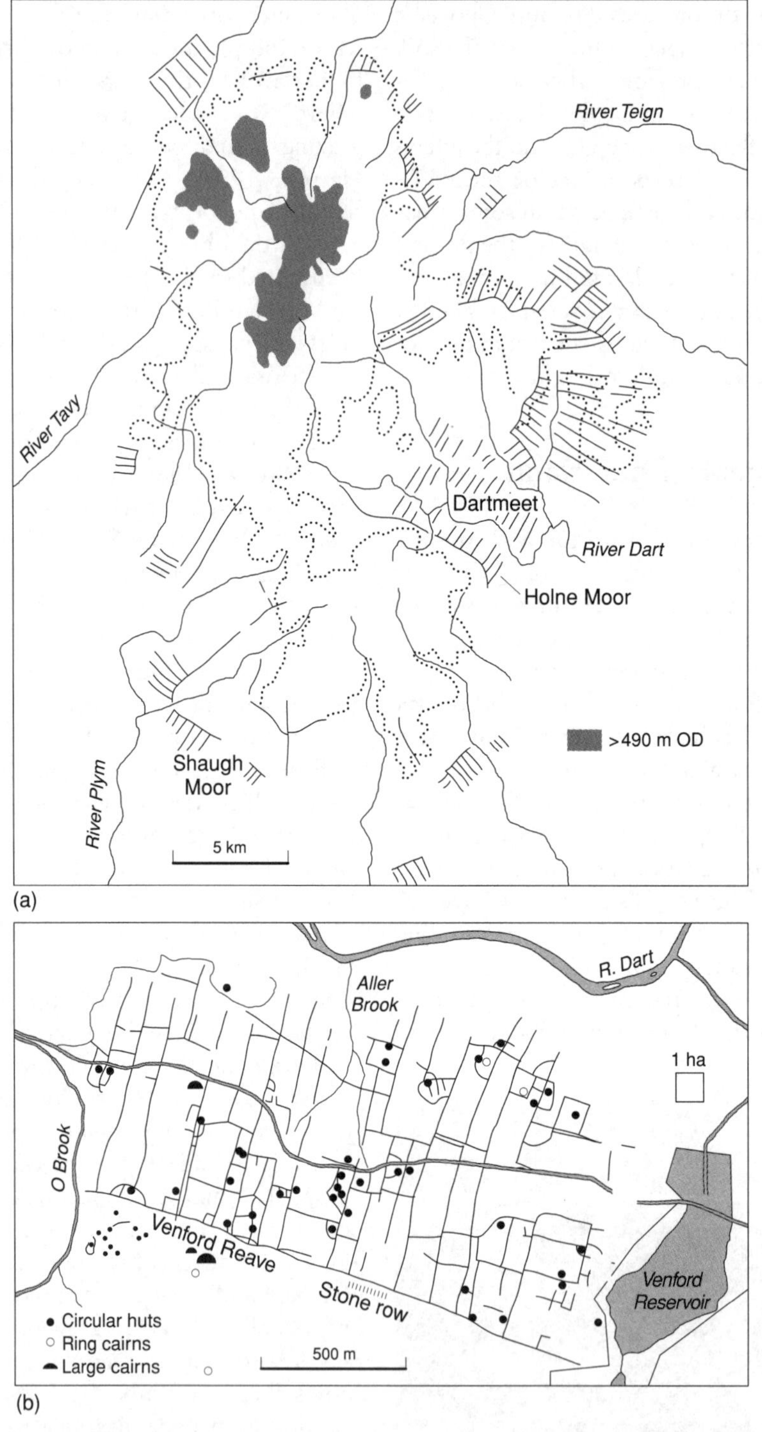

Figure 6.25 The Dartmoor reaves (a) showing the main areas of reaves round the moor; (b) the landscape of reaves, houses and cairns on Holne Moor (after Fleming, 1988, Figures 30 and 34)

have been to facilitate more intensive exploitation of the best remaining land for winter grazing and hay while leaving the higher unenclosed moor for summer grazing (French, 2003). Intensive pastoral activity only seems to have lasted for a few centuries and by the Iron Age activity was limited. As land-use became less intensive, so moorland plant communities become widespread.

Other moorlands did not experience the same intensity of Bronze Age activity. Exmoor, in south-west England, and the North York Moors in north-east England, saw greater clearance during the Iron Age. By the end of that period all the main moors existed although less extensively than later (Rackham, 1994). The fringes of all moorlands were subject to further reclamation and enclosure during the medieval period as communities sought to win back cultivable land from the moor. This occurred at a time of more favourable climatic conditions for crop growth in the uplands during the Little Optimum (p. 180).

The expansion of podzols and, with decreased permeability, peaty gleyed podzols and moorland plant communities is clearly linked to the activities of prehistoric communities, particularly pastoralism. Episodes of more intensive land-use during the middle Bronze Age and medieval period coincide with times when climate was more conducive for agriculture, but the middle Bronze Age also saw intensification and enclosure in lowland landscapes (Yates, 1999) and in both periods social factors are likely to have been at least as significant as climate. This is suggested by the fact that the dates of intensive activity and moorland formation vary geographically and on the basis of evidence for farm abandonment dates well into, and after, the Little Ice Age (p. 180). From about the Bronze Age, moorlands were mostly set aside, not as waste, but for very specific forms of exploitation, principally grazing; their present-day vegetation is to a large extent a reflection of these long-continued practices.

The development of heathland

Heathland has much in common with moorlands botanically and in terms of its origin. Heaths occur at lower elevations and in areas of lower rainfall, often on sandy podzolic soils. Trees are similarly sparse and the vegetation is dominated by evergreen dwarf shrubs, particularly members of the Ericaceae such as *Calluna* (heather) (Thompson *et al.*, 1995). Heathland extends in a coastal belt along the Atlantic seaboard from north Portugal to just beyond the Arctic Circle in Norway. Its most extensive development is in the British Isles, the Netherlands, north Germany and Denmark. Plant communities with many of the same characteristics occur in the circumpolar region and in mountains above the treeline.

Given the maritime distribution of heaths, the traditional view was that they represent a natural climax vegetation type. Indeed heaths developed in previous interglacials, particularly during the oligocratic phase, but these are crowberry (*Empetrum*) heaths, whereas *Calluna vulgaris* heaths are restricted to the Holocene (Stevenson and Birks, 1995). In some extreme oceanic situations, such as the Faeroes, St Kilda, some of the most exposed westerly parts of Orkney, Shetland, the Outer Hebrides and Caithness, pollen evidence shows few trees; there are grass and heathland communities throughout the Holocene (Walker, 1984; Birks, 1986; Jóhansen, 1996). In the Outer Hebrides crowberry heath may precede *Calluna* although on one site the latter has been present since 9.7 ka BP (Edwards *et al.*, 1995). Charcoal occurrence suggests that here fire was a key factor in the spread of heath species although, as we have already noted, the origins of burning are problematic since it predates by millennia the earliest recorded human activity (p. 196). Some palaeoecologists argue that burning is the key factor in the extension of heathland, others demonstrate that some heaths form before significant burning and suggest that the contribution of grazing may be more important (Stevenson and Birks, 1995). Neither disturbance factor is restricted to places where people are present, but both are more frequent and widespread when they are. During the Holocene these processes led to increasing acidity and expansion of acid-tolerant plants (French, 2003).

Heaths in Brittany occur in the areas of some of the great Neolithic monument complexes such as Carnac, which have a predominantly coastal distribution (Scarre, 2002). Buried soils below the Neolithic monuments show that they were mostly constructed in grazed grassy clearings in open secondary woodland with only small proportions of heath taxa (Marguerie, 1992). Although there may have been small-scale heath development in the Neolithic, this mostly occurred following extensive clearance during the Iron Age.

Inland heaths were also formerly wooded and there is evidence from both Denmark and southern England that heathland formation occurred at a wide range of dates, largely dependent on the pattern of human activity (Dimbleby, 1985). Hampstead Heath, London, and parts of the Breckland heaths in East Anglia originated from the loss of woodland through pastoral activity during the Neolithic (Greig, 1996). Field examination and micromorphological analysis of soils below Bronze Age barrows in present-day heathlands show that some were constructed on brown earth soils, but the majority sealed soils that had already begun to podzolize (Courty *et al.*, 1989). Heathland barrows often have a core of clearly recognisable turves picked out by bleached and overlying dark humus horizons which characterise a podzol (Plate 6.3). In the northern Netherlands some Neolithic barrows and many more Bronze Age barrows buried podzol profiles containing heath pollen spectra, whereas in the central and southern Netherlands widespread heath formation mostly occurred from the late Bronze Age/early Iron Age onwards (Casparie and Groenman-van Waateringe, 1982).

The heaths of western Jutland always carried more open forest than was present in eastern Denmark (Odgaard, 1988; Odgaard and Rostholm, 1987; Andersen *et al.*, 1996). Some *Calluna* was present throughout the Holocene, although there is a dramatic rise in *Calluna* pollen when farming starts after the elm decline, and *Calluna* becomes more abundant at times of high charcoal frequency (Odgaard, 1992). Pollen records from soil profiles beneath barrows dating to around 4.6 ka BP show evidence of clearance by grazing and burning at a time when podzolisation was already under way, and an increase in heath species towards the top of the old land surface (Andersen, 1993c). By the fourth millennium BP large areas of Jutland were heath, with further expansion in the second millennium BP. Other Danish heaths, however, formed during the Viking period and some of those in southern Sweden as late as the Middle Ages and sixteenth century AD (Berglund, 1991). Near Bergen, Norway, heath forms a 25 km wide coastal belt. The dates of heathland formation correlate with archaeological and place-name evidence for the development of the settlement pattern. Dates as early as 5 ka BP occur on the coast and later dates down to 1.5 ka BP moving progressively inland (Kaland, 1986). Norwegian heaths have been maintained by grazing and burning as part of a traditional agricultural system which has only declined in the present century, leading to woodland regeneration on some heaths.

An important factor in heathland expansion in the Netherlands, Flanders, north Germany and parts of Denmark from early medieval times to the nineteenth century AD was the creation of 'man-made' plaggen soils (Gimingham and de Smidt, 1983; Groenman-van Waateringe and Robinson, 1988). These are artificially augmented soils characterised by a dark humic topsoil more than 0.5 m thick. They were created by adding organic-rich material, often mucked out from animal byres. This material included turves, grass sods, forest litter and straw which pollen analysis shows were derived from diverse habitats (Groenman-van Waateringe, 1992). The progressive removal of these materials from heathland and the activities of grazing animals resulted in a substantial export of nutrients and organic matter from heath soils to infield (intensively cultivated arable). This extended heathlands and sharpened the landscape contrast between heathy grazing land and the intensively husbanded and enclosed agricultural infield. Such practices played an important part in sustaining long-term agriculture on poor sandy soils such as Pleistocene coversands.

Some coastal heaths do seem to represent climax communities, but the Norwegian evidence suggests that this may only apply to restricted local situations. The dates of heathland formation vary geographically over millennia. Burning and/

or grazing have contributed to the creation of plant communities distinctive from those of earlier interglacials. Heaths became extensive in the earlier Bronze Age, and burial mounds of this date are particularly concentrated on heaths in Britain, Denmark and parts of the Netherlands. It may be that by this time these more open areas had emerged as ancestral seasonal grazing lands, particularly used in winter. Aaby (1997) argues that their particular value was that they could be grazed in winter.

The origins of grasslands

In areas climatically too dry or cold for the development of climax woodland, extensive tracts of natural grassland exist such as the African savannah, the American prairies and the Asiatic steppe; natural grasslands also exist above the treeline in mountainous areas. Some grasslands such as the African savannah have existed for millions of years (p. 144); today their ecology is often influenced by natural fire regimes and they have been extended geographically by human activity.

The first Europeans in North America who ventured beyond the forests of the eastern states encountered vast tracts of prairie extending from beyond the Mississippi to the Rockies. Holocene pollen sequences show that the position of the prairie forest boundary had advanced eastward between 10 and 7.5 ka BP and then retreated to the west in the later Holocene. The earlier Holocene prairie expansion corresponds to a drier phase also marked by low lake levels (see Figure 3.19b). Contrasting with this climatic pattern is the substantial body of already reviewed evidence that areas of grassland in North America were created and extended by native Americans' use of fire (p. 193). A further important factor in the maintenance of open prairie conditions was the grazing by bison. There is evidence for the regeneration of woodland at the prairie margin and in other grasslands following the cessation of burning and the decimation of the bison (Figure 6.16), both consequences of European conquest.

The Eurasian steppe environments have existed continuously from those of the Lateglacial in areas of low precipitation and winter cold. In Russia the Holocene trend has been for forest to advance over steppe with modest reversal in the last millennium which is attributed to human activity (Peterson, 1993). In eastern Europe smaller core areas of natural steppe and steppe woodland existed through the Holocene in the Hungarian Plain and the Black Sea but these have expanded since 6.5 ka BP (Huntley and Prentice, 1993; Berglund *et al.*, 1996a). Steppe now extends over areas once wooded and in some parts of the Hungarian Plain this change took place as late as the seventeenth century (Behre, 1988).

In western Europe lowland temperate grasslands occur in some coastal areas and more extensively on calcareous strata such as chalk and limestone geologies which have thin rendzina soils. These calcareous grasslands are often species rich and contain species which occurred in the Lateglacial steppe. What is controversial is whether the species have maintained a presence in these areas through the Holocene, or have subsequently expanded from very localised refugia, for example on sea cliffs.

Three recent palaeoecological studies in Britain suggest the continued existence, through the Holocene, of at least some more open areas, on a steep slope on the South Downs (Waller and Hamilton, 2000), in the Allen Valley on Cranbourne Chase (French, 2003) and in the Yorkshire Wolds (Bush, 1993). This new evidence accords with the thesis of Vera (2000) that the early Holocene environment may not have been so uniformally wooded as was once thought. That said, there is also extensive evidence on the chalk for former woodland in the first half of the Holocene. Archaeological excavations frequently reveal bowl-shaped hollows containing woodland molluscs, and interpreted as tree-throw pits (see Figure 1.5). Buried soils below prehistoric monuments also have woodland molluscs in their base and grassland species in the upper part of the profile, as for instance at Avebury (Figure 6.26). Around Stonehenge there is evidence for former woodland but also for a significant area of long-

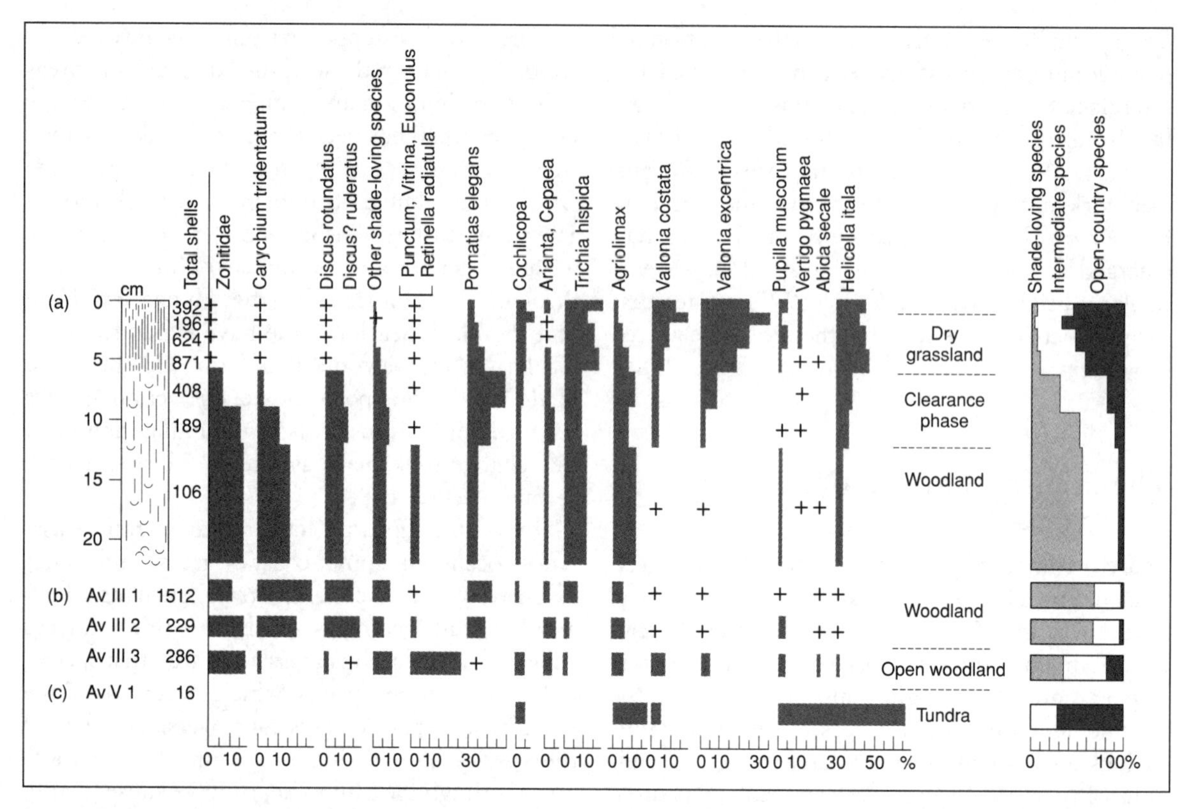

Figure 6.26 Land snail diagram from the buried soil below the bank of Avebury Neolithic henge, this shows the decreasing abundance of shade-loving species and their replacement by taxa of open conditions such as grassland (after Evans, 1972). See Plate 2.4 for an illustration of the Avebury buried soil

standing grassland going back to at least the earlier Neolithic. Plate 6.4 is a modelled reconstruction of two phases in the development of the Neolithic environment (Evans, 1993; M.J. Allen, 1997). The best Late Quaternary environmental sequence from the chalk at Holywell Combe, investigated during construction of the Channel Tunnel, also indicates woodland in the early Holocene (Preece and Bridgeland, 1998, 1999). Recent ecological trends also support the idea of former woodland. Sheep grazing on these grasslands declined in the second half of the twentieth century and the other main grazer, the rabbit, was decimated in 1954 by myxomatosis, a disease introduced by people to contain them. Reduced grazing resulted in scrub and woodland invasion of the grassland showing that it was a plagioclimax, a succession prevented by grazing and human pressure from proceeding to full climax (Smith, 1980).

Disturbance, human agency and the structuration of landscape

In several of the landscape types considered we are presented with apparently contradictory strands of evidence. Some sites on chalkland, moorland and heathland point to closed woodland in the first half of the Holocene. Others hint at the possible persistence of some more open areas. This apparent contradiction is only problematic because we implicitly apply the generalising principle that if area A supported woodland in the mid-Holocene then area B on that bedrock with closely similar environmental parameters will likewise have been wooded. The concept of patch dynamics explains why some areas may never have achieved the ecological climax we expect. In these areas disturbance factors were particularly concentrated due to shallow soils, steep slopes, places where animals or people were most active, etc. Such places are likely to have remained more open than others and thus to have served as refugia from which open country taxa subsequently spread as people created more clearings.

Environmental patchiness also offers a possible explanation for the emergence of concentrations of prehistoric monuments in some areas. At Stonehenge (Plate 6.4), for instance, there were things going on long before Stonehenge itself (Cleal *et al.*, 1995; M.J. Allen, 1997). Large posts were erected in a possible clearing as early as 9.5 ka BP. By the early to middle Neolithic (6–5 ka BP) there was a larger grassland area and a cluster of long barrows. Only later, around 5 ka BP, was the first phase of Stonehenge itself created. The existence of enigmatic linear cursus monuments (*c.*5 ka BP) and the later addition of a ceremonial avenue (4.4–3.5 ka BP) linking Stonehenge to the River Avon hints that there could be a relationship here between routeways and the factors which, over millennia, created, or maintained, open areas which made these places special. Something similar may be seen at Avebury where the great henge lies at the convergence points of two routes marked by avenues of standing stones. Here again the mollusc evidence (Figure 6.26) shows that the henge was preceded by a grassy clearing. Similar arguments might perhaps apply to barrow cemetery concentrations on some areas of heathland, or those places on moorland which were selected for stone rows, stone circles, etc.

It is certainly not claimed that all concentrations of prehistoric monuments are determined by antecedent environmental conditions but rather that these conditions and, particularly, routes through landscape, deserve greater consideration than is often given. Some monuments will have been placed to make reference to earlier landscape structures and activities. Others may have been intended to defy and oppose existing monuments and landscape structures. The Dartmoor reaves marching across landscapes in defiance of local topography (p. 219) could be seen in these terms. The argument presented for the importance of antecedent environmental patchiness is not in essence environmentally deterministic because the factors responsible for that patchiness will very often be a product of an intimate combination of human agency and natural factors. It is not necessarily a very profitable exercise to try to weigh up which of the disturbance factors was the most important in relation to a given ecological change. It is often not the factors individually that are important and interesting, but rather the way they interact together to structure landscape and ecological communities in the long term.

7 People, climate and erosion

Introduction

In the past some scientists have emphasised the role of climate as the dominant influence on later Holocene erosion history, while others have attributed greater influence to the role of people. Today there is a growing recognition of the complex interactions between factors, leading to the advocacy of more sophisticated multi-causal explanations (A.G. Brown, 1997; Endfield, 1997; Macklin, 1999). Erosion is a universal geomorphological process. Under natural conditions erosion rates are governed by such factors as climate (especially precipitation levels and distribution, as well as temperature range), vegetation, slope angle and aspect, soil and bedrock type. Erosion occurs particularly on unvegetated, or sparsely vegetated, slopes where soil and sediment are exposed to subaerial weathering. In those semi-arid or arid areas with less than *c*.600 mm of rainfall (e.g. parts of the Mediterranean, western United States and central Asia) erosion rates will be closely related to the extent of partial vegetation cover at particular times (French, 2003). Even in those areas which had once been fully vegetated, progressive Holocene clearance and the creation of agricultural landscapes have meant that anthropogenic factors became increasingly important in determining the rate and pattern of soil erosion. Many studies demonstrate greater runoff and erosion with decreasing vegetation cover (Lockwood, 1983). Data from North America suggest that the river sediment load doubles for every 20 per cent loss of forest cover. On sandy soils in Bedfordshire, UK, for example, sediment yield reaches 17.7 tonnes per hectare per year (abbreviated hereafter as t ha^{-1} y^{-1}) on bare ground as compared to 2.4 t under grass and zero under woodland (Morgan, 1995; Goudie, 2001). Under natural conditions erosion rates are generally below 1 t ha^{-1} y^{-1}, whereas soil losses under agriculture in the temperate zone may exceed 100 t ha^{-1} y^{-1} (Boardman, 2002).

Erosion is generally an episodic process concentrated in events of varying magnitude. Some may occur on a regular and gradual basis, for instance during rain of a certain frequently achieved intensity. But weather is often highly variable and wet years may produce 10 times the erosion of dry years (Boardman, 1998). Of particular importance in terms of their sediment yield are events of medium frequency and magnitude. Studies in England and central Europe have shown that up to 80 per cent of erosion occurs in major storms which take place two to five times a year (Richter, 1986; Morgan, 1995). Catastrophic storms of a magnitude recurring every 100, or 1000, years can also make a major contribution to erosion. Rare events can also have a transformatory effect on landscape by crossing critical thresholds and creating unstable conditions which set in train a new cycle of erosion and sedimentation (Starkel, 2002). There is increasing evidence that extreme events are more concentrated during periods of climatic transition (p. 139). Rare events can also be devastating in human terms, especially because they will often transgress the boundaries of expectation and existing coping strategies (p. 140). The 1952 Exmoor flood carried 100 000 t of boulders, soil and uprooted trees and devastated the small town of Lynmouth, England, killing 34 people (Kidson, 1953).

There is an important distinction to be made between periods of secular climatic change, which dominated thinking in the older palaeoenvironmental literature, and the greater emphasis currently given to the transformatory effects of rare events and episodic processes. Increased erosion and deposition may occur during major secular episodes of higher rainfall but this is by no means necessarily the case. Storm intensity and thus erosion may be greater in periods of lower mean rainfall. Much depends on the distribution of rainfall within the year and the frequency of high-rainfall events, or those falling on saturated or frozen ground, etc. Furthermore, it may not be the event in isolation which is significant but its relationship to a unique chain of events (**event sequence**) of varying frequency and magnitude (Brunsden, 2001). Much depends on the sensitivity of an environment to change at a given time (p. 8). Of the range of factors affecting environmental sensitivity, including the disturbance factors outlined in Chapter 6, the role of human agency is particularly important in the Holocene. Thus by reducing, or removing, vegetation cover people sensitise slopes to the effects of episodic climatic processes such as high-rainfall events.

Given the importance of issues of timescale, studies of past and present erosion are complementary (Bell and Boardman, 1992). Present studies help us to understand the processes operating, but monitoring is mostly very short term and provides little evidence of rare events. The past dimension provides information on longer-term erosion rates and the impact of societies and agricultural systems very different from our own. As always, however, the present does not provide a precise analogue (Boardman, 1992): generally today fields are larger, organic matter levels are reduced, organic manures have given way in many areas to chemical fertilisers, fallow periods are fewer, soil fauna are likely to be impoverished and with it soil structural stability. The passage of heavy wheeled vehicles also contributes to modern erosion. These factors mean that, in general, present erosion rates may be expected to be higher than those of the past. However, particular past climatic episodes and/or land-use regimes may well have given rise to rates higher than at present, as indeed the evidence for prehistoric erosion suggests. Clearly, from examples outlined below, the serious problems which modern agriculturalists recognise in many areas are not just recent phenomena.

Processes of soil erosion are reviewed by Morgan (1995), Lal (2002) and Toy *et al.*, (2002). The main types of erosion are as follows:

1. **Rainsplash.** Raindrop impact causes soil aggregates to shatter and produces a hard surface crust which reduces infiltration. Soil particles are flung into the air by impact and, on a slope, move downslope.
2. **Overland flow/sheet flow.** The soil's infiltration capacity is exceeded and water flows across the surface.
3. **Rill erosion.** Overland flow becomes channelled and forms ephemeral erosion channels.
4. **Gully erosion.** Relatively permanent channels formed by running water.
5. **Mass movement.** Mud flows, landslides and debris avalanches.
6. **Subsurface flow.** Movement of fine particles through voids in the soil and movement of minerals in ionic solution.
7. **Wind erosion.** Movement of silt particles in suspension (e.g. loess) and coarser sand grade particles by saltation and surface creep (e.g. dunes and coversand).

In addition to the above, most disturbance factors that affect soil on a slope will, for reasons of gravity, induce more downslope than upslope movement: such factors include frost heave, cultivation and the effects of animals. Both wild and domestic animals affect slope processes locally, for example around animal burrows, or routes frequented by herds. Of special importance, in anthropogenic terms, is the direct effect of cultivation itself, which experiments show leads to significant net downslope movement altering the gradient and redistributing soil within individual fields (Govers *et al.*, 1994; Govers, 2002). It is particularly as a result of cultivation processes, acting with the other factors, that the upslope parts of fields show evidence of soil loss, often in the

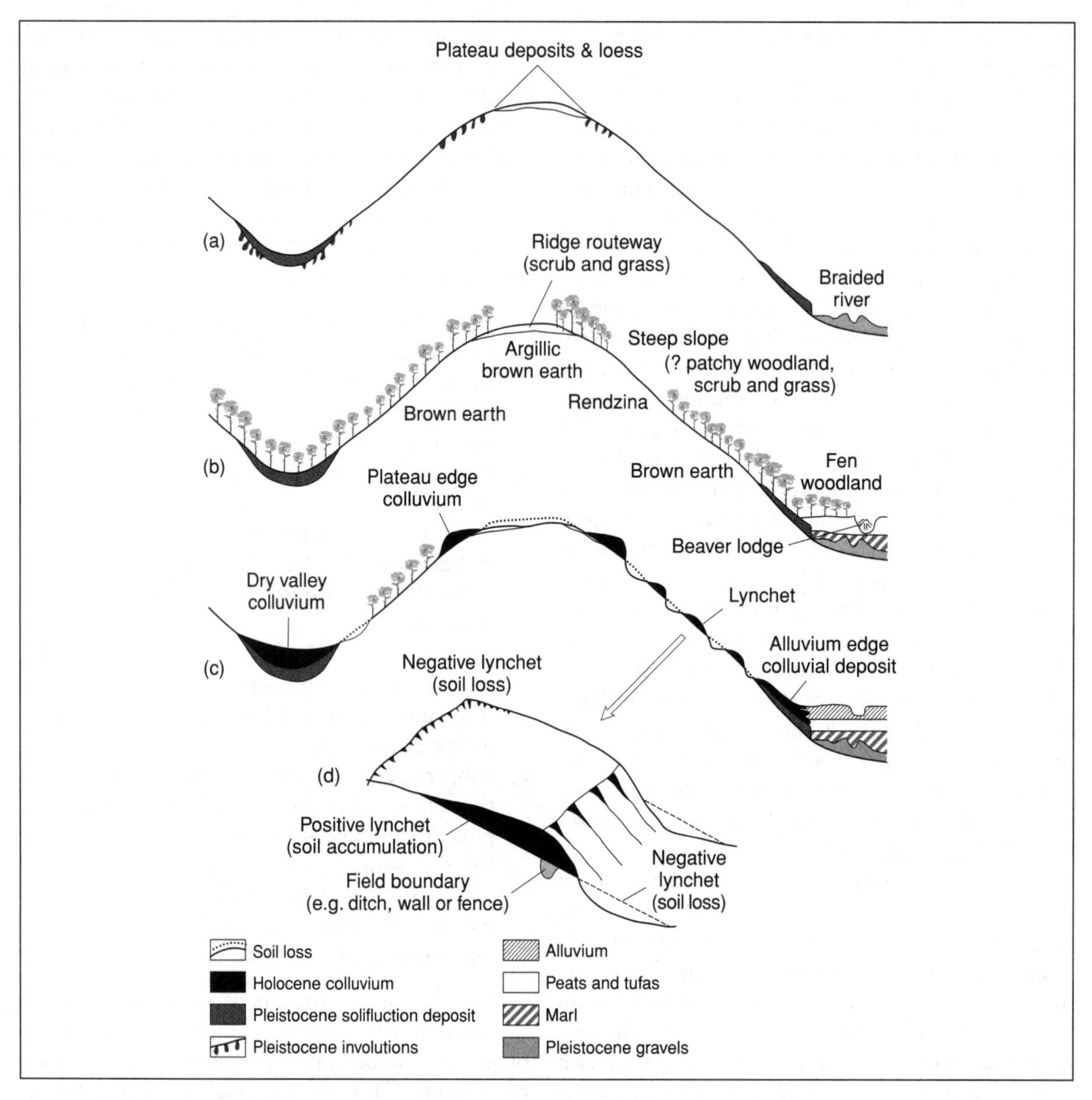

Figure 7.1 Soil and sediment landscape relationships, a diagram based on the chalk of south-east England: (a) in the Lateglacial; (b) at mid-Holocene woodland maximum; (c) in later Holocene prehistory, showing the situations in which eroded sediments occur; (d) detail showing the structure of a field boundary lynchet

form of a pronounced step, a negative **lynchet**, while the downslope boundary is marked by sediment accumulation, a positive lynchet (Figures 7.1d and 7.2). Below these accumulations there may be evidence of former boundaries such as fence lines, ditches or stone walls. In many areas, such as the Mediterranean and Andes, people have deliberately constructed terraces as a means of conserving soil and moisture (Miller and Gleason, 1994). Cultivation itself also results in soil changes including loss of nutrients with crops, reduction in soil organic matter and fauna such as earthworms, all of which reduce soil structural stability. These factors, together with the increased size of voids created by

Figure 7.2 A colluvial lynchet bank of Bronze Age to Romano-British date. At the base there is a boundary ditch and fence (marked by posts), Bishopstone, England (photo Brenda Westley)

tillage, increase silt translocation down-profile which may produce 'agric horizons', the formation of which is exacerbated by the plough which turns a furrow and causes smearing at the base of the plough zone. Thus cultivation reduces infiltration capacity and soil structural stability and leads to increased erosion. Where traces of ancient fields or cultivation marks (see Figure 2.18) are present, then past agriculture can be readily identified. It may also be identified in micromorphological thin sections (p. 39), although this generally requires recognition of a combination of features to distinguish the effects of tillage from other soil disturbance factors (Macphail, 1998; French, 2003).

The relationship between vegetation cover, land-use and erosion highlights the need for a seamless whole landscape approach which integrates both the slope process and those operating in riverine environments (Figure 7.1). Of particular importance in terms of human activity are the slope-related, soil erosion processes which produce colluvial sediments. **Colluvium** is poorly sorted or unsorted sediment laid down by slope processes including slopewash and downslope creep, sometimes augmented by cultivation. Significant colluviation occurs on unvegetated or poorly vegetated slopes which are widespread in arid and semi-arid areas; in temperate areas colluviation mainly occurs in unvegetated or partly vegetated agricultural landscapes. **Alluvium** is sediment laid down by running water and consequently sorted. Coarser sands and gravels reflect high energy conditions, such as a river's bed load, whereas silts and clays comprise the suspended sediment laid down, for instance, during overbank flooding. Alluvial processes have been reviewed from a geoarchaeological perspective by A.G. Brown (1997), with particular reference to north-west Europe, and by Waters (1996), with particular reference to North America. Riverine processes and relationships are complex since sediments often derive from substantial catchments. Erosion upstream, whether by channel incision, bank erosion or colluvial slope processes, may be associated with deposition downstream. Sediment **storage** also occurs, for varying timescales, on field boundaries (lynchets), at the base of slopes and in valleys (Figure 7.1). A proportion of the eroded sediment will be flushed out of the catchment, for example to the oceans. The key factor is the relationship between sediment supply and the competence of a river to transport the eroded sediment; this is affected by climate, runoff, vegetation, etc.

The occurrence of erosion constitutes a problem in human terms if its rate is greater than that of soil formation. Information on soil formation rates is very limited; studies typically suggest rates of around 2.5 cm of soil forming in 300–1000 years depending on bedrock, rainfall and the other environmental factors (Schertz and Nearing, 2002). Present erosion rates may be measured in the field using a variety of apparatus (Morgan, 1995), or calculated from the volume of erosion features. Field monitoring generally provides data over relatively short timescales of *c.*1–10 years. Longer-term rates may be calculated from the volume of alluvium and colluvium. This, however, is a calculation of sediment in storage and needs to be supplemented by information on losses to the catchment by river or stream action, including the activities of episodic and seasonal streams.

In long-term studies, dating of the products of erosion may be based on archaeological artefacts, especially pottery, when midden material and manure have been spread on fields to maintain fertility (Bell, 1983). Radiocarbon dating of organic material contained in eroded sediments can also provide an excellent chronology (Preece and

Bridgland, 1998). In river valleys dendrochronology is also increasingly important as a dating technique for timbers incorporated by flood or landslip events and where people have created wooden structures. Optical dating (p. 55) is also applied to the development of alluvial and colluvial chronologies (Lang and Nolte, 1999; Lang and Hönscheidt, 1999). The short-lived isotope ^{137}caesium (p. 56) from nuclear weapons testing post AD 1945, makes it possible to calculate the amount of soil redistributed since that date (Quine and Walling, 1992; Foster, 2000). Such studies show that between 14 and 73 per cent may stay within a field, the remainder being exported, for example by stream transport. In the catchment of the River Severn, UK, about three times as much sediment is removed as becomes stored on floodplains (Quine and Walling, 1992). Cosmogenic isotopes, such as ^{36}Cl (p. 56), provide a measure of bedrock surface lowering over longer timescales (Granger *et al.*, 1996; Harbor, 2002). The long-term record is particularly important in providing evidence of natural erosion rates, the so-called '**geologic norm**'.

A simple model of soil and landscape change in Figure 7.1 identifies some of the key Holocene changes and soil erosion features showing landscape change at three stages of development: the Lateglacial, mid-Holocene woodland and later prehistoric agricultural landscape. It is based on chalk landscapes in south-east England but many of the trends and features apply more widely. This also demonstrates the concept of a **catena**: the variations in soil type which occur in relation to slope because of weathering, slope processes and drainage. Residual older soils survive in plateau areas, the soils of steep slopes are thin and eroded and there are accumulations of soil at the base of slopes and in valleys; where the water-table is high these soils may be wet gleys. Figure 7.1a shows the situation in the Late Pleistocene with unstable slopes subject to freeze-thaw and solifluction. River valleys at times of snowmelt were characterised by high discharges and coarse sedimentation by braided streams. High rates of erosion and sedimentation also marked the unstable and sparsely vegetated environments that existed in the first few centuries of the Holocene following the rapid warming *c.*11.5 ka BP. During the period of maximum woodland development (Figure 7.1b) in the mid-Holocene, brown earth woodland soils developed and erosion rates were low. As hypothesised in Chapter 6, some steep slopes, or places where animals congregated, may have had patchy woodland and some local erosion could have occurred. River valleys at this stage were wooded, experienced low rates of minerogenic sedimentation, deposits were organic-rich and peats and tufas formed. In the agricultural landscapes of later prehistory, areas with widespread cultivation often show evidence of soil erosion and colluvial accumulations in the situations indicated in Figure 7.1c: the perennial boundaries of cultivation at plateau edges, in dry valleys, on slopes as lynchets and at the base of slopes on the edge of alluvium. By this time many river valleys were cleared and drained and the increasing erosion gives rise to later Holocene accumulations of minerogenic alluvium.

Valley sediments in North America

Valley sediments in North America have been extensively investigated from a geoarchaeological perspective because a high proportion of prehistoric sites are stratified in alluvial sequences (Waters, 1996; Holliday, 1992, 1997). Many of the most well-preserved sites were close to water bodies, such as rivers and oxbow lakes, and are found on successive palaeosol horizons, representing episodes of environmental stability, which are separated by episodes of alluviation and colluviation, representing episodes of reduced stability. Holocene river valley sequences show evidence of broadly contemporary sedimentary changes over wide geographical areas (see Figure 4.25). Only since the late eighteenth or nineteenth century have anthropogenic factors significantly influenced the sedimentary patterns of North American rivers. Prior to that the main drivers of sedimentary change are widely considered to be climatic (Knox, 2000).

Climatic factors are especially marked in the semi-arid and arid south-west where changes in the amount and seasonal distribution of precipitation are a major influence on vegetation cover and river regime. Variations in the magnitude and frequency of floods are seen as particularly

Figure 7.3 Pueblo Bonito, Chaco Canyon, New Mexico: a ceremonial and population centre of the eleventh and twelfth centuries AD. An arroyo in the background was incised after AD 1100 and may have impeded floodplain cultivation (photo Harold D. Walter, courtesy Museum of New Mexico, Negative No. 128725)

important forcing factors disrupting equilibria and initiating new erosion cycles (Knox, 2000). Major flood events are increasingly documented during periods of climatic instability. A catastrophic flood in California in AD 1605 (Schimmelmann *et al.*, 2003), is the highest magnitude event in a sequence of flood episodes recurring about every 200 years over the last 2 millennia which, it is thought, were driven by climatic events on a continental or larger scale. These episodic events do not appear to be linked to El Niño events, but had significant social impact on communities in middle and South America.

The archaeologically rich area of the Colorado Plateau, where pueblo settlements are wonderfully preserved and settlement histories are often precisely dated by dendrochronology, suggests a rather more subtle interplay between aridity changes, flooding and social dynamics. Cyclical patterns of aggradation and degradation have been identified lasting *c.*550 and 275 years, and degradation phases correlate with dendrochronological evidence for periods of high climatic variability (Gumerman, 1988). During degradation phases the incision of deep erosion gullies, **arroyos**, were associated with episodic flash floods (Figure 7.3). Rainfall variability was a key influence on the way of life of Hohokam and ancestral Pueblo (Anasazi) communities in these semi-arid landscapes (Cordell, 1997). In many places half of the precipitation was in high-intensity summer storms and rainfall was often insufficient for crop growth. The solution was the development of highly sophisticated strategies for water and soil capture and conservation (Bettis and Hajic, 1995; Waters, 1996). Optimal sites were selected for floodwater farming, for instance on alluvial fans at arroyo mouths. The strategies included check dams, contour terraces, irrigation canals, bordered gardens and stone mulching. The progressive development of these water-control strategies would have demanded, and created, high levels of social cohesion and organisation. Together with the continued use of wild resources to buffer against the significant risk of crop failure (2 out of 5 years in places), these represent a novel suite of coping strategies (Chapter 5) which facilitated the development of sizeable communities at sites such as Chaco Canyon, New

Mexico (Figure 7.3) and Snaketown, Arizona. However, high population levels proved not to be sustainable because of the fluctuating rainfall pattern, and during periods of drought there was progressive settlement abandonment between AD 1150 and 1450. The extensive manipulation of drainage by floodwater farming and irrigation may have exacerbated the impact of climatic instability because, during the period of Hohokam communities AD 300–1500, there was a marked increase in floodplain entrenchment. Thus a particular event sequence created a situation in which critical thresholds were crossed, triggering erosion and downcutting thus in turn rendering inoperable existing irrigation networks and fields fed by floodwater (Bettis and Hajic, 1995; Waters, 1996). Further widespread arroyo formation followed the onset of cattle ranching from the 1880s. The sedimentary effects of land-use changes after European contact (Knox, 2001) are, however, far more marked and widespread than the generally modest and local effects attributed to prehistoric communities.

Mediterranean valleys

A long-held theory (Vita-Finzi, 1969) was that Mediterranean valleys have one phase of Holocene alluviation, which occurred in post-classical times and was climatically driven, perhaps by the changes of the Little Optimum or Little Ice Age (p. 93). This view is no longer tenable as a result of a generation of Mediterranean landscape archaeological surveys, many of which have included geoarchaeological investigation (Figure 7.4). Recent research includes international programmes which provide time depth for issues of current environmental concern such as landscape degradation, erosion and desertification (van der Leeuw, 1994; Leveau *et al.*, 1999; Castro *et al.*, 2000). The relative contributions of climate, human agency and tectonics in the Mediterranean river systems are increasingly under investigation (Lewin *et al.*, 1995b; Vermeulen and de Dapper, 2000; Grove and Rackham, 2001).

Human activity has been a key factor in Mediterranean vegetation change. Evergreen forests of **sclerophyll** species (those drought-adapted by leaves with thick epidermis and a waxy coat) have been replaced by **maquis** and **garrigue** communities, low shrubby communities with many aromatic plants. In the driest areas these communities have been further degraded to steppe with patchy vegetation cover (Castro *et al.*, 1998). Progressive vegetation change will have increased runoff and erosion.

Many valley floors show thick alluvial and colluvial sequences (Figure 7.5). The history of these landscapes shows marked punctuations: during stable episodes soils form and rivers are incised, and these are separated by phases of instab-

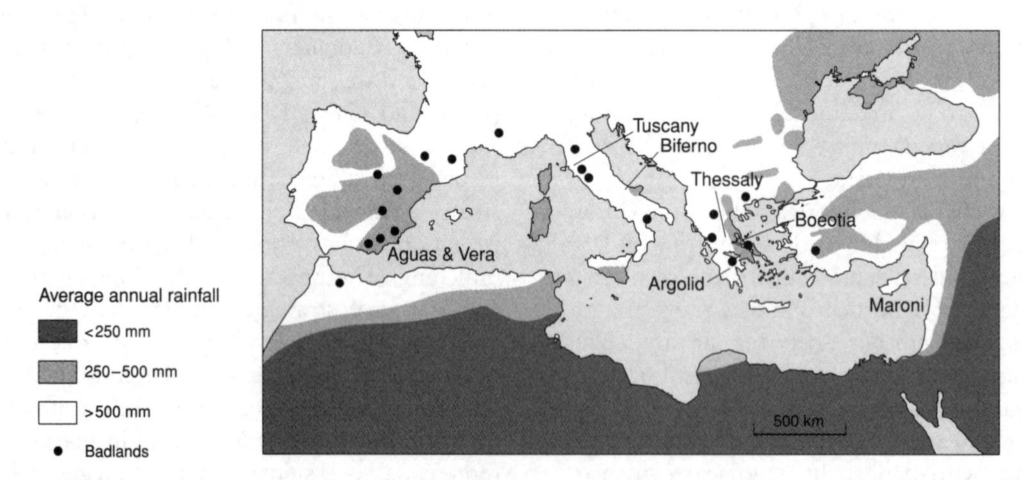

Figure 7.4 The Mediterranean showing areas of low annual rainfall and the locations of selected archaeological/geoarchaeological surveys. Examples of Mediterranean badlands are also shown (badlands after Grove and Rackham, 2001, Figure 15.1)

Figure 7.5 Marone, Cyprus: a sediment sequence in a coastal cliff showing a truncated basal soil overlain by colluvium containing prehistoric pottery (photo Martin Bell)

ility when slope erosion and valley sedimentation are pronounced (French, 2003). Landscape archaeological survey often shows phases of dense settlement and intensive land-use punctuated by phases when sites are few. Increasingly, however, it is clear that the relationships between these two distinct forms of punctuated sequence are not straightforward. Analysis of the dates of alluviation at 85 sites in the north Mediterranean by Grove and Rackham (2001: Figure 16.17) shows a very wide range of dates from the mid-Holocene *c.*5 ka BP, when a trend to increasing aridity around the Mediterranean has been recognised (Macklin *et al.*, 1995). There is a particular concentration of alluviation from *c.* AD 600. Grove and Rackham conclude that weather was a more important driver of alluviation than human agency. In south-east Spain some alluviation occurs prior to major human impact, highlighting the contribution of natural disturbance factors, in this, the driest part of Europe (Figure 7.4; Castro *et al.*, 2000).

Many of the episodes of instability are not, however, in phase with known secular Holocene climate changes and dates vary greatly between survey areas. Around the Mediterranean evidence of Neolithic alluviation is limited to areas such as Thessaly, Greece, which was densely settled (van Andel and Runnels, 1995); in many areas vegetation clearance and environmental impact seem to have been quite limited. In south-east Spain alluviation is particularly marked during the period of intensive land-use during the Argaric Bronze Age 4.2–3.4 ka BP (Castro *et al.*, 1998, 2000). As a result of this, woodland disappeared, steppe-like vegetation developed and there is evidence of plants that demonstrate **salinisation** (increasing soil salinity). Many areas, such as the Biferno Valley, Italy (Barker, 1995) and the Argolid, Greece (Jameson *et al.*, 1994), show pronounced landscape instability and alluviation during classical times. It is frequently the case that far higher rates of erosion occurred in recent centuries than in earlier millennia, strengthening concerns about the sustainable use of Mediterranean landscapes (Hunt and Gilbertson, 1995; Castro *et al.*, 1998; French, 2003).

An important question concerns the role of the stone-fronted terraces which are widespread on Mediterranean hillslopes. These retain soil and water and extend cultivable land. During periods of terrace abandonment and collapse, slope instability and erosion will have increased. This, it is suggested, would have led to increased alluviation; in the Argolid following the collapse of classical agriculture and in south-east Spain following the abandonment of Moorish terraces and irrigation after the medieval Christian reconquest of the area. In the last case abandonment is manifestly the product of wider social conflict rather than environmentally damaging practices. The problem is that information about the origins and history of terracing is very limited. Some Bronze Age terraces have been dated, but many examples appear to be relatively recent, at least in their present form (Grove and Rackham, 2001).

Despite close correlations between land-use intensity and alluviation in many recent survey areas, there are concerns about uncritical over-emphasis on anthropogenic factors to the exclusion of other disturbance factors (Bintliff, 1992, 2000; Endfield, 1997). Some alluviations are not sufficiently precisely dated for confident correlation with the archaeological record and there is a need to guard against circular reasoning. In some areas erosion has been attributed both to intensive land-use *and* periods of abandonment without independent biological evidence to justify the apparent contradiction.

Figure 7.6 Badlands seen from the walls of the Etruscan city of Volterra, Italy (photo Martin Bell)

The importance of giving balanced consideration to natural as well as anthropogenic factors is illustrated by the most extreme form of Mediterranean erosion, namely **badlands**: areas so intensely dissected and gullied that they are unsuitable for agriculture (Figure 7.6; Woodward, 1995). These generally occur on poorly consolidated strata of Tertiary geological age, and, despite the semi-arid and partially vegetated appearance of many badlands, they are not confined to the areas of lowest rainfall (Figure 7.4). In south-east Spain major gullies of the badlands have, in some cases, been dated before significant human impact in prehistory and some are pre-Holocene (Gilman and Thornes, 1985). Grove and Rackham (2001) conclude that Mediterranean badlands are essentially natural, geologically conditioned landforms, noting that there is no documented Mediterranean instance where they have developed from normal landscape in historic times.

Anthropogenically focused interpretations of Mediterranean alluviation history have sometimes given insufficient emphasis to two key points: the role of extreme events and the interactions between factors (Bintliff, 2000). Environments that have been sensitised by a reduction in vegetation, however caused, will be particularly susceptible to erosion. In addition to the important role of human agency, a range of other factors creates perturbations within the Mediterranean environment: erratic rainfall, grazing, fire, etc. The last is a potentially critical factor in a landscape where many vegetation communities are fire adapted, but fire history has been little investigated by comparison with northern Europe and America (Grove and Rackham, 2001: 193).

Central and eastern continental Europe

The occurrence in continental Europe of river valley sequences with episodes of downcutting and aggradation which are broadly coeval from one basin to another, and are in some cases dendrochronologically dated, has established the important contribution of climatic factors in the changing riverine geomorphology of this region (Starkel, 1985, 2002). In the earlier Holocene there was an extended period when Alpine slopes had not yet fully stabilised following the last cold stage; here a chronology for episodic alluviation episodes is provided by the radiocarbon dating of trees buried by accumulating sediments (Miramount *et al.*, 2000). These alternating alluviation and stabilisation phases occur between 10.5 and 7 ka BP during which period the driver of instability must surely be climatic. After this, there is widespread evidence for greater mid-Holocene slope stability, channel incision and limited alluviation.

Anthropogenic factors come into play rather later in this area than in the Mediterranean and western Europe, reflecting the generally later dates of extensive clearance (Berglund *et al.*, 1996a). Erosion episodes are increasingly well dated using optical dating (p. 55; Lang *et al.*, 1998) and extensive 'archaeological-type' trenching relates sediments to archaeological and historical evidence for land-use and climate (Bork *et al.*, 1998). Evidence of Neolithic colluviation is generally small-scale and local to specific sites (Lang and Wagner, 1996). Increased colluviation occurs during the Bronze Age in some areas (Maier and Vogt, 2001) and greater clearance during this period and the Iron Age may account for the onset of flood loam accumulations in these periods (Lang and Nolte, 1999). In many river systems it is during the medieval period that extensive deposition of flood loams occurs, frequently burying archaeological sites of earlier date (Dreslerová, 1995). Anthropogenic factors are increasingly seen as key influences on the pattern of alluviation particularly in the last millennium (Klimek, 1999).

As dating methods improve, so it becomes increasingly evident that both slope instability and alluviation are highly episodic processes (e.g. Richter, 1986; Bork *et al.*, 1998). As in a Mediterranean context, this has led to a growing interest in the role of extreme events and their changing frequency. In Germany and its surroundings there is a well-documented example of a rare event which provides a most valuable insight to the subtle interplay between land-use and climatic factors. During July AD 1342 exceptional rainfall occurred; this is calculated to have a recurrence interval of 1000 years and to have generated overland flows 50–100 times higher than any recorded during the twentieth century (Bork *et al.*, 1998). Massive gullies and extensive fans formed in some areas; so severe was the erosion in parts of Niedersachsen and the Wolfsschlucht, Germany, that the soil was totally eroded away and what was once agricultural land was abandoned to forest for centuries. The event is the most extreme manifestation of an episode of wetter, stormy conditions which marked the transition at the beginning of the Little Ice Age (p. 94; Lamb, 1995). Had this event occurred in the sixth century AD it would have had only limited impact, as at that time 90 per cent of Germany was forested; however, by the fourteenth century AD this had reduced to *c.*15 per cent, with the result that extensively cultivated landscapes were highly sensitised to an event of this magnitude.

British Isles: colluviation in chalk landscapes

Chalk areas contain rich archaeological landscapes, and many were more intensively used in the past than they are today. The light soils were suitable for early cultivation, and in some areas soil depth and fertility have declined through time. There is much sediment in storage as lynchets and dry valley fills (Figure 7.1), because on permeable strata surface streams are few. A particularly important Lateglacial and Holocene sequence has been investigated at Holywell Combe, Folkestone, in advance of construction of the Channel Tunnel (Preece and Bridgland, 1998, 1999). There were surviving soils of Lateglacial Interstadial age, early Holocene tufas and evidence for a wooded landscape, the clearance of which in the Bronze Age resulted in slope instability and colluviation. Erosion at this date is part of a wider pattern seen also on the South Downs and in Wessex (Bell, 1983; Allen, 1992). It is during the Bronze Age that recognisable agricultural landscapes with field boundaries marked by lynchets appear on the chalk (Fowler, 2000). Colluvial increments of Neolithic date are not widely represented, perhaps because at this time cultivation remained small-scale and shifting (p. 199). In some areas original woodland soils appear to have been eroded leaving only truncated fossil tree holes (Figure 7.1).

Some sequences show clear evidence of progressive soil changes with basal sediments rich in loess, which was laid down as a blanket in the Late Pleistocene (p. 38) and survived in some areas into the Holocene. Subsequent colluvium of Iron Age and later date contains a higher proportion of stones and, later, chalk granules, a sequence reflecting progressively thinning soils (Plate 7.1). Other study areas show very little evidence for colluviation, including those rich in prehistoric archaeology such as the Stonehenge landscape (see Plate 6.4; M.J. Allen, 1997) and Cranbourne Chase in Dorset. In the latter area colluvial increments are small and mostly post-prehistoric (French *et al.*, 2000; French, 2003). Evidence of former woodland in parts of that landscape is lacking and one factor which contributed to the change from shallow brown earth to rendzinas was the stripping of turf to make burial mounds. Marked contrasts between areas with little erosion and those with extensive colluvia are important in helping us to identify spatial diversity within prehistoric landscapes. They also support the emerging view, discussed in the context of grassland history (p. 233), that even in wildwood times these landscapes may have been more patchy than once believed, with extensive

Figure 7.7 Severe erosion on fields recently drilled with winter cereals at Rottingdean, England, October 1987. The rills and gullies have cut through thin topsoil into the underlying chalk and fans of coarse material including chalk have been deposited on the valley floor (photo John Boardman)

woodland but also scrubby and grassy areas with thinner soils (French, 2003; Figure 7.1b).

Comparisons in southern England between present-day soil erosion monitoring and earlier Holocene sediment sequences have helped to elucidate the processes operating in these landscapes (Bell and Boardman, 1992). Erosion today (Figure 7.7) is almost entirely on arable land and is greatest on autumn-sown fields. Interestingly, there is plant macrofossil evidence for autumn sowing in the Bronze Age and Iron Age when erosion rates were also high (Jones, 1981). Rills are responsible for most of the erosion and they deposit fans of chalky granules on the valley floors. Removal of fine sediment by temporary runoff streams leaves stony valley fill sediments (Plate 7.1). Ten years' monitoring showed erosion was highly episodic with most in just two wet periods. Erosion from one rain-fall event with a 25-year return period was so severe that if the same land-use persists for 300–500 years the soil would be totally eroded away. Erosion rates up to 200 m^3 ha^{-1} y^{-1} were recorded (Boardman, 1993). Present erosion is also seen to be particularly concentrated in specific topographic locations (Boardman, 1990), which may help to explain the spatial variation and patchiness of prehistoric occurrences.

Data from present-day erosion studies have been combined with archaeological evidence and assumptions about past conditions to model soil change over a timescale of 7 ka (Favis-Mortlock *et al.*, 1997; Figure 7.8). This showed that, given present assumptions, it is theoretically possible for a thick loess cover to have been eroded. The model replicates the heavy erosion which is seen on some sites during the Bronze and Iron Ages and the reduced erosion in later periods as stoniness increased and land-use became less intensive. Despite the challengeable assumptions involved in modelling it does enable us to conceptualise the effects of episodic processes in the long term, using, in this case, a stochastic weather generator in daily time steps over 7 ka. By adjusting the modelled parameters it is possible to identify the factors that have a major effect on erosion rates and those which are of less significance. In this chalkland example the nature of land-use and the stoniness of the soil emerged as particularly important factors governing the sensitivity of the soil to erosion.

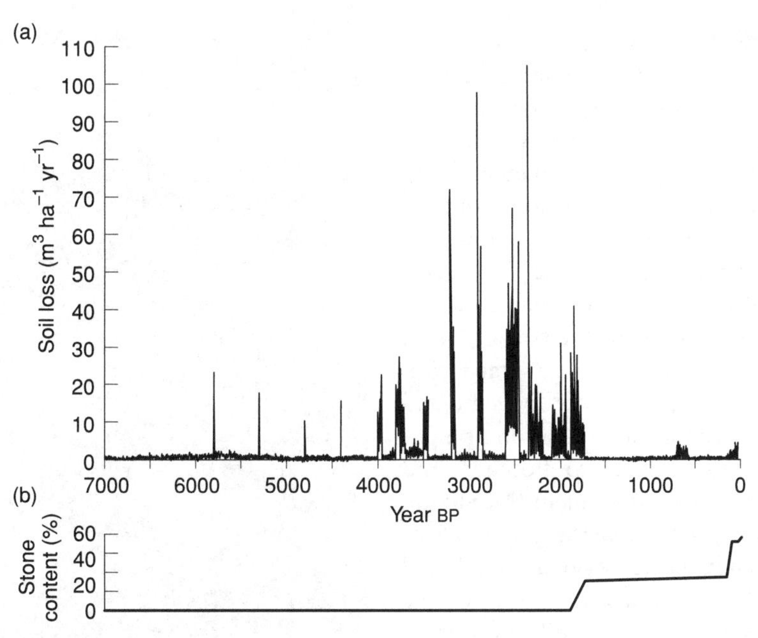

Figure 7.8 Modelled erosion rates of loess cover on chalk soils of the South Downs over the last 7 ka: (a) soil loss per year; (b) increasing stone content (after Favis-Mortlock *et al.*, 1997, Figure 1)

British Isles: river valleys

Recent years have seen a growing appreciation of the richness of British archaeological landscapes buried within alluvial sequences, the contribution that archaeological evidence can make to dating of these sequences and the importance of an integrated geoarchaeological approach (Needham and Macklin, 1992; A.G. Brown, 1997). Many of these contexts are under pressure from large-scale gravel, and other aggregate, extraction. Changing river regimes in the Lateglacial and Holocene (p. 137) have been outlined by Rose (1994) and A.G. Brown (1997). Rose's model identifies contrasting geomorphic contexts laterally along a river's course and there are also marked contrasts as between the upland catchments of the north and west and the lowland catchments of the south and east (Howard and Macklin, 1999; Lewin and Macklin, 2003). A typical sequence of sedimentary changes in the lowlands is shown in Plate 7.2. The latest extensive gravel deposition in many lowland catchments occurred during the Loch Lomond Stadial between 12.6 and 11.5 ka BP. Drainage remained disrupted, water bodies were extensive, groundwater was calcareous and in places calcareous marls formed. Woodland subsequently became extensive, drainage improved, mature soils formed and sedimentation rates declined. Thus, through much of the early Holocene, river channels remained generally stable and rates of minerogenic sedimentation were low. Where sedimentation occurs it reflects the high levels of biological activity within river valleys at this time. Peats developed in areas of waterlogging, encouraged locally perhaps by the activities of beavers (Coles, 2001), and tufa formation occurred where there were calcareous springs. Once people started clearing landscapes, environmental stability declined and minerogenic alluviation increased. Neolithic alluvium is limited but interestingly does occur in the major concentration of Neolithic activity and clearance at the headwaters of the river Kennet around Avebury henge (Evans *et al.*, 1993). In the Thames Valley whole prehistoric landscapes are associated with former channels of the Thames and its tributaries, emphasising the contribution of river valley resources to the lives of past communities (Allen *et al.*, 1997). The most detailed sedimentary investigation from an archaeological perspective is at Runnymede (Needham, 2000), where there is evidence for a major flood episode and gravel deposition *c*.4 ka BP. There was human activity at this time but no specific evidence suggests anthropogenic causation. Deposition of fine-grained alluvium in the Bronze Age does, however, occur. In this period extensive clearance took place and an agricultural landscape was demarcated by extensive ditch systems on the river terraces (Yates, 1999). The major late Bronze Age waterfront settlement at Runnymede, with waterlogged wood structures and evidence for the deposition of high-status metalwork, was subject to flooding and erosion at the time of its abandonment (Needham, 2000). In the Thames as a whole the grasslands of the floodplain, which had been seasonally exploited as pasture in the Bronze and Iron Ages, became subject to increasing flooding and alluviation through the later Iron Age and into the Romano-British period, apparently reflecting the increasing scale of arable farming in the catchment (Robinson, 1992). Renewed alluviation occurs in the medieval period in the Thames and other Midland valleys, corresponding to a time when there were extensive landscapes of ridge-and-furrow cultivation, parts of which are now buried under alluvium.

At Hemington in the Trent Valley, very different sedimentary conditions obtained; the confluence of rivers, some from upland areas with snowmelt, gave rise to a high-energy river in the medieval period which deposited extensive gravels. Within these is a wealth of waterlogged wood structures (Salisbury, 1992): fishing weirs, mills and three bridges swept away by successive catastrophic floods dendrochronologically dated AD 1140, 1210 and 1403 (Figure 7.9; A.G. Brown, 1997; Ellis and Brown, 1998).

Macklin and Lewin (2003) have sought a climatic signal within Holocene river valley sequences by identifying contemporary sedimentary change in different catchments. When the dates are plotted out (Figure 7.10) they show marked vertical alignments or steps on which basis 14 episodes of marked Holocene flooding were defined. Many of these episodes are close in time to climatic deteriorations independently identified on the basis of wetness shifts in mire sequences (p. 32, Figure 3.20). Some also correlate with North Atlantic drift ice and other indicators of global climate change. The two most marked events are at 2 and 2.5 ka BP.

Figure 7.9 Medieval bridges in gravels of the River Trent at Hemington (photo Susan Ripper, University of Leicester Archaeological Services)

Figure 7.10 shows the greater sensitivity of upland rivers to flooding events in recent centuries, maybe reflecting the effects of the Little Ice Age (A.G. Brown, 1998), or more intensive land-use including mining which may have sensitised these upland catchments to the effects of climate vagary. Before the Little Ice Age Figure 7.10 shows no marked difference in the sensitivity of upland and lowland rivers. It is suggested that, as dating precision improves, we are beginning to glimpse an underlying pattern of climate/weather-related

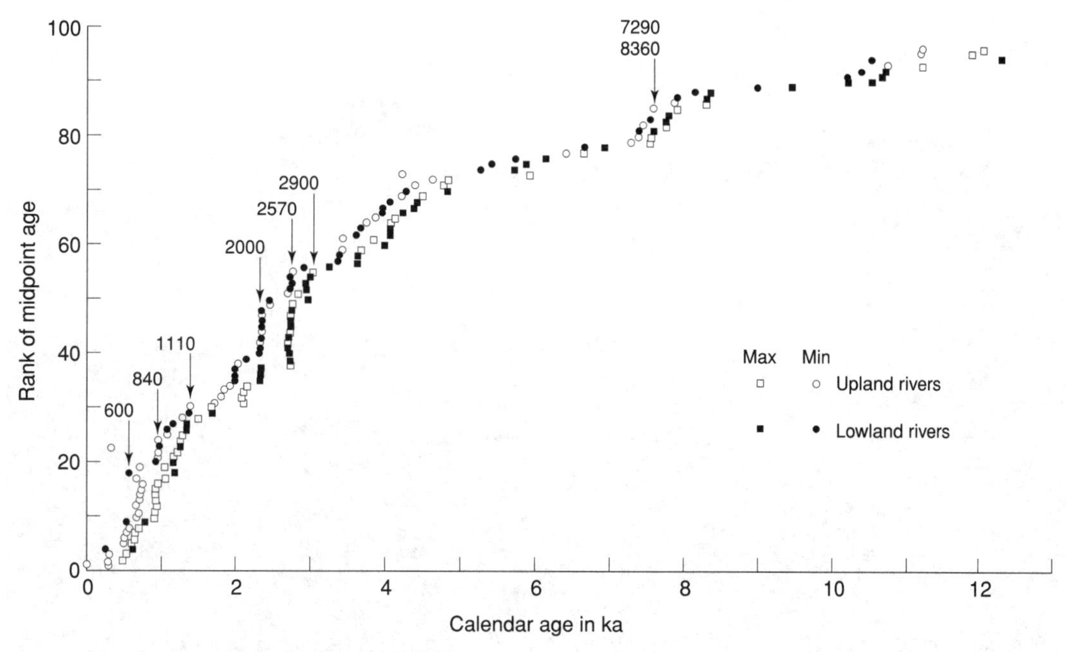

Figure 7.10 British alluvial sequences, the radiocarbon dates of geomorphologically significant changes in river activity for upland and lowland rivers (after Edwards and Whittington, 2001; Macklin and Lewin, 2003, Figure 1a). Sites are ranked according to the mid-point of the calibrated age range. Arrows mark major flood episodes.

instability which is particularly picked out by environments made most sensitive by human agency.

Aeolian sediments

The effects of wind erosion were most spectacularly demonstrated by the American dust bowl on the Great Plains in the 1930s which resulted from a drought and depleted vegetation cover, exacerbated by years of overgrazing and the rapid extension of cereal growing facilitated by mechanisation (Goudie, 2001; Toy *et al.*, 2002). In 1934, one four-day storm transported 300 million tonnes of sediment, some of it up to 3300 km! Less dramatic dust storm episodes recurred here about every 10–20 years in the twentieth century. Equally severe twentieth-century dust storms have occurred in the Russian steppe (Warren, 2002).

Earlier episodes of aeolian erosion on a more local scale are recorded in Europe. Deflation from cultivated fields resulted in peaks of dust deposition in Danish mire sequences between the late Bronze Age and the late pre-Roman Iron Age and again in the Viking period (Aaby, 1997). This complements the pollen record by providing an additional source of evidence concerning the extent of arable land. In western Europe the most dramatic effects of wind blow are in areas of coversand deposited during the Pleistocene (p. 38). When cleared for cultivation in the later Holocene, soil organic matter became depleted, podzols and heaths formed and, when subsequently cultivated, became subject to deflation in dry periods. On the Drenthe Plateau in the northern Netherlands this happened locally in a few small areas in the Neolithic but deflation increased in the period 2–3 ka BP when substantial areas of 'Celtic fields' were buried by remobilised coversand (van Gijn and Waterbolk, 1984; Fokkens, 1998). In the southern Netherlands, the Veluwe Plateau remained wooded until the eighth century AD when heathland began to expand with increasing settlement and charcoal production. Extensive aeolian deposition began in the tenth century AD and buried the early medieval settle-

Figure 7.11 Kootwijk, Netherlands, central area of a former pond adjacent to an early medieval settlement. The dark layer represents the extension of arable land over the former pond in the tenth century AD. A later well (beside scale) cuts through the aeolian sand which buried the settlement and its fields (photo Amsterdams Archaeologisch Centrum (AAC), University of Amsterdam, courtesy of H.A. Heidinga)

ment of Kootwijk (Heidinga, 1987; Groenman-van Waateringe and van Wijngaarden-Bakker, 1987). The effects of deforestation and agriculture were exacerbated by a dry episode which first dried up the settlement's pond and then lowered water-tables, as reflected in the levels of successive wells (Figure 7.11).

Wind erosion in Britain occurs in the drier east: East Anglia, Lincolnshire and Yorkshire (Morgan, 1995). In pre-modern times remobilised coversands were again mainly affected. Early medieval settlements, cemeteries and fields at West Heslerton, Yorkshire, and West Stow, Suffolk, were buried by sands in this way and in the latter heathland there was a further episode of dune mobilisation during the Little Ice Age *c.* AD 1570, which buried buildings and fields (Macphail, 1987; Rackham, 1994).

Lakes

Lake margins are key contexts for the preservation of waterlogged archaeological sequences (Coles and Coles, 1996). Such ecotonal situations from which communities could exploit a diversity of resources were so attractive that people learnt to cope with the hazard of floods and fluctuating lake levels. Lakes provide particular opportunities for comparative multi-proxy environmental investigations using pollen, charcoals, sediment accumulation rates, chemistry, diatoms and magnetic properties (p. 32). They are often well dated by radiocarbon, dendrochronology (often on humanly made structures) and archaeomagnetics. The increasing identification of sections of annually laminated sediments is also making an important contribution to the development of more precise

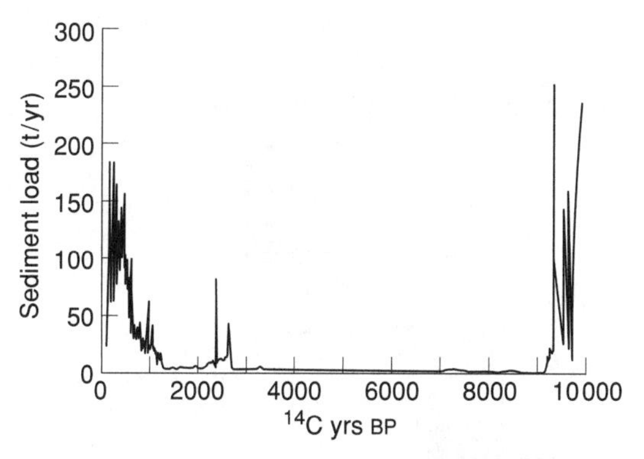

Figure 7.12 Holocene sedimentation in Lake Bussjösjön, southern Sweden (after Dearing *et al.*, 1990, Figure 12.8a)

land-use and erosion chronologies (Hicks *et al.*, 1994).

The high sedimentation rates of the Lateglacial continued into the early Holocene until environmental stability under woodland cover was achieved (Figure 7.12). During this stage mollusc-rich calcareous marls were sometimes deposited (Walker *et al.*, 1993). It was in the latter part of this phase, as woodland extended, that occupation took place at Star Carr on the side of temporary Lake Pickering, Yorkshire. Here, as in many early Holocene lakes, calcareous marls accumulated in the lake centre and peats in the vegetated zone round its margins (Day, 1995; Mellars and Dark, 1998). Vegetation disturbance by people is recorded around the lake by pollen and charcoals (p. 195). By 7.6 ka BP sedimentation reached the stage when vegetation spread to cover the former lake, bringing to an end the favourable environmental conditions which had attracted a concentration of early Mesolithic sites.

The 12 000-year record from Lake Gosciaz, Poland, includes a long sequence of annually laminated sediments between *c*.7600 and 3900 BP with evidence of disturbance which is likely to be related to human activity both before, and after, the elm decline, the latter being marked by a particularly thick sediment band (Ralska-Jasiewiczowa and van Geel, 1992). Annually laminated sediments at Diss Mere, eastern England, similarly showed evidence of human activity and charcoal, somewhat in advance of the elm decline, and were of great importance in identifying the timescale of the decline (p. 199; Peglar, 1993b). Subsequent changes in the mere sediments and chemistry reflect increasing, but fluctuating, levels of human activity (Fritz, 1989; Peglar, 1993b).

From the middle Neolithic to the Bronze Age the margins of Alpine lakes were especially attractive for 'lake village' communities (p. 205; Schlichtherle, 1997). Synchronous changes to the levels of some German and Swiss lakes reflect climatic influences probably related to the extent of Alpine glaciers. Fluctuating levels of some lakes and the occurrence of laminations correlate with changing intensities of lake shore settlement (Joos, 1982; Merkt and Muller, 1994).

As a result of the major Ystad survey in southern Sweden, the settlement and environmental history of this area is perhaps more thoroughly investigated than any comparable area of Europe (Berglund, 1991). Sedimentary changes in three lakes can be related to changes in lake palynology and magnetic properties, the latter providing evidence of the relative proportions of topsoil and subsoil eroded (Dearing *et al.*, 1990; Dearing, 1991). Sedimentation rates were high in the very early Holocene, very low in the early and mid-Holocene but increased markedly from *c*.2.7 ka BP when extensive deforestation and cultivation occurred in the catchments concerned (Figure 7.12). Under woodland the estimated rate of soil formation exceeded the erosion rate which was *c*.0.2 t ha^{-1} y^{-1}. The maximum erosion rate at 300–500 BP is calculated to be about 50 times the rate of soil formation, which, if those conditions continued, could expose subsoil within 200 years. The Little Ice Age also apparently contributed to this period of high erosion, since it is observed that most erosion here today occurs following periods of frozen ground.

Analysis of sediment accumulations in 50 British and Irish lakes (Figure 7.13) demonstrates that accelerated sedimentation rates, as seen for instance at Braeroddach Loch, Scotland (Figure 7.13b), occur in those basins along with palynological evidence of human activity (Edwards and Whittington, 2001). Lakes with no evidence of accelerated sedimentation, or where it occurs very late, are all in the west and north of the British Isles (Figure 7.13a). Some basins show a

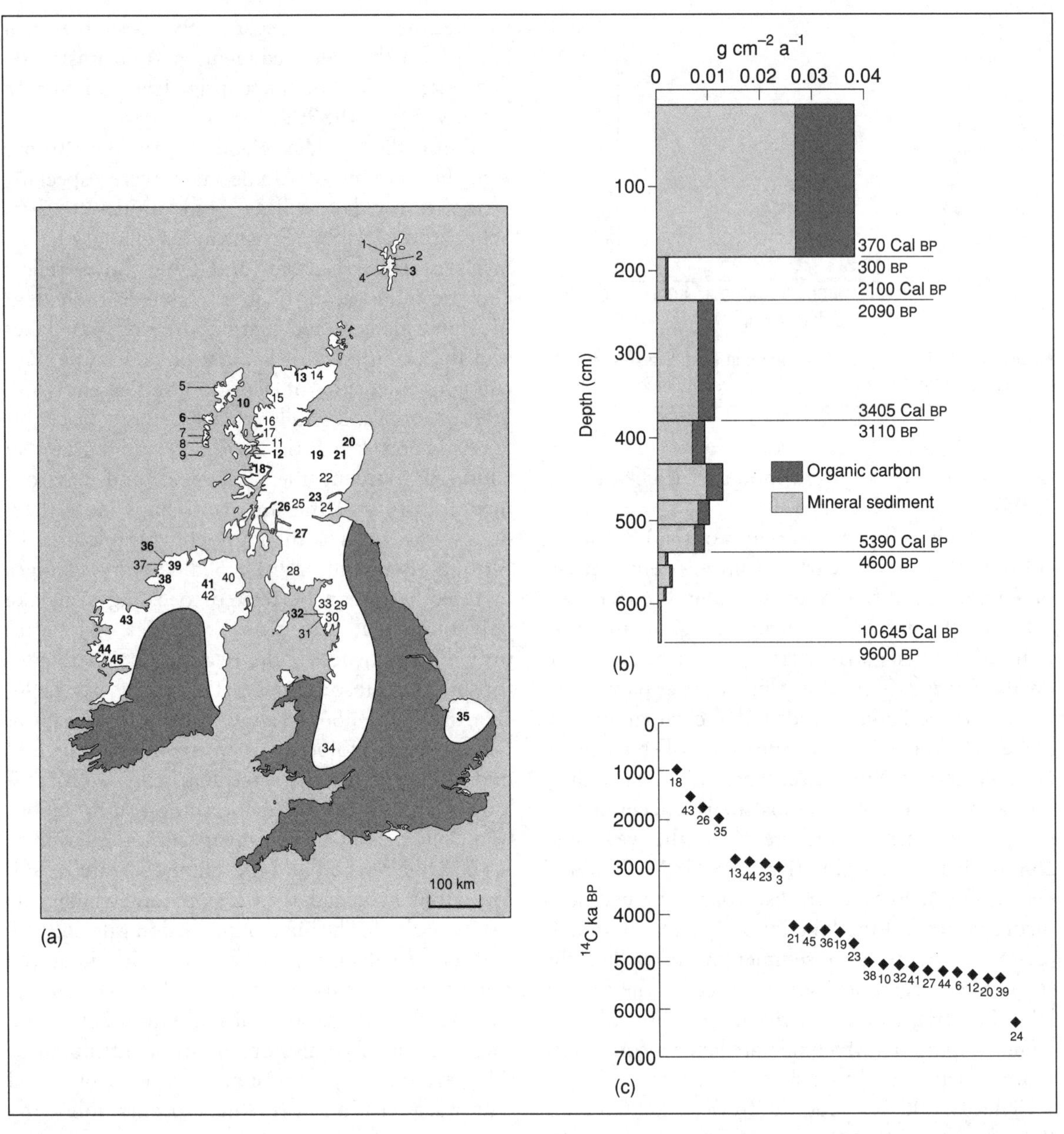

Figure 7.13 Lake sedimentation in the British Isles: (a) locations of dated lakes distinguishing those with (in bold), and those without, accelerated later Holocene sedimentation; (b) sediment deposition rates in Braeroddach Loch, Scotland; (c) dates at which accelerated sedimentation occurs at sites on (a) (a and c after Edwards and Whittington, 2001; b after Edwards and Rowntree, 1980, Figure 15.6)

time lag between the onset of activity and increased sedimentation. Increased sedimentation rates are clustered at 5.9, 4.9 and 3 ka BP (Figure 7.13c). It is striking that two of these phases of increased sedimentation are within the Neolithic, the first at the elm decline. At this time evidence for colluviation and river floods (Figure 7.10) is limited. This may be because the generally smaller catchment scale of British lakes will be more sensitive to the effects of early agricultural activity than larger river catchments. It may also be that, as around the Alps, lakes in the British Isles were particularly attractive foci for early agricultural activity. The third phase of accelerated lake sedimentation at 3 ka BP corresponds to the most marked phase of river flooding (Figure 7.10).

Lakes in North America show little sign of increased input related to human activity until after European contact. Before contact, sedimentation rates and lake sediment characteristics reflect the operation of natural processes, and lake levels provide evidence of climatic fluctuations (Haworth, 1972; Cwynar, 1978; Webb *et al.*, 1993). Dramatic changes occur, however, with the arrival of Europeans which is typically marked by an increase in pollen of the agricultural indicator *Ambrosia* (ragweed). A further useful marker horizon in recent lake sediments is the chestnut decline (p. 203) which occurred in eastern North America during the 1930s. European agriculture is also reflected in changes in lake sediment chemistry, in diatom assemblages (Davis and Norton, 1978) and in a change from organic to minerogenic sediment with increased ash from clearance burning. In Frains Lake, Michigan, for example, the establishment of settlement and associated clearance in 1830 was accompanied by an initial dramatic increase in sedimentation to 30 times its previous level; erosion then stabilised at 10 times previous levels (Davis, 1976).

Erosion and flood: perception and response

Even today many agricultural communities are unaware that they are suffering rates of soil erosion that are unsustainable. Part of the problem is that the phenomenon is stochastic. Each of our sources of evidence relates to different timescales and provides only partial knowledge (see Figure 1.7). It may only become apparent that serious erosion is occurring when evidence from a range of timescales is integrated. How a community responds will depend on their rationalisation of what they have observed and the timescale over which they have information. Sometimes changes were so serious that upheavals of settlement pattern or economy resulted. Iron Age communities on the Drenthe Plateau of the Netherlands and medieval communities in several areas had to abandon whole landscapes in the face of advancing remobilised coversand. Similarly, increased flooding during the Iron Age led to the abandonment of seasonal settlement on the Thames floodplain.

It is evident from the quantities of artefacts in many agricultural soils that prehistoric communities responded to decreasing soil fertility by manuring, and by medieval times there are plaggen, or man-made, soils (p. 222). What is often less clear, however, is the extent to which field boundaries, such as those marked by lynchets in north-west Europe (Figure 7.2), or the terraces of the Mediterranean, are deliberate soil conservation measures. Organised field systems with lynchets become widespread from the middle Bronze Age and a few Mediterranean terraces may also go back to the Bronze Age, but it remains very unclear when they became established as a major aspect of Mediterranean land-use.

Human responses to erosion events are not, however, always as predicted by a deterministic perspective. The native American coastal site at Ozette, Washington, was buried by a catastrophic mudslide, possibly earthquake triggered, in *c.* AD 1750 prior to European contact (Figure 5.20; Ames and Maschner, 1999). A memory of the disaster is part of the oral history of the Makah tribe and within the mudslide are preserved buildings and one of the finest assemblages of decorated organic artefacts ever discovered. Despite the mudslide, and a number of subsequent lesser slides, the settlement was rebuilt and continued until the 1940s. Much earlier at Brean Down, England (Plate 5.2), Bronze Age communities continued to occupy a settlement intermittently for 600 years after all the soil in their immediately adjacent land was eroded away down to underlying Pleistocene breccia (Bell, 1990). At Brean Down, as at Ozette, communities found ways of coping with catastrophic erosion, presumably because in both cases a combination of the rich resources of particular coastal situations and the social significance of place were perceived as outweighing the erosion hazards.

8 The role of the past in a sustainable future: environment and heritage conservation

Introduction

The last three chapters have reviewed the history of human–environment relationships and have established how the activities of people have contributed to the development of many of the landscapes that are of conservation importance today. In 1992 the United Nations Conference on Environment and Development at Rio set a new international sustainability-based environmental agenda continued by follow-up meetings, including Kyoto (1997) and Johannesburg (2002).[1] The emphasis is on international cooperation on climate change issues (Chapter 9), sustainable development and biodiversity. The UN Convention on Biological Diversity, which came into force in December 1993, required states to develop a national strategy for biodiversity conservation. Biodiversity increases the capacity of ecosystems to adapt to future environmental perturbations, however caused. The genetic library of biodiversity is essential to the future development of pharmaceuticals and new crop plants, particularly those tolerant of arid and saline conditions in parts of the world where people are underfed (Myers, 1997).

A time perspective for sustainability and biodiversity

Time-depth is central to concepts of sustainability and biodiversity; both require an understanding of past environmental history (Redman, 1999) and are the rationale for what might be called 'green archaeology' (Bell, 2004). The timescale of data required will vary according to the nature of the environmental phenomena in question and the timescale of stochastic processes (e.g. disturbance factors) affecting ecosystems and populations. One of the main lessons from the palaeoenvironmental record is that we live in a changeable, non-equilibrium, world (Chapter 1). This has led some to question the general applicability of the sustainability concept (van der Leeuw, 1994; Simmons, 1999). Despite this acknowledged problem, the term remains a widely understood shorthand for an evaluation of whether particular ways of life and activities can be continued at their present level in the short, medium and long term. Among the perturbing influences which give rise to non-equilibrium conditions, human agency is one of the most important, as Chapters 6 and 7 have shown. People represent a key ecological factor in almost all terrestrial ecosystems. Time-depth must therefore include an understanding of the contribution of people to the formation and development of environments.

The archaeological and anthropological record provides examples of societies that were conserving of resources and others which were responsible for major environmental perturbations. First Nation Americans of the Pacific north-west coast practised a varied and sophisticated fishing technology (Stewart, 1977). An abundance of salmon provided the basis for a hunter-gatherer culture of exceptional artistic richness, yet fish stocks only declined after the arrival of Europeans (Fowler and Turner, 1999). Statistical information on fish stocks in the North Atlantic has only been gathered since *c.* AD 1900. Today large cod more than 1 m long are exceedingly rare in Icelandic waters, whereas middens of the medieval period show that then they were common (Amorosi *et al.*, 1994).

Quite apart from baseline information on ecology and biodiversity in periods preceding the ecological impacts of recent centuries, there is increasing recognition of the contribution that indigenous knowledge and cultural tradition can make to biodiversity and sustainability (Turner, 1997). The role of indigenous people in sustainable development is now enshrined as Principle 22 of the Rio declaration of 1992.[1] There is abundant evidence that cultural traditions and practices, such as transplanting and clearing, create and maintain biodiversity, as for instance in the well-documented planting activities of First Nation Americans in California (Blackburn and Anderson, 1993) or the Kayopo in Amazonia (McNeely and Keeton, 1995). In this way the maintenance of culturally diverse ways of life may be seen as contributing to ecological diversity. This is one of the many ways, as noted in Chapter 1, in which scientific and social perspectives may be seen as increasingly convergent.

Societies of the past did not, however, necessarily occupy self-sustaining Gardens of Eden. Ecologically destructive past activity has been identified on many Polynesian islands, particularly Easter Island (Flenley and Bahn, 2003). However, in this last case others have argued that insufficient emphasis may have been given to the ecologically and socially destructive effects of the very first contact with Europeans (Rainbird, 2002). Significant erosion, of clearly pre-Columbian origin, has been identified in the Mexican Highlands (O'Hara *et al.*, 1993). Even societies in which philosophy and religion emphasise respect for animals, mutualism and interdependency with the natural world, such as Aboriginal communities in Australia and First Nations in the Americas, have been shown in Chapter 6 to have had significant ecological effects. By no means all of these could be considered ecologically harmful; some have enhanced biodiversity and many have contributed to the sustainability of human populations.

Archaeological sites in the landscape

Heritage conservation tends to be focused on individual settlement sites, although arguably it is often the wider context that is significant. This includes the cemetery, trackways and fields, as well as the neighbouring bogs and ponds containing biological and sedimentary evidence essential for an understanding of that site's environmental context and history. Sites preserved in isolation from their context, surrounded by deep-ploughed arable, have often lost much of their value. Cultural landscapes (Figure 8.1) in many parts of Europe preserve much evidence of former land-use and economy and are increasingly valued for the contribution they make to local identity, a sense of place and thus tourism (Birks *et al.*, 1988; Fairclough and Rippon, 2002).

The geographical relationship between landscape features contributes to interpretation of environmental evidence and an understanding of the cosmology of past societies. In a cultural landscape the particular conjunction of banks, tombs, fields, natural vegetation and other features including geology and topography may all have been used to convey meaning to a contemporary observer moving through that landscape. So far phenomenological approaches have mainly emphasised monuments, topography and natural features (Figure 1.3; Tilley, 1994; Bradley, 2000a). There is clearly the potential to make greater use of the palaeoenvironmental record in considering how the landscapes of the past were perceived (Chapman and Gearey, 2000; Evans, 2003). The communication of knowledge and experience through perceived landscape relationships is likely to have been of special significance to pre-literate societies. The 'dreamtime' of Australian aboriginals is a tradition incorporating the natural world. The landscape is 'charged with awe, every rock, spring, and waterhole represents a concrete trace of a sacred drama carried out in mythical times' (Eliade, 1973). Young initiates learn of the deeds of their ancestors from inscriptions on the landscape, sometimes the literal inscriptions of rock art, sometimes myths of how landscapes were made. It follows that within landscapes the significance of individual features may relate not so much to their own specific properties but to their relationship to one another and the total landscape, natural as well as cultural. That constitutes a powerful case for working towards an integrated cultural and natural perspective on landscape and conservation issues.

Palaeoenvironmental studies and nature conservation

There is a growing awareness of the common interests of archaeology and nature conservation (Lambrick, 1985; Macinnes and Wickham-Jones, 1992; Cox *et al.*, 1995). This is not always reflected in the legislative and administrative arrangements of individual nation states but it is a growing theme at international level (Coles and Olivier, 2001). Many of the key contexts for the investigation of past people/environment relationships are areas of relict landscape (Figure 8.1). The habitats discussed in Chapter 6 provide examples: heathland, moorland, old grassland and woodland areas. These have escaped intensive agriculture and disturbance in recent times and may be seen as marginal in terms

Figure 8.1 A relict landscape on moorland, Horridge Common, Dartmoor, England, showing middle Bronze Age fields forming part of a reave system, trackways and circular huts (photo BMC 50, copyright reserved Cambridge University Collection of Air Photographs)

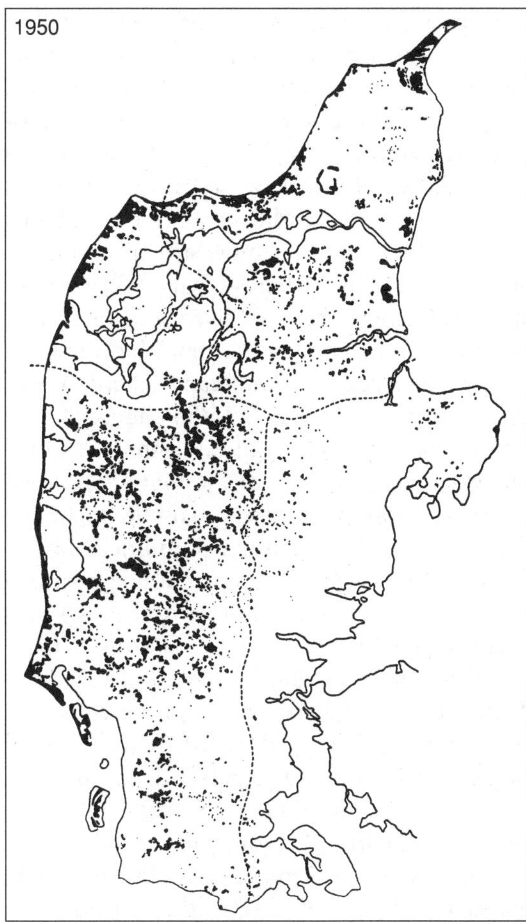

Figure 8.2 Habitat loss illustrated by the reduction of heathland in Jutland, Denmark, between 1800 and 1950 (after Nielsen, 1953 and Kristiansen, 1985, Figure 5a–b)

of the rather specific and exacting requirements of modern farming. Many such landscapes have been progressively encroached on as a result of agricultural intensification in the post-medieval period. Figure 8.2 shows the loss of heathland habitat in Denmark. In many parts of Europe the Common Agricultural Policy of the European Union led to further loss of cultural landscapes (Willems, 1998). However, increasing yields, the agricultural surpluses of food mountains and a growing move towards habitat conservation have now created a reverse trend; substantial tracts of land throughout Europe are now being returned to 'nature'.

Areas of relict landscape often contain rare plant and animal communities of conservation importance. Archaeological sites and natural environments of deposition (bogs, lakes, dunes, valley sediments and coastal sediment sequences) are together valuable sources of evidence for past environmental conditions. Of particular significance is the relationship between this evidence, whether artefactual, sedimentary or biological, and the context in which it occurs. It is a non-renewable resource, deserving of greater emphasis in conservation debates (Greeves, 1990).

Living organisms and their habitats understandably attract the most widespread attention in conservation. We need to identify those species which are most at risk and require special protection. These are contained in Red Data Books listing species of extreme national rarity.[2] The conditions which these require are important determinants of conservation management policy. Less evident is the role that cultural and palaeoenvironmental perspectives have to play in the development of effective conservation strategies. In many parts of the world there are few habitats that are truly natural, as most are the products of long histories of human activity (Chapter 6). People and nature have evolved together in a coevolutionary way so that current biodiversity is to a significant degree a landscape inheritance of what went before (Boswijk and Whitehouse, 2002). Effective conservation requires both a detailed knowledge of the ecology of the biota concerned and the past trajectory of the habitat, including the history of human land-use (Lambrick, 1985). No longer can many habitats containing rare species be seen as the remnants of natural ecosystems that can be conserved simply by putting a fence round them and excluding grazing and other disturbance. Early in the history of nature conservation that approach frequently led to successional ecological change, such as scrub and tree invasion of grassland, or heathland environments, not infrequently resulting in reduction, or loss, of the species it was intended to conserve. Inappropriate management of parts of the Drenthe Plateau in the Netherlands has seen areas turn from heathland to grassland (Bottema, 1988). Invariably positive management is needed, the development of particular grazing strategies, the coppicing of wood, the cutting of reeds or setting back the effects of hydroseral succession to recreate wet areas favourable to certain taxa.

Understanding trends in biodiversity also requires an historical perspective to which palaeoenvironmental research contributes (Brown and Caseldine, 1999). This must take into account different spatial and temporal resolution and levels of taxonomic precision of palaeoenvironmental studies as compared to surveys of present vegetation. Such research contributes to ongoing debates about the character of the wildwood (Svenning, 2002). In the Netherlands, for instance, nature conservationists have sought to recreate a wildwood which is park-like and ecologically diverse based on the theories of Vera (2000). This envisages higher levels of grazing than has normally been contemplated in palaeoenvironmental interpretations and is generating lively debate about issues long taken for granted (Bottema, 1988; Louwe-Kooijmans, 1995). It is clear that in the past, people enhanced biodiversity by creating situations in which a range of species could flourish; this is illustrated by palynological case studies in Scotland (Tipping *et al.*, 1999). In Denmark palynology shows that taxon richness increases up to the Iron Age as diverse environments are created by human agency. Biodiversity then declines as landscapes became more uniform. The most marked decline is associated with the agricultural landscapes of the twentieth century (Andersen, 1993a). The reduction in biodiversity through extinctions has already been reviewed (pp. 86–192 and 210–212).

The palaeoenvironmental record can also contribute to debates concerning the enhancement of biodiversity by reintroducing species which became locally extinct in the past. There has been a successful reintroduction of the white tailed sea eagle to Britain where it was well represented in the prehistoric faunal record and had only been extinct since AD 1918 (Love, 1983). The beaver has been successfully reintroduced to Denmark and there are controversial plans to reintroduce it to Britain. Bones and beaver-gnawed wood are well represented in prehistoric contexts and the last historical record is by Gerald of Wales in AD 1188. A current archaeologically-focused project on the effects of modern populations of European beavers in parts of France (Coles, 2001) is contributing to an understanding of the ecological effects of this species. In appropriate topographic situations reintroduction would significantly extend wetland habitats, and increase biodiversity and would help to restore the health of rivers.

Knowledge of the native, or introduced, status of endangered taxa likewise contributes to conservation strategies. The small-leaved lime (*Tilia cordata*) was at one time regarded as a minor, or even non-native, species in many parts of Europe. That was until the implications of its low pollen productivity became fully appreciated (Greig, 1982). It is now recognised as having been one of the most abundant species in the Climatic

Optimum wildwood of Britain, Germany, Belgium and south Scandinavia. Its occurrence today provides a good indicator of ancient woodland of high conservation value (Rackham, 2003). A converse case is represented by two plants now rare in Britain due to the abandonment of the agricultural regime on which they depended. The corncockle (*Agrostemma githago*) and cornflower (*Centurea cyanus*) were, however, only introduced in the last two millennia (Robinson, 1985). The fact that they are introduced should not necessarily diminish their conservation value, but it may encourage reflection on why they are being conserved: because of their contribution to the ecosystem as a whole (e.g. for insect life) or as vestiges of a former agricultural system. There is also the question of the extent to which introductions are sustainable in the long term given the natural variability of climate, land-use and many other factors.

Environmental management and archaeological assessment

At one time the main archaeological strategy in the face of development was excavation. Salvage archaeology developed in the USA particularly from the period of major dam construction. Rescue archaeology in Britain was particularly in response to motorway construction in the 1960s and 1970s, and similar developments occurred throughout Europe. Today rescue archaeology takes place on a much reduced scale, where development cannot be avoided for social, or economic, imperatives. Major transport infrastructure programmes, for instance, frequently reveal archaeological sites in deep sedimentary sequences. Some recent examples have revealed major archaeological and palaeoenvironmental sequences, noted in previous chapters: the Storebaelt fixed link bridge/tunnel between Denmark and Sweden (Pedersen *et al.*, 1997); the Betuweroute transport line in the southern Netherlands (Louwe Kooijmans, 2003); and the Channel Tunnel between England and France (Preece and Bridgeland, 1998). In general, however, excavation in advance of development, so-called preservation by record, is today seen as a second-best option where *in situ* preservation is not practical. This is because it is never possible to make a perfect record of original contextual relationships. Future generations will have improved methods and techniques at their disposal.

There is general recognition today that the unbridled building and industrial development of previous generations is not sustainable. Environmental and planning policy exists to evaluate the effects of development at the planning stage. Important in the evolution of this approach was the USA Natural Environmental Policy Act of 1969 which created foundations for an environmental protection policy. Environmental assessment is required of federal activities that affect human health and environment including the historical, cultural and natural heritage (O'Donnell, 1995; McManamon, 2000). An allied concept is that of **cultural resource management** which is concerned with all those features, both natural and created, which are associated with human activity (Fowler, 1982). The concept was first developed (*c.*1971) within the USA National Park Service[3] as part of the development of an integrated conservation strategy. Environmental assessment was adopted by the EEC (now European Union) in a directive of 1985 which applied to major engineering projects. Since then the approach has become more widely adopted in planning policy and a further 1997 directive specifically includes archaeology in its definitions (Willems, 1998). In 1992 the environmental assessment concept was enshrined as Principle 17 in the declaration of the Rio conference.[1]

A logical development of the idea of environmental assessment is that of **mitigation strategy**: works designed to counteract the detrimental environmental effects of development. Mitigation strategies include modifying developments so that they affect less sensitive areas, or designing developments so that important deposits are preserved. Although the principles of heritage conservation are widely accepted internationally, practice does vary significantly, even within Europe, and there is a particular problem dealing with threats to heritage which lie outside the normal planning/development control process.

In situ preservation

The move away from excavation and towards *in situ* preservation means that it is important to develop ways of evaluating the significance of archaeological sites non-destructively without excavation. It has also created a need to improve understanding of the burial environment on archaeological sites, the processes of decay and preservation and the ways in which appropriate conditions for preservation may be maintained (Corfield *et al.*, 1996). A somewhat controversial case is the reburial, following excavation, of Shakespeare's Rose Theatre (built AD 1587) in sediments below an office block in London (Bowsher, 1998). The experimental earthwork project established in the early 1960s on two sites, on chalk at Overton Down, Wiltshire (Figure 8.3) and acidic sandy soils at Wareham Heath, Dorset, is the longest-running experiment which is contributing to questions of *in situ* preservation (Bell *et al.*, 1996). Excavations of these earthworks have been made at intervals of 2, 4, 8, 16 and 32 years and will continue at 64 and 128 years, providing information on how the earthworks weather and change over time, the effects of burial on biological evidence (e.g. pollen and snails) and the decay and preservation processes affecting a range of identical organic and inorganic artefacts buried in successive sections. Results so far demonstrate the rapid nature of initial (<10 years) post-burial change, after which a state approaching quasi-equilibrium seems to be achieved. At Lejre, Denmark the making of experimental barrows (burial mounds; Figure 8.4) has established how the development of an iron pan by redox (reduction/oxidation processes) within the mound can, over a period as short as months, create wet anaerobic conditions within a barrow. This explains the exceptional conditions within some barrows, most notably the burial of Egtved girl found with a preserved head of hair, a complete set of Bronze Age clothing and other preserved organic artefacts (Breuning-Madsen and Holst, 1998; Holst *et al.*, 2001), in an oak coffin dendrochronologically dated 1370 BC. Both earthwork experiments demonstrate that relatively short-term experiments can form the basis of better understanding of the nature of changes over much longer archaeological timescales.

Figure 8.3 The experimental earthwork on old chalk grassland at Overton Down, England, showing the excavation of the earthwork 32 years after construction (photo Edward Yorath)

An increasingly important dimension of heritage mitigation strategy is **green development:** the enhancement of the conservation value of a site to compensate for the loss of habitat through development elsewhere. Nature reserves are being created on reclaimed land, for instance abandoned peat cuttings, quarries or, in the case of a reserve in Flevoland in the Netherlands, the recreation of native woodland on land only reclaimed from the sea between 1957 and 1968. A rather different case of green development is current restoration of heathland at Greenham Common, UK, on land that served as a Cruise missile base during the closing years of the Cold War, and was the site of

Figure 8.4 Experimental barrows at Lejre, Denmark: oak 'coffin' after excavation showing the preservation of animal burials wrapped in textiles designed to investigate the exceptional preservation conditions found within some Danish Bronze Age barrows (photo Marianne Rasmussen, Historiks-Arkaeologisk Forsøgscenter, Lejre)

a peace camp protest for two decades. Restoration of that site has involved decontamination, and the concrete bunkers which housed missiles are now listed buildings (those with statutory protection) because of their significance to the history of the Cold War. Green developments frequently involve the making of ponds, lagoons and the creation of scrapes to make wetter areas in peat bogs; unfortunately such processes can be destructive to buried archaeology. This highlights an important difference between the conservation of archaeology, which cannot regenerate, and nature, which, given the right carefully created conditions, sometimes can.

A new approach to nature conservation in the USA has been **mitigation banking**: a developer buys a share of habitat (e.g. wetland) in some different location as compensation for environmental destruction on a development site. Such an approach to the commodification of habitat conservation is controversial and has obvious dangers. Its advocates argue, however, that one large nature reserve is frequently preferable to many small areas. Environmental scientists in the USA have also advocated an approach to conservation based on evaluation of ecosystem or nature's services. The services include pest control, insect pollination, fisheries, climatic effects, vegetation, flood control, recycling of matter, composition of the atmosphere, genetic diversity, soil retention, formation and fertility (Mooney and Ehrlich, 1997). A financial calculation is made of the value of these services, so that this may be balanced against the calculated benefits (e.g. jobs, wealth creation) of development proposals. The notion is that we need to attribute value in order to achieve a full understanding of what we stand to lose through developments. Cultural heritage may be similarly valued in terms of its contribution to tourism, leisure activities and the quality of life.

National strategies for conservation

Reviews of national arrangements for heritage conservation (Cleere, 1984, 1989) reveal considerable diversity, particularly in the extent to which heritage conservation and the conservation of environments, plants and animals are integrated. In part this is reflective of differing administrative traditions, national circumstances, perceptions and priorities. In a changing world each nation clearly has something to learn from best practice internationally.

In the USA the first National Park was created at Yellowstone in 1872, followed by establishment in 1916 of the National Park Service,[3] the government's lead preservation agency; today the service administers 32 million ha of land (Jameson and Hunt, 1999). The movement for National Parks reflected a desire to protect wilderness. Those sites in the vast holdings of federal land that exist west of the Mississippi, including the National Parks and areas administered by the Forestry Service, have a greater level of protection than in much of the eastern United States where a much greater proportion of the land is in private ownership (McGimsey and Davis, 1984). In the United States there is also a growing awareness of the need to take cognisance of the significance of landscape features for First Nation Americans (Laidlaw, 1989). Likewise Australian heritage legislation originally protected only those sites with physical Aboriginal remains, but subsequently this has been extended to cover natural features with significance for Aboriginal communities (Flood, 1989).

In Britain separate arrangements exist for heritage and wildlife conservation. Here the first National Parks were created as late as 1949 as areas of natural beauty, although, as we have seen (Chapter 6), the environment of most of these areas (e.g. Dartmoor and the North York Moors) has been greatly affected by long histories of human activity (Figure 8.1). The lead body for wildlife and geological conservation in England is English Nature.[4] The most important sites are protected as National Nature Reserves and there is a much larger number of Sites of Special Scientific Interest which cannot be impacted upon without appropriate consultation and safeguards. English Heritage is the lead body for heritage conservation in England,[5] and sites are protected under the Ancient Monuments and Archaeological Areas Act of 1979. Some of the most important monuments are in state guardianship, a much larger number are on a Schedule of Ancient Monuments and a much larger number still are on the Sites and Monuments records of local planning authorities. Sites are afforded protection through the planning process, particularly the application of Planning Policy Guidance Note 16; this codifies a preference for *in situ* preservation and assessment in advance of any development proposals (Wainwright, 2000). A somewhat schizoid tendency as between arts and environmental emphasis is illustrated by the successive government departments responsible for archaeology in England over the last 30 years: Ministry of Works; Department of the Environment; Department of National Heritage; Department of Culture, Media and Sport. For most of this period nature conservation has come under different government departments, and is currently under the Department for Environment, Food and Rural Affairs.

In Germany the emphasis of conservation has historically been on the mainly cultural or arts aspects. Here the extent of destruction affecting both archaeological sites and wildlife habitats has, in the past, been significant: only 5–8 per cent of the monument heritage recorded since 1830 survives (Reichstein, 1984).

Probably the country with the most unified conservation strategy for archaeology and nature is Denmark. Both are protected under the Protection of Nature Act (1992) administered by the Danish Forest and Nature Agency[6] of the Ministry of the Environment (Dehn and Kristiansen, 1993). This integrated strategy reflects the perceived importance of archaeological sites within a Danish landscape that is widely understood to be almost entirely humanly created. Denmark is a country with a particularly high level of public awareness of archaeology, natural history and past environmental conditions. In all 5 per cent of the country has been designated for conservation purposes on the basis of the value of areas in terms of their biology, cultural history and recreational value. Individual archaeological sites are surrounded by a 100 m protection zone in order to preserve the relationship between monuments and their landscape. This is an approach reflective of an almost unique appreciation of the relationship between culture and landscape in both the past and the present.

International dimensions and World Heritage

The UNESCO World Heritage convention[7] was signed in 1972 and has since played an increasing role in the designation and protection of those sites of outstanding importance (Prott, 1992). Sites are proposed by states party to the convention (176 countries); this can be on the basis of their cultural or natural significance, or a combination of the two. By 2004 there were 582 cultural sites, 149 natural sites and 23 sites with dual designations distributed as shown in Figure 8.5. This shows that cultural sites are concentrated in Europe, while many other parts of the world, particularly Africa and Australia, have a more limited coverage in which natural sites predominate. The number of European proposals allowed is currently limited in an attempt to achieve a more equitable distribution. Increasingly there is recognition that the nature/culture distinction is too often unhelpful (von Droste *et al.*, 1995). Ayers Rock in Australia is a natural site but one with profound cultural significance to Aboriginals. What could be more natural than remote Henderson Island in the Pacific, unoccupied at the time of the first

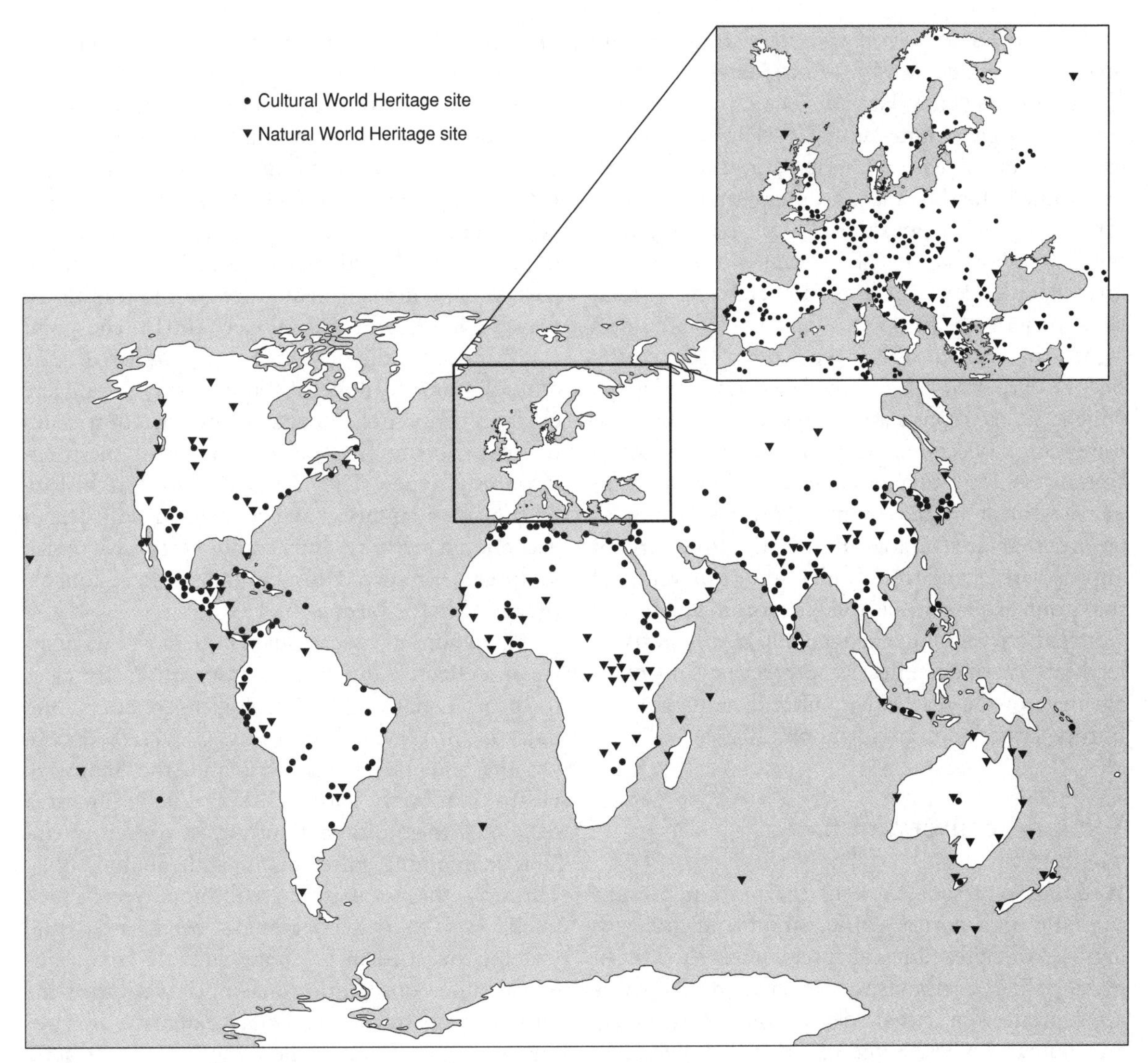

Figure 8.5 The distribution of World Heritage sites. This map shows sites designated up to June 2002 (based on data from UNESCO World Heritage Centre website)

European contact? Now, however, it is apparent that the island had earlier been settled by Polynesian communities who affected its ecology (Preece, 1998a). The states responsible for each World Heritage site make a commitment to develop management plans for the long-term conservation of those sites and these plans are subject to international scrutiny. Another aspect of UNESCO's heritage work has been to organise international action in situations where threats to heritage were beyond the resources of individual states, as for instance the Aswan Dam in Egypt, recording of the classical site of Carthage and the conservation of Sri Lankan temples.

Also in the custody of UNESCO is the Convention on Wetlands signed at Ramsar in Iran in 1971.[8] This early international move to protect wetlands reflects their obvious cross-border significance for migrating birds. By 2004 there were 138 states signatory to the convention and these have undertaken to work towards the wise and sustainable use of wetlands.

Within Europe, conservation strategies are increasingly influenced by a European perspective. The European Union Habitats Directive developed since 1994 gives states until 2004 to designate a list of Special Areas of Conservation. The British government for instance is designating 340 sites for this higher European level of protection covering 1.7 million ha, or 7.4 per cent of the UK land area. Parallel developments are occurring in relation to heritage. In 1992 many European states signed the Council of Europe Valetta Convention on the Protection of Archaeological Heritage, which as the convention says is a 'source of European collective memory'.[9] States make a commitment to take archaeology into account in assessment of development proposals and to ensure that archaeological investigations are of appropriate standard. Under the umbrella of the Council of Europe the European Landscape Convention was launched in 2001; this aims to facilitate the more effective protection and management of the distinctive cultural landscapes of Europe (Fairclough and Rippon, 2002).

Wetland conservation

Wetlands represent some of the most important sites for both nature conservation and archaeology as the preceding chapters have shown. In many periods, wetlands were seen as special places appropriate for ritual deposition of artefacts and human bodies. Some of the finest objects in the National Museums of Ireland, Denmark and other countries come from such contexts. The Gundestrup Cauldron, a remarkable piece of Iron Age iconography, was deposited on a small Danish bog, the Ravensmose, which today is a nature reserve within an agricultural landscape. Bogs in the north-west European countries have preserved bog bodies which, when investigated using the techniques of modern palaeoenvironmental and forensic science, can produce a wealth of evidence for diet, living conditions and ritual practice as illustrated by Lindow Man in the UK and many other bog bodies (Turner and Scaife, 1995; van der Sanden, 1996). The threat to this resource is all too apparent, however, and 96 per cent of the raised bogs known in England in 1850 have now been lost through peat cutting, drainage, etc.

Today there is a growing recognition of the importance of social and cultural dimensions in wetland conservation within the framework of the Ramsar convention (p. 253). This includes the development of strategies that conserve the archaeological heritage, and develop an understanding of past wetland management practices and traditional ways of life (Coles and Olivier, 2001). The point is illustrated by the rich estuarine mangrove forests and sea grass meadows of south-west Florida. Here evidence from archaeological contexts contributes to an understanding of the origins of these environment types. They have a 7000-year history of human adaptation which created cultures of exceptional artistic richness represented by wooden sculptures preserved in the anaerobic sediments (Purdy, 1991; Marquardt, 1996).

Wetlands have seen some of the most imaginative and effective integrated conservation strategies to protect nature, archaeological remains and palaeoenvironmental resources (Cox *et al.*, 1995). On the Somerset Levels, England, the Shapwick Heath National Nature Reserve is in an area reclaimed from old peat cuttings, with fen woodland containing rare plant and animal taxa. Through this runs the Neolithic Sweet Track (p. 164). The water-table has been raised and water is pumped in to create suitable conditions for wildlife conservation and preservation of the buried wood trackway. Today extensive areas of the Somerset Levels where peat was formerly cut for garden use has been given over to nature conservation and water-tables are being raised, in particular to meet the needs of bird conservation. An even more ambitious scheme is that by the Danish National Forest and Nature Agency in Åmose, where there is a chain of bogs in a river valley, many of which were once drained for agricultural land. Field walking had produced exceptionally abundant finds of Mesolithic and Neolithic artefacts. The area is now being progressively restored to wetland with the eventual aim of raising the water-tables over some 40 km^2 (Fischer, 1999). The Federsee in Germany (Plate 8.1) is also a wetland of exceptional archaeological richness: 18 waterlogged prehistoric settlements are known and the area is rich in birds and has many Red Data Book species (see note 2). Much of this area was once farmed but now a project part-funded by the European Life Programme has acquired much of the 3300 ha wetland for conservation. Public accessibility has been facilitated by a trail through part of the restored wetland, which includes partial reconstructions of some of the previously excavated waterlogged sites. This is linked to a museum of finds from the settlement excavations with reconstructions of houses of various prehistoric periods (Schlichtherle and Strobel, 1999).

An area of England which has attracted particular concern is Thorne and Hatfield Moors (Figure 8.6), 3000 ha of raised bog from which large-scale peat extraction has been undertaken for garden compost. One trackway has been found and there is evidence of votive artefact and body deposition in wetlands of the wider region. There are extensive areas of buried ancient woodland (p. 211) of great palaeoenvironmental importance, which contain many beetle species now extinct in Britain (Whitehouse, 2004). During 2002 an agreement was concluded to end peat extraction on this site, and with it the virtual end of commercial peat cutting in the UK, within 2–5 years. This creates the opportunity for the restoration of this raised mire as a nature reserve. Similar experiments in raised mire restoration are under way in the Somerset Levels and the Corlea trackway site in Ireland (see Figure 6.23). It remains to be seen whether, and over what timescale, such ambitious restoration schemes may be achieved. Clearly a detailed understanding of bog history represents an important part of achieving effective restorations.

Figure 8.6 Thorne Moors, Yorkshire: trees of a dendrochronologically dated pine forest exposed by commercial peat milling (photo Martin Bell)

From a wetland archaeological perspective, monitoring of water-tables and burial environment is essential to establish the effectiveness of *in situ* preservation. The Sweet Track preservation programme was implemented in 1982 and subsequent monitoring by small strategic excavations and analysis has demonstrated that so far the strategy had been highly effective (Brunning *et al.*, 2000). Similarly at Åmose in Denmark,, the Mesolithic site complex, and at Nydam, where ritual deposits of weapons and wooden artefacts including boats of the Roman Iron Age lie buried in a bog, techniques of monitoring the water-table, water quality and burial environment are being pioneered (Fischer, 1999).

Coastal environments represent a particular and increasingly important set of wetland management issues. Only in recent years has the archaeological richness of this environment become apparent (Fulford *et al.*, 1997; McErlean *et al.*, 2002). Coastal wetlands share the archaeological and natural historical value of other wetlands but, added to this, archaeological sites and palaeoenvironmental sequences are frequently exposed by erosion, and these areas offer particular opportunities for work on past coastal change and the ways in which past communities were adapted to a highly changeable environment (p. 159). Coastal environments were particularly attractive to past communities and these areas are also disproportionately impacted upon today by developments including harbours, industry, etc. Many of these coastal wetlands are also subject to the effects of **coastal squeeze** where, seaward of seawalls, wetland habitats of conservation importance such as salt marshes, reed swamps and mangroves are being eroded at an accelerated rate as a result of sea-level rise related to global warming (Chapter 9). In some areas the policy adopted is **managed retreat:** abandonment of coastal defences which are not considered economically sustainable in the context of current sea-level rise. This creates new areas of salt marsh and wetland habitat on land formerly reclaimed for agriculture. An analogous

Figure 8.7 Gwent Levels Wetland Reserve, Goldcliff, South Wales, UK. Saline lagoons have been created as part of a mitigation strategy for loss of bird habitat in Cardiff Bay. To seaward there are intertidal exposures of Mesolithic and Iron Age settlements, the latter on the surface of peat exposures. (Copyright Royal Commission on Ancient and Historic Monuments of Wales)

case is the artificial creation of the Gwent Levels Wetland Reserve in Wales (Bell, 2004). Here saline lagoons (Figure 8.7) were excavated behind the seawall as a mitigation strategy to compensate for the loss of intertidal mudflats 20 km away in Cardiff Bay, where a barrage had been created to permanently flood the bay as part of a programme to regenerate the docks area. The area chosen for the lagoons was one of extreme archaeological sensitivity and the nature conservation strategy had to be carefully linked to archaeological assessments to ensure that there was minimal impact on the buried archaeological resource, particularly a Roman period land surface with drainage ditches, which was just 0.8 m below the surface.

Presenting the past

A recurrent theme throughout this chapter has been the growing relationship between heritage and natural historical aspects of conservation, the two contributing to sustainable strategies for the future. Education and tourism also have a role to play in promoting a more integrated approach. However, in some countries tradition militates in varying degrees against an integrated perspective. Under these circumstances people may be presented as apart from nature, rather than part of it. In London the British Museum has a strongly cultural focus with an emphasis on objects, while the Natural History Museum has an emphasis on the natural world, albeit with reference to the role of people. More integrated perspectives are seen in Denmark, for instance in the displays of the National Musuem and Moesgard Museum, which emphasise the environmental context of human activity, particularly in the Mesolithic and Neolithic periods. Søllerød Museum presents artefacts and ecofactual evidence from the complex of Mesolithic sites at Vedbaek alongside dioramas depicting contemporary environments. In the Danish National Museum it is similarly emphasised that a major part of the Bronze Age and Iron Age collections derives from ritual deposition in wetland contexts. The origins of part of the British Museum collection derived from ritual deposition in, for instance, riverine contexts does not receive comparable emphasis, with the result that the objects are to some extent decontextualised.

Integrated presentations of natural historical and heritage aspects are increasingly found in the interpretation centres and display boards of organisations like National Parks, nature reserves and amenity sites. An innovative approach is the French concept of the ecomuseum (Blockley, 1999) which presents multiple sites in a landscape setting. An example has been created by a local community initiative at Plougerneau in north Brittany. This celebrates the landscape and history which gives a distinctive identity to that community, particularly its maritime relationships. The ecomuseum comprises linked walks which take in points of historical, ecological and archaeological interest. Sites include standing stones, a church and cemetery buried by blown sand in the Little Ice Age (see Figure 5.25), a museum of the ethnohistory of seaweed exploitation, local churches, lighthouses, etc. Such an initiative contributes to the sustainable development of a rural community through tourism. The contribution of heritage to the identity of place, the ideal of community involvement in conservation and a drive to make the past accessible to a wider public, beyond affluent and intellectual elites, are currently key themes in the development of heritage policy in the UK (English Heritage, 2000; Department of Culture, Media and Sport, 2001).

The physical reconstruction of sites and landscapes is one way of going beyond the limitations of traditional museum buildings and making the past come alive for a wider public, as the reconstruction of colonial Williamsburg in the USA shows (Hume, 1999). Many reconstructed sites exist throughout the world, although most have a mainly educational and presentational emphasis (Stone and Planel, 1999). Two which have been particularly engaged in experiment are Lejre, Denmark (founded 1964; Rasmussen and Grønnow, 1999) and Butser, England (founded 1972; Reynolds, 1999). Such sites provide an opportunity to test theories about how things were done in the past and thus develop an approach based on explicit hypothesis testing (p. 17). Experiments at Butser have shown that seed grain can be most successfully stored in pits and that primitive cereal types can give unexpectedly high yields under prehistoric conditions of cultivation. Lejre has contributed to an understanding of crop husbandry and woodland management (Meldgaard and Rasmussen, 1996) and current research on *in situ* preservation below barrows has already been noted (p. 250).

Reconstruction *in situ* on archaeological sites has frequently proved contentious, as for instance in the work of the USA National Park Service (Jameson and Hunt, 1999). Reconstructions away from actual sites are less problematic. They may also be open to question, however, as of necessity they are based on imperfect knowledge and there is the danger that one particular interpretation is perceived as fact and becomes almost written in stone for decades.

Integrated perspectives

It has been argued that heritage and nature conservation mutually contribute to the development of sustainable futures. Examples have been noted of imaginative habitat restoration schemes, some of which link archaeological and natural historic objectives, particularly in wetlands. In some countries, however, separate agencies, legal frameworks, a lack of coordination and lack of empathy between the different aspects of conservation weaken the case. Increasingly, however, the international dimension is influential. The Rio summit and its successors, World Heritage sites and Ramsar designation all show a growing recognition of the importance of the past dimension and the contribution which social factors can make to nature conservation. The role of indigenous groups and traditional ways of life in the maintenance of biodiversity and sustainable futures is also increasingly recognised. If conservation strategies are to work then they cannot simply be imposed at national or international level, there must be local ownership and commitment. This underlines the importance of education and presentation, in an accessible way, of the past and people's place as part of nature. On the world stage it is important that heritage concepts are not simply dominated by Western academic agendas. The promotion of global perspectives has been an objective of the World Archaeological Congress since 1986. Others have similarly argued for a multivocal approach which is receptive to different views of the past (Hodder, 1999).

The post-Rio agendas of sustainability, biodiversity and the valuing of indigenous identity and place currently exist in an uneasy tension alongside other forms of international organisation (e.g. International Monetary Fund; World Bank; World Trade Organisation) with an emphasis on monetarist values, development and free trade as the solution to the world's problems. In charting a course between such powerful influences, the time-depth provided by palaeoenvironmental studies serves to alert us to the dangers of unbridled environmental manipulation. The knowledge we gain from archaeology and anthropology about the achievements of societies very different from our own may, likewise, make us more receptive to the importance of maintaining and celebrating cultural diversity alongside the biodiversity which it helps to maintain.

Notes (www resources)

International conventions, heritage organisations and legal conservation frameworks are constantly evolving, and information on current developments can be obtained at the following websites:

1. Rio: United Nations Conference on Environment and Development – http://www.igc.apc.org/habitat/agenda21/rio-dec.html
2. Red Data Book lists – http://www.iucn.org.themes/ssc/redlists
3. USA National Park Service – http://www.nationalparks.org
4. English Nature – http://www.english-nature.org.uk There are separate nature conservation agencies in the devolved regions of Britain: Wales, Scotland and Northern Ireland.
5. English Heritage – http://www.english-heritage.org.uk There are separate heritage agencies in the devolved regions of Britain noted above.
6. Danish Forest and Nature Agency – http://www.sns.dk/internat
7. World Heritage (UNESCO) http://whc.unesco.org/heritage.html
8. Ramsar Convention on Wetlands – http://www.ramsar.org
9. Valetta Convention on the Protection of Archaeological Heritage – http://conventions.coe.int/treaty/en

9 The impact of people on climate

Introduction

In previous chapters, the interactive relationship of people and the environment has been explored against a background of environmental change. A central theme throughout has been the historical scientist's adherence to the uniformitarian principle (Chapter 2), with environmental reconstructions based on geological and biological proxy records relying heavily on contemporary analogues. Our understanding of human environmental perception and response to change is also heavily dependent on the recent ethnohistorical record. Thus the present provides a methodological key which can be used to unlock the mysteries of the past. In Chapter 8, we examined some of the strategies that have been put in place to sustain and conserve the record of the past. However, as concern has grown about the future of the planet, research in the historical earth and archaeological sciences has taken on an additional dimension, with the search for clues as to future climatic trends and their likely impact on the global environment. Analogues from the past are now being sought to provide climatic scenarios for the future, while historical proxy data are being widely employed in increasingly sophisticated simulation models that seek to predict the course of climatic change (Houghton *et al.*, 2001). Hence, while the present may well be regarded as the key to the past, it might equally be said that the past is the key to the future. The point, of course, is that environmental scientists and archaeologists are working along a temporal continuum. Consequently, any discussion of the historical interaction between people, climate and landscape would be incomplete without some reference to the future. It is this prospective view which constitutes the basis for this final chapter.

The greenhouse effect

Although natural forcing factors (solar output variations, volcanic aerosols, etc.) may be responsible for some of the global climatic changes that are apparent in recent meteorological records (see below), there is now a substantial body of evidence pointing to the increasing influence of human activity on the world's climate. Well over a century ago it was suggested that atmospheric gases could retard heat output and hence raise surface temperatures of the planet and, by the turn of the century, a quantitative relationship had been established between atmospheric carbon dioxide and global temperatures (Revelle, 1985). It was not until the 1950s, however, that the full significance of what has now become known as the greenhouse effect, namely the warming of the atmosphere arising from the accumulation of anthropogenically produced gases, began to be fully appreciated (Kellogg, 1987). Although some still remain sceptical about the nature and extent of the greenhouse effect, a combination of empirical data, including terrestrial and ocean temperature records (Bradley, 2000b; Barnett *et al.*, 2001) and measurements of outgoing long-wave radiation spectra (Harries *et al.*, 2001), coupled with the results of climate simulation modelling (Tett *et al.*, 1999), provide a powerful confirmation of the hypothesis that recent climate change can be attributed, in large measure, to the effects of human activity.

Increasing concern over the potentially adverse effects of anthropogenic activity on global climate led to the establishment in 1988 of the **Intergovernmental Panel on Climate Change (IPCC)** under the aegis of the World Meteorological Organisation and the United Nations Environment Programme. The aims of the IPCC were (a) to assess available scientific information on climate change; (b) to assess the environmental and socio-economic impacts of climate change; and (c) to formulate response strategies. The IPCC First Assessment Report was completed in 1990 (Houghton *et al.*, 1990), and a Supplementary Report (Houghton *et al.*, 1992) was produced in 1992. A Second Assessment Report appeared in 1995 (Houghton *et al.*, 1996), which updated the information contained in the First Assessment, and which also involved an extensive review of issues related to the economic and social aspects of climate change (Bruce *et al.*, 1995; Watson *et al.*, 1996). These included the likely impacts of climate change on a range of different environments (drylands, coastal areas, mountain regions, etc.) and human activities (e.g. agriculture, water resources management, fisheries, forestry), as well as possible adaptation and mitigation options. A Third Assessment was published in July 2001 (Houghton *et al.*, 2001), and was accompanied by reports on the impacts of, adaptations to, and vulnerability of, society to climate change (McCarthy *et al.*, 2001), and possible mitigation strategies (Metz *et al.*, 2001).

The IPCC Reports have provided much of the factual basis for international agreements on controlling greenhouse gas emissions. The 1990 Assessment and the updated information in the 1992 Supplementary Report laid the groundwork for the United Nations Framework Convention of Climate Change (UNFCCC) which was agreed at the Earth Summit in Rio de Janeiro in June 1992. At this meeting, the Convention was signed by over 150 countries and came into force after the fiftieth ratification was achieved in 1994. The Convention was the forerunner for the World Climate Summit held in Kyoto, Japan, in December 1997, where it was agreed that industrialised nations would reduce net CO_2 emissions by ~5 per cent by 2010 relative to 1990 levels (Table 9.1). The Protocol, however, is complex, and allows participant countries to meet all or part of their emission obligations by cutting back on gases other than CO_2 (Manne and Richels, 1999; Reilly *et al.*, 1999). Moreover, the Protocol does not come into force as a legally binding document until at least 55 per cent of the parties to the Convention have ratified it. In addition, enough of the 'Annex 1' parties (the industrialised countries) should be included to account for at least 55 per cent of their total CO_2 emissions in 1990. The United States was responsible for 38 per cent of these emissions, the European Union for ~22 per cent and Japan for ~8 per cent. This condition means that until a number of key developed countries ratify the Protocol, it cannot become operational (Bolin, 1998). The implementation of mitigation measures to control the emission of greenhouse gases is now, therefore, very much caught up in the world of international politics (O'Riordan and Jäger, 1996), and it seems likely that future emission control scenarios will be dictated as much by what can be successfully negotiated as by what is required on the basis of the scientific evidence.

Atmospheric carbon dioxide

A major element contributing to the greenhouse effect is carbon dioxide (CO_2) which absorbs thermal long-wave infrared radiation. This leads to atmospheric warming and hence an increase in global surface temperatures. The most valuable archive of past changes in atmospheric CO_2 and other greenhouse gases are the polar ice sheets (Chapter 2) from which a record of both long-term and more recent (post-industrial) atmospheric trace gas changes has been obtained (Raynaud *et al.*, 1993, 2000). Measurements of atmospheric concentrations of CO_2 in ice cores from Greenland and Antarctica show that the mean concentration in late Holocene pre-industrial levels was around 280 ppmv (parts per million by volume). Thereafter, the ice-core data reflect a progressive increase in CO_2 concentrations with values of 315–320 ppmv being recorded for the 1950s (Figure 9.1a). Continuous monitoring of atmospheric CO_2 began in 1958 and these observations show an

Table 9.1 Commitments to limit or reduce emissions of equivalent CO_2 from 1990 to 2010 by Annex 1 (Industrialised nations) parties as agreed to at Kyoto, compared with changes in CO_2 emissions from 1990 to 1995 (reprinted with permission from Bolin, B. 1998, The Kyoto negotiations on climate change: a scientific perspective, *Science*, **279**, 16 January, pp. 330–31. Copyright 1998 AAAS)

Party	*Allowed 1990–2010*	*Observed 1990–1995*
European Union*	**–8%**	**–1%**
Austria	–8	–3
Belgium/Luxembourg	–8	+1
Denmark	–8	+18
Finland	–8	+3
France	–8	–4
Germany	–8	–9
Greece	–8	+7
Ireland	–8	–1
Italy	–8	–1
Netherlands	–8	+7
Portugal	–8	+49
Spain	–8	+14
Sweden	–8	+7
UK and N. Ireland	–8	–4
OECD, except EU	**(–6)**	**+8**
Australia	+8	+8
Canada	–6	+9
Iceland	+10	–4
Japan	–6	+8
New Zealand	0	+16
Norway	+1	+9
Switzerland	–8	–5
United States	–7	+7
Countries in trans.**	**(–1)**	**–29**
Bulgaria	–6	n.a.
Croatia	–5	n.a.
Czech Republic	–8	–23
Estonia	–8	n.a.
Hungary	–6	–15
Latvia	–8	n.a.
Poland	–6	n.a.
Romania	–8	n.a.
Russian Federation	0	n.a.
Slovakia	–8	n.a.
Slovenia	–8	n.a.
Ukraine	0	n.a.
Non-Annex I parties	–	**+25**

* Members of the European Union will implement their respective commitments in accordance with the provisions of Article 4 of the Convention.

** Countries that are undergoing the process of transition to a market economy.

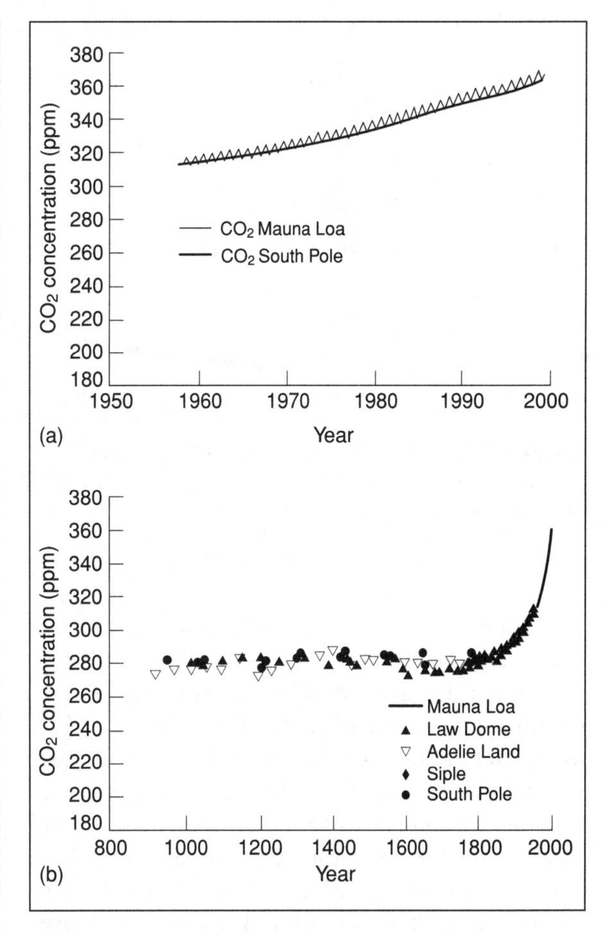

Figure 9.1 (a) Direct measurements of atmospheric CO_2 concentration at Mauna Loa, Hawaii, and at the South Pole over the period 1958 to 1999. (b) CO_2 concentration in Antarctic ice cores for the past millennium. Recent atmospheric measurements (Mauna Loa) are shown for comparison (after Houghton *et al.*, 2001, Figures 10a and b)

increase from around 315 to *c.*367 ppmv in the period up to 1999 (Figure 9.1b). The average rate of increase since 1980 is 0.4 per cent per year, and this is due to a rise in CO_2 emissions, the majority of which are due to fossil fuel burning, while the rest (10–30 per cent) are predominantly due to land-use change, especially deforestation. Data from the 1990s show that the annual rates of CO_2 increase varied from 0.9 to 2.8 ppmv/yr, at an average rate of ~1.5 per cent per year. These annual changes reflect short-term climatic variability, which alters the rate at which atmospheric CO_2 is taken up and released by the oceans and land

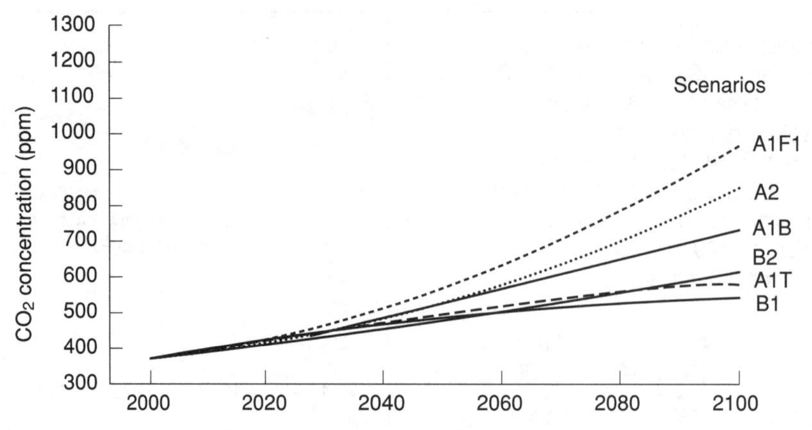

Figure 9.2 Atmospheric concentrations of CO_2 resulting from the six SRES scenarios. These range from a scenario which envisages a continued emphasis on the use of fossil fuels (A1F1), through a scenario which embodies economies based on non-fossil energy sources (A1T), to a scenario which involves, *inter alia,* the introduction of clean and resource-efficient technologies (B1) (after Houghton *et al.*, 2001, Figure 18a)

(Houghton *et al.*, 2001). Overall, however, the data indicate an increase in atmospheric CO_2 of 20–30 per cent in less than 200 years, attributable principally to an increase in fossil fuel combustion. When set in the context of the longer ice-core record, it is apparent that the industrial increases in CO_2 are unique in terms of growth rate, and have resulted in atmospheric concentrations which are unprecedented over the course of the last 420 ka (Raynaud *et al.*, 2000) and, maybe, even over the past 20 million years (Prentice *et al.*, 2001)!

Predicted changes in CO_2 concentration are shown in Figure 9.2. Analysis by the IPCC (Houghton *et al.*, 2001) suggests six possible scenarios for future CO_2 emissions. These so-called SRES scenarios (Special Report on Emissions Scenarios) are based on future energy requirements and energy use, but also include a range of factors such as economic development, demographic trends, and social and environmental attitudes. The calculation of future concentrations of atmospheric CO_2, along with other greenhouse gases, entails modelling the processes that transform and remove different gases from the atmosphere. Hence, in terms of CO_2, future concentrations are calculated using models of the carbon cycle, which model the exchanges of CO_2 between the atmosphere and the oceans and terrestrial biosphere. All of the SRES emissions scenarios imply increases in CO_2 concentrations to 2100, although the trajectories are significantly different (Figure 9.2). The carbon cycle models predict atmospheric CO_2 concentrations of between 540 and 970 ppmv by the end of the twenty-first century (90–250 per cent above the pre-industrial level of 280 ppmv), although if uncertainties about the magnitude of the climatic feedback from the terrestrial biosphere are taken into account, the total range is between 490 and 1250 ppmv (75–350 per cent above the 1750 concentration). An additional uncertainty is that the SRES scenarios do not acknowledge future governmental initiatives, so that, for example, no scenarios are included that explicitly assume implementation of the emissions targets in the Kyoto Protocol. Should increasing awareness of, and international concern about, the effects of rising atmospheric CO_2 levels lead to a significant reduction of the quantities of CO_2 released from industrial and related sources, then future atmospheric CO_2 increases could be towards the lower end of the range of the SRES scenarios.

Other atmospheric trace gases

Carbon dioxide is not the only constituent of the atmosphere that is important in determining the global heat budget, however, for it has long been recognised that other less abundant trace gases are also increasing and are likely to have an effect on climate comparable with that of CO_2. These gases

Table 9.2 Greenhouse gases that are affected by human activities and which contribute to the greenhouse effect (after Houghton *et al.*, 2001, Table 1)

	CO_2 (carbon dioxide)	CH_4 (methane)	N_2O (nitrous oxide)	CFC-11 (chlorofluoro carbon-11)	HFC-23 (Hydrofluoro carbon-23)	CF_4 (Perfluoromethane)
Pre-industrial concentration	about 280 ppm	about 700 ppb	about 270 ppb	zero	zero	40 ppt
Concentration in 1998	365 ppm	1745 ppb	314 ppb	268 ppt	14 ppt	80 ppt
Rate of concentration change	1.5 ppm/yr	7.0 ppb/yr	0.8 ppb/yr	−1.4 ppt/yr	0.55 ppt/yr	1 ppt/yr
Atmospheric lifetime	5 to 200 yr	12 yr	114 yr	45 yr	260 yr	>50 000 yr

which, like CO_2, act to block the escape from the atmosphere of thermal infrared radiation, include methane (CH_4), nitrous oxide (N_2O), tropospheric ozone (O_3) and the chlorofluorocarbons or 'Freons' (CFCs) CFC-11 and CFC-12 (CF_2Cl_2). All of these gases are affected by human activities (Table 9.2). Houghton *et al.* (2001) have estimated that, in the period 1750–2000, these gases collectively account for ~40 per cent of the observed *radiative forcing* (the change in energy available to the global earth–atmosphere system due to changes in these various forcing agents), the remaining 60 per cent being attributable to the effects of increasing atmospheric CO_2. However, other scientists have argued that the influence of non-CO_2 greenhouse gases has been greater than this, and that the rapid warming of recent decades, particularly since 1850, can be attributed in large measure to the increase in atmospheric concentrations of CFCs, CH_4 and N_2O (Hansen *et al.*, 2000).

Methane

Of the trace gases listed above, methane is perhaps the most significant, with estimates suggesting that it is responsible for around 20 per cent of the overall greenhouse effect, approximately one-third that of CO_2 (Houghton *et al.*, 2001). CH_4 is produced by microbial activities during the mineralisation of carbon under anaerobic conditions, e.g. in waterlogged soils or in the intestines of animals. It is also released by anthropogenic activities, such as exploitation of natural gas, biomass burning, and fossil fuel mining (A.T. Smith, 1995). Slightly more than half of the current CH_4 emissions are estimated to be anthropogenic. Ice-core studies show that atmospheric CH_4 concentrations have risen steadily since the middle of the eighteenth century (Figure 9.3a) from a pre-industrial level of about 700 ppbv (parts per billion volume) to a value in

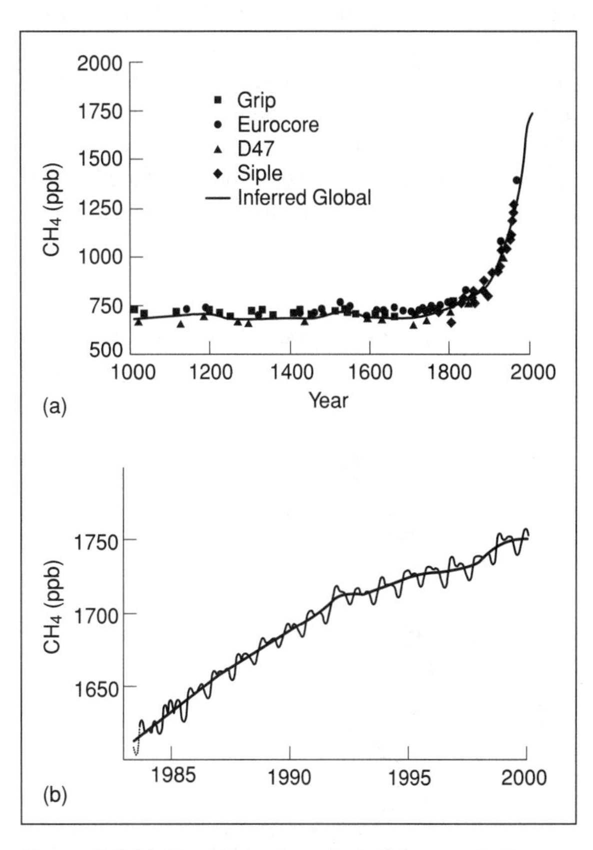

Figure 9.3 (a) Changes in atmospheric CH_4 concentration determined from ice-core and air samples for the past 1000 years. (b) Globally averaged CH_4 (monthly varying) and deseasonalised CH_4 (smooth line) abundance 1983–1999 (after Houghton *et al.*, 2001, Figures 11a and 11b)

excess of 1700 ppbv in the 1990s, an increase of ~150 per cent (Houghton *et al.*, 2001). Direct measurement of atmospheric CH_4 began in 1978, and the current estimate for CH_4 concentrations is around 1745 ppbv. Over the last 20 years there has been a slight decline in the methane growth rate (Figure 9.3b), although the increase has been highly variable, ranging from near zero in 1992 to 13 ppbv in 1998 (Houghton *et al.*, 2001).

As with CO_2, these high post-industrial CH_4 concentrations as revealed in the ice-core records have not been exceeded over the course of the last 420 000 years (Raynaud *et al.*, 2000). A curious feature of the CH_4 record, however, is the gradual increase in CH_4 concentration (of *c.*100 ppbv) from around 5 ka BP onwards (Figure 9.4). Ruddiman and Thomson (2001) have suggested that this may reflect the onset of large-scale rice farming in Asia and the consequent expansion of wetlands, which acted as methane sources. However, other factors, for example the increase in organic-rich wetlands following mid-Holocene sea-level stabilisation, and the growing numbers of domestic herbivores, might also have contributed to higher atmospheric CH_4 levels in the second half of the Holocene. If this hypothesis is correct, it implies that as much as 25 per cent of atmospheric CH_4 in Late Holocene pre-industrial times may also be attributable to human activity.

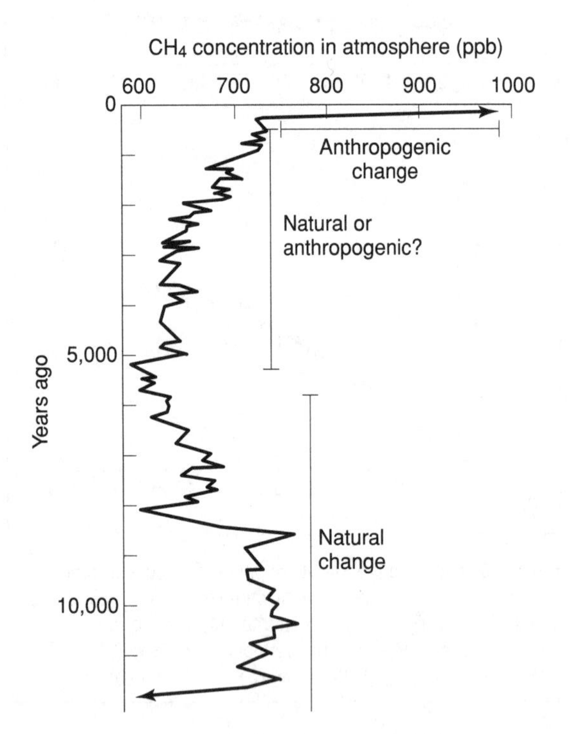

Figure 9.4 Atmospheric CH_4 variations from 12 ka BP to present, showing the gradual increase in methane concentrations from around 5 ka BP which may be partly due to anthropogenic activity (after Ruddiman and Thomson, 2001, Figure 1a)

Nitrous oxide

Although N_2O is also increasing in concentration the rate is considerably less than that of CH_4, recent measurements (1980–98) pointing to a rise of around 0.25 per cent per year (Houghton *et al.*, 2001). The flux of N_2O into the atmosphere is due primarily to microbial processes in soil and water and is part of the nitrogen cycle. Principal sources of atmospheric N_2O include fossil fuel burning, biomass burning, and mineral fertilisers, although these are often difficult to quantify. Ice-core data suggest pre-industrial levels of around 275 ppbv (Figure 9.5), rising to 314 ppbv in 1998 (Flückiger *et al.*, 1999). It has been estimated that the radiative

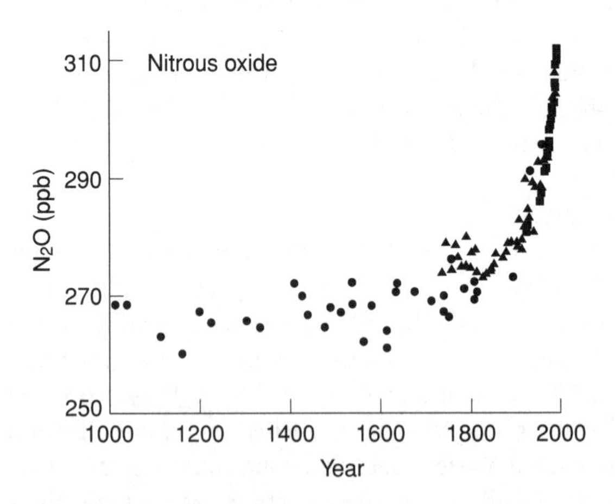

Figure 9.5 Measurements of the concentrations of nitrous oxide (N_2O) over the last 1000 years in cores from the Greenland and Antarctic ice sheets. Different symbols show different recording sites. Note the significant increase in N_2O values during the twentieth century (after Houghton *et al.*, 2001, Figure 8a)

forcing from N_2O is ~6 per cent of the total of all of the long-lived and globally mixed greenhouse gases (Houghton *et al.*, 2001).

Ozone

In the troposphere, ozone is produced through the oxidation of methane, and from various short-lived precursor gases, mainly carbon monoxide (CO), nitrogen oxides (NO_x), and non-methane hydrocarbons (NMHC). There are indications that concentrations of tropospheric O_3 are increasing as a result of the enhanced emission of CH_4, CO and NO_x (Houghton *et al.*, 1996). Observational evidence points to a rise in tropospheric O_3 over the Northern Hemisphere (north of 20° N) during the past three decades, with measurements from Europe suggesting a doubling in lower troposphere O_3 concentrations since earlier in the twentieth century (Schimel *et al.*, 1996). In other Northern Hemisphere areas, however, O_3 levels have declined (Tarasick *et al.*, 1995). According to Houghton *et al.* (2001), a combination of observational and modelling evidence suggests that tropospheric O_3 has increased by about 35 per cent since pre-industrial times, with some regions experiencing larger increases and some smaller. It has been estimated that O_3 currently contributes around 15 per cent to greenhouse-induced radiative forcing, which means that it is the third most important greenhouse gas after CO_2 and CH_4. It must be emphasised that these increases in O_3 occur only in the lower atmosphere, for in some parts of the world, a significant decrease in O_3 content has been recorded in the overlying stratosphere. As will be shown below, this decline in O_3 levels in the middle atmosphere, which again appears to be attributable largely to anthropogenic activity, poses a different set of problems for life on earth.

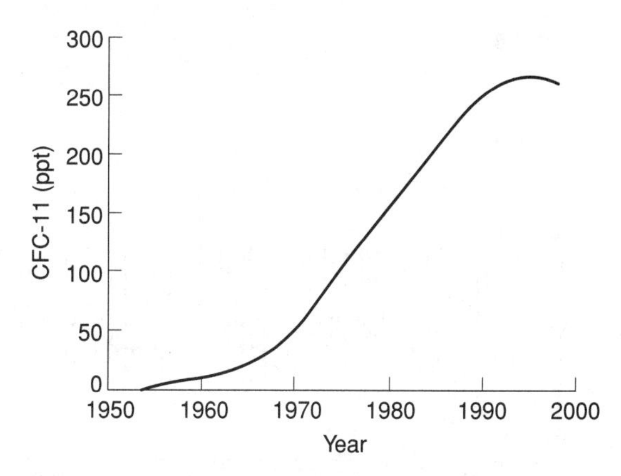

Figure 9.6 Global mean CFC-11 tropospheric abundance in pptv (parts per trillion by volume) from 1950 to 1998 (after Houghton *et al.*, 2001, Figure 12). This shows the slowdown and ultimate decline in chlorofluorocarbon concentrations following the implementation of the Montreal Protocol (see p. 275)

Halocarbons and halogenated compounds

These are carbon compounds which contain chlorine, fluorine, bromine or iodine, and many are effective greenhouse gases as they directly absorb infrared radiation. Those that contain chlorine (CFCs: chlorofluorocarbons; and HCFCs: halogenated chlorofluorocarbons) and bromine (halons) also destroy ozone in the lower stratosphere (Houghton *et al.*, 1996). Measurement of trace gases in polar ice cores indicates that natural sources of these complex molecules are minimal or non-existent, and hence their presence in the atmosphere is due entirely to anthropogenic activity (Butler *et al.*, 1999). Manufactured CFCs, which were introduced over 60 years ago, have a variety of uses, including solvents, refrigerator coolant fluids and propellants for aerosol sprays, and while they are ideal industrial chemicals in that they are highly stable, unreactive and non-toxic, they do not degrade readily in the atmosphere. Hence, by 1980 atmospheric concentrations of these compounds was increasing by around 6 per cent per year (Figure 9.6), particularly over the industrialised regions of the northern middle latitudes. In the 1990s, however, there was a significant slowdown in growth rates of atmospheric chlorofluorocarbons, especially CFC-11 and CFC-12, as CFC consumption was phased out under the terms of the Montreal Protocol (p. 275; Elkins *et al.*, 1993). Observational data suggest that the atmospheric abundances of the two principal CFCs peaked in the early to mid 1990s (Figure 9.6), and that both are now in decline. Nevertheless, they still contribute around 14 per cent of the radiative forcing of all of the global greenhouse gases (Houghton *et al.*, 2001).

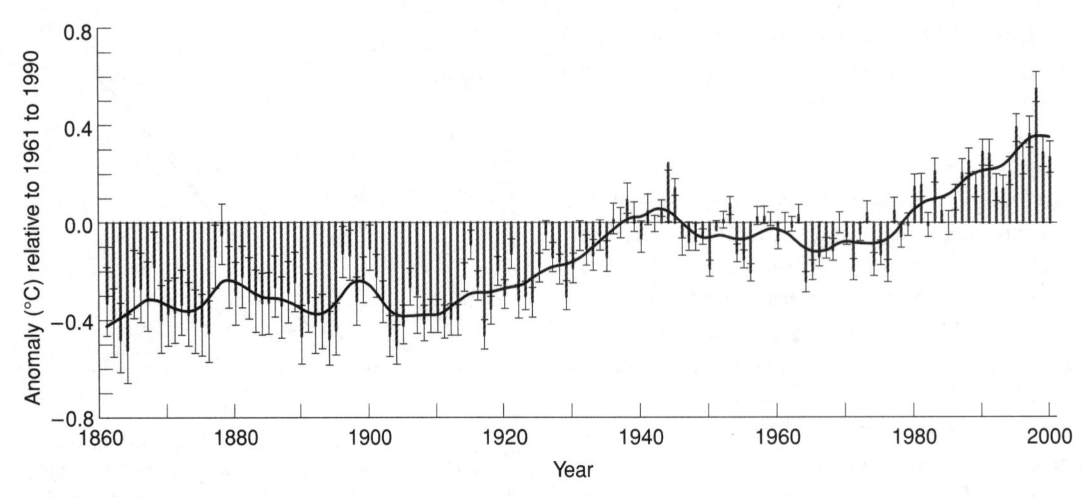

Figure 9.7 Smoothed annual anomalies of combined land-surface, air and sea-surface temperatures (°C) for the period 1861 to 2000 relative to the 1961–1990 average (after Folland *et al.*, 2001, Figure 2.7)

The role of aerosols

Aerosols are particles and very small droplets of natural and human origin that occur in the atmosphere. They include dust and other particles comprising a range of different chemicals, and are produced by a variety of processes, both natural (dust storms, volcanic activity, etc.) and anthropogenic. The latter category includes sulphate aerosols, biomass-burning aerosols, and fossil fuel black carbon (or soot) (Houghton *et al.*, 2001). Aerosols in the atmosphere influence the radiation balance of the earth by scattering and absorbing radiation, and by modifying the optical properties, amount and lifetime of clouds. Some aerosols, such as soot, tend to warm the surface, but the net climatic effect is believed to be negative radiative forcing, leading to a cooling of the earth's surface (Schimel *et al.*, 1996). Volcanic aerosols, particularly stratospheric sulphur, have long been considered as a possible forcing factor in short-term climate change during the Late Quaternary (Chapter 3). However, it is now believed that anthropogenic sulphate aerosols from fossil fuel combustion may also have exerted an influence on global climate by offsetting some of the warming induced by the increase in greenhouse gases. For example, recent climatic modelling results suggest that in the twentieth-century temperature record (Figure 9.7), the reduction in the overall warming trend between 1946 and the mid 1970s could reflect sulphate aerosols balancing the effect of greenhouse gases (Tett *et al.*, 1999).

Consequences of the greenhouse effect

The IPCC has evaluated the extent of the radiative impact of the greenhouse effect. Globally averaged estimates of radiative forcing due to changes in greenhouse gas concentrations since pre-industrial times are +2.43 W m^{-2} (watts per square metre) for the direct effect of the well-mixed greenhouse gases: 1.46 W m^{-2} from CO_2, 0.48 W m^{-2} from CH_4, 0.34 W m^{-2} from the halocarbons, and 0.15 W m^{-2} from N_2O. There is also a positive radiative forcing of 0.35 W m^{-2} for tropospheric ozone. These increases are offset by decreases of 0.15 W m^{-2} for stratospheric ozone, and 0.7 W m^{-2} for anthropogenic aerosols, mainly sulphates (80 per cent), but also fossil fuel organic carbon and organic aerosols from biomass burning (Houghton *et al.*, 2001). It must be emphasised, however, that these are first-order estimates, and there will be considerable spatial variation in patterns of forcing between the globally well-mixed greenhouse gases, the regionally varying tropospheric ozone, and the even more regionally concentrated aerosols. The environmental impact of these changes in radiative forcing is considered in the following section.

Global temperature changes

There is a considerable body of empirical evidence for recent climatic warming. Global surface air temperature data for the present century (Figure 9.7) show an initial episode of maximum warmth around 1940, but a significant increase in temperature from the late 1970s onwards. The 1990s in particular have been exceptionally warm, with 1998 exceeding all annual temperatures for at least 1000 years (Bradley, 2000b), while 2002 and 2009 were the second and third warmest years since temperature measurements began. Indeed, the 10 hottest years in the modern instrumental record, which began in 1855, have all been since 1990. In Britain, four of the five hottest years in the central England temperature record, which goes back to 1659, have occurred since 1990, with the record English temperature of 38.5 °C (101.3 °F) being achieved at Faversham, Kent, on 10 August 2003. Germany has a new record of 40.8 °C, Switzerland one of 41.5 °C and Portugal 47.3 °C. The IPCC assessments indicate an increase in mean global surface temperature of between 0.3 and 0.6 °C since the late nineteenth century, and an increase of 0.2–0.3 °C over the past 40 years (Nichols *et al.*, 1996). The rate of temperature rise at the surface has been of the order of 0.2 °C per decade since 1979, and this has been accompanied by an increase in tropospheric temperature of *c.*0.1 °C per decade. This general warming trend has not been globally uniform, however. The recent warmth has been greatest over the continents between 40 and 70° N, while other areas, such as the Caribbean, northern South America and West Africa have actually cooled. Marine data show a significant increase in heat content in the world's oceans between the mid-1950s and the mid-1990s, equivalent to a volume mean warming of 0.06 °C. In the North Atlantic, the mean temperature increase for the 0–300 m layer was 0.3 °C (Levitus *et al.*, 2000). This apparent warming trend is entirely consistent with model predictions for greenhouse-gas-induced climatic change in both the terrestrial and oceanographic realms (e.g. Tett *et al.*, 1999; Barnett *et al.*, 2001).

Despite the broad measure of agreement that appears to exist between empirical data and modelling results, there is still the possibility that the temperature trends that are apparent in the records *could* be explained almost entirely in terms of natural variability within the climatic system (Mitchell, 1989). This was acknowledged in the first IPCC Assessment where it was concluded that the observed warming was 'broadly consistent with predictions of climate models, but it is also of the same magnitude as natural climate variability' (Houghton *et al.*, 1990). In the 10 years that have elapsed between then and the publication of the third IPCC Assessment, however, significant progress has been made in reducing uncertainty, particularly with respect to distinguishing and quantifying the magnitude of climate response to different external influences. As a result, the third IPCC Assessment concluded that 'in the light of new evidence and taking into account the remaining uncertainties, most of the observed warming over the last 50 years is likely (*66–90 per cent chance*) to have been due to the increase in greenhouse gas concentrations' (Houghton *et al.*, 2001).

Future climatic trends are derived from atmospheric general circulation models (AGCMs) which simulate global climatic patterns and hence enable predictions to be made about the influence of atmospheric trace gases (Kattenberg *et al.*, 1996). The IPCC projections for future global warming are based on an estimate of 'climatic sensitivity', which is the predicted equilibrium response of global surface temperature to a doubling of equivalent CO_2 concentration. This has been estimated to be in the range 1.5–4.5 °C (Houghton *et al.*, 2001). Using the SRES emissions scenarios, which include emissions of both greenhouse gases and aerosols, an increase in global mean temperature of between 1.4 and 5.8 °C is predicted (Figure 9.8). The IPCC models suggest, however, a considerable range in predicted temperature increases under the different SRES scenarios. For example, for the end of the twenty-first century, the mean change in global average surface temperatures, relative to the period 1961–90, is 3.0 °C (range 1.3–4.5 °C) for the A2 scenario, and 2.2 °C (range 0.9–3.4 °C) for the B2 scenario (Cubasch *et al.*, 2001). Under all

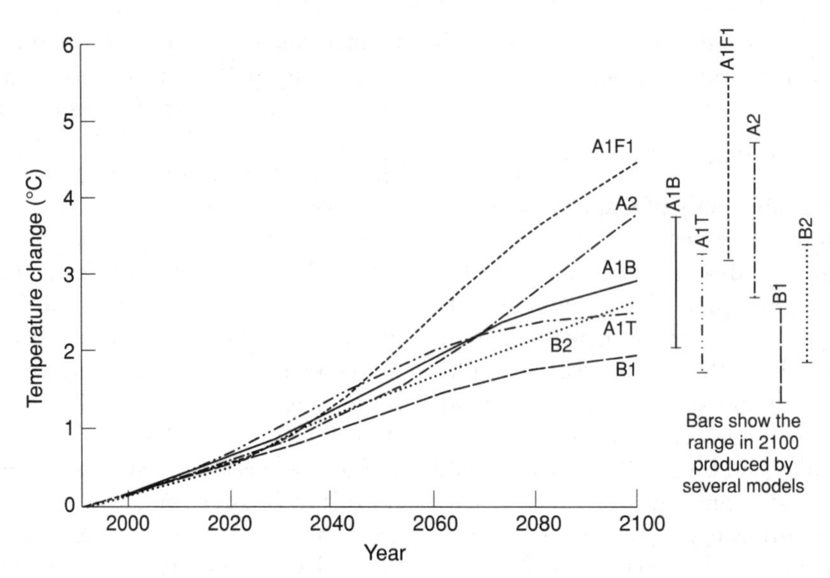

Figure 9.8 Global mean temperature projections for the six SRES emissions scenarios. The bars on the right show the temperature range for 2100 for each scenario as predicted by the climate models (after Houghton *et al.*, 2001, Figure 22)

of the scenarios in Figure 9.8, however, the average rate of warming is very likely (90–99 per cent chance) to be greater than any seen during the last 10 ka, although there would be considerable natural variability in the annual to decadal changes.

It is very likely that land areas will warm more rapidly than the global average, particularly at northern high latitudes in the winter months, where higher temperatures will lead to a more restricted sea-ice cover. Indeed, the IPCC model simulations suggest that in winter the warming for all Northern Hemisphere high-latitude regions exceeds the global mean warming by more than 40 per cent (1.3–6.3 °C for the range of SRES scenarios). In summer, warming is predicted to be in excess of 40 per cent above the global mean change in central and northern Asia. Only in south Asia and southern South America in the period June–August, and in South-East Asia for both summer and winter, do the models consistently show warming that is less than the global average (Houghton *et al.*, 2001).

Predictions of future climate change have often used the year 2100 as the target baseline, but policy-makers need short-term climate predictions (and predictions in which they can have confidence) if they are to develop strategies for coping with climate change over the two- to three-decade planning period (Zwiers, 2002). The results of two recent, but quite different, modelling experiments are of particular significance in this respect, for they show a very broad measure of agreement about the range of likely temperature change over the next 20–30 years. Estimates of mean global temperature increase for the decade 2020–30 relative to 1990–2000 (with a likelihood range of 5–95 per cent) are 0.3–1.3 °C (Stott and Kettleborough, 2002) and 0.5–1.1 °C (Knutti *et al.*, 2002). These estimates are unaffected by the choice of IPCC emissions scenarios. Over the longer timescale, these modelling exercises suggest that the potential warming for 2100 could even exceed the IPCC predicted temperature increase of 5.8 °C.

These predicted climate changes resulting from anthropogenic activity must be set alongside the natural climatic forcing factors discussed in Chapter 3. For example, future changes in solar irradiance have been estimated for the next 20 years using a combination of statistical and geophysical techniques (Lean, 2001). The results suggest peak irradiance in 2010, with levels comparable to or slightly lower than in previous maxima in 2000, 1989 and 1981. Minima will occur in 2006 and 2016. Total irradiance forcing of climate between 1996 and 2016 is in the range ±0.1 W m^{-2}. This compares with a forecast net anthropogenic climate forcing over the next 20 years of 0.5–0.9 W m^{-2}.

Global precipitation changes

Overall, global land precipitation has increased by about 2 per cent since the beginning of the twentieth century (Figure 9.9a), and while this increase is statistically significant, it is neither spatially nor temporally uniform (Folland *et al.*, 2001). For example, although precipitation in the higher northern latitudes is seen to have increased markedly over the last 30–40 years (Figure 9.9b), in the Northern Hemisphere subtropics a gradual fall in precipitation values (*c.*0.2–0.3 per cent per decade) occurred throughout much of the twentieth century, although this trend has been reversed from the late 1980s (Figure 9.9c). Indeed, the

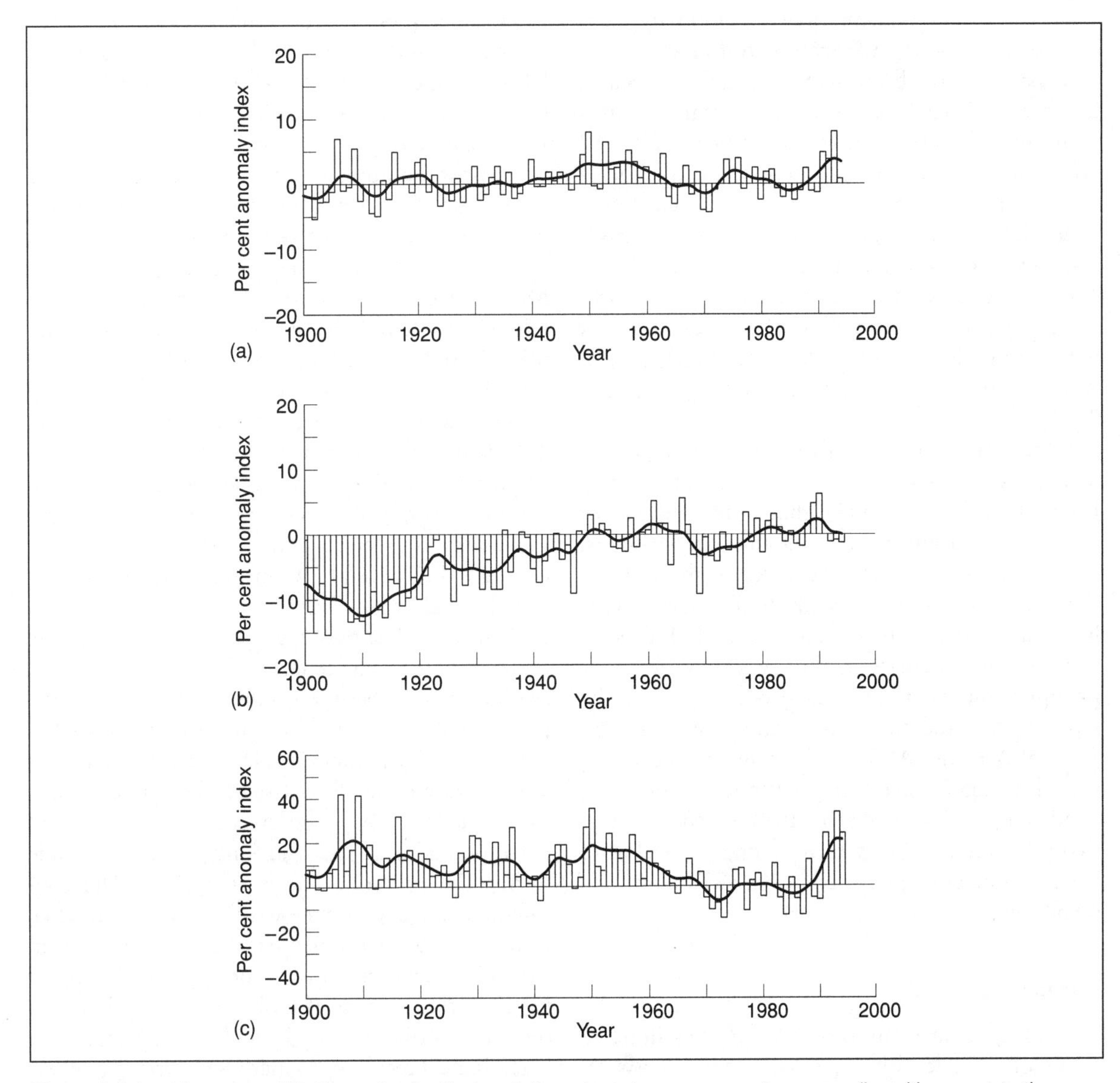

Figure 9.9 Annual mean precipitation series for the twentieth century shown as percentage anomalies with respect to the 1961–90 mean: (a) global land areas; (b) northern high latitudes; (c) Northern Hemisphere subtropics (after Jones and Hulme, 1996, Figures 6b, 8b and 9b)

mid-latitudes of the Northern Hemisphere have had precipitation totals exceeding the 1961–90 mean every year since 1995. The present rate of increase in precipitation in the mid- and high-latitude regions of the Northern Hemisphere is of the order of 0.5–1.0 per cent per decade, with the wettest year on record being recorded in 1998. In the Southern Hemisphere, twentieth-century increases in precipitation have also been recorded in Australia and Argentina (Folland *et al.*, 2001).

The IPCC assessment envisages globally averaged water vapour, evaporation and precipitation increasing, with both increases and decreases being experienced at the regional scale. Results from modelling simulations under both SRES A2 and B2 emissions scenarios indicate the likelihood of increased summer and winter precipitation over high-latitude regions, with winter increases also over northern mid-latitudes, tropical Africa and Antarctica, and summer increases over southern and eastern Asia. By contrast, lower levels of winter rainfall are projected for Australia, central America and southern Africa (Houghton *et al.*, 2001).

One feature of the IPCC projections is the increase in extreme weather and climate events. Data suggest, for example, that in many areas of the Northern Hemisphere mid- and high latitudes, significant increases have occurred in the proportion of total annual precipitation derived from heavy and extreme precipitation events. Indeed, a 2–4 per cent increase in the frequency of heavy precipitation events may have occurred over the latter half of the twentieth century. Similarly, in parts of Asia and Africa, the frequency and intensity of droughts have increased in recent decades. Modelling results suggest that the frequency of such extreme events is likely to accompany the future rise in temperature predicted by the SRES emissions scenarios.

Sea-level changes

Tide-gauge data from a number of sites indicate that global sea level has risen at a rate of between 1.0 and 2.0 mm yr^{-1} over the course of the twentieth century, a more rapid rate of rise than occurred during the nineteenth century (Houghton *et al.*, 2001). No significant acceleration of sea-level rise has been detected in the tide-gauge records during the course of the twentieth century. However, satellite altimeter data suggest a rate of sea-level rise for the 1990s greater than the mean rate of rise for much of the twentieth (possibly as high as 3.1 mm yr^{-1}), although it is too early to say whether or not these new records indicate a recent acceleration in sea-level rise, or whether they reflect systematic differences between the two measuring techniques (Church and Gregory, 2001). The estimated contributions to the observed rise in sea level over the period 1910–90 are: thermal expansion of ocean waters (contribution 0.3–0.7 mm yr^{-1}), melting of glaciers and ice caps (0.2–0.4 mm yr^{-1}), isostatic adjustments since the Last Glacial Maximum (0–0.5 mm yr^{-1}), with further smaller contributions from ice-sheet melting, permafrost melting and the products of terrestrial storage. Modelling estimates suggest that those components related to climate change (Figure 9.10a) contribute 0.3–0.8 mm yr^{-1} to global sea-level rise, and that these terms do show an acceleration during the course of the twentieth century (Figure 9.10b). The data suggest that it is very likely that greenhouse-induced warming has made a significant contribution to the observed rise in sea level during the course of the twentieth century, primarily through the thermal expansion of ocean waters and the loss of land ice.

Figure 9.11 shows the future rise in sea level under the six SRES scenarios. The IPCC projects a sea-level rise of between 0.09 and 0.88 m for the period 1990–2100, with a central value of 0.48 m (Church and Gregory, 2001). The latter gives an average rate of sea-level rise of about two to four times the rate over the twentieth century, and corresponds to a rise of around 5 cm per decade during the course of the next century. The considerable range within these estimates (Figure 9.11), however, reflects the degree of uncertainty within the scientific community not only with regard to the future course of global warming, but also over the terrestrial, hydrological and especially glaciological response to increased surface temperatures. For example, while greenhouse warming is likely to lead to melting of the smaller ice caps and glaciers and a rise in sea level, the IPCC

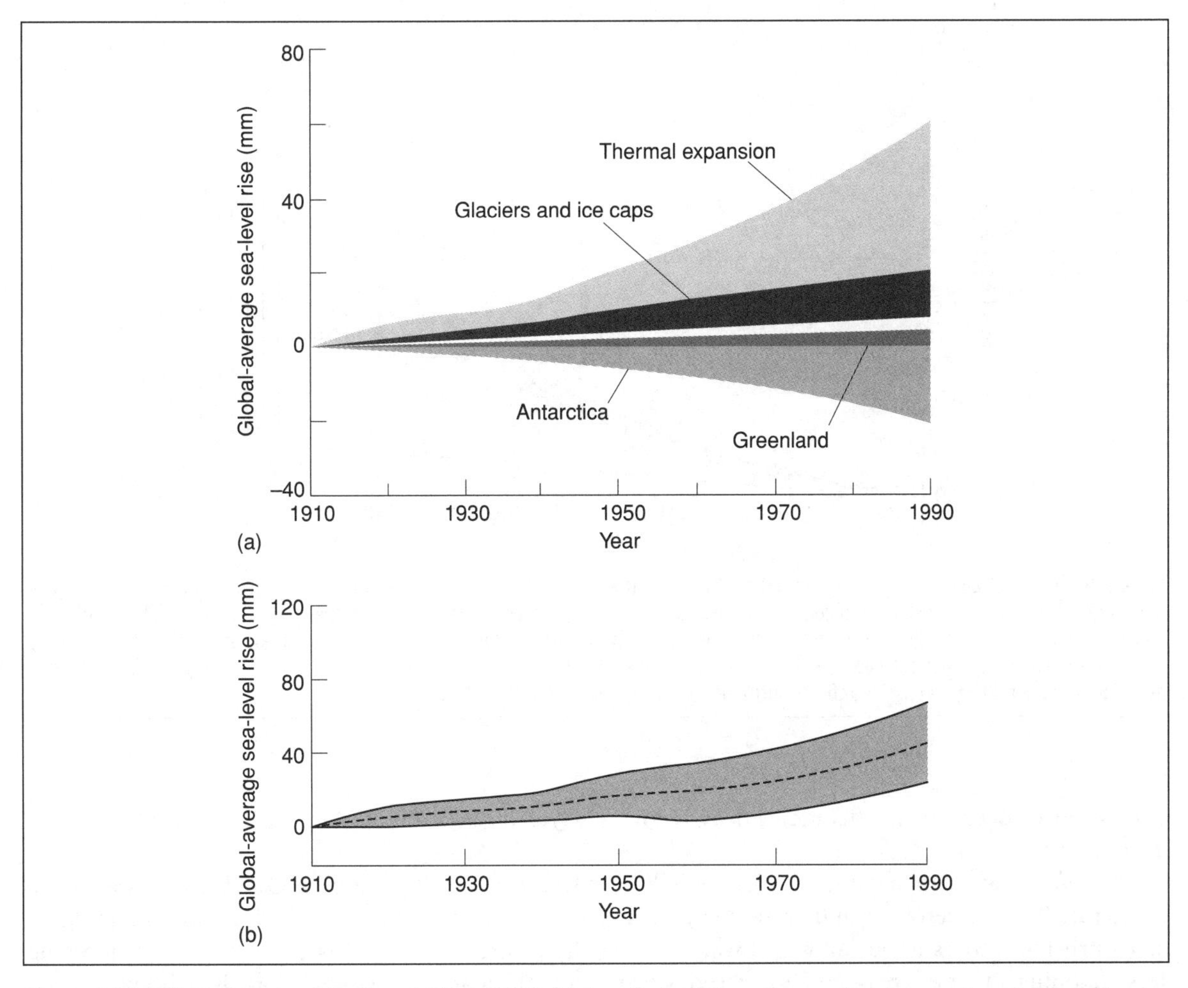

Figure 9.10 Estimated sea-level rise from 1910 to 1990: (a) the effects of thermal expansion of sea water, glacier and ice cap, and Greenland and Antarctic contributions resulting from climate change and calculated from modelling exercises; (b) the mid range and upper and lower bounds for the computed response of sea level to climate change. These curves represent our best estimates of the impact of anthropogenic climate change on global sea level during the course of the twentieth century (after Church and Gregory, 2001, Figures 11.10a and 11.10b)

assessment suggests that increased precipitation in the high-latitude regions (Figure 9.9) could, to some extent, offset this effect, with the Antarctic ice sheet in particular gaining in mass and thereby contributing to a lowering of global sea level (Houghton *et al.*, 2001). Concern has been expressed over the stability of the marine-based West Antarctic ice sheet, for it has been suggested that the rapid destruction of this ice mass as a result of greenhouse warming during the course of the next century could raise global sea levels by 4–6 m (Oppenheimer, 1998). Current opinion, however, is that collapse of the West Antarctic ice sheet is unlikely, at least over the course of the twenty-first century (Vaughan and Spouge, 2001), and hence the projected future sea-level rise in the IPCC assessment makes no allowance for dynamic instability of the West Antarctic ice sheet. Accordingly, it seems likely that the principal mechanisms by which greenhouse warming will raise global sea level during the course of the next century are the melting of low-latitude glaciers (with the Greenland and Antarctic ice sheets probably playing minor, or even negative roles), and the thermal

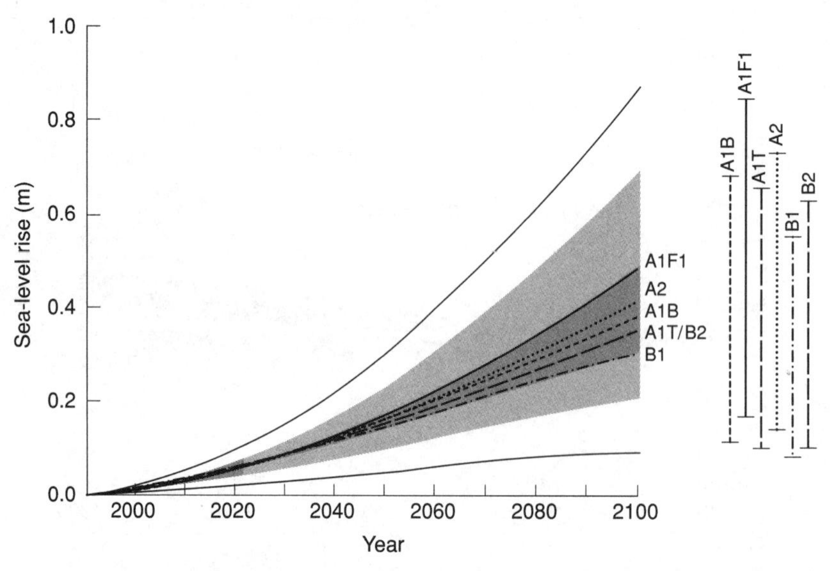

Figure 9.11 Global average sea-level rise 1990–2100 for the six SRES emissions scenarios. The region in dark shading shows the range of the average for all modelled SRES emission scenarios; the light shading shows the range of all modelled SRES emissions scenarios; the region delimited by the outer lines shows the range of all modelled scenarios, and includes uncertainty in land-ice changes, permafrost changes and sediment deposition. The vertical bars to the right show the range in 2100 of all models for the six SRES scenarios (after Church and Gregory 2001, Figure 11.12)

expansion of ocean water (Church and Gregory, 2001).

The effects of a gradual rise in sea level will be accelerated coastal erosion and shoreline retreat, salt intrusion into estuaries and freshwater aquifers, flooding of new areas and increased storm damage, and the progressive dislocation of human activity in coastal regions (Viles, 1989). Coastal inundation from extreme sea-level events, notably storm surges, is likely to pose a particular threat (Hubbert and McInnes, 1999). Areas particularly at risk are the thousands of hectares of coastal wetlands of western Europe and North America (p. 255) where marsh development has kept pace with the course of Holocene sea-level changes but where salt marsh is now being increasingly lost. Crustal subsidence around the North Sea coasts of southern England and along the Atlantic seaboard of the United States will tend to increase the threat of coastal inundation in these areas. In low-latitude regions, storm surges resulting from the increased frequency and intensity of tropical cyclones could become an increasing danger under enhanced greenhouse conditions (Pittock *et al.*, 1996).

Hydrological changes

Both the 1996 and 2001 IPCC assessments conclude that the impact of global warming on hydrological systems is likely to be considerable, and a major research effort has been directed towards the development of hydrological models for estimating the effect of climatic change on, for example, stream and river regimes (Arnell *et al.*, 1996; Arnell and Liu, 2001). Evidence is now emerging for a significant increase in the frequency of great floods in the world's major river systems during the course of the twentieth century, and modelling results suggest that this trend is likely to continue under a scenario of global warming (Milly *et al.*, 2002). This would not only have a profound effect on river systems themselves, but also on land and ecosystems adjoining the rivers. Moreover, it is not only quantity of water in river networks that is likely to be influenced by future climate change, but also water quality. In the mid- and high latitudes, for example, increased summer dryness is likely to lead to a reduction in stream flow which could increase the risk of water pollution through, for example, lower dissolved oxygen concentrations or increased levels of chemical compounds (Carmichael *et al.*, 1996; Cruise *et al.*, 1999). In mountain regions, accelerated glacier retreat will probably result in increased summer flow as water is released from long-term storage, but the duration of these enhanced flow regimes will depend on glacier size and the rate of glacier melt (Arnell and Liu, 2001). The hydrological effects of future climate change are going to require the implementation of new integrated water management strategies, although the capacity to implement effective management responses is unevenly distributed around the world, and remains low in many transition (i.e. from a communist to a market economy) and developing countries (McCarthy *et al.*, 2001).

Effects on agriculture

A considerable amount of work, involving experimental research, detailed modelling of basic processes, and a knowledge of both physical and biological processes, is beginning to provide an understanding of the direct and indirect effects of climate change on agricultural production. For example, experimental work into the effects of elevated atmospheric CO_2 concentrations has shown that the mean value yield response of C_3 crops (most crops except maize, sugar cane, millet and sorghum) to a doubling of CO_2 is +30 per cent (Reilly *et al.*, 1996). However, increased CO_2 is only one parameter likely to influence future agricultural production, and the 2001 IPCC assessment indicates that the response of crop yields to climate change is likely to vary widely, depending on the species, cultivar, soil conditions and other locational factors (Gitay *et al.*, 2001). Hence, although increased CO_2 concentration can stimulate crop growth, that benefit may not always overcome the adverse effects of excessive heat and drought that are likely to result from greenhouse-induced climate change. Crop modelling assessments suggest that a warming of a few degrees Celsius will result in a generally positive response for mid-latitude crop yields, but a temperature increase of more than a few degrees Celsius will result in negative responses in those areas. In low-latitude regions, by contrast, a warming of only a few degrees Celsius would be likely to induce a negative response in crop yields, as many are growing near their maximum temperature tolerance. This negative effect would be exacerbated by any significant decline in rainfall.

At the continental scale, grain yields in Africa are projected to decrease for many of the future climatic scenarios, diminishing food security in small food-importing countries. In many of the countries of arid, tropical and temperate Asia, the IPCC predicts a decrease in agricultural productivity and aquaculture due to thermal and water stress, sea-level rise, floods and droughts. In northern areas of Asia, by contrast, agriculture is likely to expand and increase in productivity. In Australia and New Zealand, the net impact on some temperate crops of climate and CO_2 changes may initially be beneficial, but this balance is expected to become negative for some areas and crops with further climate change. In Europe, there are likely to be broadly positive effects in northern areas, but productivity is predicted to decrease in southern and eastern Europe. In Latin America, yields of several important crops are projected to decrease in many areas, and subsistence farming in some regions is likely to be threatened. Finally, in North America, some crops would benefit from modest warming accompanied by increasing CO_2, but effects would vary with crops and regions, including declines due to drought in some areas of the Canadian Prairies and US Great Plains and potentially increased food production in central and northern Canada. However, benefits for crops would decline at an increasing rate and possibly become a net loss with further warming (McCarthy *et al.*, 2001).

Effects on forest ecosystems

Although evidence for the ecological impacts of recent climate change can be found in both terrestrial and marine environments (Walther *et al.*, 2002), on current evidence it is difficult to predict the impact of future climate change on the world's forests, with modelling results suggesting subtle and often non-linear responses of forest ecosystems to global warming (Prentice *et al.*, 1993; Neilson and

Drapek, 1998). Forests are composed of long-lived organisms, and responses to climate change and resulting impacts may take a long time to propagate through the system. Moreover, forest responses to climate change and the resulting impacts may extend longer than the change in climate (Gitay *et al.*, 2001). Nevertheless, both the 1996 and 2001 IPCC assessments suggest that climatic changes arising from the greenhouse effect are likely to affect a number of the world's forested regions, with the largest and earliest impacts (reflected particularly in changes in productivity) occurring in the high-latitude boreal forests. Northern treelines are projected to advance slowly into regions currently occupied by tundra, while at the southern boundary boreal coniferous forests are likely to give way to grassland. In lowland humid–tropical regions, tropical forests are more likely to be affected by changes in land-use (deforestation) than by climate change, although decreases in soil moisture may accelerate forest loss in many areas where water availability is already marginal (Cannell *et al.*, 1996; Gitay *et al.*, 2001). It is possible that future increases in CO_2 might also be a factor influencing forest ecosystems, for recent experimental work has shown a significant increase in tree growth rates under elevated CO_2 conditions. However, the extent to which trees are likely eventually to become acclimatised to higher CO_2 levels remains to be established (DeLucia *et al.*, 1999).

The ozone layer

During the 1970s it was discovered that not only did the increase in atmospheric CFC concentrations contribute to the greenhouse effect, but that continued emission of CFCs also posed a possible threat to the ozone layer. Stable CFC compounds rise through the troposphere into the stratosphere where they are exposed to the intense UV radiation that is absorbed by O_3 at lower altitudes. Exposure to radiation leads to a breakdown of the normally stable CFCs into more reactive forms such as chlorine (Cl), a chemical which is known from laboratory studies to destroy O_3 (Stolarski, 1988). In 1985, a significant decline (by 40–50 per cent) in springtime atmospheric O_3 levels was first reported over Antarctica (Farman *et al.*, 1985), and a similar reduction in atmospheric O_3 was detected over the Arctic during the winter months of the late 1980s and 1990s (Hofmann *et al.*, 1989; Müller *et al.*, 1997; Figure 9.12). Although such levels of O_3 depletion could reflect natural atmospheric variations, it is now generally accepted that this seasonal thinning of the O_3 layer is directly attributable to anthropogenic Cl pollution (Drake, 1995).

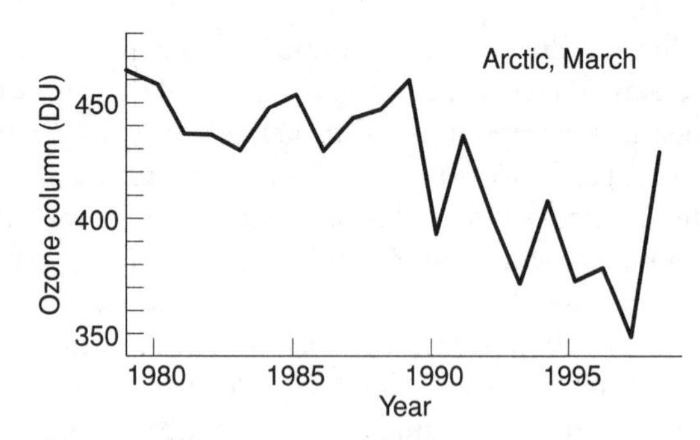

Figure 9.12 Time series of the average total ozone column over the Arctic during March from 1979 to 1998. The vertical scale is in Dobson units (one DU is the thickness, measured in units of hundredths of a millimetre, that the ozone column would occupy at standard temperature and pressure). The progressive decline in late winter–early spring ozone concentration can be clearly seen (reprinted with permission from *Nature*, after Salawitch, R.J., 1998, A greenhouse warming connection, *Nature*, **392**, p. 551, Figure 1a. Copyright 1998 Macmillan Magazines Ltd.)

Progressive depletion of stratospheric O_3 levels could have far-reaching implications for life on earth. Although O_3 constitutes less than 1 ppm of the atmospheric gases, it absorbs much of the potentially harmful incoming UV radiation and prevents it from reaching the earth's surface. An increase in UV radiation could lead to a higher incidence of, and morbidity from, eye diseases, skin cancer and infectious diseases, as well as affecting animal health, crops, aquatic ecosystems, biogeochemical cycles and air quality (van der Luen *et al.*, 1995). Particularly vulnerable to increased fluxes of UV radiation are polar regions, where components of both terrestrial and marine ecosystems are highly susceptible to increased UV exposure. In Antarctica, measurements of UV radiation have shown significantly enhanced UV levels associated with the springtime O_3 reductions, while data from Argentina, Chile, New Zealand and

Australia have revealed relatively high UV levels compared with corresponding latitudes in the Northern Hemisphere (Madronich *et al.*, 1995). However, significant reductions in atmospheric O_3 have been observed over northern Europe, particularly in the mountains, raising fears about the possible harmful effects of increased UV (especially UV-B) radiation in those areas also (Björn *et al.*, 1998).

Increasing concern over the depletion of atmospheric O_3 led to the signing of the Montreal Protocol on Substances that Deplete the Ozone Layer. This was formulated under the aegis of the United Nations Environment Programme in 1987, agreed by 170 countries, and revised several times during the 1990s. Under the terms of the Protocol, O_3-depleting chemicals are to be progressively phased out, and progress towards this objective is to be reviewed at least every four years (e.g. UNEP/WMO, 1998). The positive effects of the Montreal Protocol are already being seen in the reduction in atmospheric levels of certain O_3-depleting halogen compounds (chloro- and bromocarbons). However, others, such as CFC-11, are falling more slowly whereas CFC-12 values are still rising (Montzka *et al.*, 1999). Moreover, as CFCs are not due to be phased out (under the terms of the Protocol) until 2030, emissions (and hence atmospheric concentrations) are predicted to increase substantially, and legally, over the next decade (Fraser and Prather, 1999). Greenhouse gases may further exacerbate the situation, for while they warm the lower atmosphere, they cool the stratosphere leading to the formation of ice crystals which catalyse the destruction of O_3 by CFCs (Schrope, 2000). Indeed, modelling results show Arctic O_3 losses increasing to a maximum in the decade 2010–2019, roughly a decade after the projected maximum in stratospheric chlorine abundance. The severity and duration of the Antarctic O_3 hole are also predicted to increase because of greenhouse-gas-induced stratospheric cooling over the coming decades (Shindell *et al.*, 1998). Hence, while some scientists are optimistic that if current CFC policies continue, the Antarctic and Arctic O_3 holes will be gone by around 2050, others are less convinced that this will be the case.

Acid deposition

Acid deposition (often referred to as 'acid rain') is a form of atmospheric pollution that has affected, and continues to affect, many of the industrialised regions of the world. Most rainfall is naturally acidic (i.e. $pH < 7$), but pollution can also occur through snow, hail, gas clouds, fog, mist and dry dust. The basic elements involved in acid deposition are sulphur dioxide (SO_2) and the two nitrogen oxides (NO_x), nitric oxide (NO) and nitrogen dioxide (NO_2). These chemical pollutants are released into the atmosphere through the burning of fossil fuels and petroleum products, from the smelting of metallic ores, from petrochemical and related industries, and from vehicle exhausts. Ice-core data from Greenland show that, prior to 1900, atmospheric sulphur and nitrate levels were generally low, but since the turn of the century concentrations of nitrates have doubled and sulphates have trebled (Wolff and Peel, 1985). This dramatic increase in atmospheric acidity is reflected in twentieth-century precipitation measurements, and is also apparent in the changing composition of diatom assemblages (p. 28) which show increasing acidity of lake waters in upland areas of western Europe over the course of the past two centuries (e.g. Jones *et al.*, 1993).

The effects of acid deposition are considerable. Increased acidification of lake ecosystems has resulted in a decline in fish stocks and other aquatic biota in lakes and rivers throughout North America and northern Europe (Minns *et al.*, 1990; Hesthagen *et al.*, 1999), due partly to the increased acidity of lake waters, and partly to higher concentrations of dissolved metals (such as aluminium) which are often found in waters of low pH. These metals are not only toxic to many forms of aquatic life, but may also constitute a threat to human health in more remote areas where water treatment is rudimentary (Moghissi, 1986). Mountain lakes, which are often in areas of low natural base status, are especially sensitive to inputs of atmospheric pollutants (Skjelkvåle and Wright, 1999). In North America, a combination of climate warming and increased lake acidification from acid deposition has led to a decrease in dissolved organic carbon concentration in some boreal lakes, resulting in a marked increase in exposure of the upper water column to solar UV radiation (Schindler *et al.*, 1996). Enhanced acidification of soils leads to nutrient depletion, which is exerting increasing environmental stress on forested regions of North America and Europe (Cowling, 1989). Data from southern Sweden, for example, indicate that over the past 35 years, soil pH has fallen by up to 1.5 units and a long-term decline in tree growth seems inevitable (Falkengren-Grerup, 1989). Building damage, crop damage and human health problems are further consequences of acid deposition. In the atmosphere, higher concentrations of SO_2 and NO_x will affect the global radiation balance, influence atmospheric chemical reactions, and contribute to the scattering of solar radiation as the aerosol load of the troposphere increases (see above).

In recent years, there has been a significant decline in the rate of acid deposition, especially of sulphur, across large areas of Europe and North America. This has come about largely as a result of national and international efforts to limit or to reduce significantly the output of the airborne pollutants that lead to acid deposition. The first legislation aimed at reducing sulphur emissions was signed in 1985 under the auspices of the United Nations Economic Commission for Europe (UN-ECE) Convention on Long-Range Transboundary Air Pollution (the First Sulphur Protocol) under the terms of which signatory countries committed themselves to a 30 per cent reduction in sulphur emissions (relative to 1980) by 1993. This was followed by the Second Sulphur Protocol in 1994, with an agreement to reduce emissions by 60 per cent by the year 2010, again relative to 1980 levels. The First Nitrogen Protocol (1988) agreed to stabilise nitrogen emissions at 1987 levels by 1994, and a further Protocol was signed in 1999 (Jenkins, 1999). Measurements have shown that in northern and central Europe SO_2 concentrations in air decreased by 63 per cent between 1985 and 1996, while in the USA and Canada, SO_2 emissions declined by 28 per cent between 1980 and 1995. These declines are supported by empirical data from lakes and streams in both Europe and North America which show a significant reduction in levels of acidification (Stoddard *et al.*, 1999). However, while it would seem from the results of recent monitoring that international legislation is beginning to have a positive effect on the natural environment, the recovery from acid pollution is likely to be a long-term process. This is borne out, for example, by data from forest ecosystems which suggest that the response to any reduction in acid deposition will be slow (Likens *et al.*, 1996), and by evidence from mountain regions which shows the slow release into mountain streams of acid deposition frozen into snow and ice (Schindler, 1999). Moreover, in some areas (central Europe, for example), many sites are showing a significant delay in aquatic recovery from acidification, despite the marked reduction that has occurred in anthropogenic acid deposition (Alewell *et al.*, 2000). In addition, although sulphur emissions are declining in Europe and North America, they are still rising in many developing nations across the world, and continue to pose a major threat to ecosystems in tropical and subtropical climates (Kuylenstierna *et al.*, 2001). Futhermore, much remains to be learned about the biochemistry of the recovery process itself, and hence it is likely to be some considerable time before global surface waters and ecosystems are fully recovered from the effects of human-induced acid deposition.

Bibliography

Aaby, B. (1976) Cyclic climatic variations over the past 5500 years reflected in raised bogs. *Nature*, **263**, 281–4.

Aaby, B. (1986) Palaeoecological studies of mires. In *Handbook of Holocene Palaeoecology and Palaeohydrology* (edited by B.E. Berglund), John Wiley & Sons, Chichester and New York, 145–64.

Aaby, B. (1997) Mineral dust and pollen as traces of agricultural activity. In *Environment and Vikings: With special reference to Birka* (edited by V. Miller), PACT 52, 115–21.

Aaris-Sørensen, K. (1980) Depauperation of the mammalian fauna of the island of Zealand during the Atlantic period. *Videnskabelige Meddelelser Dansk Naturhistorisk Forening*, **142**, 131–8.

Aaris-Sørensen, K. (1998) *Danmarks Forhistoriske Dyreverden*. Gyldendal, Copenhagen.

Aberet, O. and Jacomet, S. (1997) Analysis of plant macrofossils in goat/sheep faeces from the Neolithic lake shore settlement of Horgen Scheller – an indication of prehistoric transhumance. *Vegetation History and Archaeobotany*, **6**, 235–9.

Adams, R.M. (1988) Introductory remarks: spatial and temporal contexts. In *Conceptual Issues in Environmental Archaeology* (edited by J.L. Bintliff, D.A. Davidson and E.G. Grant), Edinburgh University Press, Edinburgh.

Adovasio, J.M. and Pedler, D.R. (1997) Monte Verde and the antiquity of humankind in the Americas. *Antiquity*, **71**, 573–80.

Aharon, P. (1984) Implications of the coral reef record from New Guinea concerning the astronomical theory of climatic change. In *Milankovitch and Climate* (edited by A. Berger, J. Imbrie, J. Hays, G. Kukla and B. Saltzman), Reidel, Dordrecht, 379–90.

Aitken, M.J. (1985) *Thermoluminescence Dating*. Academic Press, London.

Aitken, M.J. (1998) *An Introduction to Optical Dating*. Oxford University Press, Oxford.

Alcoforado, M-J., Nunes, M. de F., Garcia, J.C. and Taborda, P. (2000) Temperature and precipitation reconstruction in southern Portugal during the late Maunder Minimum. *The Holocene*, **10**, 333–40.

Aldhouse-Green, S.H.R., Whittle, A., Allen, J.R.L., Caseldine, A.E., Culver, S.J., Day, M.H., Lundqvist, J. and Upton, D. (1992) Prehistoric human footprints from the Severn Estuary at Uskmouth and Magor Pill, Gwent, Wales. *Archaeologia Cambrensis*, **CXLI**, 14–55.

Alewell, C., Manderscheid, B., Meesenburg, H. and Bittersohl, J. (2000) Is acidification still an ecological threat? *Nature*, **407**, 856–7.

Allen, J.R.L. (1997) Subfossil mammalian tracks (Flandrian) in the Severn Estuary, S.W. Britain: mechanics of formation, preservation and distribution. *Philosophical Transactions of the Royal Society of London B*, **352**, 481–518.

Allen, J.R.L. (2000) Morphodynamics of Holocene salt marshes: a review sketch from the Atlantic and southern North Sea coasts of Europe. *Quaternary Science Review*, **19**, 1155–231.

Allen, J.R.M., Brandt, U., Brauer, A., Hubberten, H-W., Huntley, B., Keller, J., Kraml, M., Mackensen, A., Mingram, J., Negendank, J.F.W., Nowaczyk, N.R., Oberhänsli, H., Watts, W.A., Wulf, S. and Zolitschka, B. (1999) Rapid environmental changes in southern Europe during the last glacial period. *Nature*, **400**, 740–2.

Allen, J.R.M., Huntley, B. and Watts, W.A. (1996) The vegetation and climate of northwest Iberia over the last 14,000 yr. *Journal of Quaternary Science*, **11**, 125–47.

Allen, M.J. (1992) Products of erosion and the prehistoric land-use of the Wessex Chalk. In *Past and Present Soil Erosion* (edited by M.G. Bell and J. Boardman), Oxbow Monograph 22, Oxford, 37–52.

Allen, M.J. (1997) Environment and land-use: the economic development of the communities who built Stonehenge (an economy to support the stones). In *Science and Stonehenge* (edited by B. Cunliffe and C. Renfrew), Oxford University Press (for the British Academy: 92), Oxford, 115–44.

Allen, T., Hey, G. and Miles, D. (1997) A line of time: approaches to archaeology in the Upper and Middle Thames Valley, England. *World Archaeology*, **29** (1), 114–29.

Alley, R.B. (2000) The Younger Dryas cold interval as viewed from central Greenland. *Quaternary Science Reviews*, **19**, 213–26.

Alley, R.B. and Clark, P.U. (1999) The deglaciation of the northern hemisphere: a global perspective. *Annual Review of Earth and Planetary Science Letters*, **27**, 149–82.

Alley, R.B., Mayewski, P.A., Sowers, T., Stuiver, M., Taylor, K.C. and Clark, P.U. (1997) Holocene climatic instability: a prominent, widespread event 8200 yrs ago. *Geology*, **25**, 483–6.

Alley, R.B., Meese, D.A., Shuman, C.A., Gow, A.J., Taylor, K.C., Grootes, P.M., White, J.W.C., Ram, M., Waddington, E.D., Mayewski, P.A. and Zielinski, G.A. (1993) Abrupt increase in snow accumulation at the end of the Younger Dryas event. *Nature*, **362**, 527–9.

Alley, R.B., Shuman, C.A., Meese, D.A., Gow, A.J., Taylor, K.C., Cuffey, K.M., Fitzpatrick, J.J., Grootes, P.M., Zielinski, G.A., Ram, M., Spinelli, G. and Elder, B. (1997) Visual-stratigraphic dating of the GISP2 ice core: basis, reproducibility, and application. *Journal of Geophysical Research*, **102**, 26367–82.

Allison, T.D., Moeller, R.E. and Davis, M.B. (1986) Pollen in laminated sediments provides evidence for a mid-Holocene pathogen outbreak. *Ecology*, **67**, 1101–5.

Almquist, H., Diffenbacher-Krall, A.C., Flanagan-Brown, R. and Sanger, D. (2001) The Holocene record of lake levels at Mansell Pond, central Maine, USA. *The Holocene*, **11**, 189–201.

Alvarez, L.W., Alvarez, A., Asaro, F. and Michel, H.V. (1980) Extraterrestrial cause for the Cretaceous–Tertiary extinction. *Science*, **208**, 1095–108.

Ames, K.M. and Maschner, H.D.G. (1999) *Peoples of the Northwest Coast: Their archaeology and prehistory*. Thames and Hudson, London.

Ammann, B. (1988) Palynological evidence of prehistoric anthropogenic forest changes on the Swiss Plateau. In *The Cultural Landscape – Past, Present and Future* (edited by H.H. Birks, H.J.B. Birks, P.E. Kaland and D. Moe), Cambridge University Press, Cambridge, 289–99.

Ammann, B., Birks, H.J.B., Brooks, S.J., Eicher, U., von Grafenstein, J., Tobolski, K. and Wick, L. (2000) Quantification of biotic responses to rapid climatic changes around the Younger Dryas – a synthesis. *Palaeogeography, Palaeoclimatology, Palaeoecology*, **159**, 313–47.

Amorosi, T., Buckland, P.C., Magnússon, K., McGovern, T.H. and Sadler, J.P. (1994) An archaeozoological examination of the midden at Nesstofa, Reykjavik, Iceland. In *Whither Environmental Archaeology?* (edited by R. Luff and P. Rowley-Conwy), Oxbow Monograph 38, Oxford, 69–80.

Andersen, S.H. (1985) Tybrind Vig: a preliminary report on a submerged Ertebølle settlement on the west coast of Fyn. *Journal of Danish Archaeology*, **4**, 52–69.

Andersen, S.H. (1991) Norsminde. A 'køkkenmødding' with Late Mesolithic and Early Neolithic occupation. *Journal of Danish Archaeology*, **8**, 13–40.

Andersen, S.H. (1995) Coastal adaptation and marine exploitation in late Mesolithic Denmark – with some special emphasis on the Limfjord Project. In *Man and Sea in the Mesolithic* (edited by A. Fischer), Oxbow, Oxford, 41–66.

Andersen, S.H. (2000) 'Køkkenmøddinger' (shell middens) in Denmark: a survey. *Proceedings of the Prehistoric Society*, **66**, 361–84.

Andersen, S.T. (1993a) History of vegetation and agriculture at Hassing House Mose, Thy, northwest Denmark since the Ice Age. *Journal of Danish Archaeology*, **11**, 57–79.

Andersen, S.T. (1993b) Early agriculture. In *Digging into the Past: 25 years of archaeology in Denmark* (edited by S. Hvass and B. Storgaard), The Royal Society of Northern Antiquaries, Copenhagen, 88–91.

Andersen, S.T. (1993c) Early and middle Neolithic agriculture in Denmark: pollen spectra from soils in burial mounds of the Funnel Beaker Culture. *Journal of European Archaeology*, **1**, 153.

Andersen, S.T. and Rasmussen, P. (1993) Radiocarbon wiggle dating of elm declines in northwest Denmark and their significance. *Vegetation History and Archaeobotany*, **2**, 125–35.

Andersen, S.T., Aaby, B. and Odgaard, B. (1996) Denmark. In *Palaeoecological events during the last 15,000 years* (edited by B.E. Berglund, H.J.B. Birks, M. Ralska-Jasiewiczowa and H.E. Wright), John Wiley, Chichester, 215–32.

Anderson, A.J. (1988) Coastal subsidence economies in prehistoric southern New Zealand. In *The Archaeology of Prehistoric Coastlines* (edited by G. Bailey and J. Parkington), Cambridge University Press, Cambridge, 93–101.

Anderson, A.J. (1989) *Prodigious Birds: Moas and moa-hunting in prehistoric New Zealand*. Cambridge University Press, Cambridge.

Anderson, H.A., Berrow, M.L., Farmer, V.C., Hepburn, A., Russel, J.D. and Walker, A.D. (1982) A reassessment of podzol formation processes. *Journal of Soil Science*, **33**, 125–36.

Anderson, I.W. (1974) The chestnut pollen decline as a time horizon in lake sediments in eastern North America. *Canadian Journal of Earth Sciences*, **11**, 678–85.

Anderson, T. and Macpherson, J.B. (1994) Wisconsinan Lateglacial environmental change in Newfoundland: a regional review. *Journal of Quaternary Science*, **9**, 171–8.

Andres, W., Bos, J.A.A., Houben, P., Kalis, A.J., Nolte, S., Rittweger, H. and Wunderlich, J. (2001) Environmental change and fluvial activity during the Younger Dryas in central Germany. *Quaternary International*, **79**, 89–100.

Andrews, J.T. (1979) The present ice age. In *The Winters of the World* (edited by B.S. John), David & Charles, London and North Pomfret, Vt, 173–218.

Andrews, J.T. (1989) Quaternary geology of the northeastern Canadian Shield. In *Quaternary Geology of Canada and Greenland* (edited by R.J. Fulton), Geological Survey of Canada, Geology of Canada No. 1, 276–302.

Andrews, J.T. (1998) Abrupt changes (Heinrich events) in late Quaternary North Atlantic marine environments: a history and review of data and concepts. *Journal of Quaternary Science*, **13**, 3–16.

Anklin, M., Schwander, J., Stauffer, B., Tschumi, J. and Fuchs, A. (1997) CO_2 record between 40 and 8 kyr B.P. from the Greenland Ice Core Project ice core. *Journal of Geophysical Research*, **102**, 26539–45.

Appelt, M., Berglund, J. and Gulløv, H.C. (eds) (2000) *Identities and Cultural Contacts in the Arctic*. Dansk Polar Center, Copenhagen.

Arneborg, J. and Gulløv, H.C. (eds) (1998) *Man, Culture and Environment in Ancient Greenland*. Dansk Polar Center, Grønningen.

Arneborg, J., Heinemeier, J., Lynnerup, N., Nielsen, H.L., Rud, N. and Sveinbjörnsdóttir, A.E. (1999) Change in diet of the Greenland Vikings determined from stable carbon isotope analysis and ^{14}C dating of their bones. *Radiocarbon*, **41** (2), 157–68.

Arnell, N. and Liu, C. (2001) Hydrology and water resources. In *Climate Change 2001: Impacts, adaptation, and vulnerability* (edited by J.J. McCarthy, O.F. Canzini, N.A. Leary, D.J. Dokken and K.S. White), Cambridge University Press, Cambridge, 191–233.

Arnell, N., Bates, B., Lang, H., Magnuson, J.J. and Mulholland, P. (1996) Hydrology and freshwater ecology. In *Climate Change 1995. Impacts, Adaptations and Mitigation of Climate Change: Scientific-Technical analysis* (edited by R.T. Watson, M.C. Zinyowera, R.H. Moss and D.J. Dokken), Cambridge University Press, Cambridge, 325–63.

Arnold, T.G. (2002) Radiocarbon dates from the ice-free corridor. *Radiocarbon*, **44** (2), 437–54.

Ashworth, A.C., Buckland, P.C. and Sadler, J.P. (eds) (1997) *Studies in Quaternary Entomology*. Quaternary Proceedings, 5, John Wiley & Sons, Chichester and New York.

Asioli, A., Trincardi, F., Lowe, J.J. and Oldfield, F. (1999) Short-term changes during the Last Glacial–Holocene transition: comparison between Mediterranean records and the GRIP event stratigraphy. *Journal of Quaternary Science*, **14**, 373–81.

Atkinson, T.C., Briffa, K.R. and Coope, G.R. (1987) Seasonal temperatures in Britain during the last 22,000 years, reconstructed using beetle remains. *Nature*, **325**, 587–92.

Atkinson, T.C., Lawson, T.J., Smart, P.L., Harmon, R.S. and Hen, J.W. (1986) New data on speleothem deposition and palaeoclimate in Britain over the last forty thousand years. *Journal of Quaternary Science*, **1**, 67–72.

Atwater, B.F. (1987) Evidence for Great Holocene earthquakes along the outer coast of Washington State. *Science*, **236**, 942–4.

Atwater, B.F. and Yamaguchi, D.K. (1991) Sudden, probably coseismic submergence of Holocene trees and grass in coastal Washington State. *Geology*, **19**, 706–9.

Austin, D. (1985) Dartmoor and the upland village of the South-West of England. In *Medieval Villages: A review of current work* (edited by D. Hooke), Committee for Archaeology Monograph 5, Oxford, 71–9.

Austin, W.E.N. and Kroon, D. (1996) Lateglacial sedimentology, Foraminifera and stable isotope stratigraphy of the Hebridean continental shelf, northwest Scotland. In *Late Quaternary Palaeoceanography of the North Atlantic Margins* (edited by J.T. Andrews, W.E.N. Austin, H. Bergsten and A.E. Jennings), Geological Society Special Publication No. 111, 187–214.

Bahn, P.G. and Flenley, J. (1992) *Easter Island, Earth Island*. Thames and Hudson, London.

Bahn, P.G. and Vertut, J. (1997) *Journey through the Ice Age*. Wiedenfeld and Nicolson, London.

Bahn, P.G., Pettitt, P. and Ripoll, S. (2003) Discovery of Palaeolithic cave art in Britain. *Antiquity*, **77** (296), 227–31.

Bailey, S.D., Wintle, A.G., Duller, G.A.T. and Bristow, C.S. (2001) Sand deposition during the last millennium at Aberffraw, Anglesey, North Wales, as determined by OSL dating of quartz. *Quaternary Science Reviews*, **20**, 701–4.

Baillie, M.G.L. (1995) *A Slice through Time: Dendrochronology and precision dating*. Routledge, London.

Baillie, M.G.L. (1999) *Exodus to Arthur*. Batsford, London.

Baillie, M.G.L. and Munro, M.A.R. (1988) Santorini and volcanic dust veils. *Nature*, **332**, 344–6.

Baker, A., Smart, P.L., Edwards, R.L. and Richards, D.A. (1993) Annual growth banding in a cave stalagmite. *Nature*, **364**, 518–20.

Baker, D.G., Watson, B.F. and Skaggs, R.H. (1985) The Minnesota long-term temperature record. *Climatic Change*, **7**, 225–36.

Baker, N.V. and Payne, G.K. (1978) G.K. Gilbert and modern geomorphology. *American Journal of Science*, **278**, 97–123.

Baker, R.G. (1984) Holocene vegetational history of the western United States. In *Late Quaternary Environments of the United States. Volume 2. The Holocene* (edited by H.E. Wright, Jnr), Longman, London, 109–27.

Baker, R.G. (2000) Holocene environments reconstructed from plant macrofossils in stream deposits from south-eastern Nebraska, USA. *The Holocene*, **10**, 357–65.

Baker, V.R. and Bunker, R.C. (1985) Cataclysmic Late Pleistocene flooding from glacial lake Missoula: a review. *Quaternary Science Reviews*, **4**, 1–41.

Baker, V.R. and Komar, P.D. (1987) Cataclysmic flood processes and landforms. In *Geomorphic Systems of North America* (edited by W.L. Graf), Geological Society of America centennial special Volume 2, Boulder, Colorado, 423–43.

Balaam, N.D., Smith, K. and Wainwright, G.J. (1982) The Shaugh Moor Project: fourth report. *Proceedings of the Prehistoric Society*, **48**, 203–78.

Ball, T.F. and Kingsley, R.A. (1984) Instrumental temperature records at two sites in central Canada: 1768–1910. *Climatic Change*, **6**, 39–56.

Ballantyne, C.K. (2002) The Loch Lomond Readvance on the Isle of Mull, Scotland: glacier reconstruction and palaeoclimatic implications. *Journal of Quaternary Science*, **17**, 759–71.

Ballantyne, C.K. and Harris, C. (1994) *The Periglaciation of Great Britain.* Cambridge University Press, Cambridge.

Ballantyne, C.K., McCarroll, D., Nesje, A., Dahl, S.O. and Stone, J.O. (1998b) The last ice sheet in north-west Scotland: reconstructions and implications. *Quaternary Science Reviews*, **17**, 1149–84.

Ballantyne, C.K., Stone, J.O. and Fifield, L.K. (1998a) Cosmogenic Cl-36 dating of postglacial landsliding at The Storr, Isle of Skye, Scotland. *The Holocene*, **8**, 347–51.

Balouet, J.-Y. (1990) *Extinct Species of the World.* Charles Letts, London.

Barber, D.C., Dyke, A., Hillaire-Marcel, C., Jennings, A.E., Andrews, J.T., Kerwin, M.W., Bilodeau, G., McNeely, R., Southon, J., Morehead, M.D. and Gagnon, J-M. (1999) Forcing of the cold event of 8,200 years ago by catastrophic drainage of Laurentide lakes. *Nature*, **400**, 344–8.

Barber, K.E. (1981) *Peat Stratigraphy and Climatic Change*. Balkema, Rotterdam.

Barber, K.E. and Coope, G.R. (1987) Climatic history of the Severn Valley during the last 18,000 years. In *Palaeohydrology in Practice* (edited by K.J. Gregory), John Wiley, Chichester and New York, 201–16.

Barber, K.E., Chambers, F.M. and Maddy, D. (2003) Holocene palaeoclimates from peat stratigraphy: macrofossil proxy climatic records from three oceanic raised bogs in England and Ireland. *Quaternary Science Reviews*, **22**, 521–39.

Barber, K.E., Chambers, F.M., Maddy, D., Stoneman, R. and Brew, J.S. (1994) A sensitive high-resolution record of late Holocene climatic change from a raised bog in northern England. *The Holocene*, **4**, 198–205.

Barber, K.E., Maddy, D., Rose, N., Stevenson, A.C., Stoneman, R. and Thompson, R. (2000) Replicated proxy-climate signals over the last 2000 yr from two distant UK peat bogs: new evidence for regional palaeoclimatic teleconnections. *Quaternary Science Reviews*, **19**, 481–8.

Barclay, D.J., Wiles, G.C. and Calkin, P.E. (1999) A 1119-year tree ring-width chronology from western Prince William Sound, southern Alaska. *The Holocene*, **9**, 79–84.

Bard, E., Arnold, M., Maurice, P., Duprat, P., Moyes, J. and Duplessy, J-C. (1987) Retreat velocity of the North Atlantic polar front during the last deglaciation determined by ^{14}C accelerator mass spectrometry. *Nature*, **328**, 791–4.

Bard, E., Arnold, M., Hamelin, B., Tisnerat-Laborde, N. and Cabioch, G. (1998) Radiocarbon calibration by means of mass spectrometric $^{230}Th/^{234}U$ and ^{14}C ages of corals: an updated database including samples from Barbados, Mururoa and Tahiti. *Radiocarbon*, **40**, 1085–92.

Barfield, L. and Chippindale, C. (1997) Meaning in the later prehistoric rock-engravings of Mont Bégo, Alpes-Maritimes, France. *Proceedings of the Prehistoric Society*, **63**, 103–28.

Barker, G. (ed.) (1995) *A Mediterranean Valley.* Leicester University Press, London.

Barlow, D.N. and Thompson, R. (2000) Holocene sediment erosion in Britain as calculated from lake-basin studies. In *Tracers in Geomorphology* (edited by I.D.L. Foster), John Wiley & Sons, Ltd, Chichester, 455–72.

Barlow, L.K. (2001) The time period A.D. 1400–1980 in Central Greenland ice cores in relation to the North Atlantic Sector. In *The Iceberg in the Mist: Northern research in pursuit of a 'Little Ice Age'* (edited by A.E.J. Ogilvie and T. Jónsson), Kluwer Academic Publishers, Dordrecht, 101–20.

Barlow, L.K., Sadler, J.P., Ogilvie, A.E.J., Buckland, P.C., Amorosi, T., Ingimundarson, J.H., Skidmore, P., Dugmore, A.J. and McGovern, T.H. (1997) Interdisciplinary investigations of the end of the Norse Western settlement in Greenland. *The Holocene*, 7 (4), 489–99.

Barnekow, L. (1999) Holocene tree-line dynamics and inferred climatic reconstructions in the Abisko area, northern Sweden, based on macrofossil and pollen records. *The Holocene*, **9**, 253–67.

Barnes, I., Cooper, A., Shapiro, B. and Jensen, D. (2002) Dynamics of Pleistocene population extinctions in Beringian brown bears. *Science*, **295**, 2267–70.

Barnett, T.P., Pierce, D.W. and Schnur, R. (2001) Detection of anthropogenic climate change in the world's oceans. *Science*, **292**, 70–4.

Barnosky, A.D. (1986) Big game extinctions caused by late Pleistocene climatic change: Irish elk (*Megaloceros giganteus*) in Ireland. *Quaternary Research*, **25**, 128–35.

Bartlein, P.J. and Whitlock, C. (1993) Paleoclimatic interpretations of the Elk Lake pollen record. *Geological Society of America Special Paper*, **276**, 275–93.

Barton, R.N.E. (1991) Technological innovation and continuity at the end of the Pleistocene in Britain. In *The Lateglacial in North-west Europe* (edited by R.N.E. Barton, A.J. Roberts and D.A. Roe), Council for British Archaeology, London, 234–45.

Barton, R.N.E., Berridge, P.J., Walker, M.J.C. and Bevins, R.E. (1995) Persistent places in the Mesolithic landscape: an example from the Black Mountain Uplands of south Wales. *Proceedings of the Prehistoric Society*, **61**, 81–116.

Barton, R.N.E., Roberts, A.J. and Roe, D.A. (eds) (1991) *The Lateglacial in North-west Europe*. Council for British Archaeology, London.

Barton, R.N.E., Jacobi, R.N., Stapert, D. and Street, M. (2003) The Late-glacial reoccupation of the British Isles and the Cresswellian. *Journal of Quaternary Science*, **18** (7), 631–43.

Bar-Yosef, O. (1995) The role of climate in the identification of human movement and cultural transition in western Asia. In *Palaeoclimate and Evolution* (edited by E.S. Vrba, G.H. Denton, T.C. Partridge and L.H. Burkle), Yale University Press, Yale, 507–23.

Bar-Yosef, O. (2000) The impact of radiocarbon dating on Old World archaeology: past achievements and future expectations. *Radiocarbon*, **42** (1), 23–39.

Bassinot, F.E., Labayrie, L.D., Vincent, G., Quidelleur, X., Shackleton, N.J. and Lancelot, Y. (1994) The astronomical theory of climate and the Brunhes-Matuyama magnetic reversal. *Earth and Planetary Science Letters*, **126**, 91–108.

Battarbee, R.W. (2000) Palaeolimnological approaches to climate change, with special regard to the biological record. *Quaternary Science Reviews*, **19**, 107–24.

Battarbee, R.W., Flower, R.J., Stevenson, J. and Rippey, B. (1985) Lake acidification in Galloway: a palaeoecological test of competing hypotheses. *Nature*, **314**, 350–2.

Bauch, H.A., Erlenkeuser, H., Spielhagen, R.F., Struck, U., Mattheissen, J., Thiede, J. and Heinemeier, J. (2001) A multiproxy reconstruction of the evolution of deep and surface waters in the subarctic Nordic seas over the last 30,000 yr. *Quaternary Science Reviews*, **20**, 659–78.

Bean, L.J. and Lawton, H.W. (1993) Some explanations for the rise of cultural complexity in Native California with comments on proto-agriculture and agriculture. In *Before the Wilderness* (edited by T.C. Blackburn and K. Anderson), Balkema Press, Menlo Park, California, 27–54.

Beattie, O. (2000) The Kwäday Dan Ts'ī̄nchi Discovery from a glacier in British Columbia. *Canadian Journal of Archaeology*, **24**, 129–48.

Beaulieu, J-L. de and Reille, M. (1984) A long Upper Pleistocene pollen record from Les Echets, near Lyon, France. *Boreas*, **13**, 111–32.

Beaulieu, J-L. de and Reille, M. (1992) The last climatic cycle at La Grande Pile (Vosges, France): a new pollen profile. *Quaternary Science Reviews*, **11**, 431–8.

Beer, J., Johnsen, S.J., Bonani, G., Finkel, R.C., Langway, C.C., Oeschger, H., Stauffer, B., Suter, M. and Woelfli, W. (1992) ^{10}Be peaks as time markers in polar ice cores. In *The Last Deglaciation: Absolute and radiocarbon chronologies* (edited by E. Bard and W.S. Broecker), NATO ASI Series, 1, 2, Springer-Verlag, Berlin, 141–53.

Beer, J., Mende, W. and Stellmacher, R. (2000) The role of the sun in climatic forcing. *Quaternary Science Reviews*, **19**, 403–15.

Behre, K.-E. (1988) The role of man in European vegetation history. In *Vegetation History* (edited by B. Huntley and T. Webb), Kluwer, Dordrecht, 633–72.

Behre, K.-E. (1995) Die ursprüngliche Vegetation in den deutschen Marschgebieten und deren Veränderung durch prähistorische Besiedlung und Meeresspiegelbewegungen. In *Verhandlungen Gesellschaft für Ökologie XIII* (edited by G. Weidemann), Bremen, 85–96.

Behre, K.-E. and van der Plicht, J. (1992) Towards an absolute chronology for the last glacial period in Europe: radiocarbon dates from Oerel, northern

Germany. *Vegetation History and Archaeobotany*, **1**, 111–17.

Behre, K.-E., Brande, A., Küster, H. and Rösch, M. (1996) Germany. In *Palaeoecological Events during the Last 15,000 Years* (edited by B.E. Berglund, H.J.B. Birks, M. Ralska-Jasiewiczowa and H.E. Wright), John Wiley, Chichester, 507–52.

Behringer, W. (1999) Climatic change and witch-hunting: the impact of the Little Ice Age on mentalities. *Climatic Change*, **43**, 335–51.

Belknap, D.F. and Kraft, J.C. (1977) Holocene relative sea-level changes and coastal stratigraphic units on the northwest flank of the Baltimore canyon trough geosyncline. *Journal of Sedimentary Petrology*, **47**, 610–29.

Bell, M.G. (1981) Seaweed as a prehistoric resource. In *Environmental Aspects of Coasts and Islands* (edited by D.R. Brothwell and G. Dimbleby), British Archaeological Reports IS 94, Oxford, 117–26.

Bell, M.G. (1983) Valley sediments as evidence of prehistoric land-use on the South Downs. *Proceedings of the Prehistoric Society*, **49**, 119–50.

Bell, M.G. (1990) *Brean Down Excavations 1983–1987*. English Heritage Archaeological Report 15, London.

Bell, M.G. (1992) The prehistory of soil erosion. In *Past and Present Soil Erosion* (edited by M.G. Bell and J. Boardman), Oxbow Books, Oxford, 21–36.

Bell, M.G. (2000) Intertidal peats and the archaeology of coastal change in the Severn Estuary, Bristol Channel and Pembrokeshire. In *Coastal and Estuarine Environments: Sedimentology, geomorphology and geoarchaeology* (edited by K. Pye and J.R.L. Allen), Geological Society Special Publication No. 175, London, 377–92.

Bell, M.G. (2003) *Making One's Mark in the World: Trackways from a wetland and dryland perspective*. Wet Site Connections Conference, Olympia, Washington, Wetland Archaeology Research Project.

Bell, M.G. (2004) Archaeology and green issues. In *A Companion to Archaeology* (edited by J. Bintliff), Blackwell, Oxford, 509–31.

Bell, M.G. and Boardman, J. (eds) (1992) *Past and Present Soil Erosion*. Oxbow Monograph 22, Oxford.

Bell, M.G., Allen, J.R.L., Buckley, S., Dark, P. and Haslett, S.K. (2002) Mesolithic to Neolithic coastal environmental change: excavations at Goldcliff East, 2002. *Archaeology in the Severn Estuary*, **13**, 1–29.

Bell, M.G., Caseldine, A.E. and Neumann, H. (2000) *Prehistoric Intertidal Archaeology in the Welsh Severn Estuary*. Council for British Archaeology Report 120, York.

Bell, M.G., Fowler, P.J. and Hillson, S.W. (1996) *The Experimental Earthwork Project 1960–1992*. Council for British Archaeology Report 100, York.

Benito, G., Baker, V.R. and Gregory, K.J. (1998) *Palaeohydrology and Environmental Change*. John Wiley & Sons, Chichester and New York.

Benito, G., Machado, M.J. and Pérez-González, A. (1996) Climate change and flood sensitivity in Spain. In *Global Continental Changes: The context of palaeohydrology* (edited by J. Branson, A.G. Brown and K.J. Gregory), Geological Society of London, Special Publication No. 115, London, 85–98.

Benn, D.I. and Evans, D.J.A. (1998) *Glaciers and Landscape*. Edward Arnold, London, 85–98.

Bennett, K.D. (1989) A provisional map of forest types for the British Isles 5000 years ago. *Journal of Quaternary Science*, **4**, 141–4.

Bennett, K.D. (1990) Postglacial history of alder (*Alnus glutinosa* (L.) Gaertn.) in the British Isles. *Journal of Quaternary Science*, **5**, 123–34.

Bennett, K.D. (1997) *Evolution and Ecology: The pace of life*. Cambridge University Press, London.

Bennett, K.D. and Fuller, J.L. (2002) Determining the age of the mid-Holocene *Tsuga canadensis* (hemlock) decline, eastern North America. *The Holocene*, **12** (4), 421–9.

Bennett, K.D., Boreham, S., Sharp, M.J. and Switsur, V.R. (1992) Holocene history of environment, vegetation and human settlement on Catta Ness, Lunnasting, Shetland. *Journal of Ecology*, **80**, 173–241.

Bennett, K.D., Tzedakis, P.C. and Willis, K.J. (1991) Quaternary refugia of north European trees. *Journal of Biogeography*, **18**, 103–15.

Berendsen, H.J.A. and Stouthamer, E. (2001) *Palaeogeographic Development of the Rhine–Meuse Delta, The Netherlands*. Koninklijke van Gorcum, Assen.

Beresford, G. (1981) Climatic change and its effects upon the settlement and desertion of medieval villages in Britain. In *Consequences of Climatic Change* (edited by C. Delano Smith and M. Parry), Department of Geography, Nottingham, 30–9.

Berger, A. (1980) The Milankovitch astronomical theory of palaeoclimates: a modern review. In *Vistas in Astronomy* (edited by A. Beer, K. Pounds and P. Beer), Pergamon Press, Oxford, 103–22.

Berger, A. (1988) Milankovitch theory and climate. *Review of Geophysics*, **26**, 624–57.

Berger, A. (1989) Pleistocene climatic variability at astronomical frequencies. *Quaternary International*, **2**, 1–14.

Berger, A. (1992) Astronomical theory of palaeoclimates and the last glacial–interglacial cycle. *Quaternary Science Reviews*, **11**, 571–82.

Berger, W.H. and Jansen, E. (1995) Younger Dryas episode: ice collapse and super fjord heat pump. In *The Younger Dryas* (edited by S.R. Troelstra, J. van Hinte and G.M. Ganssen), Koninklikje Nederlandse Akademie van Wetenschappen, Amsterdam, 61–105.

Berglund, B.E. (1985) Early agriculture in Scandinavia: research problems related to pollen analytical studies. *Norwegian Archaeological Review*, **18**, 77–105.

Berglund, B.E. (ed.) (1991) *The Cultural Landscape during 6000 Years in Southern Sweden; the Ystad Project*. Munksgaard International Booksellers, Copenhagen.

Berglund, B.E., Birks, H.J.B., Ralska-Jasiewiczowa, M. and Wright, H.E. (eds) (1996a) *Palaeoecological Events during the last 15,000 Years*. John Wiley, Chichester.

Berglund, B.E., Digerfeldt, G., Engelmark, R., Gaillard, M.-J., Karlsson, S., Miller, U. and Risberg, J. (1996b) Sweden. In *Palaeoecological Events during the last 15,000 Years* (edited by B.E. Berglund, H.J.B. Birks, M. Ralska-Jasiewiczowa and H.E. Wright), John Wiley, Chichester, 233–80.

Bergsten, H. (1994) A high-resolution record of Lateglacial and early Holocene marine sediments from southwestern Sweden; with special emphasis on environmental changes close to the Pleistocene–Holocene transition and the influence of fresh water from the Baltic basin. *Journal of Quaternary Science*, **9**, 1–12.

Bergthórsson, P. (1969) An estimate of drift ice and temperature in Iceland in 10,000 years. *Jökull*, **19**, 94–101.

Berti, A.A. (1975) Palaeobotany of Wisconsinan Interstadials, eastern Great Lakes region, North America. *Quaternary Research*, **5**, 591–619.

Bettis, E.A. and Hajic, E.R. (1995) Landscape development and the location of evidence of Archaic cultures in the Upper Midwest. *Geological Society of America, Special Paper*, **297**, 87–129.

Beven, K. (1996) Equifinality and uncertainty in geomorphological modelling. In *The Scientific Nature of Geomorphology* (edited by B.L. Rhoads and C.E. Thorn), John Wiley & Sons, Chichester and New York, 289–313.

Bianchi, G.G. and McCave, N. (1999) Holocene periodicity in North Atlantic climate and deep-ocean flow south of Iceland. *Nature*, **397**, 515–17.

Bickerton, R.W. and Matthews, J.A. (1993) 'Little Ice Age' variations of outlet glaciers from the Jostedalsbreen ice-cap, southern Norway: a regional lichenometric-dating study of ice-marginal moraine sequences and their climatic significance. *Journal of Quaternary Science*, **8**, 45–66.

Billamboz, A. (1992) Tree-ring analysis from an archaeodendrochronological perspective; the structural timber from the South West German lake dwellings. *Lundqua*, **34**, 34–40.

Bilsborough, A. (1999) Contingency, patterning and species in hominid evolution. In *Structure and Contingency: Evolutionary processes in life and human society* (edited by J. Bintliff), Leicester University Press, London, 43–101.

Bintliff, J. (1992) Erosion in the Mediterranean lands: a reconsideration of pattern, process and methodology. In *Past and Present Soil Erosion* (edited by M.G. Bell and J. Boardman), Oxbow Monograph 22, Oxford, 125–32.

Bintliff, J. (ed.) (1999) *Structure and Contingency: Evolutionary processes in life and human society*. Leicester University Press, London.

Bintliff, J. (2000) Landscape change in classical Greece. In *Geoarchaeology of the Landscapes of Classical Antiquity* (edited by F. Vermeulen and M. de Dapper), Stichting Babasch, Leiden, 49–70.

Bintliff, J. (2004) Time, structure and agency: the Annales, emergent complexity and archaeology. In *A Companion to Archaeology* (edited by J. Bintliff), Blackwell, Oxford, 174–94.

Birchfield, G.E. and Weertman, J. (1983) Topography, albedo-temperature feedback and climatic sensitivity. *Science*, **219**, 284–5.

Birks, H.H. (2003) The importance of plant macrofossils in the reconstruction of Lateglacial vegetation and climate: examples from Scotland, western Norway and Minnesota, USA. *Quaternary Science Reviews*, **22**, 453–73.

Birks, H.H. and Ammann, B. (2000) Two terrestrial records of rapid climatic change during the glacial–Holocene transition (14,000–9000 calendar years B.P.) from Europe. *Proceedings of the National Academy of Sciences*, **97**, 1390–4.

Birks, H.H., Birks, H.J.B., Kaland, P.E. and Moe, D. (1988) *The Cultural Landscape: Past, present and future*. Cambridge University Press, Cambridge.

Birks, H.J.B. (1986) Late Quaternary biotic changes in terrestrial and lacustrine environments, with particular reference to north-west Europe. In *Handbook of Holocene Palaeoecology and Palaeohydrology* (edited by B.E. Berglund), John Wiley, Chichester and New York, 3–65.

Birks, H.J.B. (1989) Holocene isochrone maps and patterns of tree spreading in the British Isles. *Journal of Biogeography*, **16**, 503–40.

Birks, H.J.B. (1995) Quantitative palaeoenvironmental reconstructions. In *Statistical Modelling of Quaternary Science Data* (edited by D. Maddy and J.S. Brew), Quaternary Research Association, Cambridge, 161–254.

Birks, H.J.B. and Birks, H.H. (1980) *Quaternary Palaeoecology*. Edward Arnold, London.

Björck, S. (1995a) Late Weichselian to early Holocene development of the Baltic Sea – with implications for coastal settlements in the southern Baltic region. In *Man and Sea in the Mesolithic* (edited by A. Fischer), Oxbow Books, Oxford, 23–34.

Björck, S. (1995b) A review of the history of the Baltic Sea, 13.0–8.0 ka BP. *Quaternary International*, **27**, 19–40.

Björck, S., Kromer, B., Johnsen, S., Bennike, O., Hammarlund, D., Lemdahl, G., Possnert, G., Rasmussen, T.L., Wohlfarth, B., Hammer, C.U. and Spurk, M. (1996) Synchronised terrestrial–atmospheric deglacial records around the North Atlantic. *Science*, **274**, 1155–60.

Björck, S., Walker, M.J.C., Cwynar, L.C., Johnsen, S., Knudsen, K-L., Lowe, J.J., Wohlfarth, B. and INTIMATE Members (1998) An event stratigraphy for the Last Termination in the North Atlantic region based on the Greenland ice-core record. *Journal of Quaternary Science*, **13**, 283–92.

Björn, L.O., Callaghan, T.V., Gehrke, C., Johanason, U., Sonesson, M. and Gwynne-Jones, D. (1998) The problem of ozone depletion in northern Europe. *Ambio*, **27**, 275–9.

Blackburn, T.C. and Anderson, K. (eds) (1993) *Before the Wilderness: Environmental management by native Californians*. Balkema Press, Menlo Park, California.

Blackford, J.J. (2000) Palaeoclimatic records from peat bogs. *Trends in Ecology and Evolution*, **15**, 193–8.

Blackford, J.J. and Chambers, F.M. (1991) Proxy records of climate from blanket mires: evidence for a Dark Age (1400BP) climatic deterioration in the British Isles. *The Holocene*, **21**, 63–7.

Blackford, J.J. and Chambers, F.M. (1995) Proxy climate record for the last 1000 years from Irish blanket peat and a possible link to solar variability. *Earth and Planetary Science Letters*, **133**, 145–50.

Blackford, J.J., Edwards, K.J., Dugmore, A.J., Cook, G.T. and Buckland, P.C. (1992) Icelandic volcanic ash and the mid-Holocene Scots Pine (*Pinus sylvestris*) pollen decline in northern Scotland. *The Holocene*, **2**, 260–6.

Blockley, M. (1999) Archaeological reconstructions and the community in the UK. In *The Constructed Past: Experimental archaeology, education and the public* (edited by P.G. Stone and P.G. Planel), Routledge, London, 15–34.

Blong, R.J. (1982) *The Time of Darkness*. Australian National University Press, Canberra.

Blong, R.J. (1984) *Volcanic Hazards*. Academic Press, London.

Bloom, A.L. (1983) Sea level and coastal morphology of the United States through the Late Wisconsin glacial maximum. In *Late Quaternary Environments of the United States*. Volume 1. *The Pleistocene* (edited by S.C. Porter), Longman, London, 215–29.

Bloom, A.L. (1984) Sea level and coastal changes. In *Late Quaternary Environments of the United States*. Volume 2. *The Holocene* (edited by H.E. Wright, Jnr), Longman, London, 42–51.

Bluth, G.J.S., Schnetzler, C.C., Krueger, D.A.J. and Walter, L.S. (1993) The contribution of explosive volcanism to global sulfur dioxide concentrations. *Nature*, **366**, 327–30.

Boardman, J. (1990) Soil erosion on the South Downs: a review. In *Soil erosion on Agricultural Land* (edited by J. Boardman, I.D.L. Foster and J.A. Dearing), Wiley, Chichester, 87–105.

Boardman, J. (1992) Current erosion on the South Downs: implications for the past. In *Past and Present Soil Erosion* (edited by M.G. Bell and J. Boardman), Oxbow Books, Oxford, 9–20.

Boardman, J. (1993) The sensitivity of downland arable land to erosion by water. In *Landscape Sensitivity* (edited by D.S.G. Thomas and R.J. Allison), John Wiley and Sons, Ltd, Chichester, 211–28.

Boardman, J. (1998) An average soil erosion: myth or reality? *Journal of Soil and Water Conservation*, **53** (1), 46–50.

Boardman, J. (2002) Erosion assessment. In *Encyclopedia of Soil Science* (edited by R. Lal), Marcel Dekker, New York, 399–401.

Bobe, R., Behresmeyer, R.E. and Chapman, R.E. (2002) Faunal change, environmental veracity and late Pliocene hominid evolution. *Journal of Human Evolution*, **42** (4), 475–97.

Bohlen, P.J. (2002) Earthworms. In *Encyclopedia of Soil Science* (edited by R. Lal), Marcel Dekker, New York, 370–7.

Bohncke, S.J.P. and Vandenberghe, J. (1991) Palaeohydrological development in the southern Netherlands during the last 15,000 years. In *Temperate Palaeohydrology* (edited by L. Starkel, J.B. Thornes and K.J. Gregory), John Wiley, Chichester and New York, 253–81.

Bokelmann, K. (1991) Humans and reindeer in the Ahrensburgian tunnel valley, Schleswig-Holstein. In *The Lateglacial in North-west Europe* (edited by R.N.E. Barton, A.J. Roberts and D.A. Roe), Council for British Archaeology, London, 72–81.

Bolin, B. (1998) The Kyoto negotiations on climate change: a scientific perspective. *Science*, **279**, 3310–31.

Bonadonna, F. and Leone, G. (1995) Palaeoclimatological reconstructions using stable isotope data on continental molluscs from Valle di Castiglione, Roma, Italy. *The Holocene*, **5**, 461–9.

Bond, G. and Lotti, R. (1995) Iceberg discharge into the North Atlantic on millennial timescales during the Last Glaciation. *Science*, **267**, 1005–10.

Bond, G., Broecker, W., Johnsen, S., McManus, J., Labeyrie, L., Jouzel, J. and Bonani, G. (1993) Correlations between climate records from North Atlantic sediments and Greenland ice. *Nature*, **365**, 143–7.

Bond, G., Heinrich, H., Broecker, W., Labeyrie, L., McManus, J., Andrews, J., Huon, S., Jantschik, R., Clasen, S., Simet, C., Tedesco, K., Klas, M., Bonani, G. and Ivy, S. (1992) Evidence for massive discharges of icebergs into the North Atlantic during the last glacial period. *Nature*, **360**, 245–9.

Bond, G., Showers, W., Cheseby, M., Lotti, R., Almasi, P., de Menocal, P., Priore, P., Cullen, H., Hajdas, I. and Bonani, G. (1997) A pervasive millennial-scale cycle in North Atlantic Holocene and glacial climates. *Science*, **278**, 1257–66.

Bondestam, K., Vasaru, A., Vasari, Y., Lemdahl, G. and Eskonen, K. (1994) Younger Dryas and Preboreal in Salpausselkä Foreland, Finnish Karelia. *Dissertationes Botanicae*, **234**, 161–206.

Bondevik, S., Svendsen, J.I., Johnsen, G., Mangerud, J. and Kaland, P.E. (1997) The Storegga tsunami along the Norwegian coast, its age and run up. *Boreas*, **26**, 29–53.

Boon, G.C. (1980) Caerleon and the Gwent Levels in early historic times. In *Archaeology and Coastal Change* (edited by F.H. Thompson), Society of Antiquaries, London, 24–36.

Bork, H.-R., Bork, H., Dalchow, C., Faust, B., Piorr, H.-P. and Schatz, T. (1998) *Landschaftsentwicklung in Mitteleuropa*. Klett-Perthes, Gotha and Stuttgart.

Bork, H.-R., Dotterweich, M. and Schmidtchen, G. (forthcoming) Landscape changes caused by soil erosion in Germany. In *Geoarchaeology: Landscape change over archaeological timescales* (edited by M.G. Bell and J. Boardman).

Borlenschlager, S. and Oeggl, K. (eds) (2000) *The Iceman and his Natural Environment*. Springer, Vienna.

Bos, J.A.A. and Janssen, C.R. (1996) Local impact of Palaeolithic man on the environment during the end of the last glacial in The Netherlands. *Journal of Archaeological Science*, **23**, 731–9.

Boswijk, G. and Whitehouse, N.J. (2002) *Pinus* and *Prostomis*: a dendrochronological and palaeoentomological study of a mid-Holocene woodland in eastern England. *The Holocene*, **12** (5), 585–96.

Bottema, S. (1988) Back to nature? Objectives of nature management in view of archaeological research. In *Archeologie en Landschap* (edited by H. van Bierma, O.H. Harsema and W. van Zeist), Biologisch-Archeologisch Instituut Rijkuniversiteit, Groningen, 185–206.

Bottema, S. and Sarpaki, A. (2003) Environmental change in Crete: a 9000-year record of Holocene vegetation history and the effect of the Santorini eruption. *The Holocene*, **13** (5), 733–49.

Boulton, G.S., Dongelmans, P., Punkari, M. and Broadgate, M. (2001) Palaeoglaciology of an ice sheet through a glacial cycle: the European ice sheet through the Weichselian. *Quaternary Science Reviews*, **20**, 591–626.

Bourdon, B., Henderson, G.M., Lundstrom, C.C. and Turner, S.P. (eds) (2003) Uranium-series Geochemistry. *Reviews in Mineralogy and Geochemistry*, **52**, 1–656.

Bowdler, S. (1988) The archaeology of prehistoric coastlines. In *The Archaeology of Prehistoric Coastlines* (edited by G. Bailey and J. Parkington), Cambridge University Press, Cambridge, 42–52.

Bowen, D.Q. (1991) Time and space in the glacial sediment systems of the British Isles. In *Glacial Deposits in Great Britain and Ireland* (edited by J. Ehlers, P.L. Gibbard and J. Rose), Balkema, Rotterdam, 3–12.

Bowen, D.Q., Phillips, F.M., McCabe, A.M., Knutz, P.C. and Sykes, G.A. (2002) New data for the Last Glacial Maximum in Great Britain and Ireland. *Quaternary Science Reviews*, **21**, 89–101.

Bowen, D.Q., Rose, J., McCabe, A.M. and Sutherland, D.G. (1986) Correlation of Quaternary glaciations in England, Ireland, Scotland and Wales. *Quaternary Science Reviews*, **5**, 299–340.

Bowler, J.M., Johnston, H., Olley, J.M., Prescott, J.R., Roberts, R.G., Shawcross, W. and Spooner, N.A. (2003) New ages for human occupation and climatic change at Lake Mungo, Australia. *Nature*, **421**, 837–40.

Bowman, S. (1990) *Radiocarbon Dating*. British Museum Publications, London.

Bowsher, J. (1998) *The Rose Theatre: An archaeological discovery*. Museum of London, London.

Boyd, R. (ed.) (1999) *Indians, Fire and the Land*. Oregon State University Press, Corvallis, Oregon.

Bradley, R. (ed.) (1991) Conceptions of time and ancient society. *World Archaeology*, **25** (2).

Bradley, R. (1993) *Altering the Earth*. Society of Antiquaries of Scotland Monograph 8, Edinburgh.

Bradley, R. (2000a) *An Archaeology of Natural Places*. Routledge, London.

Bradley, R.S. (1999) *Quaternary Palaeoclimatology*, 2nd edition. Allen & Unwin, London.

Bradley, R.S. (2000b) Past global changes and their significance for the future. *Quaternary Science Reviews*, **19**, 391–402.

Bradshaw, R.H.W. and Mitchell, F.J.G. (1999) The palaeoecological approach to reconstructing former grazing–vegetation interactions. *Forest Ecology and Management*, **120**, 3–12.

Bradshaw, R.H.W., Hannon, G.E. and Lister, A.M. (2003) A long-term perspective on ungulate–vegetation interactions. *Forest Ecology and Management*, **181**, 267–80.

Brandt, C.J. and Thornes, J. (1996) *Mediterranean Desertification and Land Use*. John Wiley & Sons, Ltd, Chichester.

Brandt, R. and van der Leeuw, S.E. (1987) The Assendelver polders of the Netherlands and a wet perspective on the European Iron Age. In *European Wetlands in Prehistory* (edited by J.M. Coles and A.J. Lawson), Clarendon Press, Oxford, 203–26.

Brandt, R., Groenman-van Waateringe, W. and van der Leeuw, S.E. (1987) *Assendelver Polder Papers*. Van Giffen Instituut, Amsterdam.

Branson, J., Brown, A.G. and Gregory, K.J. (eds) (1996) *Global Continental Changes: The context of palaeohydrology*. Geological Society of London, Special Publication 115, London.

Bray, J.R. (1982) Alpine glacier advance in relation to proxy summer temperature index based mainly on wine harvest dates. *Boreas*, **11**, 1–10.

Brenninkmeijer, C.A.M., van Geel, B. and Mook, W.G. (1982) Variations in the D/H and $^{18}O/^{16}O$ ratios in cellulose extracted from a peat bog core. *Earth and Planetary Science Letters*, **61**, 283–90.

Breuning-Madsen, H. and Holst, M.K. (1998) Recent studies on the formation of iron pans around the oak log coffins of the Bronze Age barrows of Denmark. *Journal of Archaeological Science*, **25**, 1103–10.

Bridgland, D.R. (ed.) (1994) *Quaternary of the Thames*. Geological Conservation Review Series 7, Chapman & Hall, London.

Bridgland, D.R. (2000) River terrace systems in northwest Europe: an archive of environmental change, uplift and early human oçcupation. *Quaternary Science Reviews*, **19**, 1293–303.

Briffa, K.R. (2000) Annual climatic variability in the Holocene: interpreting the message of ancient trees. *Quaternary Science Reviews*, **19**, 87–106.

Briffa, K.R. and Matthews, J.A. (eds) (2002) Analysis of dendrochronological variability and associated natural climates in Eurasia (ADVANCE-10k). *The Holocene*, 639–794.

Briffa, K.R., Bartholin, T.S., Eckstein, D., Jones, D.D., Karlén, W., Schweingruber, F.H. and Zetterberg, P. (1990) A 1,400 year tree-ring record of summer temperatures in Fennoscandia. *Nature*, **346**, 343–9.

Briffa, K.R., Jones, P.D., Schweingruber, F.H. and Osborn, T.J. (1998) Influence of volcanic eruptions on Northern Hemisphere summer temperatures over the past 600 years. *Nature*, **393**, 450–5.

Broadbent, N.D. (1979) *Coastal Resources and Settlement Stability: A critical analysis of a Mesolithic site complex in northern Sweden*. Ayn 3, Archaeological Studies Institute of North European Archaeology, University of Uppsala, Uppsala, Borgstroms tryckeri, 268 pp.

Broadbent, N.D. and Bergqvist, K.I. (1986) Lichenometric chronology and archaeological features on raised beaches: preliminary results from the Swedish north Bothnian coast region. *Arctic and Alpine Research*, **18**, 297–306.

Broecker, W.S. and Denton, G.H. (1990) The role of ocean atmosphere reorganisations in glacial cycles. *Quaternary Science Reviews*, **9**, 305–41.

Broecker, W.S., Kennett, J.P., Flower, B.P., Teller, J.T., Trumbore, S., Bonani, G. and Wölfli, W. (1989) Routing of meltwater from the Laurentide Ice Sheet during the Younger Dryas cold episode. *Nature*, **341**, 318–21.

Broecker, W.S., Peteet, D.M. and Rind, D. (1985) Does the ocean–atmosphere system have more than one stable mode of operation? *Nature*, **315**, 21–5.

Bromley, R.G. (1996) *Trace Fossils: Biology, taphonomy and applications*. Chapman & Hall, London.

Bronger, A., Winter, R., Derevjanko, O. and Aldag, S. (1995) Loess–palaeosol sequences in Tadjikistan as a palaeoclimatic record of the Quaternary in central Asia. *Quaternary Proceedings*, **4**, 69–82.

Brooks, S.J. and Birks, H.J.B. (2000) Chironomid-inferred late-glacial and early Holocene mean July air temperatures from Kråkenes Lake, western Norway. *Journal of Paleolimnology*, **23**, 77–89.

Brooks, S.J. and Birks, H.J.B. (2000) Chironomid-inferred late-glacial air temperatures at Whiting Bog, southeast Scotland, *Journal of Quaternary Science*, **15**, 759–64.

Brothwell, D.R. and Pollard, A.M. (eds) (2001) *Handbook of Archaeological Sciences*. John Wiley & Sons, Chichester.

Brown, A.G. (1997) *Alluvial Geoarchaeology: Floodplain archaeology and environmental change*. Cambridge University Press, Cambridge.

Brown, A.G. (1998) Fluvial evidence of the Medieval warm period and the Late Medieval climatic deterioration in Europe. In *Palaeohydrology and Environmental Change* (edited by G. Benito, V.R. Baker and K.J. Gregory), John Wiley & Sons Ltd, Chichester, 43–52.

Brown, A.G. and Caseldine, C. (1999) Biodiversity from palaeoecological data. *Journal of Biogeography*, **26** (1), 3–6.

Brown, A.G. and Quine, T.A. (1999) *Fluvial Processes and Environmental Change*. John Wiley, Chichester and New York.

Brown, T. (1997) Clearances and clearings: deforestation in Mesolithic/Neolithic Britain. *Oxford Journal of Archaeology*, **16** (2), 133–46.

Brown, T.A. (1998) (ed.) Special Issue. *Ancient Biomolecules*, **2** (2–3), 97–280.

Brown, T.A. (2001) Ancient DNA. In *Handbook of Archaeological Sciences* (edited by D.R. Brothwell and A.M. Pollard), John Wiley, Chichester and New York, 213–18.

Brown, T.A., Nelson, D.E., Mathewes, R.W., Vogel, J.S. and Southon, J.R. (1989) Radiocarbon dating of pollen by Accelerator Mass Spectrometry. *Quaternary Research*, **32**, 205–12.

Brubaker, L.B. and Cooke, E.R. (1984) Tree-ring studies of Holocene environments. In *Late-Quaternary Environments of the United States*. Volume 2. *The Holocene* (edited by H.E. Wright, Jnr), Longman, London, 222–35.

Bruce, J., Lee, H. and Haites, E. (eds) (1995) *Climate Change 1995: Economic and social dimensions of climate change*. Cambridge University Press, Cambridge.

Brugam, R.B., McKeever, K. and Lolesa, L. (1998) A diatom-inferred water-depth reconstruction for an Upper Peninsula, Michigan, lake. *Journal of Paleolimnology*, **20**, 267–76.

Brunet, M., Guy, F., Pilbeam, D., Mackaye, H.T., Likius, A., Ahounta, D., Beauvilain, A., Blondel, C., Bocherens, H., Boisserie, J.-R., de Bonis, L., Coppens, Y., Dejax, J., Denys, C., Duringer, P., Elsenmann, V., Fanone, G., Fronty, P., Geraads, D., Lehmann, T., Lihoreau, F., Louchart, A., Mahamat, A., Merceron, G., Mouchelin, G., Otero, O., Pelaez Campomanes, P., Ponce de Leon, M., Rage, J.-C., Sapanet, M., Schuster, M., Sudre, J., Tassy, P., Valentin, X., Vignaud, P., Viriot, L., Zazzo, A. and Zollikofer, C. (2002) A new hominid from the Upper Miocene of Chad, Central Africa. *Nature*, **418**, 145–55.

Brunning, R., Hogan, D., Jones, J., Jones, M., Maltby, E., Robinson, M.A. and Straker, V. (2000) Saving the Sweet Track: the *in situ* preservation of a Neolithic wooden trackway, Somerset, UK. *Conservation and Management of Archaeological Sites*, **4**, 3–20.

Brunsden, D. (2001) A critical assessment of the sensitivity concept in geomorphology. *Catena*, **42**, 90–123.

Bryant, I.D. (1983) Facies sequences associated with some braided river deposits of late Pleistocene age from southern Britain. In *Modern and Ancient Fluvial Systems* (edited by J.D. Comminson and J. Lewin), Blackwell, Oxford, 267–75.

Bryant, I.D., Holyoak, D.T. and Moseley, K.A. (1983) Late Pleistocene deposits at Brimpton, Berkshire, England. *Proceedings of the Geologists' Association*, **94**, 321–43.

Bucha, V. and Bucha, V. Jr (2002) Geomagnetic forcing and climatic variations in Europe, North America and in the Pacific Ocean. *Quaternary International*, **91**, 5–15.

Buckland, P.C. (1975) Synanthropy and the deathwatch: a discussion. *The Naturalist*, **100**, 37–42.

Buckland, P.C. (1979) *Thorne Moors: a palaeoecological study of a Bronze Age site*. University of Birmingham, Department of Geography Occasional Paper, Birmingham.

Buckland, P.C. (1991) Granaries, stores and insects: the archaeology of insect synanthropy. In *La préparation alimentaire des ceréales* (edited by D. Fournier and F. Sigaut), PACT 26-5, Belgium, 69–81.

Buckland, P.C. and Coope, G.R. (1991) *A Bibliography and Literature Review of Quaternary Entomology*. J.R. Collis, Sheffield.

Buckland, P.C. and Edwards, K.J. (1984) The longevity of pastoral episodes of clearance activity in pollen diagrams: the role of post-occupation grazing. *Journal of Biogeography*, **11**, 243–9.

Buckland, P.C., Amorosi, T., Barlow, L.K., Dugmore, A.J., Mayewski, P.A., McGovern, T.H., Ogilvie, A.E.J., Sadler, J.P. and Skidmore, P. (1996) Bioarchaeological and climatological evidence for the fate of Norse farmers in medieval Greenland. *Antiquity*, **70**, 88–96.

Buckland, P.C., Dugmore, A.J. and Edwards, K.J. (1997) Bronze Age myths? Volcanic activity and human response in the Mediterranean and North Atlantic regions. *Antiquity*, **71**, 581–93.

Buckland, P.C., Dugmore, A.J. and Sadler, J. (1991) Faunal change or taphonomic problem? A comparison of modern and fossil insect faunas from south-east Iceland. In *Environmental Change in Iceland: Past and present* (edited by J.K. Maizels and C.J. Caseldine), Kluwer, Dordrecht, 127–46.

Buckland, P.C., Dugmore, A.J. and Sadler, J.P. (1998) Palaeoecological evidence for human impact on the North Atlantic islands. *Boletim do Museu Municipal do Funchal (História Naturel)*, Sup. **5**, 89–108.

Buckland, P.C., Sadler, J.P. and Sveinbjarnardóttir, G. (1992) Palaeoecological investigations at Reykholt, western Iceland. In *Norse and Later Settlement and Subsistence in the North Atlantic* (edited by C.D. Morris and D.J. Rackham), Department of Archaeology, University of Glasgow, Glasgow, 149–68.

Burgess, C. (1989) Volcanoes, catastrophe and the global crisis of the late second millennium BC. *Current Archaeology*, **117**, 325–9.

Bush, M.B. (1993) An 11,400 year palaeoecological history of a British Chalk grassland. *Journal of Vegetation Science*, **4**, 47–66.

Butler, J.H., Battle, M., Bender, M.L., Montzka, S.A., Clarke, A.D., Saltzman, E.S., Sucher, C.M., Severinghaus, J.P. and Elkins, J.W. (1999) A record of atmospheric halocarbons during the twentieth century from polar firn air. *Nature*, **399**, 749–55.

Butzer, K.W. (1977) *Geomorphology of the Lower Illinois Valley as a Spatial-Temporal Context for the Koster Archaic Site*. Illinois State Museum, Reports of Investigations No. 34.

Butzer, K.W. (1982) *Archaeology as Human Ecology*. Cambridge University Press, Cambridge.

Butzer, K.W. (1996) Ecology in the long view: settlement histories, agrosystemic strategies and ecological performance. *Journal of Field Archaeology*, **23** (2), 141–50.

Calcote, R. (2003) Mid-Holocene climate and the hemlock decline: the range limit of *Tsuga canadensis* in the western Great Lakes region, USA. *The Holocene*, **13** (2), 215–24.

Camuffo, D., Cocheo, C. and Enzi, S. (2000) Seasonality of instability phenomena (hailstorms and thunderstorms) in Padova, northern Italy, from archive and instrumental sources since AD 1300. *The Holocene*, **10**, 635–42.

Cannell, M.G.R., Cruz, R.V.O., Galinski, W. and Cramer, W.P. (1996) Climate change impacts on forests. In *Climate Change 1995. Impacts, Adaptations and Mitigations of Climate Change: Scientific-technical analyses* (edited by R.T. Watson, M.C. Zinyowera, R.H. Moss and D.J. Dokken), Cambridge University Press, Cambridge, 95–129.

Carbonel, P., Colin. J-P., Danielopol, D.L., Löffler, H. and Neustreva, I. (1988) Paleoecology of limnic ostracodes: a review of some major topics. *Palaeogeography, Palaeoclimatology, Palaeoecology*, **62**, 413–61.

Carey, S., Sigurdsson, H., Mandeville, C. and Bronto, S. (2000) Volcanic hazards from pyroclastic flow discharge into the sea. In *Volcanic Hazards and Disasters in Human Antiquity* (edited by F.W. McCoy and G. Heiken), The Geological Society of America, Boulder, Colo., 1–14.

Carmichael, J.J., Strzepek, K.M. and Minarik, B. (1996) Impacts of climate change and seasonal variability on economic treatment costs: a case study of the Nitra River Basin, Slovakia. *International Journal of Water Resources Development*, **12**, 209–27.

Carrión, J., Munuera, M., Navarro, C., Burjachs, F., Dupré, M. and Walker, M.J. (1999) The palaeoecological potential of pollen records in caves: the case of Mediterranean Spain. *Quaternary Science Reviews*, **18**, 1061–74.

Carter, S.P. (1990) The stratification and taphonomy of shells in calcareous soils: implications for land snail analysis in archaeology. *Journal of Archaeological Science*, **17**, 495–507.

Caseldine, C.J. (1984) Pollen analysis of a buried arctic-alpine soil from Vestre Memurubreen, Jotunheimen, Norway: evidence for postglacial high altitude vegetation change. *Arctic and Alpine Research*, **16**, 423–30.

Caseldine, C.J. (1999) Archaeological and environmental change on prehistoric Dartmoor – current understanding and future directions. In *Holocene Environments of Prehistoric Britain* (edited by K.J. Edwards and D. Sadler), John Wiley & Sons, Chichester, 575–83.

Caseldine, C.J. and Hatton, J. (1993) The development of high moorland on Dartmoor: fire and the influence of Mesolithic activity on vegetation change. In *Climate Change and Human Impact on the Landscape* (edited by F.M. Chambers), Chapman & Hall, London, 119–32.

Casparie, W.A. (1986) The two Iron Age wooden trackways XIV (Bou) and XV (Bou) in the raised bog of SE Drenthe (The Netherlands). *Palaeohistoria*, **28**, 169–210.

Casparie, W. (2001) Prehistoric building disasters in Derryville Bog, Ireland: trackways, floodings and erosion. In *Recent Developments in Wetland Research* (edited by B. Raftery and J. Hickey), University College, Dublin, WARP Occasional Paper 14, Dublin, 115–28.

Casparie, W. and Groenman-van Waateringe, W. (1982) Palynological analysis of Dutch barrows. *Palaeohistoria*, **22**, 7–65.

Caspers, G. and Freund, H. (2001) Vegetation and climate in the Early- and Pleni-Weichselian in northern and central Europe. *Journal of Quaternary Science*, **15**, 31–48.

Castro, P.V., Chapman, R.W., Gili, S., Lull, V., Micó, R., Rihuete, C., Risch, R. and Sanahuja, M.E. (eds) (1998) *Aguas Project: Palaeoclimatic reconstruction and the dynamics of human settlement and land-use in the area of the middle Aguas (Almeria) in the south-east of the Iberian peninsula*. Office for Official Publications of the European Communities, Luxembourg.

Castro, P.V., Gili, S., Lull, V., Micó, R., Rihuete, C., Risch, R., Sanahuja, M.E. and Chapman, R.W. (2000) Archaeology and desertification in the Vera Basin (Almeria, South-east Spain). *European Journal of Archaeology*, **3** (2), 147–66.

Catchpole, A.J.W. and Faurer, M-A. (1983) Summer sea ice severity in Hudson Strait, 1751–1870. *Climatic Change*, **5**, 115–39.

Catt, J.A. (1979) Soils and Quaternary geology in Britain. *Journal of Soil Science*, **30**, 607–42.

Catt, J.A. and Staines, S.J. (1998) Petrography of sediments and buried soils. In *Late Quaternary Environmental Change in North-west Europe: Excavations at Holywell Coombe, south-east England* (edited by R.C. Preece and D.R. Bridgland), Chapman & Hall, London, 69–85.

Caufield, S., O'Donnoll, R.G. and Mitchell, P.I. (1998) ^{14}C dating of a Neolithic field system at Céide Fields, Co Mayo, Ireland. *Radiocarbon*, **40** (2), 629–40.

Cerling, T.E. and Craig, H. (1994) Geomorphology and *in situ* cosmogenic isotopes. *Annual Review of Earth and Planetary Science*, **22**, 273–317.

Cessford, C. (2001) A new dating sequence for Çatalhöyük. *Antiquity*, **75**, 717–25.

Chalmers, A.F. (1999) *What is This Thing called Science?* 3rd edition. Open University, Buckingham.

Chamberlin, T.C. (1965) The method of multiple working hypotheses. *Science*, **148**, 745–59 (reprinted from *Science*, 1897).

Chambers, F.M. (1996) Great Britain – Wales. In *Palaeoecological Events during the Last 15,000 Years* (edited by B.E. Berglund, H.J.B. Birks, M. Ralska-Jasiewiczowa and H.E. Wright), John Wiley, Chichester, 77–94.

Chambers, F.M. and Blackford, J.J. (2001) Mid- and Late-Holocene climatic change: a test of periodicity and solar forcing in proxy-climatic data from blanket peat bogs. *Journal of Quaternary Science*, **16**, 329–38.

Chambers, F.M., Addison, K., Blackford, J.J. and Edge, M.J. (1995) Palynology of organic beds below Devensian glacigenic sediments at Pen-y-bryn, Gwynedd, North Wales. *Journal of Quaternary Science*, **10**, 157–73.

Chambers, F.M., Barber, K.E., Maddy, D. and Brew, J. (1997) A 5500-year proxy-climate and vegetation record from blanket mire at Talla Moss, Borders, Scotland. *The Holocene*, **7**, 391–9.

Chambers, F.M., Ogle, M. and Blackford, J.J. (1999) Palaeoenvironmental evidence for solar forcing of Holocene climate: linkages to solar science. *Progress in Physical Geography*, **23**, 181–204.

Champion, T., Gamble, C., Shennan, S. and Whittle, A. (1984) *Prehistoric Europe*. Academic Press, London.

Chapman, H.P. and Gearey, B.R. (2000) Palaeoecology and the perception of prehistoric landscapes: some comments on visual approaches to phenomenology. *Antiquity*, **74**, 316–31.

Chapman, M.R. and Shackleton, N.J. (2000) Evidence of 550-year and 1000-year cyclicities in North Atlantic circulation patterns during the Holocene. *The Holocene*, **10**, 287–92.

Charman, D.J. (1994) Patterned fen development in northern Scotland: developing a hypothesis from palaeoecological data. *Journal of Quaternary Science*, **9**, 285–97.

Charman, D.J. (2001) Biostratigraphic and palaeoenvironmental applications of testate amoebae. *Quaternary Science Reviews*, **20**, 1753–64.

Charman, D.J. (2003) *Peatlands and Environmental Change*. John Wiley, Chichester and New York.

Charman, D.J., Hendon, D. and Packman, S. (1999) Multiproxy surface wetness records from replicate cores on an ombrotrophic mire: implications for Holocene palaeoclimatic records. *Journal of Quaternary Science*, **14**, 451–64.

Charman, D.J., Hendon, D. and Woodland, W.A. (2000) *The Identification of Testate Amoebae (Protozoa: Rhizopoda) in Peat*. Quaternary Research Association, Technical Guide No. 6, London.

Charman, D.J., Roe, H.M. and Gehrels, W.R. (1998) The use of testate amoebae in studies of sea-level change: a case study from the Taf estuary, south Wales, UK. *The Holocene*, **8**, 209–18.

Chorley, R.J., Dunn, A.J. and Beckinsale, R.P. (1964) *The History of the Study of Landforms:* Volume 1: *Geomorphology before Davis*. Methuen, London.

Christensen, C., Fischer, A. and Mathiassen, D.R. (1997) The great sea rise in the Storebælt. In *The Danish Storebælt since the Ice Age* (edited by L. Pedersen, A. Fischer and B. Aaby), Storebælt Publications, Copenhagen, 45–54.

Christensen, K. (1997) Wood from fish weirs: forestry in the Stone Age. In *The Danish Storebælt since the Ice Age* (edited by L. Pedersen, A. Fischer and B. Aaby), Storebælt Publications, Copenhagen, 147–56.

Church, J.A. and Gregory, J.M. (2001) Changes in sea level. In *Climate Change 2001: The scientific basis* (edited by J.T. Houghton, Y. Ding, D.J. Griggs, M. Noguer, P.J. van der Linden, X. Dai, K. Maskell and C.A. Johns), Cambridge University Press, Cambridge, 640–93.

Church, M. and Ryder, J.M. (1972) Paraglacial sedimentation: a consideration of fluvial processes governed by glaciation. *Geological Society of America Bulletin*, **83**, 3059–72.

Clague, J.J. and James, T.S. (2002) History and isostatic effects of the last ice sheet in southern British Columbia. *Quaternary Science Reviews*, **21**, 71–87.

Clark, A.J., Tarling, D.H. and Noël, M. (1988) Developments in archaeomagnetic dating in Britain. *Journal of Archaeological Science*, **15**, 645–7.

Clark, J.D. and Harris, J.W.K. (1985) Fire and its roles in early hominid lifeways. *African Archaeological Review*, **3**, 3–27.

Clark, J.G.D. (1954) *Excavations at Star Carr*. Cambridge University Press, Cambridge.

Clark, J.S. (1988) Effect of climatic change on fire regimes in northwestern Minnesota. *Nature*, **334**, 233–5.
Clark, J.S. (1989) Ecological disturbance as a renewal process: theory and application to fire history. *OIKOS*, **56**, 17–30.
Clark, J.S. and Royall, P.D. (1995) Transformation of a northern hardwood forest by aboriginal (Iroquois) fire: charcoal evidence from Crawford Lake, Ontario, Canada. *The Holocene*, **5** (1), 1–9.
Clark, P.U. (1994) Unstable behaviour of the Laurentide Ice Sheet over deforming sediment and its implications for climate change. *Quaternary Research*, **41**, 19–25.
Clark, P.U., Clague, J.J., Curry, B.B., Dreimanis, A., Hicock, S.R., Miller, G.H., Berger, G.W., Eyles, N., Lamothe, M., Miller, B.B., Mott, R.J., Oldale, R.N., Stea, R.R., Szabo, J.P., Thorleifson, L.H. and Vincent, J-S. (1993) Initiation and development of the Laurentide and Cordilleran Ice Sheets following the last interglaciation. *Quaternary Science Reviews*, **12**, 79–114.
Clark, P.U., Marshall, S.J., Clarke, G.K.C., Hostetler, S.W., Licciardi, J.M. and Teller, J.T. (2001) Freshwater forcing of abrupt climate change during the last deglaciation. *Science*, **293**, 283–7.
Clark, P.U., Webb, R.S. and Keigwin, L.D. (1999) *Mechanisms of Global Climate Change at Millennial Timescales*. Geophysical Monograph 112, American Geophysical Union, Washington.
Clarke, G.K.C., Mathews, W.H. and Pack, R.T. (1984) Outburst floods from Glacial lake Missoula. *Quaternary Research*, **22**, 289–99.
Clausen, B.L. (1993) *Viking Voyages to North America*. Viking Ship Museum, Roskilde.
Clausen, H.B., Hammer, C.U., Hvidberg, C.S., Dahl-Jensen, D., Steffensen, J.P., Kipfstuhl, J. and Legrand, M. (1997) A comparison of the volcanic records over the past 400 years from the Greenland ice core project and Dye 3 Greenland ice cores. *Journal of Geophysical Research*, **102**, 26707–23.
Cleal, R.M.J., Walker, K.E. and Montague, R. (1995) *Stonehenge in its Landscape: Twentieth century excavations*. English Heritage Archaeological Report 10, London.
Cleaveland, M.K. (2000) A 963-year reconstruction of summer (JJA) streamflow in the White River, Arkansas, USA, from tree-rings. *The Holocene*, **10**, 33–41.
Cleere, H.F. (1984) *Approaches to the Archaeological Heritage*. Cambridge University Press, Cambridge.
Cleere, H. (1989) *Archaeological Heritage Management in the Modern World*. Unwin Hyman, London.
CLIMAP Project Members (1981) Seasonal reconstructions of the earth's surface at the last glacial maximum. *Geological Society of America Map and Chart Series*, MC-36.
Clottes, J. and Courtin, J. (1996) *The Cave beneath the Sea: Palaeolithic images at Cosquer*. Abrams, New York.
Clutton-Brock, J. (ed.) (1989) *The Walking Larder*. Unwin Hyman, London.
Coard, R. and Chamberlain, A.T. (1999) The nature and timing of faunal change in the British Isles across the Pleistocene/Holocene transition. *The Holocene*, **9**, 372–6.
Cohen, J. and Stewart, I. (1994) *The Collapse of Chaos*. Penguin, London.
Coles, B. (1998) Doggerland: a speculative survey. *Proceedings of the Prehistoric Society*, **64**, 45–81.
Coles, B. (2000a) Somerset and the Sweet conundrum. In *Experiment and Design* (edited by A. Harding), Oxbow, Oxford, 163–9.
Coles, B. (2000b) Doggerland: the cultural dynamics of a shifting coastline. In *Coastal and Estuarine Environments: Sedimentology, geomorphology and geoarchaeology* (edited by K. Pye and J.R.L. Allen), Geological Society Special Publication No. 175, London, 393–402.
Coles, B. (2001) The impact of Western European beaver on stream channels: some implications for past stream conditions and human activity. *Journal of Wetland Archaeology*, **1**, 55–82.
Coles, B. and Coles, J. (1986) *Sweet Track to Glastonbury*. Thames & Hudson, London.
Coles, B. and Coles, J. (1989) *People of the Wetlands*. Thames & Hudson, London.
Coles, B. and Coles, J. (1996) *Enlarging the Past*. Society of Antiquaries of Scotland, Edinburgh.
Coles, B. and Coles, J. (1998) Passages of time. *Archaeology in the Severn Estuary*, **9**, 3–16.
Coles, B. and Olivier, A. (eds) (2001) *The Heritage Management of Wetlands in Europe*. WARP, Exeter.
Coles, G. and Mills, C.M. (1998) Clinging on for grim life: an introduction to marginality as an archaeological issue. In *Life on the Edge: Human settlement and marginality* (edited by C.M. Mills and G. Coles), Oxbow Books, Oxford, vii–xii.
Coles, J. (1987) Meare Village East: the excavations of A. Bulleid and H. St George Gray, 1932–1956. *Somerset Levels Papers*, **13**, 1–254.
Coles, J. and Minnitt, S. (1995) *Industrious and Fairly Civilised: The Glastonbury lake village*. Somerset Levels Project, Exeter.
Coles, J. and Orme, B. (1983) *Homo sapiens* or *Castor fiber*? *Antiquity*, **57**, 95–102.

Comani, S. (1987) The historical temperature series of Bologna (Italy): 1716–1774. *Climatic Change*, **11**, 375–90.

Conzen, M.P. (1990) *The Making of the American Landscape*. Unwin Hyman, Boston.

Cooke, H.B.S. (1981) Age control of Quaternary sedimentary/climatic record from deep boreholes in the Great Hungarian Plain. In *Quaternary Palaeoclimatology* (edited by W.C. Mahaney), Geobooks, Norwich, 1–12.

Cooke, R.G., Norr, L. and Piperno, D.R. (1996) Native Americans and the Panamanian Landscape. In *Case Studies in Environmental Archaeology* (edited by E.J. Reitz, L.A. Newsom and S.J. Scudder), Plenum Press, London, 103–26.

Cooney, G. (1999) *Landscapes of Neolithic Ireland*. Routledge, London.

Coope, G.R. (1986) Coleoptera analysis. In *Handbook of Holocene Palaeoecology and Palaeohydrology* (edited by B.E. Berglund), John Wiley and Sons, Chichester and New York, 703–13.

Coope G.R. (2000a) The climatic significance of coleopteran assemblages from the Eemian deposits in southern England. *Geologie en Mijnbouw*, **79**, 257–68.

Coope, G.R. (2000b) Middle Devensian (Weichselian) coleopteran assemblages from Earith, Cambridgeshire (UK) and their bearing on the interpretation of 'full glacial' floras and faunas. *Journal of Quaternary Science*, **15**, 779–88.

Coope, G.R. and Lemdahl, G. (1995) Regional differences in the Lateglacial climate of northern Europe based on coleopteran analysis. *Journal of Quaternary Science*, **10**, 391–5.

Coope, G.R., Lemdahl, G., Lowe, J.J. and Walkling A. (1998) Temperature gradients in northern Europe during the last glacial–Holocene transition (14–9 ^{14}C kyr BP) interpreted from coleopteran assemblages. *Journal of Quaternary Science*, **13**, 419–34.

Cordell, L. (1997) *Archaeology of the Southwest*. Academic Press, Boston.

Corfield, M., Hinton, P., Nixon, T. and Pollard, M. (eds) (1996) *Preserving Archaeological Remains in situ*. Museum of London, London.

Cortijo, E., Labeyrie, L., Elliott, M., Balbon, E. and Tisnerat, N. (2000) Rapid climatic variability of the North Atlantic Ocean and global climate: a focus of the IMAGES program. *Quaternary Science Reviews*, **19**, 227–41.

Courty, M.A., Goldberg, P. and Macphail, R. (1989) *Soils and Micromorphology in Archaeology*. Cambridge University Press, Cambridge.

Cowling, E.B. (1989) Recent changes in chemical climate and related effects on forests in North America and Europe. *Ambio*, **18**, 167–71.

Cox, M., Straker, V. and Taylor, D. (eds) (1995) *Wetlands: Archaeology and nature conservation*. HMSO, London.

Craddock, J.M. (1976) Annual rainfall in England since 1726. *Journal of the Royal Meteorological Society*, **102**, 823–40.

Cronon, W. (1983) *Changes in the Land: Indians, colonists and the ecology of New England*. Hill and Wang, New York.

Crosby, A.W. (1986) *Ecological Imperialism: The biological expansion of Europe 900–1100*. Cambridge University Press, Cambridge.

Crowley, T.J. (2000) Causes of climate change over the past 1000 years. *Science*, **289**, 270–7.

Crowley, T.J. and Lowery, T.S. (2000) How warm was the Medieval Warm Period? *Ambio*, **29**, 51–4.

Crowley, T.J. and North, G. (1991) *Palaeoclimatology*. Oxford University Press.

Cruise, J.F., Limaye, A.S. and Al Aber, N. (1999) Assessments of impact of climate change on water quality in the southeastern United States. *Journal of the American Water Resources Association*, **35**, 1539–50.

Crumley, C.L. (ed.) (1994) *Historical Ecology: Cultural knowledge and changing landscape*. School of American Research, Sante Fe.

Cubasch, U., Meehl, G.A., Boer, G.J., Stouffer, R.J., Dix, M., Noda, A., Senior, C.A., Raper, S. and Yap, K.S. (2001) Projections of future climate change. In *Climate Change 2001: The scientific basis* (edited by J.T. Houghton, Y. Ding, D.J. Griggs, M. Noguer, P.J. van der Linden and D. Xiaosu), Cambridge University Press, Cambridge, 525–85.

Cuffey, K.M. and Clow, G.D. (1998) Temperature, accumulation, and ice sheet elevation in central Greenland through the last deglacial transition. *Journal of Geophysical Research*, **102**, 26383–96.

Cummings, V. (2000) Myth, memory and metaphor: the significance of place, space and the landscape in Mesolithic Pembrokeshire. In *Mesolithic Lifeways: Current research from Britain and Ireland* (edited by R. Young), University of Leicester Archaeology Monographs 7, Leicester, 87–96.

Cunliffe, B. (1991) *Iron Age Communities in Britain*. Routledge, London.

Currant, A. and Jacobi, R. (2001) A formal mammalian biostratigraphy for the Late Pleistocene of Britain. *Quaternary Science Reviews*, **20**, 1707–16.

Cwynar, L.C. (1978) Recent history of fire and vegetation from laminated sediment of Greenleaf

Lake, Algonquin Park, Ontario. *Canadian Journal of Botany*, **56**, 10–21.

Cwynar, L.C. and Levesque, A. (1995) Chironomid evidence for Late-Glacial climatic reversal in Maine. *Quaternary Research*, **43**, 405–13.

Cwynar, L.C., Levesque, A.J., Mayle, F.E. and Walker, I. (1994) Wisconsian Late-glacial environmental Change in New Brunswick: a regional synthesis. *Journal of Quaternary Science*, **9**, 161–4.

Czaja, A. and Frankignoul, C. (1999) Influence of the North Atlantic SST on the atmospheric circulation. *Geophysical Research Letters*, **26**, 2969–72.

Dabney, S.M. (2002) Terrace relationships: tillage erosion. In *Encyclopedia of Soil Science* (edited by R. Lal), Marcel Dekker, New York, 1308–10.

Dahl, S.O. and Nesje, A. (1992) Palaeoclimatic implications based on equilibrium-line altitude depressions of reconstructed Younger Dryas and Holocene cirque glaciers in inner Nordfjord, western Norway. *Palaeogeography, Palaeoclimatology, Palaeoecology*, **94**, 87–97.

Dahl-Jensen, D., Mosegaard, K., Gundestrup, N., Clow, G.D., Johnsen, S.J., Hansen, A.W. and Balling, N. (1998) Past temperatures directly from the Greenland ice sheet. *Science*, **282**, 268–71.

Daily, D.C. (1997) *Nature's Services: Societal dependence on natural ecosystems*. Island Press, Washington, DC.

Daily, G., Matson, P.A. and Vitousek, P.M. (1997) Ecosystem services supplied by the soil. In *Nature's Services: Societal dependence on natural ecosystems* (edited by G.C. Daily), Island Press, Washington, DC, 113–32.

Dansgaard, W., Johnsen, S.J., Clausen, H.B., Dahl-Jensen, D., Gundestrup, N., Hammer, C.U., Hvidberg, C.S., Steffensen, J.P., Sveinbjörnsdóttir, A.E., Jouzel, J. and Bond, G. (1993) Evidence for general instability of past climate from a 250-kyr ice-core record. *Nature*, **364**, 218–20.

Dansgaard, W., Johnsen, S.J., Reeh, N., Gundestrup, N., Clausen, H.B. and Hammer, C.U. (1975) Climatic changes, Norsemen and modern man. *Nature*, **255**, 24–8.

Dansgaard, W., White, J.W.C. and Johnsen, S.J. (1989) The abrupt termination of the Younger Dryas climate event. *Nature*, **339**, 532–4.

Dark, P. (2000a) *The Environment of Britain in the First Millennium AD*. Duckworth, London.

Dark, P. (2000b) Revised 'absolute' dating of the early Mesolithic site of Star Carr, North Yorkshire, in the light of changes in the early Holocene tree-ring chronology. *Antiquity*, **74**, 304–7.

Darwin, C. (1859) *The Origin of Species by Means of Natural Selection*. Murray, London.

David, C., Dearing, J. and Roberts, N. (1998) Land-use history and sediment flux in a lowland lake catchment: Groby Pool, Leicestershire, UK. *The Holocene*, **8**, 383–94.

Davidson, D.A. and Carter, S.P. (1998) Micromorphological evidence of past agricultural practices in cultivated soils: the impact of a traditional agricultural system on soils in Papa Stour, Shetland. *Journal of Archaeological Science*, **25**, 827–38.

Davidson, D.A. and Simpson, I.A. (2001) Archaeology and soil micromorphology. In *Handbook of Archaeological Sciences* (edited by D.R. Brothwell and A.M. Pollard), John Wiley & Sons, Chichester, 166–77.

Davies, K.H. and Keen, D.H. (1985) The age of Pleistocene marine deposits at Portland, Dorset. *Proceedings of the Geologists' Association*, **96**, 217–25.

Davies, S.M., Branch, N.P., Lowe, J.J. and Turney, C.S.M. (2002) Towards a European tephrochronological framework for Termination 1 and the Early Holocene. *Philosophical Transactions of the Royal Society*, **A360**, 767–802.

Davis, M.B. (1976) Erosion rates and land-use history in southern Michigan. *Environmental Conservation*, **3**, 139–48.

Davis, M.B. (1981) Outbreaks of forest pathogens in Quaternary History. *Proceedings of IV International Conference on Palynology*, Luchnow, India, **3**, 216–27.

Davis, M.B. (1984) Holocene vegetational history of the eastern United States. In *Late-Quaternary Environments of the United States*. Volume 2. *The Holocene* (edited by H.E.J. Wright), Longman, London, 166–81.

Davis, P.T. (1988) Holocene glacier fluctuations in the American cordillera. *Quaternary Science Reviews*, **7**, 129–58.

Davis, P.T. and Osborn, G. (eds) (1988) Holocene glacier fluctuations. *Quaternary Science Reviews*, **7**, 113–242.

Davis, R.B. and Norton, S.A. (1978) Paleolimnologic studies of human impact on lakes in the United States, with emphasis on recent research in New England. *Polskie Archiwum Hydrobiologii*, **25** (1/2), 99–115.

Davis, S.J.M. (1987) *The Archaeology of Animals*. Batsford, London.

Dawson, A.G. and Smith, D.E. (1997) Holocene relative sea-level changes on the margin of a glacio-isostatically uplifted area: an example from northern Caithness, Scotland. *The Holocene*, **7**, 59–77.

Dawson, A.G., Long, D. and Smith, D.E. (1988) The Storegga Slides: evidence from eastern Scotland for a possible tsunami. *Marine Geology*, **88**, 271–6.

Dawson, A.G., Smith, D.E. and Long, D. (1990) Evidence for a tsunami from a Mesolithic site in Inverness, Scotland. *Journal of Archaeological Science*, **17** (5), 509–12.

Day, P. (1995) Devensian Late-glacial and early Flandrian environmental history of the Vale of Pickering, Yorkshire, England. *Journal of Quaternary Science*, **11** (1), 9–24.

Day, S.P. (1993) Woodland origin and 'ancient woodland indicators': a case-study from Sidlings Copse, Oxfordshire, UK. *The Holocene*, **3** (1), 45–53.

Dean, J.S. (1986) *Dating and Age Determination of Biological Material*. Croom Helm, Beckenham.

Dearing, J.A. (1991) Erosion and land-use. In *The Cultural Landscape during 6000 Years in Southern Sweden*. (edited by B.E. Berglund), *Ecological Bulletins*, **41**, 283–92.

Dearing, J.A., Alstrom, K., Reynell, J. and Sandgren, P. (1990) Recent and long-term records of soil erosion from southern Sweden. In *Soil Erosion on Agricultural Land* (edited by J. Boardman, I.D.L. Foster and J.A. Dearing), John Wiley & Sons, Chichester, 173–91.

Dehn, T. and Kristiansen, K. (1993) Conservation and preservation. In *Digging into the Past: 25 years of archaeology in Denmark* (edited by S. Hvass and B. Storgaard), The Royal Society of Northern Antiquaries, Copenhagen, 283–6.

Delcourt, H.R. and Delcourt, P.A. (1991) *Quaternary Ecology: A palaeoecological perspective*. Chapman & Hall, London.

Delcourt, P.A. and Delcourt, H.R. (1984) Late Quaternary palaeoclimates and biotic responses in eastern North America and the western North Atlantic Ocean. *Palaeogeography, Palaeoclimatology, Palaeoecology*, **48**, 263–84.

DeLucia, E.H., Hamilton, J.G., Naidu, S.L., Thomas, R.B., Andrews, J.A., Finzi, A., Lavine, M., Matamala, R., Mohan, J.E., Hendrey, G.R. and Schlesinger, W.H. (1999) Net primary production of a forest ecosystem with experimental CO_2 enrichment. *Science*, **284**, 1177–9.

Denton, G. and Hughes, T.J. (1981) *The Last Great Ice Sheets*. John Wiley & Sons, Chichester and New York.

Denton, G.H. (2000) Does an asymmetric thermohaline-ice-sheet oscillator drive 100,000-yr glacial cycles? *Journal of Quaternary Science*, **15**, 301–18.

Department of Culture, Media, and Sport (2001) *A Force for the Future*. HMSO, London.

Derbyshire, E. (ed.) (1995) Wind blown sediments in the Quaternary record. *Quaternary Proceedings*, **4**, 1–96.

Derbyshire, E. (2003) Loess, and the dust-indicators and records of terrestrial and marine palaeoenvironments (DIRTMAP) database. *Quaternary Science Reviews*, **22**, 1813–2052.

d'Errico, F. and Goñi, M.F.S. (2003) Neanderthal extinction and the millennial scale climatic variability of OIS 3. *Quaternary Science Reviews*, **22**, 769–88.

de Vernal, A. and Hillaire-Marcel, C. (2000) Sea-ice cover, sea-surface salinity and halo-thermocline structure of the northwest North Atlantic: modern versus full glacial conditions. *Quaternary Science Reviews*, **19**, 65–86.

Devoy, R.J.N. (1979) Flandrian sea-level changes and vegetation history of the lower Thames estuary. *Philosophical Transactions of the Royal Society, London*, **B285**, 355–410.

Devoy, R.J.N. (1987) Sea-level changes during the Holocene: the North Atlantic and Arctic Oceans. In *Sea Surface Studies* (edited by R.J.N. Devoy), Croom Helm, Beckenham, 294–347.

Devoy, R.J.N. (1995) Deglaciation, earth crustal behaviour and sea-level changes in the determination of insularity: a perspective from Ireland. In *Island Britain: A Quaternary perspective* (edited by R.C. Preece), Geological Society of London, Special Publication No. 96, London, 181–208.

Diaz, H.F. Andrews, J.T. and Short, S.K. (1989) Climate variations in northern North America (6000 BP to present) reconstructed from pollen and tree-ring data. *Arctic and Alpine Research*, **21**, 45–59.

di Castri, F., Hansen, A.J. and Debussche, M. (eds) (1990) *Biological Invasions in Europe and the Mediterranean Basin*. Kluwer, Dordrecht.

Dickson, J.H., Bootenschlager, S., Oeggl, K., Porley, R. and McMullen, A. (1996) Mosses and the Tyrolean Iceman's southern provenance. *Proceedings of the Royal Society of London B*, **263**, 567–71.

Dickson, J.H., Oeggl, K., Holden, T.G., Handley, L.L., O'Connell, T.C. and Preston, T. (2000) The omnivorous Tyrolean Iceman: colon contents (meat, cereals, pollen, moss and whipworm) and stable isotope analyses. *Philosophical Transactions of the Royal Society of London B*, **355**, 1843–9.

Dietler, M. and Herbich, I. (1991) Living on Luo time: reckoning sequence, duration, history and biography in a rural African society. In *Conceptions of Time and Ancient Society, World Archaeology*, **25**, 248–60.

Dillehay, T.D. (ed.) (1997) *Monte Verde, a Late Pleistocene Settlement in Chile*. Vol II. *Palaeoenvironmental and Site Context*. Smithsonian Institution Press, Washington, DC.

Dillehay, T.D. (2000) *The Settlement of the Americas: A new prehistory*. Basic Books, New York.

Dimbleby, G.W. (1976) Climate, soil and man. *Philosophical Transactions of the Royal Society of London B*, **275**, 197–208.
Dimbleby, G.W. (1985) *The Palynology of Archaeological Sites*. Academic Press, London and New York.
Dincauze, D.F. (2000) *Environmental Archaeology: Principles and practice*. Cambridge University Press, Cambridge.
Ding, Z., Liu, T. and Rutter, N.W. (1993) Pedostratigraphy of Chinese loess deposits and climatic cycles in the last 2.5 Ma. *Catena*, **20**, 73–91.
Ding, Z., Yu, Z., Rutter, N.W. and Liu, T. (1994) Towards an orbital time scale for Chinese loess deposits. *Quaternary Science Reviews*, **13**, 39–70.
Dinnin, M.H. and Sadler, J.P. (1999) 10,000 years of change: the Holocene entofauna of the British Isles. In *Holocene Environments of Prehistoric Britain* (edited by K.J. Edwards and J.P. Sadler), *Quaternary Proceedings*, **7**, 545–62.
Dodge, R.E., Fairbanks, R.G., Benninger, L.K. and Maurasse, F. (1983) Pleistocene sea levels from raised coral reefs in Haiti. *Science*, **219**, 1423–5.
Dodgson, R.A., Gilbertson, D.D. and Grattan, J.P. (2000) Endemic stress, farming communities and the influence of Icelandic volcanic eruptions in the Scottish Highlands. In *The Archaeology of Geological Catastrophes* (edited by W.J. McGuire, D.R. Griffiths, P.L. Hancock and I.S. Stewart), Geological Society Special Publication No. 171, London, 267–80.
Dodson, J.R. (1989) Late Pleistocene vegetation and environmental shifts in Australia and their bearing on faunal extinctions. *Journal of Archaeological Science*, **16**, 207–17.
Dolukhanov, P.M. and Khotinskiy, N.A. (1984) Human cultures and natural environment in the USSR during the Mesolithic and Neolithic. In *Late Quaternary Environments of the Soviet Union* (edited by A. Velichko), Longman, London.
Drake, F. (1995) Stratospheric ozone depletion – an overview of the scientific debate. *Progress in Physical Geography*, **19**, 1–17.
Dreslerová, D. (1995) The prehistory of the middle Labe (Elbe) floodplain in the light of archaeological finds. *Památky archeologické*, **86**, 105–45.
Driver, T.S. and Chapman, G.P. (eds) (1996) *Time-scales and Environmental Change*. Routledge, London.
Dubois, A.D. and Ferguson, D.K. (1985) The climatic history of pine in the Cairngorms based on radiocarbon dates and stable isotope analysis, with an account of the events leading up to its colonisation. *Review of Palaeobotany and Palynology*, **46**, 55–80.
Duff, D. (1998) *Holmes' Principles of Physical Geology*. Nelson Thornes, Cheltenham.
Dugmore, A.J. (1989) Tephrochronological studies of glacier fluctuations in Iceland. In *Glacier Fluctuations and Climatic Change* (edited by J. Oerlemans), Kluwer, Dordrecht, 37–57.
Dugmore, A.J., Larsen, G. and Newton, A.J. (1995) Seven tephra isochrones in Scotland. *The Holocene*, **5**, 257–66.
Duigan, C.A. and Birks, H.J.B. (2000) The late-glacial and early Holocene palaeoecology of cladoceran microfossil assemblages at Kråkenes, western Norway, with a quantitative reconstruction of temperature changes. *Journal of Paleolimnology*, **23**, 67–76.
Duller, G.A.T (2000) Dating methods: geochronology and landscape evolution. *Progress in Physical Geography*, **24**, 111–16.
Dumayne-Peaty, L. and Barber, K.E. (1998) Late Holocene vegetational history, human impact and pollen representivity variations in northern Cumbria. *Journal of Quaternary Science*, **13**, 147–64.
Dumond, D. (1979) People and pumice on the Alaska Peninsula. In *Volcanic Activity and Human Ecology* (edited by P.D. Sheets and D.K. Grayson), Academic Press, London, 373–92.
Duplessy, J-C., Labeyrie, L., Juillet-Leclerc, A. and Duprat, J. (1992) A new method to reconstruct sea-surface salinity: application to the North Atlantic during the Younger Dryas. In *The Last Deglaciation: Absolute and radiocarbon chronologies* (edited by E. Bard and W.S. Broecker), NATO ASI Series, 1, 2, Springer-Verlag, Berlin, 201–17.
Dupont, L.M. (1986) Temperature and rainfall variations in the Holocene based on comparative palaeoecology and isotope geology of hummock and hollow (Bourtangerveen, The Netherlands). *Review of Palaeobotany and Palynology*, **48**, 71–159.
Durham, W.H. (1978) The coevaluation of human biology and culture. In *Human Behaviour and Adaptation* (edited by V. Reynolds and N. Blurton Jones), Taylor & Francis, London, 11–32.
Dyke, A.S. and Dredge, L.A. (1989) Quaternary geology of the northwestern Canadian Shield. In *Quaternary Geology of Canada and Greenland* (edited by R.J. Fulton), Geological Survey of Canada, Geology of Canada No. 1, 178–214.
Dyke, A.S., Andrews, J.T., Clark, P.U., England, J.H., Miller, G.H., Kaplan, M.R. and Briner, J.P. (2002) The Laurentide and Innuitian ice sheets during the Last Glacial Maximum. *Quaternary Science Reviews*, **21**, 9–31.
Eastham, M. and Eastham, A. (1991) Palaeolithic parietal art and its topographical context. *Proceedings of the Prehistoric Society*, **57** (1), 115–28.

Edwards, K.J. (1982) Man, space and the woodland edge – speculations on the detection and interpretation of human impact in pollen profiles. In *Archaeological Aspects of Woodland Ecology* (edited by M. Bell and S. Limbrey), British Archaeological Reports IS 146, Oxford, 5–22.

Edwards, K.J. (1989) Meso-Neolithic vegetational impact in Scotland and beyond: palynological considerations. In *The Mesolithic in Europe* (edited by C. Bonsall), John Donald, Edinburgh, 143–55.

Edwards, K.J. (1996) A Mesolithic of the Western and Northern Isles of Scotland? Evidence from pollen and charcoal. In *The Early Prehistory of Scotland* (edited by T. Pollard and A. Morrison), Edinburgh University Press, Edinburgh, 23–38.

Edwards, K.J. (1998) Detection of human impact on the natural environment. In *Science in Archaeology: An agenda for the future* (edited by J. Bayley), English Heritage, London, 69–88.

Edwards, K.J. and Rowntree, K.M. (1980) Radiocarbon and palaeoenvironmental evidence for changing rates of erosion at a Flandrian stage site in Scotland. In *Timescales in Geomorphology* (edited by R.A. Cullingford, D.A. Davidson and J. Lewin), John Wiley & Sons Ltd, Chichester, 207–23.

Edwards, K.J. and Sadler, D. (eds) (1999) *Holocene Environment of Prehistoric Britain*. John Wiley & Sons, Chichester.

Edwards, K.J. and Sugden, H. (2003) Palynological visibility and the Mesolithic: colonisation of the Hebrides, Scotland. In *Mesolithic on the Move* (edited by L. Larsson), Oxbow, Oxford, 11–19.

Edwards, K.J. and Whittington, G. (2001) Lake sediments, erosion and landscape change during the Holocene in Britain and Ireland. *Catena*, **42**, 143–73.

Edwards, K.J., Hirons, K.R. and Newell, P.J (1991) The palaeoecological and prehistoric context of minerogenic layers in blanket peat: a study from Loch Dee, southwest Scotland. *The Holocene*, **1**, 29–39.

Edwards, K.J., Whittington, G. and Hirons, K.R. (1995) The relationship between fire and long-term wet heath development in South Uist, Outer Hebrides, Scotland. In *Heaths and Moorland: Cultural landscapes* (edited by F.H. Thompson, A.J. Hester and M.B. Usher), HMSO, Edinburgh, 240–8.

Edwards, K.J., Whittington, G. and Tipping, R. (2000) The incidence of microscopic charcoal in late glacial deposits. *Palaeogeography, Palaeoclimatology, Palaeoecology*, **164**, 263–78.

Edwards, P.C. (1989) Revising the Broad Spectrum Revolution: and its role in the origins of Southwest Asian food production. *Antiquity*, **63**, 225–46.

Eglinton, G., Bradshaw, S.A., Rosell, A., Saarnthein, M., Pflaumann, U. and Tiedemann, R. (1992) Molecular record of sea surface temperature changes on 100-year timescales for glacial terminations I, II and IV. *Nature*, **356**, 423–6.

Eiríksson, J., Knudsen, K-L., Haflidason, H. and Henriksen, P. (2000) Late-glacial and Holocene palaeoceanography of the North Icelandic shelf. *Journal of Quaternary Science*, **15**, 23–42.

Ekman, S.R. and Scourse, J.D. (1993) Early and Middle Pleistocene pollen stratigraphy from the British Geological Survey borehole 81/26, Fladen Ground, central North Sea. *Review of Palaeobotany and Palynology*, **79**, 285–95.

Eliade, M. (1973) *Australian Religions: An introduction*. Cornell University Press, Ithaca.

Elias, S.A. (1994) *Quaternary Insects and their Environments*. Smithsonian Institution Press, Washington, DC.

Elias, S.A. (1997) The Mutual Climatic Range Method of palaeoclimatic reconstruction based on fossil insects: new applications and interhemispheric comparisons. *Quaternary Science Reviews*, **16**, 1217–26.

Elias, S.A., Anderson, K.H. and Andrews, J.T. (1996) Late Wisconsin climate in northeastern USA and southeastern Canada, reconstructed from fossil beetle assemblages. *Journal of Quaternary Science*, **11**, 417–21.

Elkins, J.T., Thompson, T., Swanson, T., Butler, J., Hall, B., Cummings, S., Fisher, D. and Raffo, A. (1993) Decrease in the growth rates of atmospheric chlorofluorocarbons. *Nature*, **364**, 780–3.

Ellis, C. and Brown, A.G. (1998) Archaeomagnetic dating and palaeochannel sediments: data from the Mediaeval channel fills at Hemington, Leicestershire. *Journal of Archaeological Science*, **25**, 149–63.

Elverhøi, A., Dowdeswell, J.A., Funder, S., Mangerud, J. and Stein, R. (1998) Glacial and oceanic history of the Polar North Atlantic margins: an overview. *Quaternary Science Reviews*, **17**, 1–10.

Ely, L.L. (1997) Response of extreme floods in the southwestern United States to climate variations in the late Holocene. *Geomorphology*, **19**, 175–201.

Endfield, G.H. (1997) Myth, manipulation and myopia in the study of Mediterranean soil erosion. In *Archaeological Sciences 1975* (edited by A. Sinclair, E. Slater and J. Gowlett), Oxbow Monograph 64, Oxford, 241–8.

Endfield, G.H., O'Hara, S.L. and Metcalfe, S.E. (2000) The palaeoenvironmental reconstruction of the central Mexican highlands: a re-appraisal of traditional theory. In *People as an Agent of Environmental Change* (edited by R.A. Nicholson and T.P. O'Connor), Oxbow, Oxford, 81–91.

England, J. (1999) Coalescent Greenland and Innuitian ice during the last glacial maximum: revising the Quaternary of the Canadian high Arctic. *Quaternary Science Reviews*, **18**, 421–56.

English Heritage (2000) *The Power of Place*. English Heritage, London.

Epstein, S. and Krishnamurthy R.V. (1990) Environmental information in the isotopic record of trees. *Philosophical Transactions of the Royal Society, London*, **A330**, 427–39.

Erikstad, L. and Sollid, L. (1986) Neoglaciation in South Norway using lichenometric methods. *Norsk Geografisk Tidsskrift*, **40**, 85–105.

Eronen, M., Hyvärinen, H. and Zetterberg, P. (1999) Holocene humidity changes in northern Finnish Lapland inferred from lake sediments and submerged Scots pines dated by tree-rings *The Holocene*, **9**, 569–80.

Evans, D.J.A., Butcher, C. and Kirthisingha, A.V. (1994) Neoglaciation and an early 'Little Ice Age' in western Norway: lichenometric evidence from the Sandane area. *The Holocene*, **4**, 278–89.

Evans, J.G. (1972) *Land Snails in Archaeology*. Seminar Press, London.

Evans, J.G. (1975) *The Environment of Early Man in the British Isles*. Paul Elek, London.

Evans, J.G. (1993) The influence of human communities on the English chalklands from the Mesolithic to the Iron Age: the molluscan evidence. In *Climate Change and Human Impact on the Landscape* (edited by F.M. Chambers), Chapman & Hall, London, 147–56.

Evans, J.G. (1999) *Land and Archaeology: Histories of human environment in the British Isles*. Tempus, Stroud.

Evans, J.G. (2003) *Environmental Archaeology and the Social Order*. Routledge, London.

Evans, J.G. and O'Connor, T. (1999) *Environmental Archaeology*. Sutton Publishing, Stroud.

Evans, J.G., Limbrey, S., Máté, I. and Mount, R. (1993) Environmental history of the Upper Kennet valley, Wiltshire, for the last 10,000 years. *Proceedings of the Prehistoric Society*, **59**, 139–95.

Evershed, R.P., Dudd, S.N., Charters, S., Mottram, H., Stott, A.W., Raven, A, van Bergen, P.F. and Bland, H.A. (1999) Lipids as carriers of anthropogenic signals from Prehistory. *Philosophical Transactions of the Royal Society*, **B354**, 19–31.

Evershed, R.P., Dudd, S.N., Lockheart, M.J. and Jim, S. (2001) Lipids in archaeology. In *Handbook of Archaeological Sciences* (edited by D.M. Brothwell and A.M. Pollard), John Wiley, Chichester and New York, 331–49.

Fagan, B. (1991) *Ancient North America*. Thames & Hudson, London.

Fagan, B. (2000) *The Little Ice Age*. Basic Books, New York.

Fairbridge, R.W. (1983) Isostasy and eustasy. In *Shorelines and Isostasy* (edited by D.E. Smith and A.G. Dawson), Academic Press, London and New York, 3–28.

Fairbridge, R.W. (1984) Planetary periodicities and terrestrial climate stress. In *Climatic Changes on a Yearly to Millennial Basis* (edited by N.-A. Mörner and W. Karlén), Reidel, Dordrecht, 509–20.

Fairclough, G. and Rippon, S. (eds) (2002) *Europe's Cultural Landscape: Archaeologists and the management of change*. Europae Archaeologicae Consilium, Brussels.

Falkengren-Grerup, U. (1989) Soil acidification and its impact on ground vegetation. *Ambio*, **18**, 179–83.

Farman, J.C., Gardiner, B.G. and Shanklin, J.D. (1985) Large seasonal losses of ozone in Antarctica reveal seasonal ClO_x/NO_x interactions. *Nature*, **315**, 207–10.

Favis-Mortlock, D. (2002) Erosion by water. In *Encyclopedia of Soil Science* (edited by R. Lal), Marcel Dekker, New York, 452–6.

Favis-Mortlock, D., Boardman, J. and Bell, M.G. (1997) Modelling long-term anthropogenic erosion of a loess cover: South Downs, UK. *The Holocene*, **7**, 79–89.

Fedje, D.W. and Josenhans, H. (2000) Drowned forests and archaeology on the continental shelf of British Columbia, Canada. *Geology*, **28**, 99–102.

Fenton, A. (1978) *The Northern Isles: Orkney and Shetland*. John Donald, Edinburgh.

Ferguson, C.W. and Graybill, D.A. (1983) Dendrochronology of bristlecone pine: a progress report. *Radiocarbon*, **25**, 287–8.

Fiedel, S.J. (2002) Initial human colonization of the Americas: an overview of the issues and the evidence. *Radiocarbon*, **44**, 407–36.

Fienup-Riordan, A. (1990) *Eskimo Essays: Yup'ik lives and how we see them*. Rutgers University Press, New Brunswick.

Filion, L. and Quinty, F. (1993) Macrofossil and tree-ring evidence for a long-term forest succession and mid-Holocene hemlock decline. *Quaternary Research*, **40**, 89–97.

Finkel, R.C. and Nishiizumi, N. (1997) Beryllium 10 concentrations in the Greenland Ice Sheet Project 2 ice core from 3–40 ka. *Journal of Geophysical Research*, **102**, 26699–706.

Firth, C.R. and Stewart, I. (2000) Postglacial tectonics of the Scottish glacio-isostatic uplift centre. *Quaternary Science Reviews*, **19**, 1469–93.

Firth, C.R., Smith, D.E. and Cullingford, R.A. (1993) Late Quaternary glacio-isostatic uplift patterns for Scotland. *Neotectonics: Recent advances* (edited by L.A. Owen, I. Stewart and C. Vita Vinzi), *Quaternary Proceedings*, **3**, 1–14.

Fischer, A. (1991) Pioneers in deglaciated landscapes: the expansion and adaptation of Late Palaeolithic societies in Southern Scandinavia. In *The Lateglacial in North-west Europe* (edited by R.N.E. Barton, A.J. Roberts and D.A. Roe), Council for British Archaeology, London, 100–21.

Fischer, A. (1999) Stone Age Åmose. Stored in museums and preserved in the living bog. In *Bog Bodies, Sacred Sites and Wetland Archaeology* (edited by B. Coles, J. Coles and M.S. Jørgensen), WARP, Exeter, 85–92.

Fisher, T.G., Smith, D.G. and Andrews, J.T. (2002) Preboreal oscillation caused by a glacial Lake Agassiz flood. *Quaternary Science Reviews*, **21**, 873–87.

Fjeldskaar, W., Lindholm, C., Dehls, J.F. and Fjeldskaar, I. (2000) Postglacial uplift, neotectonics and seismicity in Fennoscandia. *Quaternary Science Reviews*, **19**, 1413–22.

Flanagan, D.C. (2002) Erosion. In *Encyclopedia of Soil Science* (edited by R. Lal), Marcel Dekker, New York, 395–8.

Flannery, K.V. (1969) Origins and ecological effects of early domestication in Iran and the Near East. In *The Domestication and Exploitation of Plants and Animals* (edited by P.J. Ucko and G.W. Dimbleby), Duckworth, London, 73–100.

Flannery, K.V. (1973) The origins of agriculture. *Annual Review of Anthropology*, **2**, 271–310.

Flannery, K.V. (ed.) (1986) *Guila Naquitz: Archaic foraging and early agriculture in Oaxaca, Mexico.* Academic Press, London.

Fleckinger, A. and Steiner, H. (1999) *The Fascination of the Neolithic Age: The ice man.* Folio Verlag, Bolzano.

Fleming, A. (1988) *The Dartmoor Reaves.* Batsford, London.

Fleming, A., Summerfield, M.A., Stone, J.O., Fifield, L.K. and Cresswell, R.G. (1999) Denudation rates for the southern Drakensberg escarpment, SE Africa, derived from *in situ*-produced cosmogenic ^{36}Cl: initial results. *Journal of the Geological Society, London*, **156**, 209–12.

Flenley, J. and Bahn, P. (2003) *The Enigmas of Easter Island.* Oxford University Press, Oxford.

Flood, J. (1983) *Archaeology of the Dreamtime: The story of prehistoric Australia and its people.* Collins, Sydney.

Flood, J. (1989) Tread softly for you tread on my bones. In *Archaeological Resource Management in the Modern World* (edited by H.F. Cleere), Unwin Hyman, London, 79–101.

Flückiger, J., Dällenbach, A., Blunier, T., Stauffer, B., Stocker, T.F., Raynaud, D. and Barnola, J-M. (1999) Variations in atmospheric N_2O concentration during abrupt climatic changes. *Science*, **285**, 227–30.

Fokkens, H. (1998) *Drowned Landscape: The occupation of the western part of the Frisian-Drentian Plateau, 4400 BC–AD 500.* Van Gorcum, Assen.

Foley, R. (1987) *Another Unique Species.* Longman, London.

Foley, R.A. (1994) Specialism, extinction and climate change in hominid evolution. *Journal of Human Evolution*, **26**, 275–89.

Foley, R.A. (1999) Pattern and process in hominid evolution. In *Structure and Contingency: Evolutionary processes in life and human society* (edited by J. Bintliff), Leicester University Press, London, 31–42.

Folland, C.K., Karl, T.R., Christie, J.R., Clarke, R.A., Gruza, G.V., Jouzel, J., Mann, M.E., Oerlemans, J., Salinger, M.J. and Wang, S-W. (2001) Observed climatic variability and change. In *Climate Change 2001: The scientific basis* (edited by J.T. Houghton, Y. Ding, D.J. Griggs, M. Noguer, P.J. van der Linden and D. Xiaosu), Cambridge University Press, Cambridge, 114–81.

Forester, R.M. (1987) Late Quaternary paleoclimate records from lacustrine ostracods. In *North America and Adjacent Oceans during the Last Deglaciation* (edited by W.F. Ruddiman and H.E. Wright, Jnr), Geological Society of America, Geology of North America, K-3, Boulder, Colo., 261–76.

Forman, R.T.T. (1995) *Land Mosaics.* Cambridge University Press, Cambridge.

Foster, I.D.L. (ed.) (2000) *Tracers in Geomorphology.* John Wiley & Sons, Ltd, Chichester.

Fowler, C.S. and Turner, N.J. (1999) Ecological/cosmological knowledge and land management among hunter gatherers. In *The Cambridge Encyclopedia of Hunters and Gatherers* (edited by R.B. Lee and R. Daly), Cambridge University Press, Cambridge, 419–25.

Fowler, P. (1982) Cultural resources management. *Advances in Archaeological Method and Theory*, **5**, 1–50.

Fowler, P. (2000) *Landscape Plotted and Pieced: Landscape history and local archaeology in Fyfield and Overton, Wiltshire.* The Society of Antiquaries of London, London.

Frankel, H. (1988) From continental drift to plate tectonics. *Nature*, **335**, 127–30.

Fraser, P.J. and Prather, M.J. (1999) Uncertain road to ozone recovery. *Nature*, **398**, 663–4.

Frechen, M. (1999) Upper Pleistocene loess stratigraphy in southern Germany. *Quaternary Science Reviews*, **18**, 243–70.

French, C. (2003) *Geoarchaeology in Action*. Routledge, London.

French, C., Lewis, H., Allen, M.J. and Scaife, R. (2000) Palaeoenvironmental and archaeological investigations on Wyke Down and in the upper Allen valley, Cranborne Chase, Dorset, England: interim summary report for 1998–9. *Dorset Proceedings*, **122**, 53–71.

French, H.M. (1996) *The Periglacial Environment*, 2nd edition. Longman, London.

Frenzel, B., Matthews, J.A. and Gläser, B. (eds) (1993) *Solifluction and Climatic Change in the Holocene*. Paläoklimaforschung No. 11, Gustav Fischer Verlag, Berlin.

Friedrich, M., Kromer, B., Kaiser, K., Spurk, M., Hughen, K.A. and Johnsen, S.J. (2001) High resolution climate signals in the Bølling-Allerød Interstadial (Greenland Interstadial 1) as reflected in European tree-ring chronologies compared to marine varves and ice-core records. *Quaternary Science Reviews*, **20**, 1223–33.

Friedrich, M., Kromer, B., Spurk, M., Hofmann J. and Kaiser, K.F. (1999) Palaeo-environment and radiocarbon calibration as derived from Lateglacial/Early Holocene tree-ring chronologies. *Quaternary International*, **61**, 27–39.

Fritz, S.C. (1989) Lake development and limnological response to prehistoric and historic land-use in Diss, Norfolk, England. *Journal of Ecology*, **77**, 182–202.

Frodeman, R. (1995) Geological reasoning: geology as an interpretative and historical science. *Geological Society of America Bulletin*, **107**, 960–8.

Fronval, T., Jansen, E., Bloemendal, J. and Johnsen, S. (1995) Oceanic evidence for coherent fluctuations in Fennoscandian and Laurentide ice sheets on millennial timescales. *Nature*, **374**, 443–6.

Fuji, N. (1988) Palaeovegetation and palaeoclimate changes around Lake Biwa, Japan, during the last *ca*. 3 million years. *Quaternary Science Reviews*, **7**, 21–8.

Fulford, M., Champion, T. and Long, A. (eds) (1997) *England's Coastal Heritage*. English Heritage Archaeological Report 15, London.

Fuller, I.C., Macklin, M.G., Lewin, J., Passmore, D.G. and Wintle, A.G. (1998) River response to high-frequency climate oscillations in southern Europe over the past 200 k.y. *Geology*, **26**, 275–8.

Funder, S., Hjort, C. and Landvik, J.Y. (1994) The last glacial cycle in East Greenland: an overview. *Boreas*, **23**, 283–93.

Gaillard, M-J., Birks, H.J.B., Emanuelsson, U. and Berglund, B.E. (1992) Modern pollen/land-use relationships as an aid in the reconstruction of past land-uses and cultural landscapes: an example from south Sweden. *Vegetation History and Archaeobotany*, **1**, 3–17.

Gale, S.J., Haworth, R.J. and Pisanu, P.C. (1995) The ^{210}Pb chronology of Late Holocene deposition in an eastern Australian lake basin. *Quaternary Science Reviews*, **14**, 395–408.

Gamble, C. (1993) *Timewalkers: The prehistory of global colonisation*. Alan Sutton, Stroud.

Gamble, C. and Soffer, O. (1991) *The World at 18,000 BP*. Unwin Hyman, London.

Gascoyne, M. (1992) Palaeoclimatic determination from cave calcite deposits. *Quaternary Science Reviews*, **11**, 609–32.

Gehrels, W.R. (1999) Middle and Late Holocene sea-level changes in eastern Maine reconstructed from foraminiferal saltmarsh stratigraphy and AMS ^{14}C dates on basal peat. *Quaternary Research*, **52**, 350–9.

Gemmell, A.M.D. (1999) IRSL from glacifluvial sediment. *Quaternary Science Reviews*, **18**, 207–17.

Gérard, J-C., (1990) Modelling the climatic response to solar variability. *Philosophical Transactions of the Royal Society, London*, **A330**, 561–74.

Gervais, B.R. and MacDonald, G.M (2001) Tree-ring and summer-temperature response to volcanic aerosol forcing at the northern tree-line, Kola Peninsula, Russia. *The Holocene*, **11**, 499–505.

Gibbard, P.L. (1985) *The Pleistocene History of the Middle Thames Valley*. Cambridge University Press, London.

Gibbard, P.L. (1994) *Pleistocene History of the Lower Thames*. Cambridge University Press, Cambridge.

Gilbertson, D.D. (1996) Explanations: environment as agency. In *Farming the Desert: The UNESCO Libyan Valleys Archaeological Survey* (edited by G. Barker, D. Gilbertson, B. Jones and D. Mattingley), UNESCO, Paris, 291–317.

Gill, A. and West, A. (2001) *Extinct*. Channel Four Books, London.

Gillespie, R. (2002) Dating the first Australians. *Radiocarbon*, **44** (2), 455–72.

Gillings, M. (1998) Embracing uncertainty and challenging dualism in the GIS based study of a palaeoflood plain. *European Journal of Archaeology*, **1** (1), 117–44.

Gilman, A. and Thornes, J. (1985) *Land Use and Prehistory in South-East Spain*. Allen & Unwin, London.

Gimingham, C.H. and de Smidt, J.T. (1983) Heaths as natural and semi-natural vegetation. In *Man's Impact on Vegetation* (edited by W. Holzner, W.M.J.A. and I. Ikusima), W. Junk, The Hague, 185–200.

Girling, M. (1988) The bark beetle *Scolytus scolytus* (Fabricius) and the possible role of elm disease in the early Neolithic. In *Archaeology and the Flora of the British Isles* (edited by M. Jones), Oxford University Committee for Archaeology Monograph, 14, Oxford, 34–8.

Gitay, H., Brown, S., Easterling, W. and Jallow, B. (2001) Ecosystems and their goods and services. In *Climate Change 2001: Impacts, adaptation and vulnerability* (edited by J.J. McCarthy, O.F. Canziani, N. Leary, D.J. Dokken and K.S. White), Cambridge University Press, Cambridge, 235–342.

Gleick, J. (1987) *Chaos*. Sphere, London.

Glover, I.C. and Higham, C.F.W. (1996) New evidence for early rice cultivation in South, Southeast and East Asia. In *The Origins and Spread of Agriculture and Pastoralism in Eurasia* (edited by D.R. Harris), UCL Press, London, 413–41.

Godwin, H. (1975) *The History of the British Flora*. Cambridge University Press, Cambridge.

Godwin, H. (1981) *The Archives of the Peat Bogs*. Cambridge University Press, Cambridge.

Goetcheus, V.G. and Birks, H.H. (2001) Full-glacial upland tundra vegetation preserved under tephra in the Beringia National Park, Seward Peninsula, Alaska. *Quaternary Science Review*, **20**, 135–47.

Goodfriend, G.A. (1992) The use of land snails in palaeoenvironmental reconstruction. *Quaternary Science Reviews*, **11**, 665–86.

Goodfriend, G.A. and Mitterer, R.M. (1993) A 45,000 year record of a tropical lowland biota: the land snail fauna from cave sediments at Coco Ree, Jamaica. *Geological Society of America Bulletin*, **105**, 18–29.

Göransson, H. (1986) Man and the forests of nemoral broad-leaved trees during the Stone Age. *Striae*, **24**, 143–52.

Göransson, H. (1987) *Neolithic Man and the Forest Environment around Alvastra Pile Dwelling*. Theses and Papers in North-European Archaeology, 20, Stockholm.

Gornitz, V.A. (1995) Sea level rise: a review of recent past and near-future trends. *Earth Surface Processes and Landforms*, **20**, 7–20.

Gosden, C. (1994) *Social Being and Time*. Blackwell, Oxford.

Gosden, C. (1999) *Anthropology and Archaeology: A changing relationship*. Routledge, London.

Gosse, J.C. and Phillips, F.M. (2001) Terrestrial in situ cosmogenic nuclides: theory and application. *Quaternary Science Reviews*, **20**, 1475–560.

Goudie, A. (1993) *Environmental Change*, 3rd edition. Oxford.

Goudie, A. (2001) *The Nature of the Environment*. Blackwell, Oxford.

Gould, S.J. (1999) Introduction: the scales of contingency and punctuation in history. In *Structure and Contingency: Evolutionary processes in life and human society* (edited by J. Bintliff), Leicester University Press, London, ix–xxii.

Gould, S.J. and Eldridge, N. (1993) Punctuated equilibrium comes of age. *Nature*, **366**, 223–7.

Gove, H. (2000) Some comments on Accelerator Mass Spectrometry. *Radiocarbon*, **42**, 127–36.

Govers, G. (2002) Tillage erosion, relationship to water erosion. In *Encyclopedia of Soil Science* (edited by R. Lal), Marcel Dekker, New York, 1330–2.

Govers, G., Vandaele, K., Desmet, P., Poesen, J. and Bunte, K. (1994) The role of tillage in soil redistribution on hillslopes. *European Journal of Soil Science*, **45**, 469–78.

Grachev, M., Vorobyova, S.S., Likhoshway, Y.V., Goldberg, E.L., Ziborova, G.A., Levina, O.V. and Khlystov, O.M. (1998) A high-resolution diatom record of the palaeoclimates of East Siberia for the last 2.5 My from Lake Baikal. *Quaternary Science Reviews*, **17**, 1101–6.

Granger, D.E., Kirchner, J.W. and Finkel, R. (1996) Spatially averaged long-term erosion rates measured from in situ-produced cosmogenic nuclides in alluvial sediment. *Journal of Geology*, **104**, 249–57.

Grattan, J.P. and Gilbertson, D.D. (1994) Acid-loading from Icelandic tephra falling on acidified ecosystems as a key to understanding archaeological and environmental stress in north and west Britain. *Journal of Archaeological Science*, **21**, 851–9.

Graumlich, L.J. (1993) A 1000-year record of temperature and precipitation in the Sierra Nevada. *Quaternary Research*, **39**, 249–55.

Gray, J. (1981) The use of stable isotope data in climatic reconstruction. In *Climate and History* (edited by T.M.L. Wigley, M.J. Ingram and B. Farmer), Cambridge University Press, London, 53–81.

Gray, J.M. and Coxon, P. (1991) The Loch Lomond Stadial glaciation in Britain and Ireland. In *Glacial Deposits in Great Britain and Ireland* (edited by J. Ehlers, P.L. Gibbard and J. Rose), Balkema, Rotterdam, 89–105.

Grayson, D.K. (1986) *The Establishment of Human Antiquity*. Academic Press, New York.

Grayson, D.K. (1987) An analysis of the chronology of late Pleistocene mammalian extinctions in North America. *Quaternary Research*, **28**, 281–9.

Grayson, D.K. (1989) The chronology of North America Late Pleistocene extinctions. *Journal of Archaeological Science*, **16**, 153–66.

Grayson, D.K. (2001) The archaeological record of human impacts on animal populations. *Journal of World Prehistory*, **15** (1), 1–69.

Grayson, D.K. and Meltzer, D.J. (2003) A requiem for North American overkill. *Journal of Archaeological Science*, **30**, 585–93.

Greeley, W.B. (1925) The relation of geography to timber supply. *Economic Geography*, **1**, 1–11.

Green, D.G. (1995) Time and spatial analysis. In *Statistical Modelling of Quaternary Science Data* (edited by D. Maddy and J.S. Brew), Technical Guide No. 5, Quaternary Research Association, Cambridge, 65–105.

Greer, S. (2002) *Ice Patch*. South Yukon First Nations, Yukon, Canada.

Greeves, T. (1990) Archaeology and the Green Movement: a case for perestroika. *Antiquity*, **63**, 659–65.

Greig, J. (1982) Past and present limewoods of Europe. In *Archaeological Aspects of Woodland Ecology* (edited by M.G. Bell and S. Limbrey), British Archaeological Reports, IS 146, Oxford, 23–55.

Greig, J. (1996) Great Britain – England. In *Palaeoecological Events during the Last 15,000 Years* (edited by B.E. Berglund, H.J.B. Birks, M. Ralska-Jasiewiczowa and H.E. Wright), John Wiley, Chichester and New York, 15–76.

Griffin, K.O. (1975) Vegetation studies and modern pollen spectra from Red Lake peatland, northern Minnesota. *Ecology*, **56**, 531–46.

Griffiths, H.I. and Holmes, J.A. (2000) *Non-Marine Ostracods and Quaternary Palaeoenvironments*. Technical Guide 8, Quaternary Research Association, London.

Griffiths, H.I., Ringwood, V. and Evans, J.G. (1994) Weichselian Late-glacial and early Holocene molluscan and ostracod sequences from lake sediments at Stellmoor, north Germany. *Archiv für Hydrobiologie*, **99**, 357–80.

Grimm, E.C., Jacobsen, G.L., Watts, W.A., Hansen, B.C.S. and Maasch, K.A. (1993) A 50,000-year record of climatic oscillations from Florida and its temporal correlation with the Heinrich Events. *Science*, **261**, 198–200.

Groenewoudt, B.J., Deeben, J., van Geel, B. and Lauwerier, C.G.M. (2001) An early Mesolithic assemblage with faunal remains in a stream valley near Zutphen, The Netherlands. *Archäologisches Korrespondenzblatt*, **31**, 329–48.

Groenman-van Waateringe, W. (1983) The early agricultural utilization of the Irish landscape: the last word on the elm decline. In *Landscape Archaeology in Ireland* (edited by T. Reeves-Smyth and F. Hamond), British Archaeological Reports, BS 116, Oxford, 217–32.

Groenman-van Waateringe, W. (1988) New trends in palynoarchaeology in Northwest Europe or the frantic search for local pollen data. In *Recent Developments in Environmental Analysis in Old and New World Archaeology* (edited by R.E. Webb), British Archaeological Reports IS 416, Oxford, 1–19.

Groenman-van Waateringe, W. (1992) Palynology and archaeology: the history of a plaggen soil from The Veluwe, The Netherlands. *Review of Palaeobotany and Palynology*, **73**, 87–98.

Groenman-van Waateringe, W. (1993) The effects of grazing on the pollen production of grasses. *Vegetation History and Archaeobotany*, **2**, 157–62.

Groenman-van Waateringe, W. and Robinson, M. (1988) *Man-made Soils*. British Archaeological Reports IS 410, Oxford.

Groenman-van Waateringe, W. and van Wijngaarden-Bakker, L.H. (eds) (1987) *Farm Life in a Carolingian Village*. van Gorcum, Assen.

Grosse-Brauckmann, G. (1986) Analysis of vegetative plant remains. In *Handbook of Holocene Palaeoecology and Palaeohydrology* (edited by B.E. Berglund), John Wiley, Chichester and New York, 591–618.

Grosswald, M.G. and Hughes, T.J. (2002) The Russian component of an Arctic ice sheet during the Last Glacial Maximum. *Quaternary Science Reviews*, **21**, 121–46.

Groube, L. (1996) The impact of diseases upon the emergence of agriculture. In *The Origins and Spread of Agriculture and Pastoralism in Eurasia* (edited by D.R. Harris), UCL Press, London, 101–29.

Grove, A.T. and Rackham, O. (2001) *The Nature of Mediterranean Europe*. Yale University Press, London.

Grove, J.M. (1997) The spatial and temporal variations of glaciers during the Holocene in the Alps, Pyrenees, Tatra and Caucasus. In *Glacier Fluctuations during the Holocene* (edited by B. Frenzel, G.S. Boulton, B. Gläser and U. Huckriede), *Palaeoclimate Research*, **24**, 67–83.

Grove, J.M. (2002) *The Little Ice Age*, 2nd edition. Routledge, London.

Grove, J.M. and Switsur, R. (1994) Glacial geological evidence for the Medieval Warm Period. *Climatic Change*, **26**, 143–69.

Grove, R.H. (1995) *Green Imperialism*. Cambridge University Press, Cambridge.

Grün, R. (2001) Trapped charge dating (ESR, TL, OSL). In *Handbook of Archaeological Sciences* (edited by D.R. Brothwell and A.M. Pollard), John Wiley, Chichester and New York, 47–62.

Guiot, J. (1987a) Late Quaternary climatic change in France estimated from multivariate pollen time series. *Quaternary Research*, **28**, 100–18.

Guiot, J. (1987b) Reconstruction of seasonal temperatures in central Canada since AD 1700 and detection of the 18.6 and 22-year signals. *Climatic Change*, **10**, 249–68.

Guiot, J., Pons, A., Beaulieu, J-L. de and Reille, M. (1989) A 140,000-year continental climate reconstruction from two European pollen records. *Nature*, **338**, 309–13.

Gumerman, G.J. (ed.) (1988) *The Anasazi in a Changing Environment.* Cambridge University Press, Cambridge.

Gupta, S.K., Sharma, P., Juyal, N. and Agrawal, D.P. (1991) Loess–palaeosol sequences in Kashmir: correlation of mineral magnetic stratigraphy with the marine palaeoclimate record. *Journal of Quaternary Science*, **6**, 3–12.

Gvozdover, M. (1995) *Art of the Mammoth Hunters: The finds from Ardeevo.* Oxbow Books, Oxford.

Haake, F.W. and Pflaumann, U. (1989) Late Pleistocene foraminiferal stratigraphy on the Vring Plateau, Norwegian Sea. *Boreas*, **18**, 343–56.

Hadorn, P. (1994) *Palynologie d'un site néolithique et histoire de la végétation des derniers 16,000 ans.* Musée cantonal d'archéologie, Archéologie neuchâteloise 18, Neuchâtel.

Haflidason, H., Eiríksson, J. and van Kreveld, S. (2000) The tephrochronology of Iceland and the North Atlantic region during the Middle and Late Quaternary: a review. *Journal of Quaternary Science*, **15**, 3–22.

Hafsten, U. (1983) Biostratigraphical evidence for late Weichselian and Holocene sea-level changes in southern Norway. In *Shorelines and Isostasy* (edited by D.A. Smith and A.G. Dawson), Academic Press, London and New York, 161–82.

Haigh, J.D. (1994) The role of stratospheric ozone in modulating the solar radiative forcing of climate. *Nature*, **370**, 544–46.

Haigh, J.D. (1996) The impact of solar variability on climate. *Science*, **272**, 981–4.

Haines-Young, R. and Petch, J. (1980) The challenge of critical rationalism for physical geography. *Progress in Physical Geography*, **4**, 63–77.

Haines-Young, R. and Petch, J. (1983) Multiple working hypotheses, equifinality and the study of landforms. *Transactions of the Institute of British Geographers*, New Series, **8**, 458–66.

Haines-Young, R. and Petch, J. (1986) *Physical Geography: Its nature and methods.* Harper & Row, London.

Hall, V.A. (2003) Assessing the impact of Icelandic volcanism on vegetation systems in the north of Ireland in the fifth and sixth millennia BC. *The Holocene*, **13** (1), 131–8.

Hall, V.A. and Pilcher J. (2002) Late-Quaternary Icelandic tephras in Ireland and Great Britain: detection, characterisation and usefulness. *The Holocene*, **12**, 223–30.

Hall, V.A., Pilcher, J.R. and McCormac, F.G. (1993) Tephra dated lowland landscape history of the north of Ireland, A.D. 750–1150. *New Phytologist*, **125**, 193–202.

Hall, V.A., Pilcher, J.R. and McCormac, F.G. (1994) Icelandic volcanic ash and the mid-Holocene Scots Pine (*Pinus sylvestris*) decline: no correlation. *The Holocene*, **4**, 79–83.

Halstead, P. and O'Shea, J. (eds) (1989) *Bad Year Economics: Cultural responses to risk and uncertainty.* Cambridge University Press, Cambridge.

Hammarlund, D. and Lemdahl, G. (1994) A Late Weichselian stable isotope stratigraphy compared with biostratigraphical data: a case study from southern Sweden. *Journal of Quaternary Science*, **9**, 13–31.

Hammarlund, D., Björck, S., Buchardt, C., Israelson, C. and Thomsen, C.T. (2003) Rapid hydrological changes during the Holocene revealed by stable isotope records of lacustrine carbonates from Lake Igelsjön, southern Sweden. *Quaternary Science Reviews*, **22**, 353–70.

Hammer, C.U., Clausen, H.B., Friedrich, W.L. and Tauber, H. (1987) The Minoan eruption of Santorini in Greece dated to 1645 BC? *Nature*, **328**, 517–19.

Hammer, C.U., Kurat, G., Hoppe, P., Grun, W. and Clausen, H.B. (2003) Thera eruption date 1645 BC confirmed by new ice core data. In *The Synchronisation of Civilisations in the Eastern Mediterranean in the Second Millennium B.C. II* (edited by M. Bietak), Verlag der Österreichischen Akademie der Wissenschaften, Vienna, 87–94.

Hammer, C.U., Mayewski, P., Peel, D. and Stuiver, M. (eds) (1997) Greenland Summit Ice Cores. Greenland Ice Sheet Project 2/Greenland Ice Core Project. *Journal of Geophysical Research*, 26315–886.

Hancock, G.S., Anderson, R.S., Chadwick, O.A. and Finkel, R.C. (1999) Dating fluvial terraces with ^{10}Be and ^{26}Al profiles: application to the Wind River, Wyoming. *Geomorphology*, **27**, 41–60.

Handler, P. and Andsager, K. (1994) El Niño, volcanism and global change. *Human Ecology*, **22**, 37–57.

Handmer, J.W. (ed.) (1987) *Flood Hazard Management: British and international perspectives.* Geobooks, Norwich.

Hann, B.J. (1990) Cladocera. In *Methods in Quaternary Ecology* (edited by B.G. Warner), Geosciences Canada Reprint Series 5, 81–91.

Hansen, J., Lacis, A., Rind, D., Russell, G., Stone, P., Fung, I., Ruedy, R. and Lerner, J. (1984) Climatic sensitivity: analysis of feedback mechanisms. In *Climate Processes and Climatic Sensitivity* (edited by J.E. Hansen and T. Takahashi), Geophysical Monogram Series 29, 130–163.

Hansen, J., Sato, M., Ruedy, R., Lacis, A. and Oinas, V. (2000) Global warming in the twenty-first century: an alternative scenario. *Proceedings of the National Academy of Sciences*, **97**, 9875–80.

Harbor, J. (2002) Erosion and landscape development, scale (space and time factors). In *Encyclopedia of Soil Science* (edited by R. Lal), Marcel Dekker, New York, 425–7.

Hardesty, D.L. (1977) *Ecological Anthropology*. John Wiley, New York.

Harlan, J.R. (1992) Indigenous African agriculture. In *The Origins of Agriculture* (edited by C.W. Cowan and P.J. Watson), Smithsonian Institution Press, Washington, 59–70.

Harries, J.E., Brindley, H.E., Sagoo, P.J. and Bantges, R.J. (2001) Increases in greenhouse forcing inferred from the outgoing longwave radiation spectra on the Earth in 1970 and 1997. *Science*, **410**, 355–7.

Harrison, S. (1999) The problem with landscape: some philosophical and practical questions. *Geography*, **84**, 355–63.

Harrison, S.P. and Digerfeldt, G. (1993) European lakes as palaeohydrological and palaeoclimatic indicators. *Quaternary Science Reviews*, **12**, 233–48.

Hart Hansen, J.P., Meldgaard, J. and Nordqvist, J. (1991) *The Greenland Mummies*. British Museum Publications, London.

Haslett, S.K. (ed.) (2002) *Quaternary Environmental Micropalaeontology*. Arnold, London.

Hastorf, C.A. (1999) Cultural implications of crop introductions in Andean prehistory. In *The Prehistory of Food: Appetites for change* (edited by C. Gosden and J. Hather), Routledge: One World Archaeology, London, 35–38.

Hather, J.G. (ed.) (1994) *Tropical Archaeobotany: Applications and new developments*. Routledge, London.

Hauptmann, H. (1999) The Urfa region. In M. Özdoğan and N. Ba gelen (eds) *Neolithic in Turkey: The cradle of civilisation*, Arkeoloji Ve Sanat Yayinlari, Istanbul, 65–86.

Haworth, E.Y. (1972) Diatom succession in a core from Pickerel Lake, Northeastern, South Dakota. *Bulletin of the Geological Society of America*, **83**, 157–72.

Haynes, G. and Eiselt, B.S. (1999) The power of Pleistocene hunter-gatherers: forward and backward searching for evidence about mammoth extinction. In *Extinctions in Near Time: Causes, contexts and consequences* (edited by R.D.E. MacPhee), Kluwer Academic Press, New York, 71–94.

Hays, J.D., Imbrie, J. and Shackleton, N.J. (1976) Variations in the earth's orbit: pacemaker of the Ice Ages. *Science*, **194**, 1121–32.

Head, L. (1994) Landscapes socialised by fire: post-colonial changes in Aboriginal fire-use in northern Australia and implications for prehistory. *Archaeology in Oceania*, **29**, 172–81.

Heckenberger, M.J., Kuikuro, A., Kuikuro, U.T., Russell, J.C., Schmidt, M., Fausto, C. and Franchetto, B. (2003) Amazonia 1492: pristine forest or cultural parkland? *Science*, **301**, 1710–14.

Hedges, R.E.M. (2001) The future of the past. *Radiocarbon*, **43**, 141–8.

Heidinga, H.A. (1987) *Medieval Settlement and Economy North of the Lower Rhine. Archaeology and History of Kootwijk and the Veluwe*. Cingula 9, Assen.

Heijnis, H. and van der Plicht, J. (1992) Uranium/thorium dating of Late Pleistocene peat deposits in NW Europe, uranium/thorium isotope systematics and open-system behaviour in peat layers. *Chemical Geology (Isotope Geoscience Section)*, **94**, 161–71.

Herz, N. and Garrison, E.G. (1998) *Geological Methods for Archaeology*. Oxford University Press, Oxford.

Hesthagen, T., Sevaldrud, I.H. and Berger, H.M. (1999) Assessment of damage to fish populations in Norwegian lakes due to acidification. *Ambio*, **28**, 112–17.

Heun, M., Schafer-Pregl, Klawan, D., Castagna, R., Accerbi, M., Borhi, B. and Salamini, F. (1997) Site of einkorn wheat domestication, identified by DNA fingerprinting. *Science*, **278**, 1312–14.

Hewitt, K. (1983) The idea of calamity in a technocratic age. In *Interpretations of Calamity* (edited by K. Hewitt), Allen & Unwin, London, 1–32.

Hicks, S., Miller, U. and Saarnisto, M. (eds) (1994) *Laminated Sediments*. PACT 41, Council of Europe, Belgium.

Hillaire-Marcel, C. (1980) Multiple component postglacial emergence, eastern Hudson Bay, Canada. In *Earth Rheology, Isostasy and Eustasy* (edited by N-A. Mörner), John Wiley, Chichester and New York, 215–30.

Hillaire-Marcel, C. and Occhietti, S. (1980) Chronology, palaeogeography and palaeoclimatic significance of late- and post-glacial events in eastern Canada. *Zeitschrift für Geomorphologie*, **24**, 373–92.

Hillam, J., Groves, C.M., Brown, D.M., Baillie, M.G.L., Coles, J.M. and Coles, B. (1990) Dendrochronology of the English Neolithic. *Antiquity*, **64**, 210–20.

Hillman, G.C. and Davis, M.S. (1990) Domestication rates in wild wheat and barley under primitive cultivation. *Biological Journal of the Linnaean Society*, **39**, 39–78.

Hillman, G., Hedges, R., Moore, A., Colledge, S. and Pettitt, P. (2001) New evidence of Lateglacial cereal cultivation at Abu Hureyra on the Euphrates. *The Holocene*, **11**, 383–93.

Hirons, K.R. (1990) The Post-glacial environment. In *Rhum: Mesolithic and later sites at Kinloch, excavations 1984–86* (edited by C.R. Wickham Jones), Society of Antiquaries of Scotland Monograph 7, 137–43.

Hodder, I. (ed.) (1996) *On the Surface: Çatalhöyük 1993–95*. British Institute of Archaeology at Ankara, London.

Hodder, I. (1999) *The Archaeological Process: An introduction*. Blackwell, Oxford.

Hofmann, D.J., Deshler, T.L., Aimedieu, P., Matthews, W.A., Johnston, P.V., Kondo, Y., Sheldon, W.R., Byrne, G.J. and Benbrook, J.R. (1989) Stratospheric clouds and ozone depletion in the Arctic during January 1989. *Nature*, **340**, 117–21.

Hofmann, W. (1986) Chironomid analysis. In *Handbook of Holocene Palaeoecology and Palaeohydrology* (edited by B.E. Berglund), John Wiley, Chichester and New York, 715–27.

Hogestijn, J.W.H. and Peeters, J.H.M. (eds) (2001) *De Mesolithische en vroeg-Neolitische vindplaats Hoge Vaart-A27 (Flevoland)*, Ammersfoorti Rapportage Archaeologische Momumentenzorg 79.

Holdaway, R.N. (1999) Introduced predators and avifaunal extinction in New Zealand. In *Extinctions in Near Time* (edited by R.D.E. MacPhee), Kluwer Academic Press, New York, 189–238.

Holliday, V.T. (1992) *Soils in Archaeology*. Smithsonian Institution Press, London and Washington.

Holliday, V.T. (1997) *Paleoindian: Geoarchaeology of the southern High Plains*. University of Texas Press, Austin.

Holmes, A. (1965) *Principles of Physical Geology*. Nelson, London.

Holst, M.K., Breuning-Madsen, H. and Rasmussen, M. (2001) The South Scandinavian barrows with well-preserved oak-log coffins. *Antiquity*, **75**, 126–36.

Hölzer, A. and Hölzer, A. (1998) Silicon and titanium in peat profiles as indicators of human impact. *The Holocene*, **8**, 685–96.

Hong, S., Candelone, J-P., Patterson, C.C. and Boutron, C.F. (1996) History of ancient copper smelting pollution during Roman and Medieval times recorded in Greenland ice. *Science*, **272**, 246–8.

Hong, Y.T., Yiang, H.B., Liu, T.S., Zhou, L.P., Beer, J., Li, H.D., Leng, X.T., Hong, B. and Qin, X.G. (2000) Response of climate to solar forcing recorded in a 6000 year $\delta^{18}O$ time series of Chinese peat cellulose. *The Holocene*, **10**, 1–7.

Hooghiemstra, H., Melice, J.L., Berger, A. and Shackleton, N.J. (1993) Frequency spectra and palaeoclimatic variability of the high-precision 30–1450 ka Funza 1 pollen record (Eastern Cordillera, Colombia). *Quaternary Science Reviews*, **12**, 141–56.

Horton, D.R. (1982) The burning question: aborigines, fire and Australian ecosystems. *Mankind*, **13**, 237–51.

Horton, D.R. (1984) Red kangaroos: last of the Australian megafauna. In *Quaternary Extinctions* (edited by P.S. Martin and R.G. Klein), University of Arizona Press, Tucson, 639–80.

Houghton, J.T., Callander, B.A. and Varney, S.K. (eds) (1992) *Climate Change 1992. The Supplementary Report to the IPCC Scientific Assessment*. Cambridge University Press, Cambridge.

Houghton, J.T., Ding, Y., Griggs, D.J., Noguer, M., van der Linden, P.J. and Xiaosu, D. (eds) (2001) *Climate Change 2001: The scientific basis*. Cambridge University Press, Cambridge.

Houghton, J.T., Jenkins, G.J. and Ephraums, J.J. (eds) (1990) *Climate Change: The IPCC scientific assessment*. Cambridge University Press, Cambridge.

Houghton, J.T., Meira Filho, L.G., Callander, B.A., Harris, N., Kattenberg, A. and Maskell, K. (eds) (1996) *Climate Change 1995: The science of climate change*. Cambridge University Press, Cambridge.

House, M. (1999) Evolution and environmental controls: Palaeozoic Black Deaths. In *Structure and Contingency: Evolutionary processes in life and human society* (edited by J. Bintliff), Leicester University Press, London, 14–30.

Housley, R.A., Gamble, C., Street, M. and Pettitt, P. (1997) Radiocarbon evidence for the late glacial recolonisation of north Europe. *Proceedings of the Prehistoric Society*, **63**, 25–54.

Hovgaard, W. (1925) The Norsemen in Greenland. *Geographical Review*, **15**, 605–16.

Howard, A.J. and Macklin, M.G. (1999) A generic geomorphological approach to archaeological interpretation and prospection in British river valleys: a guide for archaeologists investigating Holocene landscapes. *Antiquity*, **73**, 527–41.

Hoyt, D.V. and Schatten, K.H. (1997) *The Role of the Sun in Climate Change*. Oxford University Press.

Hubbert, G.D. and McInnes, K. (1999) A storm surge inundation model for coastal planning and impact studies. *Journal of Coastal Research*, **15**, 168–85.

Huddart, D., Gonzalez, S. and Roberts, G. (1999) The archaeological record and mid-Holocene marginal coastal palaeoenvironments around Liverpool Bay. *Journal of Quaternary Science*, **14**, 563–74.

Huggett, R.J. (1995) *Geoecology: A revolutionary approach*. Routledge, London.

Huggett, R.J. (1998) Soil chronosequences, soil development, and soil evolution: a critical review. *Catena*, **32**, 155–72.

Hughen, K.A., Overpeck, J.T. and Anderson, R.F. (2000) Recent warming in a 500-year temperature record from varved sediments, Upper Soper Lake, Baffin Island, Canada. *The Holocene*, **10**, 9–20.

Hughes, M.K. and Diaz, H.F. (1994) Was there a 'Medieval Warm Period', and if so, where and when? *Climatic Change*, **26**, 109–42.

Hughes, P.D.M., Mauquoy, D., Barber, K.E. and Langholm, P.G. (2000) Mire-development pathways and palaeoclimatic records from a full Holocene peat archive at Walton Moss, Cumbria, England. *The Holocene*, **10**, 465–79.

Huijzer, A.S. and Isarin, R.F.B. (1997) The multi-proxy approach to the reconstruction of past climates with an example of the Weichselian Pleniglacial in northwestern and central Europe. *Quaternary Science Reviews*, **16**, 513–33.

Huijzer, A.S. and Vandenberghe, J. (1998) Climatic reconstruction of the Weichselian Pleniglacial in north western and central Europe. *Journal of Quaternary Science*, **13**, 391–418.

Hume, I.N. (1999) Resurrection and deification at Colonial Williamsburg, USA, in *The Constructed Past: Experimental Archaeology, education and the public* (edited by P.G. Store and P.G. Planel), Routledge, London, 15–34.

Hunt, C.O. and Gilbertson, D.D. (1995) Human activity, landscape change and valley alluviation in the Feccia Valley, Tuscany, Italy. In *Mediterranean Quaternary River Environments* (edited by J. Lewin, M.G. Macklin and J.C. Woodward), A.A. Balkema, Rotterdam, 167–78.

Huntley, B. (1990) European vegetation history: palaeovegetation maps from pollen data – 13,000 yr BP to present. *Journal of Quaternary Science*, **5**, 103–22.

Huntley, B. and Birks, H.J.B. (1983) *An Atlas of Past and Present Pollen Maps of Europe 0–13,000 years ago*. Cambridge University Press, London.

Huntley, B. and Prentice, C. (1993) Holocene vegetation and climates of Europe. In *Global Climates since the Last Glacial Maximum* (edited by H.E. Wright, Jr, J.E. Kutzbach, T. Webb III, W.F. Ruddiman, F.A. Street-Perrott and P.J. Bartlein), University of Minnesota Press, Minneapolis, 136–68.

Huntley, B. and Webb, T, III. (eds) (1988) *Vegetation History*. Kluwer, Dordrecht.

Hutchinson, I. and McMillan, A.D. (1997) Archaeological evidence for village abandonment associated with Late Holocene earthquakes at the Northern Cascadia subduction zone. *Quaternary Research*, **48**, 79–87.

Hutton, J. (1788) Theory of the earth, or the investigation of the laws observable in the composition, dissolution and restoration of land upon the globe. *Transactions of the Royal Society of Edinburgh*, **1**, 209–304.

Hutton, J. (1795) *Theory of the Earth*. 2 volumes. William Creech, Edinburgh.

Imbrie, J. and Imbrie, K.P. (1979) *Ice Ages: Solving the mystery*. Macmillan, London.

Imbrie, J., Berger, A. and Shackleton, N.J. (1993a) Role of orbital forcing: a two million year perspective. In *Global Changes in the Perspective of the Past* (edited by J.A. Eddy and H. Oeschger), John Wiley, Chichester and New York, 263–77.

Imbrie, J., Berger, A., Boyle, E.A., Clemens, S.C., Duffy, A., Howard, W.R., Kukla, G., Kutzbach, J., Martinson, D.G., McIntyre, A., Mix, A.C., Molfino, B., Morley, J.J., Peterson, L.C., Pisias, N.G., Prell, W.L., Raymo, M.E., Shackleton, N.J. and Toggweiler, J.R. (1993b) On the structure and origin of major glaciation cycles. 2. The 100,000 year cycle. *Paleoceanography*, **8**, 699–735.

Imbrie, J., Boyle, E.A., Clemens, S.C., Duffy, A., Howard, W.R., Kukla, G., Kutzbach, J., Martinson, D.G., McIntyre, A., Mix, A.C., Molfino, B., Morley, J.J., Peterson, L.C., Pisias, N.G., Prell, W.L., Raymo, M.E., Shackleton, N.J. and Toggweiler, J.R. (1992) On the structure and origin of major glaciation cycles. 1. Linear responses to Milankovitch forcing. *Paleoceanography*, **7**, 701–38.

Imbrie, J., Hays, J.D., Martinson, A., McIntyre, A., Mix, A.C., Morley, J.J., Pisias, W.L. and Shackleton, N.J. (1984) The orbital theory of Pleistocene climate: support from a revised chronology of the δ^{18}Orecord. In *Milankovitch and Climate* (edited by A. Berger, J. Imbrie, J.D. Hays, G. Kukla and B. Saltzman), Reidel, Dordrecht, 269–306.

Ingold, T. (1986) *The Appropriation of Nature*. Manchester University Press, Manchester.

Ingold, T. (1990) Environment and culture in ecological anthropology. Lecture to the British Association, Science 1990, Lecture Aa9, 20–24 August 1990.

Ingold, T. (1993a) The temporality of the landscape. *World Archaeology*, **25** (2), 152–74.

Ingold, T. (1993b) Globes and spheres: the topology of environmentalism. In *Environmentalism: The view from anthropology* (edited by K. Milton), Routledge, London, 31–42.

Ingold, T. (2000) *The Perception of the Environment: Essays in livelihood, dwelling and skill*. Routledge, London.

Ingólfson, O., Björck, S., Haflidason, H. and Rundgren, M. (1997) Glacial and climatic events in Iceland reflecting regional North Atlantic climatic shifts during the Pleistocene–Holocene transition. *Quaternary Science Reviews*, **16**, 1135–44.

Ingram, M.J., Underhill, D.J. and Farmer, G. (1981) The use of documentary sources for the study of past climates. In *Climate and History* (edited by T.M.L. Wigley, M.J. Ingram and G. Farmer), Cambridge University Press, Cambridge, 180–213.

Ingstad, A.S. (1977) *The Discovery of a Norse Settlement in America: Excavations at L'Anse aux Meadows, Newfoundland 1961–1968*. Volume 1. Universitetsforlaget, Oslo.

Isarin, R.F.B. (1997) Permafrost distribution and temperatures in Europe during the Younger Dryas. *Permafrost and Periglacial Processes*, **8**, 313–33.

Isarin, R.F.B. and Bohncke, S.J.P. (1999) Mean July temperatures during the Younger Dryas in northwestern and central Europe as inferred from climatic indicator species. *Quaternary Research*, **51**, 158–73.

Isarin, R.F.B. and Renssen, H. (1999) Reconstructing and modelling Late Weichselian climates: the Younger Dryas in Europe as a case study. *Earth Science Reviews*, **48**, 1–38.

Isarin, R.F.B., Renssen, H. and Koster, E. (1997) Surface wind climate during the Younger Dryas in Europe as inferred from aeolian records and model simulations. *Palaeogeography, Palaeoclimatology, Palaeoecology*, **134**, 127–48.

Ivanovich, M. and Harmon, R.S. (1995) *Uranium-series Disequilibrium: Applications to earth, marine and environmental sciences*, 2nd edition. Clarendon Press, Oxford.

Iversen, J. (1941) Land occupation in Denmark's stone age. *Danmarks Geologiske Undersøgelse, Raekke 2*, **66**, 7–68.

Iversen, J. (1958) The bearing of glacial and interglacial epochs on the formation and extinction of plant taxa. *Uppsala University Årssk*, **6**, 210–15.

Iversen, J. (1964) Retrogressive vegetational succession in the Post-glacial. *Journal of Ecology*, **52**, 59–70.

Iversen, J. (1969) Retrogressive development of a forest ecosystem demonstrated by pollen diagrams from a fossil mor. *Oikos Suppl.*, **12**, 35–49.

Iversen, J. (1973) The development of Denmark's nature since the last glacial. *Danmarks Geologiske Undersøgelse, V Raekke*, **7-C**, 1–126.

Jackson, S.T., Overpeck, J.T., Webb III, T., Keattch, S.E. and Anderson, K.H. (1997) Mapped plant-macrofossil and pollen records of Late Quaternary vegetation change in eastern North America. *Quaternary Science Reviews*, **16**, 1–70.

Jackson, S.T., Webb, R.S., Anderson, K.H., Overpeck, J.T., Webb III, T. Williams, J.W. and Hansen, B.C.S. (2000) Vegetation and environment in eastern North America during the Last Glacial maximum. *Quaternary Science Reviews*, **19**, 489–508.

Jacobi, R.N., Tallis, J.H. and Mellars, P.A. (1976) The southern Pennine Mesolithic and the ecological record. *Journal of Archaeological Science*, **3**, 307–20.

Jacoby, G.C., Williams, P.L. and Buckley, B.M. (1992) Tree ring correlation between prehistoric landslides and abrupt tectonic events in Seattle, Washington. *Science*, **258**, 1621–3.

Jameson, J.H.J. and Hunt, W.J.J. (1999) Reconstruction versus preservation-in-place in the US National Park Service. In *The Constructed Past: Experimental archaeology, education and the public* (edited by P.G. Stone and R.G. Planel), Routledge, London, 35–62.

Jameson, M.H., Runnels, C.N. and van Andel, T.H. (1994) *A Greek Countryside: The Southern Argolid from prehistory to the present day*. Stanford University Press, Stanford.

Jannik, N.O., Phillips, F.M., Smith, G.I. and Elmore, D.A. (1991) ^{36}Cl chronology of lacustrine sedimentation in the Pleistocene Owens River system. *Bulletin of the Geological Society of America*, **103**, 1146–59.

Jashemski, W.F. (1979) Pompeii and Mount Vesuvius AD 79. In *Volcanic Activity and Human Ecology* (edited by P.D. Sheets and D.K. Grayson), Academic Press, London, 587–622.

Jelgersma, S. (1979) Sea-level changes in the North Sea basin. In *The Quaternary History of the North Sea* (edited by E. Oele, R.T.E. Schuttenheimen and A.J. Wiggers), University of Uppsala, 233–48.

Jenkins, A. (1999) End of acid reign? *Nature*, **401**, 537–38.

Jiang, H., Seidenkrantz, M-S., Knudsen, K.L. and Eiríksson, J. (2002) Late-Holocene summer sea-surface temperatures based on a diatom record from the north Icelandic shelf. *The Holocene*, **12**, 137–48.

Jochim, M. (1983) Palaeolithic cave art in ecological perspective. In *Hunter-gatherer Economy in Prehistory* (edited by G. Bailey), Cambridge University Press, Cambridge, 212–19.

Jóhansen, J. (1996) Faroe Islands. In *Palaeoecological Events during the Last 15,000 Years* (edited by B.E. Berglund, H.J.B. Birks, M. Ralska-Jasiewiczowa and H.E. Wright), John Wiley, Chichester, 145–52.

Johnsen, S., Clausen, H.B., Dansgaard, W., Fuhrer, K., Gundestrup, N., Hammer, C.U., Iversen, P., Jouzel, J., Stauffer, B. and Steffensen J.P. (1992) Irregular glacial interstadials revealed in a new Greenland ice core. *Nature*, **359**, 311–13.

Johnsen, S., Dahl-Jensen, D., Gundestrup, N., Steffensen, J.P., Clausen, H.B., Miller, H., Masson-Delmotte, V., Sveinbjörnsdottir, A.E. and White, J. (2001) Oxygen isotope and palaeotemperature records from six Greenland ice-core stations: Camp Century, Dye-3, GRIP, GISP2, Renland and NorthGRIP. *Journal of Quaternary Science*, **16**, 299–308.

Johnson, D.L., Keller, E.A. and Rockwell, T.K. (1990) Dynamic pedogenesis: new views on some key soil concepts, and a model for interpreting Quaternary soils. *Quaternary Research*, **33**, 306–19.

Johnson, W.H. (1990) Ice-wedge casts and relict patterned ground in central Illinois and their environmental significance. *Quaternary Research*, **33**, 51–72.

Jones, G. (2000) Evaluating the importance of cultivation and collecting in Neolithic Britain. In *Plants in Neolithic Britain and Beyond* (edited by A.S. Fairbairn), Oxbow, Oxford, 79–84.

Jones, G. and Halstead, P. (1995) Maslins, mixtures and monocrops: on the interpretation of archaeobotanical crop samples of heterogeneous composition. *Journal of Archaeological Science*, **22**, 103–14.

Jones, M. (1981) The development of crop husbandry. In *The Environment of Man: The Iron Age to the Anglo-Saxon period* (edited by M. Jones and G. Dimbleby), British Archaeological Report BS 87, Oxford, 95–127.

Jones, M. (2001) *The Molecule Hunt*. Penguin, London.

Jones, M.H.D. and Henderson-Sellars, A. (1990) History of the greenhouse effect. *Progress in Physical Geography*, **14**, 1–18.

Jones, M.K. and Brown, T.A. (2000) Agricultural origins: the evidence of modern and ancient DNA. *The Holocene*, **10**, 769–76.

Jones, M.K. and Colledge, S. (2001) Archaeology and the transition to agriculture. In *Handbook of Archaeological Sciences* (edited by D.R. Brothwell and A.M. Pollard), John Wiley, Chichester and New York, 393–401.

Jones, P.D. and Hulme, M. (1996) Calculating regional climatic time-series for temperature and precipitation: methods and illustrations. *International Journal of Climatology*, **16**, 361–77.

Jones, P.D., Briffa, K.R., Barnett, T.P. and Tett, S.F.B. (1998) High-resolution palaeoclimatic records for the last millennium: interpretation, integration and comparison with General Circulation Model control-run temperatures. *The Holocene*, **8**, 455–71.

Jones, R. (1969) Fire-stick farming. *Australian Natural History*, September, 224–8.

Jones, R. (1989) East of Wallace's Line: issues and problems in the colonisation of the Australian continent. In *The Human Revolution* (edited by P. Mellars and C. Stringer), Edinburgh University Press, Edinburgh, 743–82.

Jones, R.L. and Keen, D.H. (1993) *Pleistocene Environments in the British Isles*. Chapman & Hall, London.

Jones, R.T., Marshall, J.D., Crowley, S.F., Bedford, A., Richardson, N., Bloemendal, J. and Oldfield, F. (2002) A high-resolution multi-proxy Late-glacial record of climate change and intrasystem responses in northwest England. *Journal of Quaternary Science*, **17**, 329–40.

Jones, V.J., Flower, R.J., Appleby, P.G., Natkanski, J., Richardson, N., Rippey, B., Stevenson, A.C. and Battarbee, R.W. (1993) Palaeolimnological evidence for the acidification and atmospheric contamination of lochs in the Cairngorms and Lochnagar areas of Scotland. *Journal of Ecology*, **81**, 3–24.

Joos, M. (1982) Swiss midland lakes and climatic changes. In *Climatic Change in Later Prehistory* (edited by A.F. Harding), Edinburgh University Press, Edinburgh, 44–51.

Jouzel, J., Alley, R.B., Cuffey, K.M., Dansgaard, W., Grootes, P., Hoffmann, G., Johnsen, S.J., Koster, R.D., Peel, D., Shuman, C.A., Stievenard, M., Stuiver, M. and White J. (1997) Validity of the temperature reconstruction from water isotopes in ice cores. *Journal of Geophysical Research*, **102**, 26471–88.

Juniper, B.E. (1984) The natural flora of Mallorca, *Myotragus* and its possible effects, and the coming of man to the Balearics. In *The Deya Conference of Prehistory: Archaeological techniques, technology and theory* (edited by W.H. Waldren, R.E. Chapman, J. Lewthwaite and R.-C. Kennard), British Archaeological Reports IS 229, Oxford, 145–63.

Kaiser, K.F. (1994) Two Creeks Interstade dated through dendrochronology and AMS. *Quaternary Research*, **42**, 288–98.

Kaland, P.E. (1986) The origin and management of Norwegian coastal heaths as reflected by pollen analysis. In *Anthropogenic Indicators in Pollen Diagrams* (edited by K.-E. Behre), Balkema, Rotterdam, 19–36.

Kaland, P.E. (1988) The development of blanket mires in western Norway. In *Past, Present and Future* (edited by H.H. Birks, H.J.B. Birks, P.E. Kaland and D. Moe), Cambridge University Press, London, 475.

Kapsner, W.R., Alley, R.B., Shuman, C.A., Anandakrishnan, S. and Grootes, P.M. (1995) Dominant influence of atmospheric circulation on snow accumulation in Greenland over the past 18,000 years. *Nature*, **373**, 52–4.

Karlén, W. and Kuylenstierna, J. (1996) On solar forcing of Holocene climate: evidence from Scandinavia. *The Holocene*, **6**, 359–65.

Kashiwaya, K., Ochiai, S., Sakai, H. and Kawai, T. (2001) Orbit-related long-term climate cycles revealed in a 12-Myr continental record from Lake Baikal. *Nature*, **410**, 71–4.

Kattenberg, A., Giorgi, F., Grassl, H., Meehl, G.A., Mitchell, J.F.B., Stouffer, R.J., Tokioka, T., Weaver, A.J. and Wigley, T.M.L. (1996) Climate models – projections of future climate. In *Climate Change 1995: The science of climate change* (edited by J.T. Houghton, L.G. Meira Filho, B.A. Callander, N. Harris, A. Kattenberg and K. Maskell), Cambridge University Press, Cambridge, 65–132.

Kauffman, S. (1991) Antichaos and adaptation. *Scientific American*, **265** (2), 78–84.

Keating, G.M. (1978) Relation between monthly variations in global ozone and solar activity. *Nature*, **274**, 873–4.

Keigwin, L.D. and Boyle, E.A. (2000) Detecting Holocene changes in thermohaline circulation. *Proceedings of the National Academy of Sciences*, **97**, 1343–6.

Keigwin, L.D. and Lehman, S.J. (1994) Deep circulation change linked to Heinrich event 1 and the Younger Dryas in a mid-depth North Atlantic core. *Paleoceanography*, **9**, 185–94.

Keigwin, L.D., Jones, G.A. and Lehman, S.J. (1991) Deglacial meltwater discharge, North Atlantic deep circulation, and abrupt climatic change. *Journal of Geophysical Research*, **96**, 16811–26.

Kellogg, W.W. (1987) Mankind's impact on climate: the evolution of an awareness. *Climatic Change*, **10**, 113–36.

Kemp, R.A. (1998) The role of micromorphology in palaeopedological research. *Quaternary International*, **51/52**, 133–41.

Kemp, R.A. (2001) Pedogenic modification of loess: significance for palaeoclimatic reconstructions. *Earth Science Reviews*, **54**, 145–56.

Kenward, H.K. and Hall, A.R. (1995) *Biological Evidence from Anglo-Scandinavian Deposits at 16–22 Coppergate, York*. Archaeology of York Fascicule 14/7, York Archaeological Trust, York.

Kerney, M.P. (1999) *Atlas of the Land and Freshwater Molluscs of the British Isles*. Harley Books, Colchester.

Kerr, R.A. (2003) The small ones can kill you, too. *Science*, **301**, 1647.

Kershaw, A.P. (1986) Climatic change and Aboriginal burning in north-east Australia during the last two glacial/interglacial cycles. *Nature*, **322**, 47–9.

Kershaw, A.P. (1991) Palynological evidence for Quaternary vegetation and environments of mainland Southeastern Australia. *Quaternary Science Reviews*, **10**, 391–404.

Kershaw, A.P. (1993) Palynology, biostratigraphy and human impact. *The Artefact*, **16**, 12–18.

Khodri, M., Leclainche, Y., Ramstein, G., Braconnot, P., Marti, O. and Cortijo, E. (2001) Simulating the amplification of orbital forcing by ocean feedbacks in the last glaciation. *Nature*, **410**, 570–4.

Kidson, C. (1953) The Exmoor storm and the Lynmouth floods. *Geography*, **38**, 1–9.

King, F.B. and King, J.E. (1996) Interdisciplinary approaches to environmental reconstruction: an example from the Ozark Highland. In *Case Studies in Environmental Archaeology* (edited by E.J. Reitz, L.A. Newsom and S.J. Scudder), Plenum Press, London, 71–86.

Kirch, P.V. and Hunt, T.I. (1997) *Historical Ecology in the Pacific Islands: Prehistoric environmental landscape change*. Yale University Press, New Haven.

Kislev, M.E., Nadel, D. and Carmi, I. (1992) Epipalaeolithic (19,000 BP) cereal and fruit diet at Ohalo II, Sea of Galilee, Israel. *Review of Palaeobotany and Palynology*, **73**, 161–6.

Klimek, K. (1999) A 1000 year alluvial sequence as an indicator of catchment/floodplain interaction: the Ruda Valley, Sub-Carpathians, Poland. In *Fluvial Processes and Environmental Change* (edited by A.G. Brown and T.A. Quine), John Wiley & Sons Ltd, Chichester, 329–33.

Kneller, M. and Peteet, D. (1993) Late-Quaternary climate in the Ridge and Valley of Virginia, USA: changes in vegetation and depositional environment. *Quaternary Science Reviews*, **12**, 613–28.

Knox, J.C. (1984) Responses of river systems to Holocene climates. In *Late Quaternary History of the United States*. Volume 2, *The Holocene* (edited by H.E. Wright, Jnr), Longman, London, 26–41.

Knox, J.C. (1985) Responses of floods to Holocene climatic change in the Upper Mississippi valley. *Quaternary Research*, **23**, 287–300.

Knox, J.C. (1993) Large increases in flood magnitude in response to modest changes in climate. *Nature*, **361**, 430–2.

Knox, J.C. (2000) Sensitivity of modern and Holocene floods to climate change. *Quaternary Science Reviews*, **19**, 439–57.

Knox, J.C. (2001) Agricultural influence on landscape sensitivity in the Upper Mississippi River Valley. *Catena*, **42**, 193–224.

Knutti, R., Stocker, T.F., Joos, F. and Plattner, G-K. (2002) Constraints on radiative forcing and future climate change from observations and climate model scenarios. *Nature*, **416**, 719–23.

Koç, N. and Jansen, E. (1994) Response of the high latitude Northern Hemisphere to orbital climatic forcing: evidence from the Nordic Seas. *Geology*, **22**, 523–6.

Koç, N., Jansen, E. and Haflidason, H. (1993) Paleoceanographic reconstructions of surface ocean conditions in the Greenland, Iceland and Norwegian Seas through the last 14 ka based on diatoms. *Quaternary Science Reviews*, **12**, 115–40.

Koç, N., Jansen, E., Hald, M. and Labeyrie, L. (1996) Late-glacial-Holocene sea surface temperatures and gradients between the North Sea and the Norwegian Sea: implications for the Nordic heat pump. In *Late Quaternary Palaeoceanography of the North Atlantic Margins* (edited by J.T. Andrews, W.E.N. Austin, H. Bergsten and A.E. Jennings), Geological Society Special Publication No. 111, 177–85.

Kohen, J. (1995) *Aboriginal Environmental Impacts*. University of South Wales Press, Sydney.

Kolstrup, E. (1988) Late Atlantic and early Subboreal vegetational development at Trundholm, Denmark. *Journal of Archaeological Science*, **15**, 503–13.

Korhola, A. (1995) Holocene climatic variations in southern Finland reconstructed from peat initiation data. *The Holocene*, **5**, 43–58.

Krech, S. (1999) *The Ecological Indian: Myth and history*. Norton & Co, New York.

Krings, M., Stone, A., Schmitz, R.W., Krainitzki, H., Stoneking, M. and Pääbo, S. (1997) Neanderthal DNA sequences and the origin of modern humans. *Cell*, **90**, 19–30.

Kristiansen, K. (1985) Economic development in Denmark since agrarian reform. In *Archaeological Formation Processes* (edited by K. Kristiansen), Nationalmuseet, Copenhagen, 41–62.

Kristiansen, K. (2002) The birth of ecological archaeology in Denmark: history and research environments 1850–2000. In *The Neolithisation of Denmark: 150 years of debate* (edited by A. Fischer and K. Kristiansen), JR Collis, Sheffield, 11–31.

Kukla, G. and An, Z.S. (1989) Loess stratigraphy in central China. *Palaeogeography, Palaeoclimatology, Palaeoecology*, **72**, 203–25.

Kullman, L. (1988) Holocene history of the forest-alpine tundra ecotone in the Scandes Mountains (central Sweden). *New Phytologist*, **108**, 101–10.

Kuniholm, P.I. (2001) Dendrochonology and other applications of tree-ring studies in archaeology. In *Handbook of Archaeological Sciences* (edited by D.R. Brothwell and A.M. Pollard), John Wiley, Chichester and New York, 35–46.

Kuniholm, P.I., Kromer, B., Manning, S.W., Newton, M., Latini, C.E. and Bruce, M.J. (1996) Anatolian tree-rings and the absolute chronology of the east Mediterranean 2220–718 BC. *Nature*, **381**, 780–3.

Kutzbach, J.E. and Street-Perrott, F.A. (1985) Milankovitch forcing of fluctuations in the level of tropical lakes from 18–0 k yr BP. *Nature*, **317**, 130–4.

Kuylenstierna, J.C.I., Rodhe, H., Cinderby, S. and Hicks, K. (2001) Acidification in developing countries: ecosystem sensitivity and the critical load approach on a global scale. *Ambio*, **30**, 20–8.

Laflen, J.M. (2002) Erosion by water, empirical notes. In *Encyclopedia of Soil Science* (edited by R. Lal), Marcel Dekker, New York, 457–62.

Lageard, J.G.A., Chambers, F.M. and Thomas, P.A. (1999) Climatic significance of the marginalization of Scots pine (*Pinus sylvestris*) *c.*2500 BC at White Moss, south Cheshire, UK. *The Holocene*, **9** (3), 321–33.

Laidlaw, R. (1989) Cultural resource planning and management in a multiple-use agency. In *Archaeological Heritage Management in the Modern World* (edited by H.F. Cleere), Unwin Hyman, London, 232–5.

Lal, R. (ed.) (2002) *Encyclopedia of Soil Science*. Marcel Dekker, New York.

LaMarche, V.C. and Hirschboeck, K.K. (1984) Frost rings in trees as records of major volcanic eruptions. *Nature*, **307**, 121–6.

Lamb, H.H. (1972) *Climate: Present, past and future*. Methuen, London.

Lamb, H.H. (1977) *Climate: Past, present and future*. Volume 2. Methuen, London.

Lamb, H.H. (1981) An approach to the study of the development of climate and its impact on human affairs. In *Climate and History* (edited by T.M.L. Wigley, M.J. Ingram and G. Farmer), Cambridge University Press, Cambridge, 291–309.

Lamb, H.H. (1995) *Climate, History and the Modern World*, 2nd edition. Routledge, London.

Lambeck, K. (1995) Late Devensian and Holocene shorelines of the British Isles and North Sea from models of glacio-hydro-isostatic rebound. *Journal of the Geological Society, London*, **152**, 437–48.

Lambeck, K. (2002) Links between climate and sea levels for the past three million years. *Nature*, **419**, 199–206.

Lambeck, K. and Purcell, A.P. (2001) Sea-level change in the Irish Sea since the Last Glacial maximum: constraints from iostatic modelling. *Journal of Quaternary Science*, **16**, 497–506.

Lambeck, K., Esat, T.M. and Potter, E-K. (2002) Links between climate and sea levels for the past three million years. *Nature*, **419**, 199–206.

Lambrick, G. (1985) *Archaeology and Nature Conservation*. Oxford University Dept for Continuing Education, Oxford.

Landscheidt, T. (1984) Cycles of solar flares and weather. In *Climatic Changes on a Yearly to Millennial Basis* (edited by N-A. Mörner and W. Karlén), Reidel, Dordrecht, 473–81.

Lang, A. and Hönscheidt, S. (1999) Age and source of colluvial sediments at Vaihingen-Enz, Germany. *Catena*, **38**, 89–107.

Lang, A. and Nolte, S. (1999) The chronology of Holocene alluvial sediments from the Wetterau, Germany, provided by optical and ^{14}C dating. *The Holocene*, **9**, 207–14.

Lang, A. and Wagner, G.A. (1996) Infrared stimulated luminescence dating of archaeosediments. *Archaeometry*, **38** (1), 129–41.

Lang, A., Rieser, U., Habermann, J. and Wagner, G.A. (1998) Luminescence dating of sediments. *Naturwissenschaften*, **85**, 515–23.

Langdon, P.G., Barber, K.E. and Hughes, P.D.M. (2003) A 7500-year peat-based palaeoclimatic reconstruction and evidence for an 1100-year cyclicity in bog surface wetness from Temple Hill Moss, Pentland Hills, southeast Scotland. *Quaternary Science Reviews*, **22**, 259–74.

Lao, Y. and Benson, L. (1988) Uranium-series age estimates and paleoclimatic significance of Pleistocene tufas from the Lahontan Basin, California and Nevada. *Quaternary Research*, **30**, 165–76.

Larsson, L. (ed.) (2003) *Mesolithic on the Move*. Oxbow Books, Oxford.

Latham, A.G. (2001) Uranium-series dating. In *Handbook of Archaeological Sciences* (edited by D.R. Brothwell and A.M. Pollard), John Wiley, Chichester and New York, 63–72.

Lauritzen, S.-E. (1995) High-resolution palaeotemperature proxy record for the last interglaciation based on Norwegian speleothems. *Quaternary Research*, **43**, 133–46.

Lauritzen, S.-E. and Lundberg, J. (eds) (1999a) Speleothems as high-resolution palaeoclimatic archives. *The Holocene*, **9**, 643–722.

Lauritzen, S.E. and Lundberg, J. (1999b) Calibration of the speleothem delta function: an absolute temperature record for the Holocene in northern Norway. *The Holocene*, **9**, 659–70.

Le Roy Ladurie, E. (1972) *Times of Feast, Times of Famine*. Allen & Unwin, London.

Le Roy Ladurie, E. and Baulant, M. (1981) Grape harvests from the fifteenth through the nineteenth centuries. In *Climate and History: Studies in Interdisciplinary History* (edited by R.I. Rotberg and T.K. Rabb), Princeton University Press, Princeton, 259–69.

Lean, J. (1984) Solar ultraviolet irradiance variations and the earth's atmosphere. In *Climatic Changes on a Yearly to Millennial Basis* (edited by N-A. Mörner and W. Karlén), Reidel, Dordrecht, 449–71.

Lean, J. (1996) Reconstructions of past solar variability. In *Climatic Variations and Forcing Mechanisms of the Last 2000 Years* (edited by P.D. Jones, R.S. Bradley and J. Jouzel), NATO ASI Series 1, Global Environmental Change, Volume 41, Springer-Verlag, Berlin and Heidelberg, 519–32.

Lean, J. (2001) Solar irradiance and climate forcing in the near future. *Geophysical Research Letters*, **28**, 4119–22.

Lean, J., Skumanich, A. and White, O. (1992) Estimating the sun's radiative output during the Maunder Minimum. *Geophysical Research Letters*, **19**, 1591–4.

Lear, C.H., Elderfield, H. and Wilson, P.A. (2000) Cenozoic deep-sea temperatures and global ice volumes from Mg/Ca in benthic foraminiferal calcite. *Science*, **287**, 269–72.

Lee, R.B. and DeVore, I. (1968) *Man the Hunter*. Aldine, Chicago.

Leesch, D. (1997) *Un campement magdalénien au bord du lac de Neuchâtel cadre chronologique et culturel, mobilier et structures, analyse spatiale (secteur 1)*. Musée cantonal de Neuchâtel, Neuchâtel.

Legge, A.J. and Rowley-Conwy, P. (1988) *Star Carr Re-visited: A re-analysis of the large mammals*. Birkbeck College, London.

Lemdahl, G. (1991) A rapid climate change at the end of the Younger Dryas in south Sweden – palaeoclimatic and palaeoenvironmental reconstructions based on fossil insect assemblages. *Palaeogeography, Palaeoclimatology, Palaeoecology*, **83**, 313–31.

Leonard, E.M. and Reasoner, M.A. (1999) A continuous Holocene glacial record inferred from proglacial lake sediments in Banff National Park, Alberta, Canada. *Quaternary Research*, **51**, 1–13.

Leopold, E.B. and Boyd, R. (1999) An ecological history of old prairie areas in southwestern Washington. In *Fire and the Land* (edited by R. Boyd), Oregon State University Press, Corvallis, Oregon, 139–63.

Leveau, P., Trément, F., Walsh, K. and Barker, G. (1999) *Environmental Reconstruction in Mediterranean Landscape Archaeology*. Oxbow, Oxford.

Leverington, D.W., Mann, J.D. and Teller, J.T. (2000) Changes in the bathymetry and volume of Glacial Lake Agassiz between 11,000 and 9300 ^{14}C yrs BP. *Quaternary Research*, **54**, 174–81.

Levesque, A., Cwynar, L.C. and Walker, I.R. (1997) Exceptionally steep north–south gradients in lake temperatures during the last deglaciation. *Nature*, **385**, 423–6.

Levesque, A., Mayle, F.E. and Cwynar, L.C. (1993) The Amphi-Atlantic Oscillation: a proposed Late-glacial climatic event. *Quaternary Science Reviews*, **12**, 629–44.

Levitus, S., Antonov, J.I., Boyer, T.P. and Stephens, C. (2000) Warming of the world ocean. *Science*, **287**, 2225–9.

Lewin, J. and Macklin, M.G. (2003) Preservation potential for Late Quaternary river alluvium. *Journal of Quaternary Science*, **18** (2), 107–20.

Lewin, J., Macklin, M.G. and Woodward, J.C. (1995a) Quaternary fluvial systems in the Mediterranean basin. In *Mediterranean Quaternary River Environments* (edited by J. Lewin, M.G. Macklin and J.C. Woodward), A.A. Balkema, Rotterdam, 1–28.

Lewin, J., Macklin, M.G. and Woodward, J.C. (eds) (1995b) *Mediterranean Quaternary River Environments*. A.A. Balkema, Rotterdam.

Lewis, H.T. (1993) Patterns of Indian burning in California: ecology and ethnohistory. In *Before the Wilderness* (edited by T.C. Blackburn and K. Anderson), Balkema Press, Menlo Park, California, 55–116.

Lewis, H.T. and Ferguson, T.A. (1999) Yards, corridors and mosaics. In *Indians, Fire and the Land* (edited by R. Boyd), Oregon State University Press, Corvallis, Oregon, 164–84.

Lewis, J.S. (1996) *Rain of Iron and Ice: The very real threat of comet and asteroid bombardment*. Addison Wesley, Reading, Massachusetts.

Licciardi, J.M., Teller, J.T. and Clark, P.U. (1999) Freshwater routing by the Laurentide Ice Sheet during the last deglaciation. In *Mechanisms of Global Climate Change at Millennial Timescales* (edited by P.U. Clark, R.S. Webb and L.D. Keigwin), Geophysical Monograph 122, American Geophysical Union, Washington, 177–202.

Likens, G.E., Driscoll, C.T. and Buso, D.C. (1996) Long-term effects of acid rain: response and recovery of a forest ecosystem. *Science*, **272**, 244–6.

Limbrey, S. (1975) *Soil Science and Archaeology*. Academic Press, London.

Lipps, J.H. (ed.) (1993) *Fossil Prokaryotes and Protists*. Blackwell, London.

Lister, A.M. (1992) Mammalian fossils and Quaternary biostratigraphy. *Quaternary Science Reviews*, **11**, 329–44.

Lister, A.M. and Bahn, P. (1994) *Mammoths*. Macmillan, New York.

Lister, A.M. and Sher, A.V. (1995) Ice cores and mammoth extinction. *Nature*, **378**, 23–4.

Litt, T. (ed.) (2003) Environmental response to climate and human impact in central Europe during the last 15,000 years – a German contribution to PAGES-PEPIII. *Quaternary Science Reviews*, **22**, 1–124.

Liu, T., Ding, N. and Rutter, N.W. (1999) Comparison of Milankovitch periods between continental loess and deep sea records over the last 2.5 Ma. *Quaternary Science Reviews*, **18**, 1205–12.

Liversage, D. and Robinson, D. (1993) Prehistoric settlement and landscape development in the sandhill belt of south Thy. *Journal of Danish Archaeology*, **11**, 39–56.

Loader, N.J. and Switsur, V.R. (1995) Reconstructing past environmental change using stable isotopes in tree rings. *Botanical Journal of Scotland*, **48**, 65–78.

Loader, N.J., Switsur, V.R. and Field, E.M. (1995) High-resolution stable isotope analysis of tree rings: implications of 'microdendroclimatology' for palaeoenvironmental research. *The Holocene*, **95**, 457–60.

Lockwood, J.G. (1983) Modelling climate change. In *Background to Palaeohydrology* (edited by K.J. Gregory), John Wiley, Chichester and New York, 25–50.

Long, A.J., Innes, J.B., Kirby, J.R., Lloyd, J.M., Rutherford, M.M., Shennan, I. and Tooley, M.J. (1998) Holocene sea-level change and coastal evolution in the Humber Estuary, eastern England: an assessment of rapid coastal change. *The Holocene*, **8**, 229–47.

Long, A.J., Scaife, R.G. and Edwards, K.J. (2000) Stratigraphic architecture, relative sea-level, and models of estuary development in southern England: new data from Southampton Water. In *Coastal and Estuarine Environments: Sedimentology, geomorphology and geoarchaeology* (edited by K. Pye and J.R.L. Allen), Geological Society Special Publication No. 175, London, 253–80.

Lotter, A. and Kienast, F. (1990) Validation of a forest succession model by means of annually laminated sediments. In *Laminated Sediments* (edited by M. Saarnisto and A. Kahra), Geological Survey of Finland, Special Paper 14, Espoo, 25–32.

Lotter, A.F., Birks, H.J.B., Hofmann, W. and Marchetto, A. (1997) Modern diatom, cladocera, chironomid and chrysophyte cyst assemblages as quantitative indicators for the reconstruction of past environmental conditions in the Alps. I. Climate. *Journal of Paleolimnology*, **1**, 395–420.

Lotter, A.F., Eicher, U., Siebenthaler, U. and Birks, H.J.B. (1992) Late-glacial climatic oscillations as recorded in Swiss lake sediments. *Journal of Quaternary Science*, **7**, 187–204.

Lotter, A.F., Walker, I.R., Brooks, S.J. and Hofmann, W. (1999) An intercontinental comparison of chironomid palaeotemperature inference models: Europe vs North America. *Quaternary Science Reviews*, **18**, 717–35.

Lõugas, L., Ukkonen, P. and Jungner, H. (2002) Dating the extinction of European mammoths: new evidence

from Estonia. *Quaternary Science Review*, **21**, 1347–54.
Louwe Kooijmans, L.P. (1974) *The Rhine Meuse Delta*. Leiden University Press (Analecta Praehistorica Leidensia VII), Leiden.
Louwe Kooijmans, L.P. (1980) Archaeology and coastal change in the Netherlands. In *Archaeology and Coastal Change* (edited by F.H. Thompson), Society of Antiquaries, London, 106–33.
Louwe Kooijmans, L.P. (1985) *Sporen in het land: De Nederlandse delta in de prehistorie*. Meulenhoff Informaticf, Amsterdam.
Louwe Kooijmans, L.P. (1987) Neolithic settlement and subsistence in the wetlands of the Rhine–Meuse Delta of the Netherlands. In *European Wetlands in Prehistory* (edited by J.M. Coles and A.J. Lawson), Clarendon Press, Oxford, 227–52.
Louwe Kooijmans, L.P. (1993) Wetland exploitation and upland relations of prehistoric communities in the Netherlands. In *Flatlands and Wetlands* (edited by J. Gardiner), *East Anglian Archaeology*, **50**, 71–116.
Louwe Kooijmans, L.P. (1995) Prehistory or paradise? In *Wetlands: Archaeology and nature conservation* (edited by M. Cox, V. Straker and D. Taylor), HMSO, London, 3–17.
Louwe Kooijmans, L.P. (2003) The Hardinxveld sites in the Rhine/Meuse Delta, The Netherlands, 5500–4500 cal BC. In *Mesolithic on the Move* (edited by L. Larsson), Oxbow Books, Oxford, 608–24.
Love, J.A. (1983) *The Return of the Sea Eagle*. Cambridge University Press, Cambridge.
Lowe, D.J., Newnham, R.M., McFadgen, B.G. and Higham, T.F.G. (2000) Tephras and New Zealand archaeology. *Journal of Archaeological Science*, **27**, 859–70.
Lowe, J.J. (ed.) (1994) North Atlantic Seaboard Programme IGCP-253 'Climate changes in areas adjacent to the North Atlantic during the Last Glacial-Interglacial Transition'. *Journal of Quaternary Science*, **9**, 95–198.
Lowe, J.J. (2001) Quaternary geochronological frameworks. In *Handbook of Archaeological Sciences* (edited by D.R. Brothwell and A.M. Pollard), John Wiley, Chichester and New York, 9–21.
Lowe, J.J. and Walker, M.J.C. (1997) *Reconstructing Quaternary Environments*, 2nd edition. Addison-Wesley-Longman, London.
Lowe, J.J., Hoek, W. and INTIMATE Members (2001) Inter-regional correlation of palaeoclimatic records for the Last Glacial–Interglacial Transition: a protocol for improved precision recommended by the INTIMATE project group. *Quaternary Science Reviews*, **20**, 1175–88.
Lozek, V. (1986) Mollusca analysis. In *Handbook of Holocene Palaeoecology and Palaeohydrology* (edited by B.E. Berglund), John Wiley & Sons, Chichester and New York, 729–40.
Luckman, B.H., Briffa, K.R., Jones, P.D. and Schweingruber, F.H. (1997) Tree-ring based reconstruction of summer temperatures at the Columbia Icefield, Alberta, Canada. *The Holocene*, **7**, 375–90.
Lundelius, E.L. (1989) The implications of disharmonious assemblages for Pleistocene extinctions. *Journal of Archaeological Science*, **16**, 407–17.
Lundqvist, J. (1986) Late Weichselian glaciation and deglaciation in Scandinavia. *Quaternary Science Reviews*, **5**, 269–92.
Lyell, C. (1830–33) *Principles of Geology*. 3 volumes. Murray, London.
Lyman, R.L. (1994) *Vertebrate Taphonomy*. Cambridge University Press, Cambridge.
Lyman, R.L. (1996) Applied zooarchaeology: the relevance of faunal analysis to wildlife management. *World Archaeology*, **28**, 110–25.
Maarleveld, G.C. (1976) Periglacial phenomena and the mean annual temperature during the last glacial time in the Netherlands. *Biuletyn Peryglacjalny*, **26**, 57–78.
McAndrews, J.H. (1988) Human disturbance of North American forests and grasslands: the fossil pollen record. In *Vegetation History* (edited by B. Huntley and T. Webb), Kluwer, Dordrecht, 674–97.
McCabe, A.M. and Clark, P.U. (1998) Ice-sheet variability around the North Atlantic Ocean during the last deglaciation. *Nature*, **392**, 373–77.
McCabe, M., Knight, J. and McCarron, S. (1998) Evidence for Heinrich event 1 in the British Isles. *Journal of Quaternary Science*, **13**, 549–68.
McCarroll, D. and Ballantyne, C.K. (2000) The last ice sheet in Snowdonia. *Journal of Quaternary Science*, **15**, 765–78.
McCarroll, D. and Pawellek, F. (2001) Stable carbon isotope ratios of *Pinus sylvestris* from northern Finland and the potential for extracting a climate signal from long Fennoscandian chronologies. *The Holocene*, **11**, 517–26.
McCarroll, D., Knight, J. and Rijsdijk, K. (eds) (2001) The glaciation of the Irish Sea. *Journal of Quaternary Science*, **16**, 391–506.
McCarthy, J.J., Canziani, O.F., Leary, N., Dokken, D.J. and White, K.S. (eds) (2001) *Climate Change 2001: Impacts, adaptation and vulnerability*. Cambridge University Press, Cambridge.
McCave, I.N., Manighetti, B. and Beveridge, N.A.S. (1995) Circulation in the glacial North Atlantic inferred from grain-size measurements. *Nature*, **374**, 149–52.

McClung de Tapia, E. (1992) The origins of agriculture in Mesoamerica and Central America. In *The Origins of Agriculture* (edited by C.W. Cowan and P.J. Watson), Smithsonian Institution Press, Washington, 143–72.

McCoy, F.W. and Heiken, G. (2000a) The Late-Bronze Age explosive eruption of Thera (Santorini), Greece: regional and local effects. In *Volcanic Hazards and Disasters in Human Antiquity* (edited by F.W. McCoy and G. Heiken), The Geological Society of America, Boulder, Colo., 43–70.

McCoy, F.W. and Heiken, G. (eds) (2000b) *Volcanic Hazards and Disasters in Human Antiquity*. The Geological Society of America, Boulder, Colo.

McCracken Peck, R. (1990) *Land of the Eagle: A natural history of North America*. BBC Books, London.

McDermott, F., Mattey, D.P. and Hawksworth, C. (2001) Centennial-scale Holocene climate variability revealed by a high-resolution speleothem ^{18}O record from SW Ireland. *Science*, **294**, 1328–31.

McErlean, T., McConkey, R. and Forsythe, W. (2002) *Strangford Lough: An archaeological survey of the maritime cultural landscape*. Environment and Heritage Service, Belfast.

McGimsey, C.R. and Davis, H.A. (1984) United States of America. In *Approaches to the Archaeological Heritage* (edited by H.F. Cleere), Cambridge University Press, Cambridge, 116–24.

McGlade, J. (1995) Archaeology and the ecodynamics of human-modified landscapes. *Antiquity*, **69**, 113–32.

McGlade, J. (1999) The times of history: archaeology, narrative and non-linear causality. In *Time and Archaeology* (edited by T. Murray), Routledge, London, 139–63.

McGlade, J. and van der Leeuw, S.E. (eds) (1997a) *Time, Process and Structural Transformation in Archaeology*. Routledge, London.

McGlade, J. and van der Leeuw, S.E. (1997b) Introduction: archaeology and non-linear dynamics – new approaches to long-term change. In *Time Process and Structural Transformation in Archaeology* (edited by S.E. van der Leeuw and J. McGlade), Routledge, London, 1–31.

McGovern, T.H. (1992) Bones, buildings and boundaries: palaeoeconomic approaches to Norse Greenland. In *Norse and Later Settlement and Subsistence in the North Atlantic* (edited by C.D. Morris and D.J. Rackham), Department of Archaeology, University of Glasgow, Glasgow, 193–224.

McGuire, W.J., Griffiths, D.R., Hancock, P.L. and Stewart, I.S. (eds) (2000) *The Archaeology of Geological Catastrophes*. Geological Society, Special Publications 171, London.

Macinnes, L. and Wickham-Jones, C.R. (1992) *All Natural Things: Archaeology and the green debate*. Oxbow, Oxford.

Macklin, M.G. (1999) Holocene river environments in prehistoric Britain: human interaction and impact. *Journal of Quaternary Science*, **14**, 521–30.

Macklin, M.G. and Lewin, J. (1986) Terraced fills of Pleistocene and Holocene age in the Rheidol Valley, Wales. *Journal of Quaternary Science*, **1**, 21–34.

Macklin, M.G. and Lewin, J. (1993) Holocene river alluviation in Britain. *Zeitschrift für Geomorphologie* (Supplement), **88**, 109–22.

Macklin, M.G. and Lewin, J. (2003) River sediments, great floods and centennial-scale Holocene climate change. *Journal of Quaternary Science*, **18**, 101–5.

Macklin, M.G., Lewin, J. and Woodward, J.C. (1995) Quaternary fluvial systems in the Mediterranean basin. In *Mediterranean Quaternary River Environments* (edited by J. Lewin, M.G. Macklin and J.C. Woodward), AA Balkema, Rotterdam, 1–28.

McManamon, F.P. (2000) The protection of archaeological resources in the United States: reconciling preservation with contemporary society. In *Cultural Resource Management in Contemporary Society* (edited by F.P. McManamon and A. Hutton), Routledge, London, 40–54.

McManus, J.F., Oppo, D.W. and Keigwin, L.D. (2002) Thermohaline circulation and prolonged interglacial warmth in the North Atlantic. *Quaternary Research*, **58**, 17–21.

McNeely, J.A. and Keeton, M.S. (1995) The interaction between biological and cultural diversity. In *Cultural Landscapes of Universal Value* (edited by B. von Droste, H. Plachter and M. Rossler), Fischer, Stuttgart, 25–38.

Macphail, R.I. (1986) Palaeosols in archaeology: their role in understanding Flandrian pedogenesis. In *Palaeosols: Their recognition and interpretation* (edited by V.P. Wright), Blackwell, Oxford, 263–90.

Macphail, R.I. (1987) A review of soil science in archaeology in England. In *Environmental Archaeology: A regional review* (edited by H.C.M. Keeley), Historic Buildings and Monuments Commission for England, London, 332–79.

Macphail, R.I. (1998) A reply to Carter and Davidson's 'An evaluation of the contribution of soil micromorphology to the study of ancient arable agriculture'. *Geoarchaeology*, **13**, 549–64.

Macphail, R.I. and Cruise, G.M. (1996) Soil micromorphology. In *The Experimental Earthwork Project* (edited by M. Bell, P.J. Fowler and S.W. Hillson), CBA Research Report 100, York, 95–107.

MacPhee, R.D.E. (ed.) (1999) *Extinctions in Near Time: Causes, contexts and consequences*. Kluwer Academic Press, New York.

Macpherson, J. (1985) The postglacial development of vegetation in Newfoundland and eastern Labrador-Ungava: synthesis and climatic implications. *Syllogeus*, **55**, 267–280.

Maddy, D. (1997) Uplift-driven valley incision and river terrace formation in southern England. *Journal of Quaternary Science*, **12**, 539–45.

Maddy, D. and Bridgland, D.R. (2000) Accelerated uplift resulting from Anglian glacioisostatic rebound in the Middle Thames valley, UK?: evidence from the river terrace record. *Quaternary Science Reviews*, **19**, 1581–88.

Maddy, D.M., Lewis, S.G., Scaife, R.G., Bowen, D.Q., Coope, G.R., Green, C.P., Hardaker, T., Keen, D.H., Rees-Jones, J., Parfitt, S. and Scott, K. (1998) The Upper Pleistocene deposits at Cassington, near Oxford, England. *Journal of Quaternary Science*, **13**, 205–32.

Madole, R.F. (1995) Spatial and temporal patterns of Late Quaternary eolian deposition, eastern Colorado, USA. *Quaternary Science Reviews*, **14**, 155–77.

Madronich, S., McKenzie, R.L., Caldwell, M.M. and Björn, L.O. (1995) Changes in ultraviolet radiation reaching the earth's surface. *Ambio*, **24**, 143–52.

Maher, B. and Thompson, R. (eds) (1999) *Quaternary Climates, Environments and Magnetism*. Cambridge University Press, London.

Maier, U. (1999) Agricultural activities and landuse in a Neolithic village around 3900 BC: Hornstaad Hörnle I A, Lake Constance, Germany. *Vegetation History and Archaeobotany*, **8**, 87–94.

Maier, U. and Vogt, R. (2001) Reconstructing the Neolithic landscape at Western Lake Constance. In *Estuarine Archaeology: The Severn and beyond*. Archaeology in the Severn Estuary 11 (edited by S. Rippon), SELRC, Exeter, 121–30.

Maizels, J. (1997) Jökulhlaup deposits in proglacial areas. *Quaternary Science Reviews*, **16**, 793–819.

Maizels, J. and Aitken, J. (1991) Palaeohydrological change during deglaciation in upland Britain: a case study from northeast Scotland. In *Temperate Palaeohydrology* (edited by L. Starkel, J.B. Thornes and K.J. Gregory), John Wiley & Sons, Chichester and New York, 105–45.

Mangerud, J. (1991) The Scandinavian Ice Sheet through the last interglacial/glacial cycle. In *Klimageschichtliche Probleme der Letzen 130,000 Jahre* (edited by B. Frenzel), G. Fischer, Stuttgart and New York, 307–30.

Mangerud, J., Astakhov, V. and Svendsen, J-I. (2002) The extent of the Barents-Kara ice sheet during the Last Glacial Maximum. *Quaternary Science Reviews*, **21**, 111–19.

Manley, G. (1974) Central England temperatures: monthly means 1659–1973. *Quarterly Journal of the Royal Meteorological Society*, **100**, 389–405.

Mann, M.E., Bradley, R.S. and Hughes, M.K. (1998) Global-scale temperature patterns and climatic forcing over the past six centuries. *Nature*, **392**, 779–87.

Mann, M.E., Gille, E., Bradley, R.S., Hughes, M.K., Overpeck, J.T., Keimig, F.T. and Gross, W. (2000) Global temperature patterns in past centuries: an interactive presentation. *Earth Interactions*, **4** (4), 1–29.

Manne, A.S. and Richels, R.G. (1999) An alternative approach to establishing trade-offs among greenhouse gases. *Nature*, **410**, 675–7.

Manning, S.W. (1999) *A Test of Time*. Oxbow Books, Oxford.

Manning, S.W., Kromer, B., Kuniholm, P.I. and Newton, M.W. (2001) Anatolian tree rings and a new chronology for the East Mediterranean Bronze–Iron Ages. *Science*, **294**, 2532–5.

Marcus, J. and Flannery, K.V. (1996) *Zapotec Civilisation: How urban society ended in Mexico's Oaxaca Valley*. Thames & Hudson, New York.

Marguerie, D. (1992) *Evolution de la végétation sous l'impact humain en Armorique du Néolithique aux périodes historiques*. Travaux du Laboratoire d'Anthropologie, Préhistoire et Quaternaire Amoricains, 40, Rennes.

Marinatos, S. (1939) The volcanic destruction of Minoan Crete. *Antiquity*, **13**, 425–39.

Marquardt, W.H. (1996) Four discoveries: environmental archaeology in southwest Florida. In *Case Studies in Environmental Archaeology* (edited by E.J. Reitz, L.E. Newsom and S.J. Scudder), Plenum Press, London, 17–34.

Marshall, S.J., James, T.S. and Clarke, G.K.C. (2002) North American ice sheet reconstruction at the Last Glacial Maximum. *Quaternary Science Reviews*, **21**, 175–92.

Marten, G.G. (2001) *Human Ecology: Basic concepts for sustainable development*. Earthscan, London.

Martin, P.S. (1967) Prehistoric overkill. In *Pleistocene Extinctions: The search for a cause* (edited by P.S. Martin and H.E. Wright), Yale University Press, New Haven, 75–120.

Martin, P.S. (1984) Prehistoric overkill: the global model. In *Quaternary Extinctions* (edited by P.S. Martin and R.G. Klein), University of Arizona Press, Tucson, 354–403.

Martin, P.S. and Klein, R.G. (eds) (1984) *Quaternary Extinctions*. University of Arizona Press, Tucson.

Martin, P.S. and Steadman, D.W. (1999) Prehistoric extinctions on islands and continents. In *Extinctions in Near Time: Causes, contexts and consequences* (edited by R.D.E. MacPhee), Kluwer Academic Press, New York, 17–56.

Martin, P.S. and Wright, H.E. (1967) *Pleistocene Extinctions: The search for a cause.* Yale University Press, New Haven.

Martin, P.S., Thompson, R.S. and Long, A. (1985) Shasta ground sloth extinction: a test of the blitzkreig model. In *Environments and Extinctions: Man in Late-glacial North America* (edited by J.I. Mead and D.J. Meltzer), Center for the Study of Early Man, Orono, 5–14.

Martínez-Cortizas, A., Pontevedra-Pombal, X., García-Rodeja, E., Nóvoa-Muñoz, J.C. and Shotyk, W. (1999) Mercury in a Spanish peat bog: archive of climate change and atmospheric metal deposition. *Science*, **284**, 939–42.

Martinson, D.G., Pisias, N.G., Hays, J.D., Imbrie, J., Moore, T.C. and Shackleton, N.J. (1987) Age dating and the orbital theory of ice ages: development of a high-resolution 0–300,000 year chronostratigraphy. *Quaternary Research*, **27**, 1–29.

Maslin, M.A., Li, X.S., Loutre, M-F. and Berger, A. (1998) The contribution of orbital forcing to the progressive intensification of northern hemisphere glaciation. *Quaternary Science Reviews*, **17**, 411–26.

Mathews, R.W. (1993) Evidence for Younger-Dryas age cooling on the north Pacific coast of America. *Quaternary Science Reviews*, **12**, 321–32.

Matthews, J.A. (1992) *The Ecology of Recently Deglaciated Terrain.* Cambridge University Press, Cambridge.

Matthews, J.A. (1997) A preliminary history of Holocene colluvial (debris-flow) activity, Leirdalen, Jotunheimen, Norway. *Journal of Quaternary Science*, **12**, 117–30.

Matthews, J.A., Dahl, S.O., Nesje, A., Berrisford, M.S. and Andersson, C. (2000) Holocene glacier variations in central Jotunheimen, southern Norway based on distal glaciolacustrine sediment cores. *Quaternary Science Reviews*, **19**, 1625–47.

Matthews, W., French, C.A.I., Lawrence, T., Cutler, D.F. and Jones, M.K. (1997) Microstratigraphic traces of site formation processes and human activities. *World Archaeology*, **29**, 281–30.

Mauquoy, D. and Barber, K.E. (1999) Evidence for climatic deterioration associated with the decline of *Sphagnum imbricatum* Hornsch. Ex. Russ. in six ombrotrophic mires from Northern England and the Scottish Borders. *The Holocene*, **9**, 423–37.

Mayewski, P., Meeker, L.D., Twickler, M.S., Whitlow, S., Yang, Q. Lyons, W.B. and Prentice, M. (1997) Major features and forcing of high-latitude northern hemisphere circulation using a 110,000-year-long glaciological series. *Journal of Geophysical Research*, **102**, 26345–66.

Mead, J.I. and Meltzer, D.J. (1984) North American late Quaternary extinctions and the ^{14}C record. In *Quaternary Extinctions* (edited by P.S. Martin and R.G. Klein), University of Arizona Press, Tucson, 440–50.

Mead, J.I. and Meltzer, D.J. (eds) (1985) *Environments and Extinctions: Man in Late-glacial North America.* Center for the Study of Early Man, Orono.

Meese, D.A., Gow, A.J., Aley, R.B., Zielinski, G.A., Grootes, P.M., Ram, M., Taylor, K.C., Mayewski, P.A. and Bolzan, J.F. (1998) The Greenland Ice Sheet project 2 depth-age scale: methods and results. *Journal of Geophysical Research*, **102**, 26411–24.

Meldgaard, M. and Rasmussen, M. (1996) *Arkæologiske eksperimenter i Lejre.* RHODOS, Lejre.

Mellars, P. (1976) Fire, ecology, animal populations and man: a study of some ecological relationships in prehistory. *Proceedings of the Prehistoric Society*, **42**, 15–45.

Mellars, P. (1987) *Excavations on Oronsay: Prehistoric human ecology on a small island.* Edinburgh University Press, Edinburgh.

Mellars, P. and Dark, P. (1998) *Star Carr in Context.* McDonald Institute Monograph, Cambridge.

Meltzer, D.J. and Mead, J.I. (1985) Dating Late Pleistocene extinctions: theoretical issues, analytical bias and substantive results. In *Environments and Extinctions: Man in Late-glacial North America* (edited by J.I. Mead and D.J. Meltzer), Center for the Study of Early Man, Orono. 145–174.

Menounos, B. and Reasoner, M.A. (1997) Evidence for cirque glaciation in the Colorado Front Range during the Younger Dryas chronozone. *Quaternary Research*, **48**, 38–47.

Mercer, R. and Healy, F. (forthcoming) *Hambledon Hill, Dorset, England: Excavation and survey of a Neolithic monument complex and its surrounding landscape.* English Heritage, London.

Merkt, J. and Muller, H. (1994) Laminated sediments in South Germany from the Neolithic to Hallstatt period. In *Laminated Sediments* (edited by S. Hicks, U. Miller and M. Saarnisto), PACT 41, Belgium, 101–16.

Merritt, J.W. (1992) The high-level marine shell-bearing deposits of Clava, Inverness-shire, and their origin as glacial rafts. *Quaternary Science Reviews*, **11**, 759–80.

Mesolella, K.J., Matthews, R.K., Broecker, W.S. and Thurber, D.L. (1969) The astronomical theory of climatic change: Barbados data. *Journal of Geology*, **77**, 250–74.

Metz, B., Davidson, O., Swart, R. and Pan, J. (eds) (2001) *Climate Change 2001: Mitigation*. Cambridge University Press, Cambridge.

Mikolajewicz, U., Maier-Reimer, E., Crowley, T.J. and Kim, K.Y. (1993) Effects of Drake and Panamanian gateways on the circulation of an ocean model. *Paleoceanography*, **8**, 409–27.

Miller, G.H., Magee, J.W., Johnson, B.J., Fogel, M.L., Spooner, N.A., McCulloch, M.T. and Aycliffe, L.K. (1999a) Pleistocene extinction of *Genyornis newtoni*: human impact on Australian megafauna. *Science*, **283**, 205–8.

Miller, G.H., Mode, W.N., Wolfe, A.P., Sauer, P.E., Bennike, O., Forman, S.L., Short, S.K. and Stafford, T.W. Jnr (1999b) Stratified interglacial lacustrine sediments from Baffin Island, Arctic Canada: chronology and palaeoenvironmental implications. *Quaternary Science Reviews*, **18**, 789–810.

Miller, N.F. and Gleason, K.L. (1994) *The Archaeology of Garden and Field*. University of Pennsylvania Press, Philadelphia.

Miller, R.F. and Elias, S.A. (2000) Late-glacial climate in the Maritime Region, Canada, reconstructed from mutual climate range analysis of fossil Coleoptera. *Boreas*, **29**, 79–88.

Milly, P.C.D., Wetherald, R.T., Dunne, K.A. and Delworth, T.L. (2002) Increasing risk of great floods in a changing climate. *Nature*, **415**, 514–17.

Milne, G.A., Davis, J.L., Mitrovica, J.X., Scherneck, H-G., Johansson, J.M., Vermeer, M. and Koivula, H. (2001) Space-geodetic constraints on glacial isostatic adjustment in Fennoscandia. *Science*, **291**, 2381–5.

Milne, G.A., Mitrovica, J.X. and Schrag, D.P. (2002) Estimating past continental ice volume from sea-level data. *Quaternary Science Reviews*, **21**, 361–76.

Minnis, P.E. (1992) Earliest plant cultivation in the desert borderlands of North America. In *The Origins of Agriculture* (edited by C.W. Cowan and P.J. Watson), Smithsonian Institution Press, Washington, 121–42.

Minns, C.K., Moore, J.E., Schindler, D.W. and Jones, L.M. (1990) Assessing the potential extent of damage to inland lakes in eastern Canada due to acidic deposition. III. Predicted impacts on species richness and aquatic biota. *Canadian Journal of Fisheries and Aquatic Science*, **47**, 821–30.

Minor, R. and Grant, W.C. (1996) Earthquake-induced subsidence and burial of Late Holocene archaeological sites, Northern Oregon coast. *American Antiquity*, **61** (4), 772–81.

Miramount, C., Sivan, O., Rosique, T., Edouard, J.L. and Jorda, M. (2000) Subfossil tree deposits in the middle Durance (South Alps, France): environmental change Allerød to Atlantic. *Radiocarbon*, **42** (3), 423–35.

Mitchell, F.J.G. (1986) *Reading the Irish Landscape*. Country House, Dublin.

Mitchell, J.F.B. (1989) The 'greenhouse' effect and climate change. *Reviews of Geophysics*, **27**, 115–39.

Mitchell, J.F.B. (1993) Modelling of palaeoclimates: examples from the recent past. *Philosophical Transactions of the Royal Society, London*, **B341**, 267–75.

Mithen, S. (1990) *Thoughtful foragers: A study of prehistoric decision making*. Cambridge University Press, Cambridge.

Mithen, S. (1991) Ecological interpretations of Palaeolithic art. *Proceedings of the Prehistoric Society*, **57** (1), 103–14.

Mithen, S. (1996) Simulating mammoth hunting and extinctions: implications for North America. In *Time, Process and Structured Transformation in Archaeology* (edited by S.E. van der Leeuw and J. McGlade), Routledge, London, 176–214.

Mithen, S. (1998) *The Prehistory of Mind*. Orion, London.

Mithen, S. (2000) *Hunter-gatherer landscape archaeology: The Southern Hebrides Mesolithic Project 1988–98*, Volumes 1 and 2. McDonald Institute, Cambridge.

Mithen, S. (2003) *After the Ice: A global human history, 20,000–5000 BC*. Weidenfeld & Nicolson, London.

Mix, A., Bard, E. and Schneider, R. (2001) Environmental processes of the ice age: land, oceans, glaciers (EPILOG). *Quaternary Science Reviews*, **20**, 627–59.

Moe, D. and Rackham, O. (1992) Pollarding and a possible explanation of the Neolithic elm fall. *Vegetation History and Archaeobotany*, **1**, 63–8.

Moffet, L., Robinson, M.A. and Straker, V. (1989) Cereals, fruit and nuts. In *The Beginnings of Agriculture* (edited by A. Milles, D. Williams and N. Gardner), British Archaeological Reports IS 496, Oxford, 243–61.

Moghissi, A.A. (1986) Potential public health impacts of acid deposition. *Water Quality Bulletin*, **11**, 3–5.

Molfino, B., Heusser, L.H. and Woillard, G.M. (1984) Frequency components of a Grande Pile pollen record: evidence of precessional orbital forcing. In *Milankovitch and Climate* (edited by A. Berger, J. Hays, J. Imbrie, G. Kukla and B. Saltzman), Reidel, Dordrecht, 391–404.

Molloy, K. and O'Connell, M. (1993) Early land use and vegetation history at Derryinver Hill, Renvyle Peninsula, Co. Galway, Ireland. In *Climate Change and Human Impact on the Landscape* (edited by F.M. Chambers), Chapman & Hall, London, 185–204.

Molloy, K. and O'Connell, M. (1995) Palaeoecological investigations towards the reconstruction of environment and land-use changes during prehistory at Céide Fields, western Ireland. *Probleme der Küstenforschung im südlichen Nordseegebiet*, **23**, 187–225.

Montzka, S.A., Butler, J.H., Elkins, J.W., Thompson, T.M., Clarke, A.D. and Lock, L.T. (1999) Present and future trends in the atmospheric burden of ozone-depleting halogens. *Nature*, **398**, 690–4.

Moody, J. and Grove, A.T. (1990) Terraces and enclosure walls in the Cretan landscape. In *Man's Role in Shaping the Eastern Mediterranean Landscape* (edited by S. Bottema, G. Entjes-Nieborg and W. van Zeist), AA Balkema, Rotterdam, 183–91.

Mooney, H.A. and Ehrlich, P.R. (1997) Ecosystem services: a fragmentary history. In *Nature's Services: Societal dependence on natural ecosystems* (edited by G.C. Daily), Island Press, Washington, DC, 11–19.

Moore, A.M.T. and Hillman, G.C. (2000) The Pleistocene to Holocene transition and human economy in Southwest Asia: the impact of the Younger Dryas. *American Antiquity*, **57**, 482–94.

Moore, A.M.T., Hillman, G.C. and Legge, A.J. (2000) *Village on the Euphrates*. Oxford University Press, Oxford.

Moore, P.D. (1985) Forests, man and water. *International Journal of Environmental Studies*, **25**, 159–66.

Moore, P.D. (1986) Hydrological changes in mires. In *Handbook of Holocene Palaeoecology and Palaeohydrology* (edited by B.E. Berglund), John Wiley, Chichester and New York, 92–107.

Moore, P.D. (1993) The origin of blanket mire, revisited. In *Climate Change and Human Impact on the Landscape* (edited by F.M. Chambers), Chapman & Hall, London, 217–24.

Moore, P.D., Webb, J.A. and Collinson, M.D. (1991) *Pollen Analysis*, 2nd edition. Blackwell, Oxford.

Morel, P. and Müller, W. (1997) *Un campement magdalénien au bord du lac de Neuchâtel étude archéozoologique*. Musée contonal d'archéologie, Neuchâtel.

Morgan, A.V. (1987) Late Wisconsin and early Holocene palaeoenvironments of east-central North America based on assemblages of fossil Coleoptera. In *The Geology of North America*, Volume K-3, *North America and Adjacent Oceans during the Last Deglaciation* (edited by W.F. Ruddiman and H.E. Wright, Jnr), Geological Society of America, Boulder, Colo., 353–70.

Morgan, R.A., Litton, C.D. and Salisbury, C.R. (1987) Trackways and tree trunks – dating Neolithic oaks in the British Isles. *Tree-ring Bulletin*, **47**, 61–9.

Morgan, R.P.C. (1995) *Soil Erosion and Conservation*, 2nd edition. Longman, Harlow.

Mörner, N-A. (1980a) Eustasy and geoid changes as a function of core/mantle changes. In *Earth Rheology, Isostasy and Eustasy* (edited by N-A. Mörner), John Wiley, Chichester and New York, 535–53.

Mörner, N-A. (1980b) The Northwest European sea-level laboratory and regional Holocene eustasy. *Palaeogeography, Palaeoclimatology, Palaeoecology*, **29**, 218–300.

Mörner, N-A. (1991) Course and origin of the Fennoscandian uplift: the case for two separate mechanisms. *Terra Nova*, **3**, 408–13.

Mott, R.J. (1994) Wisconsinan Late-glacial environmental change in Nova Scotia: a regional synthesis. *Journal of Quaternary Science*, **9**, 155–60.

Movius, H.L. (1977) Excavation of the Abri Pataud, Les Eyzies (Dordogne): Stratigraphy. *Bulletin of the American School of Prehistoric Research*, **31**.

Mulholland, M.T. (1988) Territoriality and horticulture: a perspective for prehistoric southern New England. In *Holocene Human Ecology in Northeastern North America* (edited by G.P. Nicholas), Plenum Press, New York, 137.

Müller, R., Crutzen, P.J., Grooß, J-U., Brühl, C., Russell, J.M. III, Gernmandt, H., McKenna, D.S. and Tuck, A.F. (1997) Severe chemical ozone loss in the Arctic during the winter of 1995–6. *Nature*, **389**, 709–12.

Munaut, A.V. (1986) Dendrochronology applied to mire environments. In *Handbook of Holocene Palaeoecology and Palaeohydrology* (edited by B.E. Berglund), John Wiley & Sons, Chichester and New York, 371–85.

Murray, J.W. (1991) *Ecology and Palaeoecology of Benthic Foraminifera*. Longman, London.

Murray, T. (1999) A return to the 'Pompeii premise'. In *Time and Archaeology* (edited by T. Murray), Routledge, London, 8–27.

Myers, N. (1997) Biodiversities genetic library. In *Nature's Services: Societal dependence on natural ecosystems* (edited by G.C. Daily), Island Press, Washington, DC, 255–73.

Needham, S. (2000) *The Passage of the Thames: Holocene environment and settlement at Runnymede*. British Museum Press, London.

Needham, S. and Macklin, M.G. (eds) (1992) *Alluvial Archaeology in Britain*. Oxbow Monograph 27, Oxford.

Neilson, R.P. and Drapek, R.J. (1998) Potentially complex biosphere responses to transient global warming. *Global Change Biology*, **4**, 505–21.

Nesje, A. and Dahl, S.O. (2000) *Glaciers and Environmental Change*. Arnold, London.

Nesje, A., Dahl, S.A., Andersson, C. and Matthews, J.A. (2000) The lacustrine sedimentary sequence in

Sygneskardvatnet, western Norway: a continuous, high-resolution record of the Jostedalsbreen ice cap during the Holocene. *Quaternary Science Reviews*, **19**, 1047–66.

Newell, R.E. (1981) Further studies of the atmospheric temperature change produced by Mt Agung volcanic eruption in 1963. *Journal of Volcanology and Geothermal Research*, **11**, 61–6.

Newman, W.S., Cinqemani, L.J. Pardi, R.R. and Marcus, L.F. (1980) Holocene delevelling of the United States east coast. In *Earth Rheology, Isostasy and Eustasy* (edited by N-A. Mörner), John Wiley & Sons, Chichester and New York, 449–63.

Nichols, N., Gruza, G.V., Jouzel, J., Karl, T.R., Ogallo, L.A. and Parker, D.E. (1996) Observed climate variability and change. In *Climate Change 1995: The science of climate change* (edited by J.T. Houghton, L.G. Meira Filho, B.A. Callander, N. Harris, A. Kattenberg and K. Maskell), Cambridge University Press, Cambridge, 1285–357.

Nielsen, V. (ed.) (1953) *Legislation concerning Ancient Monuments in Denmark*. Nationalmuseet, Copenhagen.

Nilssen, J.P. and Sandøy, S. (1990) Recent lake acidification and cladoceran dynamics: surface sediment and core analysis from lakes in Norway, Scotland and Sweden. In *Palaeolimnology and Lake Acidification* (edited by R.W. Battarbee, J. Mason, I. Renberg and J.P. Talling), Royal Society, London, 73–83.

Nordtorp-Madson, M.A. (2000) The cultural identity of the Late Greenland Norse. In *Identities and Cultural Contacts in the Arctic* (edited by M. Appelt, J. Berglund and H.C. Gulløv), Dansk Polar Center, Copenhagen, 55–60.

Nydick, K.R., Bidwell, A.B., Thomas, E. and Verkamp, J.C. (1995) A sea-level rise curve from Guildford, Connecticut, USA. *Marine Geology*, **124**, 137–59.

O'Brien, S.R., Mayewski, P.A., Meeker, L.D., Meese, D.A., Twickler, M.S. and Whitlow, S. (1995) Complexity of Holocene climate as reconstructed from a Greenland ice core. *Science*, **270**, 1962–4.

O'Connell, M. (1994) *Connemara: Vegetation and land use since the last Ice Age*. The Office of Public Works, Dublin.

O'Connell, M. and Molloy, K. (2001) Farming and woodland dynamics in Ireland during the Neolithic. *Proceedings of the Royal Irish Academy*, **101B** (1–2), 99–128.

O'Connell, M., Huang, C.C. and Eicher, U. (1999) Multidisciplinary investigations, including stable-isotope studies, of thick Late-glacial sediments from Tory Hill, Co. Limerick, western Ireland. *Palaeogeography, Palaeoclimatology, Palaeoecology*, **147**, 169–208.

O'Connor, T. (2000) *The Archaeology of Animal Bones*. Sutton, Stroud.

Odgaard, B.V. (1988) Heathland history in western Jutland, Denmark. In *The Cultural Landscape: Past, present and future* (edited by H.H. Birks, H.J.B. Birks, P.E. Kaland and D. Moe), Cambridge University Press, Cambridge, 311–19.

Odgaard, B.V. (1992) The fire history of Danish heathland areas as reflected by pollen and charred particles in lake sediments. *The Holocene*, **2**, 218–26.

Odgaard, B.V. and Rostholm, H. (1987) A single grave barrow at Harreskor, Jutland. *Journal of Danish Archaeology*, **6**, 87–100.

O'Donnell, P.M. (1995) Cultural landscapes in North America: an overview of status in the United States of America. In *Cultural Landscapes of Universal Value* (edited by B. von Droste, H. Plachter and M. Rossler), Fischer Verlag, Stuttgart, 210–33.

Oeschger, H. and Langway, C.C. Jnr (eds) (1989) *The Environmental Record in Glaciers and Ice Sheets*. John Wiley & Sons, Chichester and New York.

Ogden, J., Deng, Y., Boswijk, G. and Sandiford, A. (2003) Vegetation changes since early Maori fires in Waipoua Forest, Northern New Zealand. *Journal of Archaeological Science*, **30**, 733–67.

Ogilvie, A.E. (1992) Documentary evidence for changes in the climate of Iceland, AD 1500 to 1800. In *Climate since AD 1500* (edited by R.S. Bradley and P. Jones), Routledge, London, 92–117.

O'Hara, S.L., Street-Perrott, F.A. and Burt, T.P. (1993) Accelerated soil erosion around a Mexican highland lake caused by prehistoric agriculture. *Nature*, **362**, 48–51.

O'Hear, A. (1989) *An Introduction to the Philosophy of Science*. Clarendon, Oxford.

Oldale, R.N. and O'Hara, C.J. (1980) New radiocarbon dates from the inner Continental Shelf off southeastern Massachusetts and local sea-level curve for the past 12,000 years. *Geology*, **8**, 102–6.

Oldfield, F., Richardson, N. and Appleby, P.G. (1995) Radiometric dating (^{210}Pb, ^{137}Cs, ^{241}Am) of recent ombrotrophic peat accumulation and evidence for changes in mass balance. *The Holocene*, **5**, 141–8.

Oldfield, F., Wake, R., Boyle, J., Jones, R., Nolan, S., Gibbs, Z., Appleby, P., Fisher, E. and Wolff, G. (2003) The Late Holocene history of Gormire Lake, north-east England and its catchment: a multi-proxy reconstruction of past human impact. *The Holocene*, **13**, 677–90.

Olsen, C.G., Nettleton, W.D., Porter, D.A. and Brasher, B.R. (1997) Middle Holocene aeolian activity on the High Plains of west-central Kansas. *The Holocene*, **7**, 255–62.

Olsson, I.U. (1986) Radiometric dating. In *Handbook of Holocene Palaeoecology and Palaeohydrology* (edited by B.E. Berglund), John Wiley & Sons, Chichester and New York, 273–312.

Onac, B.P. and Lauritzen, S-E. (1996) The climate of the last 150,000 years recorded in speleothems: preliminary results from north-western Romania. *Theoretical and Applied Karstology*, **9**, 9–21.

Oppenheimer, M. (1998) Global warming and the stability of the West Antarctic Ice Sheet. *Nature*, **393**, 325–32.

O'Riordan, T. and Jäger, J. (eds) (1996) *Politics of Climate Change: A European perspective*. Routledge, London.

Osborn, G. and Bevis, K. (2001) Glaciation in the Great Basin of the western United States. *Quaternary Science Reviews*, **20**, 1377–410.

Osborn, G. and Luckman, B.H. (1988) Holocene glacier fluctuations in the Canadian Cordillera (Alberta and British Columbia). *Quaternary Science Reviews*, **7**, 115–29.

O'Shea, J. (1989) The role of wild resources in smallscale agricultural systems: tales from the lakes and the plains. In *Bad Year Economics* (edited by P. Halstead and J. O'Shea), Cambridge University Press, London, 57–67.

O'Sullivan, A. (2001) *Foragers, Farmers and Fishers in a Coastal Landscape*. Royal Irish Academy, Dublin.

O'Sullivan, P.E. (1983) Annually-laminated lake sediments and the study of Quaternary environmental changes. *Quaternary Science Reviews*, **1**, 245–312.

Ovchinnikov, I.V., Götherström, A., Romanova, G.P., Kharitonov, V.M., Lidén, K. and Goodwin, W. (2000) Molecular analysis of Neanderthal DNA from the northern Caucasus. *Nature*, **404**, 490–3.

Overpeck, J., Hughen, K., Hardy, D., Bradley, R., Case, R., Douglas, M., Finney, B., Gajewski, K., Jacoby, G., Jennings, A., Lamoureux, S., Lasca, A., McDonald, G., Moore, J., Retelle, M., Smith, S., Wolfe, A. and Zielinski, G. (1997) Arctic environmental changes of the last four centuries. *Science*, **278**, 1251–6.

Overpeck, J.T., Webb, R.S. and Webb, T. III (1992) Mapping eastern North America vegetation change of the past 18 ka: no-analogs and the future. *Geology*, **20**, 1071–4.

Parker, A.G., Goudie, A.S., Anderson, D.E., Robinson, M.A. and Bonsall, C. (2002) A review of the mid-Holocene elm decline in the British Isles. *Progress in Physical Geography*, **26** (1), 1–45.

Parker, D.E., Wilson, H., Jones, P.D., Christy, J.R. and Folland, C.K. (1996) The impact of Mount Pinatubo on world-wide temperatures. *International Journal of Climatology*, **16**, 487–97.

Parry, M.L. (1978) *Climatic Change, Agriculture and Settlement*. Dawson, Folkestone.

Parry, M.L. (1981) Evaluating the impact of climatic change. In *Consequences of Climatic Change* (edited by C. Delano Smith and M. Parry), Department of Geography, University of Nottingham, Nottingham, 3–16.

Parry, M.L. and Carter, T.R. (1985) The effect of climatic variations on agricultural risk. *Climatic Change*, **7**, 95–110.

Partridge, T.C., Demenocal, P.B., Lorentz, S.A., Paiker, M.J. and Vogel, J.C. (1997) Orbital forcing of climate over South Africa: a 200,000-year rainfall record from the Pretoria saltpan. *Quaternary Science Reviews*, **16**, 1125–33.

Paterson, W.A. and Backman, A.E. (1988) Fire and disease history of forests. In *Vegetation History* (edited by B. Huntley and T. Webb III), Kluwer, Dordrecht, 603–32.

Patience, J. and Kroon, D. (1991) Oxygen isotope chronostratigraphy. In *Quaternary Dating Methods – a User's Guide*. (edited by P.L. Smart and P.D. Frances), Technical Guide 4, Quaternary Research Association, Cambridge, 199–228.

Patterson, W.A. and Backman, A.E. (1988) Fire and disease history of forests. In *Vegetation History* (edited by B. Huntley and T. Webb), Kluwer, Dordrecht, 603–32.

Patterson, W.A. III, Edwards, K.J. and McGuire, D.J. (1987) Microscopic charcoal as a fossil indicator of fire. *Quaternary Science Reviews*, **6**, 3–23.

Peacock, J.D. (1989) Marine molluscs and Late Quaternary environmental studies with particular reference to the Late-Glacial period in Northwest Europe. *Quaternary Science Reviews*, **8**, 179–92.

Peacock, J.D. (1993) Late Quaternary marine molluscs as palaeoenvironmental proxies: a compilation and assessment of basic numerical data for NE Atlantic species found in shallow water. *Quaternary Science Reviews*, **12**, 263–75.

Pearce, F. (1989) *Climate and Man*. Vision Books, London.

Pearsall, D.M. (1992) The origins of plant cultivation in South America. In *The Origins of Agriculture* (edited by C.W. Cowan and P.J. Watson), Smithsonian Institution Press, Washington, 173–206.

Pearsall, W.H. (1950) *Mountains and Moorlands*. Collins, New Naturalist, London.

Pearson, M.G. (1973) Snowstorms in Scotland 1782–1786. *Weather*, **28**, 195–201.

Pedersen, L., Fischer, A. and Aaby, B. (eds) (1997) *The Danish Storebælt since the Ice Age*. Storebælt Publications, Copenhagen.

Peel, D.A. (1994) The Greenland Ice-Core Project (GRIP): reducing uncertainties in climatic change. *NERC News*, April, 26–30.

Peglar, S.M. (1993a) The development of the cultural landscape around Diss Mere, Norfolk, UK, during the past 7000 years. *Review of Palaeobotany and Palynology*, **76**, 1–47.

Peglar, S.M. (1993b) The mid-Holocene *Ulmus* decline at Diss Mere, Norfolk: a year-by-year pollen stratigraphy from annual laminations. *The Holocene*, **3**, 1–13.

Peglar, S.M. and Birks, H.J.B. (1993) The mid-Holocene *Ulmus* fall at Diss Mere, south-east England – disease and human impact? *Vegetation History and Archaeobotany*, **2**, 61–8.

Peglar, S., Fritz, S.C., Alapieti, T., Saarnisto, M. and Birks, H.J.B. (1984) The composition and formation of laminated lake sediments in Diss Mere, Norfolk, England. *Boreas*, **13**, 13–28.

Peltier, W.R. (2002) On eustatic sea level history: Last Glacial Maximum to Holocene. *Quaternary Science Reviews*, **21**, 377–96.

Penney, D.N. (1987) Applications of Ostracoda to sea-level studies. *Boreas*, **16**, 237–47.

Perry, I. and Moore, P.D. (1987) Dutch elm disease as an analogue of Neolithic elm decline. *Nature*, **326**, 72–3.

Peteet, D.M. (ed.) (1993) Global Younger Dryas? *Quaternary Science Reviews*, **12**, 277–355.

Peteet, D.M. (1995) Global Younger Dryas? *Quaternary International*, **28**, 93–104.

Peteet, D.M., Daniels, R.A., Heusser, L.E., Vogel, J.S., Southon, J.R. and Nelson, D.E. (1994) Wisconsinan Late-glacial environmental change in southern New England: a regional synthesis. *Journal of Quaternary Science*, **9**, 151–4.

Peterson, G.M. (1993) Vegetational and climate history of the western former Soviet Union. In *Global Climates since the Last Glacial Maximum* (edited by H.E. Wright, J.E. Kutzbach, T. Webb, W.F. Ruddiman, F.A. Street-Perrott and P.J. Bartlein), University of Minneapolis Press, Minneapolis, 169–93.

Petit, J.R., Jouzel, J., Raynaud, D., Barkov, N.I., Basile, I., Bender, M., Chappellaz, J., Davis, J., Delaygue, G., Delmotte, M., Kotlyakov, V.M., Legrand, M., Lipenkov, V., Lorius, C., Pépin, L., Ritz, C., Saltzman, E. and Stievenard, M. (1999) 420,000 years of climate and atmospheric history revealed by the Vostok deep Antarctic ice core. *Nature*, **399**, 429–36.

Petrequin, A.-M. and Petrequin, P. (1988) *Le Néolithique des Lacs: Préhistoire des lacs de Chalain et de Clairvaux (4000–2000 av.J.-C.)*. Editions Errance, Paris.

Petzelberger, B. (2000) Coastal development and human activities in NW Germany. In *Coastal and Estuarine Environments: Sedimentology, geomorphology and geoarchaeology* (edited by K. Pye and J.R.L. Allen), Geological Society Special Publication No. 175, London, 365–76.

Péwé, T.L. (1983) Alpine permafrost in the contiguous United States: a review. *Arctic and Alpine Research*, **15**, 145–56.

Péwé, T.L. (1984) The periglacial environment in North America during Wisconsin time. In *Late Quaternary Environments of the United States*. Volume 1. *The Late Pleistocene* (edited by S.C. Porter), Longman, London, 157–89.

Pfister, C. (1984) The potential of documentary data for the reconstruction of past climates. In *Climatic Changes on a Yearly to Millennial Basis* (edited by N-A. Mörner and W. Karlén), Reidel, Dordrecht, 331–7.

Pfister, C. (1992) Monthly temperature and precipitation in central Europe 1525–1979: quantifying documentary evidence on weather and its effects. In *Climate since AD 1500* (edited by R.S. Bradley and P. Jones), Routledge, London, 118–42.

Pfister, C. and Bareiss, W. (1994) The climate in Paris from 1675 to 1713 after the meteorological journal of Louis Morin. In *Climate in Europe 1675–1715* (edited by B. Frenzel, C. Pfister and B. Gläser), Special Issue, ESF Project European Palaeoclimate and Man, Fischer, Stuttgart, 323–76.

Pfister, C., Luterbacher, J., Schwarcz-Zanetti, G. and Wegmann, M. (1998) Winter air temperature variations in western Europe during the Early and High Middle Ages (AD 750–1300). *The Holocene*, **8**, 535–52.

Pfister, C., Zhongwei, Y. and Schüle, H. (1994) Climatic variations in western Europe and China, AD 1645–1715: a preliminary continental-scale comparison of documentary evidence. *The Holocene*, **4**, 206–11.

Phillips, F.M., Zreda, M.G., Flinsch, M.R., Elmore, D. and Sharma, P. (1996) A re-evaluation of cosmogenic ^{36}Cl production rates in terrestrial rocks. *Geophysical Research Letters*, **23**, 949–51.

Phillips, F.M., Zreda, M.G., Smith, S.S., Elmore, D., Kubik, P.W. and Sharma, P. (1990) Cosmogenic chlorine-36 in rocks: a method for surface exposure dating. *Science*, **231**, 41–3.

Phillips, J.D. (1996) Deterministic complexity, explanation, and predictability in geomorphic systems. In *The Scientific Nature of Geomorphology* (edited by B.L. Rhoads and C.E. Thorn), John Wiley & Sons, Chichester and New York, 315–35.

Phillips, J.D. (1999) *Earth Surface Systems: Complexity, order and scale*. Blackwell, Oxford.

Pickett, S.T.A. and White, P.S. (1985) *The Ecology of Natural Disturbance and Patch Dynamics*. Academic Press, Orlando.

Pienitz, R., Smol, J.P. and MacDonald, G. (1999) Paleolimnological reconstruction of the Holocene climatic trends from two boreal treeline lakes, Northwest Territories, Canada. *Arctic and Alpine Research*, **31**, 82–93.

Pilcher, J.R. and Hall, V. (2001) *Flora Hibernica: The wild flowers, plants and trees of Ireland*. The Collins Press, Cork.

Pilcher, J.R., Hall, V.A. and McCormac, F.G. (1996) An outline tephrochronology for the Holocene of the north of Ireland. *Journal of Quaternary Science*, **11**, 48–94.

Piperno, D.R. and Pearsall, D.M. (1998) *The Origins of Agriculture in the Lowland Neotropics*. Academic Press, London.

Pitkänen, A., Lehtonen, H. and Huttunen, P. (1999) Comparison of sedimentary microscopic charcoal particle records in a small lake with dendrochronological data: evidence for the local origin of microscopic charcoal produced by forest fires of low intensity in eastern Finland. *The Holocene*, **9**, 559–68.

Pittock, B., Walsh, K. and McInnes, K. (1996) Tropical cyclones and coastal inundation under enhanced greenhouse conditions. *Water, Air and Soil Pollution*, **92**, 159–69.

Playfair, J. (1802) *Illustrations of the Huttonian Theory of the Earth*. Cadell, Davies and Creech, Edinburgh.

Ponel, P. and Coope, G.R. (1990) Lateglacial and early Flandrian coleoptera from La Taphanel, Massif Central, France: climatic and ecological implications. *Journal of Quaternary Science*, **5**, 235–50.

Ponel, P. and Lowe, J.J. (1992) Coleopteran, pollen and radiocarbon evidence from the Prato Spilla 'D' succession, N. Italy. *Comptes Rendues de l'Académie des Sciences, Paris*, Serie II, **315**, 1425–31.

Pons, A, Guiot, J., de Beaulieu, J-L. and Reille, M. (1992) Recent contributions to the climatology of the last glacial-interglacial cycle based on French pollen sequences. *Quaternary Science Reviews*, **11**, 439–48.

Popper, K.R. (1972) *Objective Knowledge*. Oxford University Press, Oxford.

Popper, K.R. (1974) *Conjectures and Refutations*, 5th edition. Routledge and Kegan Paul, London.

Porter, S.C. (1986) Pattern and forcing of Northern Hemisphere glacier variations during the last millennium. *Quaternary Research*, **26**, 27–48.

Porter, S.C. and Swanson, T.W. (1998) Radiocarbon age constraints on rates of advance and retreat of the Puget Lobe of the Cordilleran ice sheet during the last glaciation. *Quaternary Research*, **50**, 205–13.

Potts, R. (1996) Evolution and climate variability. *Science*, **273**, 922–3.

Preece, R.C. (ed.) (1995) *Island Britain: A Quaternary perspective*. Geological Society Special Publication No. 96, London.

Preece, R.C. (1998a) Impact of early Polynesian occupation on the land snail fauna of Henderson Island, Pitcairn group (South Pacific). *Philosophical Transactions of the Royal Society of London B*, **353**, 347–68.

Preece, R.C. (1998b) Synthesis. In *Late Quaternary Environmental Change in North-west Europe: Excavations at Holywell Coombe, southeast England* (edited by R.C. Preece and D.R. Bridgland), Chapman & Hall, London, 359–82.

Preece, R.C. (2001) Non-marine Mollusca and archaeology. In *Handbook of Archaeological Sciences* (edited by D.R. Brothwell and A.M. Pollard), John Wiley & Sons, Chichester and New York, 135–45.

Preece, R.C. and Bridgland, D.R. (eds) (1998) *Late Quaternary Environmental Change in North-west Europe: Excavations at Holywell Coombe, southeast England*. Chapman & Hall, London.

Preece, R.C. and Bridgland, D.R. (1999) Holywell Coombe, Folkestone: a 13,000 year history of an English chalkland valley. *Quaternary Science Reviews*, **18**, 1075–125.

Preece, R.C., Kemp, R.A. and Hutchinson, J.N. (1995) A Late-glacial colluvial sequence at Watcombe Bottom, Ventnor, Isle of Wight, England. *Journal of Quaternary Science*, **10**, 107–22.

Prentice, I.C., Farquar, G.D., Fasham, M.J.R., Goulden, M.L., Meimann, M., Jaramillo, V.J., Kheshgi, H.S., Le Quéré, C., Scholes, R.J. and Wallace, D.W.R. (2001) The carbon cycle and atmospheric carbon dioxide. In *Climate Change 2001: The scientific basis* (edited by J.T. Houghton, Y. Ding, D.J. Griggs, M. Noguer, P.J. van der Linden and D. Xiaosu), Cambridge University Press, Cambridge, 185–235.

Prentice, I.C., Sykes, M.T., Lautenschlager, M., Harrison, S.P., Denissenko, O. and Bartlein, P.J. (1993) A simulation model for the transient effects of climate change on forest landscape. *Ecological Modelling*, **65**, 51–70.

Price, T.D. (2000) *Europe's First Farmers*. Cambridge University Press, Cambridge.

Pringle, H. (2002) Out of the Ice. *Canadian Geographic*, **122**, 56–64.

Probert-Jones, J.R. (1984) On the homogeneity of the annual temperature of central England since 1659. *Journal of Climatology*, **4**, 241–53.

Prott, L.V. (1992) A common heritage: The World Heritage Convention. In *All Natural Things: Archaeology and the green debate* (edited by L. Macinnes and C.R. Wickham-Jones), Oxbow, Oxford, 65–88.

Purdy, B.A. (1991) *The Art and Archaeology of Florida's Wetlands*. CRC Press, Boston.

Pye, K. and Allen, J.R.L. (2000) *Coastal and Estuarine Environments: Sedimentology, geomorphology and geoarchaeology*. Geological Society Special Publication 175, London.

Quine, T.A. and Walling, D.E. (1992) Patterns and rates of contemporary soil erosion derived using caesium-137: measurement, analysis and archaeological significance. In *Past and Present Soil Erosion* (edited by M.G. Bell and J. Boardman), Oxbow Monograph 22, Oxford, 185–96.

Rackham, O. (1986) *The History of the Countryside*. Dent & Sons, London.

Rackham, O. (1994) *The Illustrated History of the Countryside*. Dent & Sons, London.

Rackham, O. (2003) *Ancient Woodland*, 2nd edition. Castlepoint Press, Colvend.

Raftery, B. (1990) *Trackways through Time: Archaeological investigations on Irish bog roads 1985–1989*. Headline Publishing, Dublin.

Raftery, B. (1996) *Trackway excavations in the Mountdillon Bogs, Co Longford, 1985–1991*. Department of Archaeology, University College Dublin, Dublin.

Rahmstorf, S. (1995) Bifurcations of the North Atlantic thermohaline circulation in response to changes in the hydrological cycle. *Nature*, **378**, 145–9.

Rahmstorf, S. (2002) Ocean circulation and climate during the past 120,000 years. *Nature*, **419**, 207–13.

Rainbird, P. (2002) A message for our future? The Rapa Nui (Easter Island) ecodisaster and Pacific island environments. *World Archaeology*, **33** (3), 436–51.

Ralska-Jasiewiczowa, M. and van Geel, B. (1992) Early human disturbance of the natural environment recorded in annually laminated sediments of Lake Gosciaz, central Poland. *Vegetation History and Archaeobotany*, **1**, 33–42.

Ramis, D., Alcover, J.A., Coll, J. and Trias, M. (2002) The chronology of the first settlement of the Balearic Islands. *Journal of Mediterranean Archaeology*, **15** (1), 3–24.

Rampino, M.R. and Ambrose, S.H. (2000) Volcanic winter in the Garden of Eden: the Toba super-eruption and the late Pleistocene human population crash. In *Volcanic Hazards and Disasters in Human Antiquity* (edited by F.W. McCoy and G. Heiken), The Geological Society of America., Boulder, Colo., 70, 71–82.

Rampino. M.R. and Self, S. (1982) Historic eruptions of Tambora (1815), Krakatau (1883) and Agung (1963): their stratospheric aerosols and climatic impact. *Quaternary Research*, **18**, 127–43.

Rampino, M.R. and Self, S. (1993) Climate–volcanism feedback and the Toba eruption of ~74,000 years ago. *Quaternary Research*, **40**, 269–80.

Rapp, G. and Hill, C.L. (1998) *Geoarchaeology: The earth science approach to archaeological interpretation*. Yale University Press, New Haven.

Rasmussen, M. (1989) Leaf foddering of livestock in the Neolithic: archaeobotanical evidence from Weier, Switzerland. *Journal of Danish Archaeology*, **8**, 51–77.

Rasmussen, M. and Grønnow, B. (1999) The historical–archaeological experimental centre at Lejre, Denmark: 30 years of experimenting with the past. In *The Constructed Past: Experimental archaeology, education and the public* (edited by P.G. Stone and P.G. Planel), Routledge, London, 136–45.

Rasmussen, P. (1995) Mid-Holocene vegetation development at the inland Ertebølle settlement Ringkløster, East Jutland. *Journal of Danish Archaeology*, **12**, 65–86.

Raymo, M.E. (1997) The timing of major climatic terminations. *Paleoceanography*, **12**, 577–85.

Raymo, M.E. and Ruddiman, W.F. (1992) Tectonic forcing of late Cenozoic climate. *Nature*, **359**, 117–22.

Raynaud, D., Barnola, J-M., Chappellaz, J., Blunier, T., Indermühle, A. and Stauffer, B. (2000) The ice record of greenhouse gases: a view in the context of future changes. *Quaternary Science Reviews*, **19**, 9–18.

Raynaud, D., Jouzel, J., Barnola, J-M., Chappellaz, J., Delmas, R.J. and Lorius, C. (1993) The ice record of greenhouse gases. *Science*, **259**, 926–34.

Reasoner, M.A., Osborn, G. and Rutter, N.W. (1994) Age of the Crowfoot Advance in the Canadian Rocky Mountains: a glacial event coeval with the Younger Dryas oscillation. *Geology*, **22**, 439–42.

Redman, C.L. (1999) *Human Impact on Ancient Environments*. University of Arizona Press, Tucson.

Regnell, J., Gaillard, M.-J., Bartholin, T.S. and Karsten, P. (1995) Reconstruction of environment and history of plant use during the late Mesolithic/Ertebolle culture at the inland settlement of Bikeberg III, south Sweden. *Vegetation History and Archaeobotany*, **4**, 67–91.

Reichstein, J. (1984) Federal Republic of Germany. In *Archaeological Heritage Management in the Modern World* (edited by H. Cleere), Unwin Hyman, London, 37–47.

Reille, M., de Beaulieu, J-L., Svobodova, H., Andrieu-Ponel, V. and Goeury, C. (2000) Pollen analytical biostratigraphy of the last five climate cycles from a long continental sequence from the Velay region (Massif Central, France). *Journal of Quaternary Science*, **15**, 665–86.

Reilly, J., Baethgen, W., Chege, F.E., van de Geijn, S.C., Erda, L., Iglesias, A., Kenny, G., Patterson, D., Rogasik, J., Rötter, R., Rosenzweig, C., Sombroek, W. and Westbrook, J. (1996) Agriculture in a changing climate: impacts and adaptations. In *Climate Change 1995: The science of climate change* (edited by J.T. Houghton, L.G. Meira Filho, B.A. Callander, N. Harris, A. Kattenberg and K. Maskell), Cambridge University Press, Cambridge, 427–67.

Reilly, J., Prinn, R., Harnisch, J., Fitzmaurice, J., Jacoby, H., Kicklighter D., Melillo, J., Stone, P., Sokolov, A. and Wang, C. (1999) Multi-gas assessment of the Kyoto Protocol. *Nature*, **401**, 549–55.

Reitz, E.J., Newsom, L.A. and Scudder, S.J. (eds) (1996) *Case Studies in Environmental Archaeology*. Plenum Press, London.

Reksten, T. (2001) *The Illustrated History of British Columbia*. Douglas and MacIntyre, Vancouver.

Renberg, I. (1981) Formation, structure and visual appearance of iron-rich varved lake sediments. *Verhandlungen Internationale Vereinigung für Limnologie*, **21**, 94–101.

Renberg, I., Korsman, T. and Birks, H.J.B. (1993) Prehistoric increases in the pH of acid-sensitive Swedish lakes caused by land-use changes. *Nature*, **362**, 824–6.

Rendall, H., Worsley, P., Green, D. and Parks, D. (1991) Thermoluminescence dating of the Chelford Interstadial. *Earth and Planetary Science Letters*, **103**, 182–9.

Renfrew, C. (1993) Cognitive archaeology: some thoughts on the archaeology of thought. *Cambridge Archaeological Journal*, 3 (2), 248–50.

Renfrew, C. and Bahn, P. (2000) *Archaeology: Theories, methods and practice*. Thames & Hudson, London.

Renssen, H. and Isarin, R.F.B. (1998) Surface temperature in NW Europe during the Younger Dryas: a GCM simulation compared with temperature reconstructions. *Climate Dynamics*, **14**, 33–44.

Renssen, H., van Geel, B., van der Plicht, J. and Magny, M. (2000) Reduced solar activity as a trigger for the start of the Younger Dryas. *Quaternary International*, 68–71, 373–83.

Revelle, R. (1985) Introduction: the scientific history of carbon dioxide. In *The Carbon Cycle and Atmospheric CO_2: Natural variations Archaeoan to present* (edited by E.T. Sundquist and W.S. Broecker), Geophysical Monograph 32, American Geophysical Union, Washington, DC, 1–4.

Reynolds, P. (1974) Experimental Iron Age storage pits: an interim report. *Proceedings of the Prehistoric Society*, **40**, 118–31.

Reynolds, P. (1999) Butser Ancient Farm, Hampshire, UK. In *The Constructed Past: Experimental archaeology, education and the public* (edited by P.G. Stone and P.G. Planel), Routledge, London, 124–35.

Rial, J.A. (1999) Pacemaking the ice ages by frequency modulation of the earth's orbital eccentricity. *Science*, **285**, 55–9.

Rial, J.A. and Anaclerio, C.A. (2000) Understanding non-linear response of the climate system to orbital forcing. *Quaternary Science Reviews*, **19**, 1709–22.

Rice, R.J. (1988) *Fundamentals of Geomorphology*, 2nd edition. Longman, London.

Richards, M.P. and Hedges, R. (1999) A Neolithic revolution? New evidence of diet in the British Neolithic. *Antiquity*, **73**, 891–7.

Richards, M.P. and Sheridan, J.A. (2000) New AMS dates on human bone from Mesolithic Oronsay. *Antiquity*, **74**, 313–15.

Richards, M.P., Schulting, R.J. and Hedges, R. (2003a) Sharp shift in diet at onset of Neolithic. *Nature*, **425**, 38–66.

Richards, M.P., Price, T.D. and Koch, E. (2003b) Mesolithic and Neolithic subsidence in Denmark: new stable isotope data. *Current Anthropology*, **44**, 287–95.

Richter, G. (1986) Investigations of soil erosion in Central Europe. In *Soil Erosion* (edited by C.P. Burnham and J.I. Pitman), SEESOIL, 3, Silsoe, 14–27.

Ridge, J.C. and Toll, N.J. (1999) Are late-glacial climatic oscillations recorded in varves of the upper Connecticut Valley, northeastern United States? *Geologiska Föreningens i Stockholm Förhandlingar*, **121**, 187–93.

Riehle, J., Dumond, D.E., Meyer, C.E. and Schaaf, J.M. (2000) Tephrochronology of the Brooks River Archaeological District, Katmai National Park and Preserve, Alaska: what can and cannot be done with tephra deposits. In *The Archaeology of Geological Catastrophes* (edited by W.J. McGuire, D.R. Griffiths, P.L. Hancock and I.S. Stewart), Geological Society Special Publication No. 171, London, 245–66.

Rigaud, J.-P. and Simek, J.F. (1990) The last Pleniglacial in the South of France (24,000–14,000 years ago). In *The World at 18,000 BP* (edited by O. Soffer and C. Gamble), Unwin Hyman, London, 69–88.

Rind, D. and Overpeck, J. (1993) Hypothesised causes of decade-to-century-scale climatic variability: climate model results. *Quaternary Science Reviews*, **13**, 357–74.

Rindos, D. (1989) Darwinism and its role in the explanation of domestication, In *Foraging and Farming: The evolution of plant exploitation* (edited by D.R. Harris and G.C. Hillman), Unwin Hyman, London, 27–41.

Ringberg, B. and Erlström, M. (1999) Micromorphology and petrography of Late Weichselian glaciolacustrine varves in southeastern Sweden. *Catena*, **35**, 147–77.

Rippon, S. (1997) *The Severn Estuary: Landscape evolution and wetland reclamation*. Leicester University Press, London.

Rippon, S. (2000) *The Transformation of Coastal Wetlands*. British Academy, Oxford.

Rittenour, T.M., Goble, R.J. and Blum, M.D. (2003) An optical age chronology of Late Pleistocene fluvial deposits in the northern Mississippi valley. *Quaternary Science Reviews*, **22**, 1105–10.

Roberts, H.M. and Wintle, A.G. (2001) Equivalent close determinations for polymineralic fine grains using the SAR protocol: applications to a Holocene sequence of the Chinese loess plateau. *Quaternary Science Reviews*, **20**, 859–63.

Roberts, N. (2002) Did prehistoric landscape management retard the post-glacial spread of woodland in Southwest Asia? *Antiquity*, **76**, 1002–10.

Roberts, N. and Wright, H.E. (1993) Vegetational, lake level and climatic history of the Near East and Southwest Asia. In *Global Climates since the Last Glacial Maximum* (edited by H.E. Wright, J.E. Kutzbach, T. Webb, W.F. Ruddiman, F.A. Street-Perrott and P.J. Bartlein), University of Minneapolis Press, Minneapolis, 194–220.

Robertsson, A.M., Hicks, S., Åkerlund, A., Risberg, J. and Hackens, T. (eds) (1996) *Landscapes and Life*. PACT 50 11.13, Belgium.

Robinson, M. (1985) Nature conservation and environmental archaeology. In *Archaeology and Nature* (edited by G. Lambrick), Oxford University Dept for Continuing Education, Oxford, 11–17.

Robinson, M. (1992) Environment, archaeology and alluvium on the river gravels of the South Midlands. In *Alluvial Archaeology in Britain* (edited by S. Needham and M.G. Macklin), Oxbow Monograph 27, Oxford, 197–208.

Robinson, M. (2000) Further consideration of Neolithic charred cereals, fruit and nuts. In *Plants in Neolithic Britain and Beyond* (edited by A.S. Fairbairn), Oxbow, Oxford, 85–90.

Robinson, M. (2001) Insects as palaeoenvironmental indicators. In *Handbook of Archaeological Sciences* (edited by D.R. Brothwell and A.M. Pollard), John Wiley & Sons, Chichester and New York, 121–33.

Robinson, S.G., Maslin, M.A. and McCave, N. (1995) Magnetic susceptibility variations in Late Pleistocene deep-sea sediments of the northeast Atlantic: implications for ice-rafting and palaeocirculation at the last glacial maximum. *Paleoceanography*, **10**, 221–50.

Robock, A. and Mao, J. (1992) Winter warming from large volcanic eruptions. *Geophysical Research Letters*, **12**, 2405–8.

Rodhe, H. (1990) A comparison of the contribution of various atmospheric gases to the greenhouse effect. *Science*, **248**, 1217–19.

Rodwell, M.J., Rowell, D.P. and Folland, C.K. (1999) Oceanic forcing of the wintertime North Atlantic Oscillation and European climate. *Nature*, **398**, 320–3.

Rösch, M. (1996) New approaches to prehistoric land-use reconstruction in south-western Germany. *Vegetation History and Archaeobotany*, **5**, 65–79.

Rose, J. (1994) Lateglacial and Early Holocene river activity in lowland Britain. *Paläoklimaforschung*, **14**, 51–74.

Rose, J., Boardman, J., Kemp, R.A. and Whiteman, C. (1985) Palaeosols and the interpretation of the British Quaternary stratigraphy. In *Geomorphology and Soils* (edited by K. Richards, R.R. Arnett and S. Ellis), George Allen & Unwin, London, 348–75.

Rose, J., Lee, J.A., Kemp, R.A. and Harding, P.A. (2000) Palaeoclimate, sedimentation and soil development during the last glacial stage (Devensian), Heathrow Airport, London, UK. *Quaternary Science Reviews*, **19**, 827–48.

Rousseau, D-D., Limondin, N. Magnin, F. and Puissegur, J-J. (1994) Temperature oscillations over the last 10,000 years in western Europe estimated from terrestrial Mollusca assemblages. *Boreas*, **23**, 66–73.

Rousseau, D-D., Limondin, N. and Puissegur, J-J. (1993) Holocene environmental signals from mollusc assemblages in Burgundy (France). *Quaternary Research*, **40**, 237–53.

Rowe, P.J., Atkinson, T.C. and Turner, C. (1999) U-series dating of Hoxnian interglacial deposits at Marks Tey, Essex, England. *Journal of Quaternary Science*, **14**, 693–702.

Rowley-Conwy, P. (1981) Slash and burn in the temperate European Neolithic. In *Farming Practice in British Prehistory* (edited by R. Mercer), Edinburgh University Press, Edinburgh, 85–96.

Rowley-Conwy, R. (1982) Forest grazing and clearance in temperate Europe with special reference to Denmark: an archaeological view. In *Archaeological Aspects of Woodland Ecology* (edited by M. Bell and S. Limbrey), British Archaeological Reports IS 146, Oxford, 199–216.

Ruddiman, W.F. and Duplessy, J-C. (1985) Conference on the last deglaciation: timing and mechanism. *Quaternary Research*, **23**, 1–17.
Ruddiman, W.F. and Kutzbach, J.E. (1991) Plateau uplift and climatic change. *Scientific American*, **264**, 66–75.
Ruddiman, W.F. and McIntyre, A. (1981) The North Atlantic Ocean during the last deglaciation. *Palaeogeography, Palaeoclimatology, Palaeoecology*, **35**, 145–214.
Ruddiman, W.F. and Raymo, M. (1988) Northern Hemisphere climate regimes during the past 3 Ma: possible tectonic connections. *Philosophical Transactions of the Royal Society, London*, **B318**, 411–30.
Ruddiman, W.F. and Thomson, J.S. (2001) The case for human causes of increased atmospheric CH_4 over the last 5000 years. *Quaternary Science Reviews*, **20**, 1769–77.
Ruddiman, W.F., McIntyre, A. and Shackleton, N.J. (1986a) North Atlantic sea-surface temperatures for the last 1.1 million years. In *North Atlantic Palaeoceanography* (edited by C.P. Summerhayes and N.J. Shackleton), Geological Society Special Publication 21, 155–73.
Ruddiman, W.F., Raymo, M. and McIntyre, A. (1986b) Matuyama 41,000 year cycles: North Atlantic Ocean and northern hemisphere ice sheets. *Earth and Planetary Science Letters*, **80**, 117–29.
Rummery, T.A. (1983) The use of magnetic measurements in interpreting the fire histories of lake drainage basins. *Hydrobiologia*, **103**, 53–8.
Rumsby, B.T. and Macklin, M.G. (1996) River response to the last neoglacial (the 'Little Ice Age') in northern, western and central Europe. In *Global Continental Changes: The context of palaeohydrology* (edited by J. Branson, A.G. Brown and K.J. Gregory), Geological Society of London, Special Publication 115, London, 217–33.
Russell, B. (1961) *A History of Western Philosophy*, 2nd edition. Allen & Unwin, London.
Rymer, L. (1978) The use of uniformitarianism and analogy in palaeoecology, particularly pollen analysis. In *Biology and Quaternary Environments* (edited by D. Walker and J.C. Guppy), Australian Academy of Sciences, Canberra, 245–58.
Saarnisto, M. (1986) Annually laminated lake sediments. In *Handbook of Holocene Palaeoecology and Palaeohydrology* (edited by B.E. Berglund), John Wiley & Sons, Chichester and New York, 343–70.
Saarnthein, M., Winn, K., Jung, S.J.A., Duplessy, J-C., Labeyrie, L. Erlenkauser, H. and Ganssen, G. (1994) Changes in east Atlantic deepwater circulation over the past 30,000 years: eight time slice reconstructions. *Paleoceanography*, **9**, 209–67.
Sadler, J.P. (1991) Archaeological and biogeographical implications of palaeoentomological studies in Orkney and Iceland. Unpublished Ph.D. thesis, University of Sheffield.
Salawitch, R.J. (1998) A greenhouse warming connection. *Nature*, **392**, 551.
Salisbury, C.R. (1992) The archaeological evidence for palaeochannels in the Trent Valley. In *Alluvial Archaeology in Britain* (edited by S. Needham and M.G. Macklin), Oxbow Monograph 27, Oxford, 155–62.
Salisbury, E. (1964) *Weeds and Aliens*. Collins, London.
Samuels, S.R. (1991) *Ozette Archaeological Project Research Reports*. Volume 1: *House Structure and Floor Midden*. Washington State University, Seattle.
Santilli, R., Ormö, J., Rossi, A.P. and Komatsu, G. (2003) A catastrophe remembered: a meteorite impact of the fifth century AD in the Abruzzo, central Italy. *Antiquity*, **77**, 313–20.
Saurer, M., Robertson, I., Siegwolf, R. and Levenberger, M. (1998) Oxygen isotope analysis of cellulose: an inter-laboratory comparison. *Analytical Chemistry*, **70**, 2074–80.
Scarre, C. (1993) *Timelines of the Ancient World*. Dorling Kindersley, London.
Scarre, C. (2002) *Monuments and Landscape in Atlantic Europe*. Routledge, London.
Schertz, D.L. and Nearing, M.A. (2002) Erosion tolerance/soil loss tolerances. In *Encyclopedia of Soil Science* (edited by R. Lal), Marcel Dekker, New York, 448–51.
Schibler, J., Jacomet, S., Hüster-Plogman, H. and Brombacher, C. (1997) Economic crash in the 37th and 36th centuries cal BC in Neolithic lake shore sites in Switzerland. *Anthropozoologica*, **25/26**, 553–70.
Schimel, D., Alves, D., Enting, I., Heimann, M., Joos, F., Raynaud, D., Wigley, T., Prather, M., Derwent, R., Ehhalt-D., Fraser, P., Sanhueza, E., Zhou, X., Jonas, P., Charlson, R., Rodhe, H, Sadasivan, S., Shine, K.P., Fouquart, Y., Ramaswamy, V., Solomon, S., Srnivasan, J., Albritton, D., Derwent, R., Isaksen, I., Lal, M. and Wuebbles, D. (1996) Radiative forcing of climate change. In *Climate Change 1995: The science of climate change* (edited by J.T. Houghton, L.G. Meira Filho, B.A. Callander, N. Harris, A. Kattenberg and K. Maskell), Cambridge University Press, Cambridge, 65–132.
Schimmelmann, A., Lange, C.B. and Meggers, B.J. (2003) Palaeoclimate and archaeological evidence for a ~200-yr recurrence of floods and droughts linking

California, Mesoamerica and South America over the past 2000 years. *The Holocene*, **13** (5), 763–78.

Schimmelmann, A., Zhao, M., Harvey, C.C. and Lange, C.B. (1998) A large California flood and correlative global climatic events 400 years ago. *Quaternary Research*, **49**, 51–61.

Schindler, D.W. (1999) From acid rain to toxic snow. *Ambio*, **28**, 350–5.

Schindler, D.W., Curtis, P.J., Parker, B. and Stainton, M.P. (1996) Consequences of climate warming and lake acidification for UV-B penetration in North American boreal lakes. *Nature*, **379**, 705–8.

Schlichtherle, H. (1990) *Siedlungsarchäologie im Alpenvorland I*. Konrad Theiss Verlag, Stuttgart.

Schlichtherle, H. (1997) *Pfahlbauten rund um die Alpen*. Konrad Theiss Verlag, Stuttgart.

Schlichtherle, H. and Strobel, M. (1999) *Archaeology and Protection of Nature in the Federsee Bog*. Landesdenkmalamt Baden-Württemberg, Stuttgart.

Schlüchter, C. (2000) The Quaternary stratigraphy of the Alps. *Quaternary International*, **63/64**, 129.

Schönwiese, C-D. (1988) Volcanism and air temperature variations in recent centuries. In *Recent Climatic Change* (edited by S. Gregory), Belhaven Press, London and New York, 20–9.

Schrope, M. (2000) Successes in fight to save ozone layer could close holes by 2050. *Nature*, **408**, 627.

Schweingruber, F.H. (1988) *Tree Rings: Basics and applications of dendrochronology*. Reidel, Dordrecht.

Scott, D.B., Brown, K., Collins, E.S. and Medioli, F.S. (1995) A new sea-level curve from Nova Scotia: evidence for a rapid acceleration of sea-level rise in the late mid-Holocene. *Canadian Journal of Earth Sciences*, **32**, 2071–80.

Scuderi, L.A. (1987) Glacier variations in the Sierra Nevada, California, as related to a 1200-year tree-ring chronology. *Quaternary Research*, **27**, 220–31.

Sealy, J. (2001) Body tissue chemistry and palaeodiet. In *Handbook of Archaeological Sciences* (edited by D.R. Brothwell and A.M. Pollard), John Wiley & Sons Ltd, Chichester and New York, 269–79.

Seidl, M.A., Finkel, R.C., Caffee, M.W., Hudson, G.B. and Dietrich, W.E. (1997) Cosmogenic isotope analysis applied to river longitudinal profile evolution: problems and interpretations. *Earth Surface Processes and Landforms*, **22**, 195–209.

Sejrup, H-P., Larsen, E., Landvik, J., King, E.L., Haflidason, H. and Nesje, A. (2000) Quaternary glaciations in southern Fennoscandia: evidence from southwestern Norway and the northern North Sea regions. *Quaternary Science Reviews*, **19**, 667–86.

Sejrup, H-P., Sjøholm, J., Furnes, H., Beyer, J., Eide, L., Jansen, E. and Mangerud, J. (1989) Quaternary tephrachronology on the Iceland Plateau, north of Iceland. *Journal of Quaternary Science*, **4**, 109–14.

Seppala, M. (1987) Periglacial phenomena of northern Fennoscandia. In *Periglacial Processes and Landforms in Britain and Ireland* (edited by J. Boardman), Cambridge University Press, Cambridge, 45–55.

Severinghaus, J.P., Sowers, T., Brook, E.J., Alley, R.B. and Bender, M.L. (1998) Timing of abrupt climate change at the end of the Younger Dryas interval from thermally fractionated gases in polar ice. *Nature*, **391**, 141–6.

Shackleton, N.J. (1987) Oxygen isotopes, ice volume and sea level. *Quaternary Science Reviews*, **6**, 183–90.

Shackleton, N.J. (2000) The 100,000-year ice-age cycle identified and found to lag temperature, carbon dioxide and orbital eccentricity. *Science*, **289**, 1897–902.

Shackleton, N.J. and Opdyke, N.D. (1973) Oxygen isotope and palaeomagnetic stratigraphy of equatorial Pacific core V28-238: oxygen isotope temperatures and ice volumes on a 10^5 and 10^6 year scale. *Quaternary Research*, **3**, 39–55.

Shackleton, N.J., Berger, A. and Peltier, W.R. (1990) An alternative astronomical calibration of the lower Pleistocene timescale. *Transactions of the Royal Society of Edinburgh: Earth Sciences*, **81**, 251–61.

Shackleton, N.J., Chapman, M.R., Sánchez-Goñi, M.F., Pailler, D. and Lancelot, Y. (2002) The classic marine isotope substage 5e. *Quaternary Research*, **58**, 14–16.

Shackleton, N.J., Le, J., Mix, A. and Hall, M.A. (1992) Carbon isotope records from Pacific surface waters and atmospheric carbon dioxide. *Quaternary Science Reviews*, **11**, 387–400.

Shane, L.K.C. and Anderson, K.H. (1993) Intensity gradients and reversals in Lateglacial environmental change in east-central North America. *Quaternary Science Reviews*, **12**, 307–20.

Shaw, E.M. (1985) Some aspects of rainfall records with selected computational examples from northern England. In *The Climatic Scene* (edited by M.J. Tooley and G.M. Sheail), Allen & Unwin, London, 60–92.

Shaw, E.M. (1988) *Hydrology in Practice*. Van Nostrand Reinhold International, London.

Shennan, I. (1983) Flandrian and Late Devensian sea-level changes and crustal movements in England and Wales. In *Shorelines and Isostasy* (edited by D.E. Smith and A.G. Dawson), Academic Press, London and New York, 255–84.

Shennan, I. and Andrews, J.E. (eds) (2000): *Holocene Land–Ocean Interaction and Environmental Change around the North Sea*. Geological Society Special Publication No. 166, London.

Shennan, I. and Horton, B. (2002) Holocene land- and sea-level changes in Great Britain. *Journal of Quaternary Science*, **17**, 511–26.

Shennan, I., Innes, J.B., Long, A.J. and Zong, Y. (1994) Late Devensian and Holocene relative sea-level at Loch nan Eala, near Arisaig, northwest Scotland. *Journal of Quaternary Science*, **9**, 261–83.

Shennan, I., Lambeck, K., Horton, B., Innes, J., Lloyd, J., McArthur, J., Purcell, T. and Rutherford, M. (2000a) Late Devensian and Holocene records of relative sea-level changes in northwest Scotland and their implications for glacio-hydro-isostatic modelling. *Quaternary Science Reviews*, **19**, 1103–35.

Shennan, I., Lambeck, K., Flather, R, Horton, B., McArthur, J., Innes, J., Lloyd, J., Rutherford, M. and Wingfield, R. (2000b) Modelling western North Sea palaeogeographies and tidal changes during the Holocene. In *Holocene Land–Ocean Interaction and Environmental Change around the North Sea* (edited by I. Shennan and J.E. Andrews), Geological Society Special Publication No. 166, 299–319.

Shennan, I., Lambeck, K., Horton, B., Innes, J., Lloyd, J., McArthur, J. and Rutherford, M. (2000c) Holocene isostasy and relative sea-level changes on the east coast of England. In *Holocene Land–Sea Interaction and Environmental Change around the North Sea* (edited by I. Shennan and J.E. Andrews), Geological Society Special Publication No. 166, 275–98.

Shennan, I., Peltier, W.R., Drummond, R. and Horton, B. (2002) Global to local scale parameters determining relative sea-level changes and the post-glacial isostatic adjustment of Great Britain. *Quaternary Science Reviews*, **21**, 397–408.

Shennan, I., Rutherford, M.M., Innes, J.B. and Walker, K.J. (1996) Late glacial sea level and ocean margin environmental changes interpreted from biostratigraphic and lithostratigraphic studies of isolation basins in northwest Scotland. In *Late Quaternary Palaeoceanography of the North Atlantic Margins* (edited by J.T. Andrews, W.E.N. Austin, H. Bergsten and A.E. Jennings), Geological Society Special Publication No. 111, 229–44.

Shermer, M. (1995) Exorcising Laplace's demon: chaos and antichaos, history and metahistory. *History and Theory*, **34**, 59–83.

Sherratt, A. (1997) Climatic cycles and behavioural revolutions: the emergence of modern humans and the beginning of farming. *Antiquity*, **71**, 271–87.

Shindell, D.T., Rind, D. and Lonergan, P. (1998) Increased polar stratospheric ozone losses and delayed eventual recovery owing to increasing greenhouse-gas concentrations. *Nature*, **392**, 589–92.

Shotyk, W., Weiss, D., Appleby, P.G., Cherbukin, A.K., Frei, R., Gloor, M., Kramers, J.D., Reese, S. and van der Knaap, W.O. (1998) History of atmospheric lead deposition since 12,370 ^{14}C yr BP from a peat bog, Jura Mountains, Switzerland. *Science*, **281**, 1635–40.

Shuman, B., Bartlein, P., Logar, N., Newby, P. and Webb III, T. (2002) Parallel climate and vegetation responses to the early Holocene collapse of the Laurentide Ice Sheet. *Quaternary Science Reviews*, **21**, 1793–805.

Sibrava, V., Bowen, D.Q. and Richmond, G.M. (1986) Quaternary glaciations of the Northern Hemisphere. *Quaternary Science Reviews*, **5**.

Siegert, M.J. (2001) *Ice Sheets and Late Quaternary Environmental Change*. John Wiley, Chichester and New York.

Simmons, A.H. (1998) Of tiny hippos, large cows and early colonists in Cyprus. *Journal of Mediterranean Archaeology*, **11**, 232–41.

Simmons, I.G. (1989) *Changing the Face of the Earth*. Blackwell, Oxford.

Simmons, I.G. (1993a) *Environmental History: A concise introduction*. Blackwell, Oxford.

Simmons, I.G. (1993b) *Interpreting Nature: Cultural constructions of the environment*. Routledge, London.

Simmons, I.G. (1996) *The Environmental Impact of Later Mesolithic Cultures*. Edinburgh University Press, Edinburgh.

Simmons, I.G. (1999) History, ecology, contingency, sustainability. In *Structure and Contingency* (edited by J. Bintliff), Leicester University Press, London, 118–31.

Simmons, I.G. (2001) Ecology into landscape: some English moorlands in the later Mesolithic. *Landscapes*, **2**, 42–55.

Simola, H. (1994) Sedimentary records of human occupation in the Eastern Finnish Lake District. In *Laminated Sediments* (edited by S. Hicks, U. Miller and M. Saarnisto), PACT, Belgium, 117–23.

Simola, H., Coard, M.A. and O'Sullivan, P.E. (1981) Annual laminations in the sediment of Looe Pool, Cornwall. *Nature*, **290**, 238–41.

Simpson, I.A., van Bergen, P.F., Elhmmali, M., Roberts, D.J. and Evershed, P.P. (1999) Lipid biomarkers of manuring practice in relict anthropogenic soils. *The Holocene*, **9**, 223–9.

Singh, G. and Geissler, E.A. (1985) Late Cainozoic history of vegetation, fire, lake levels and climate, at Lake George, New South Wales, Australia. *Philosophical Transactions of the Royal Society of London B*, **311**, 379–447.

Skjelkvåle, B.L. and Wright, R.F. (1999) Mountain lakes: sensitivity to acid deposition and global climate change. *Ambio*, **27**, 280–6.

Skog, G. and Regnell, J. (1995) Precision calendar year dating of the elm decline in a *Sphagnum* peat bog in Southern Sweden. *Radiocarbon*, **37** (2), 197–201.

Slaymaker, O. (ed.) (1996) *Geomorphic Hazards*. John Wiley & Sons, Ltd, Chichester and New York.

Small, J. and Witherick, M. (1995) *A Modern Dictionary of Geography*, 3rd edition. Hodder Arnold, London.

Smart, P.L. (1991) Uranium series dating. In *Quaternary Dating Methods – A User's Guide* (edited by P.L. Smart and P.D. Frances), Technical Guide 4, Quaternary Research Association, Cambridge, 45–83.

Smart, P.L. and Frances, P.D. (eds) (1991) *Quaternary Dating Methods – A User's Guide*. Quaternary Research Association, Cambridge.

Smironova, T.Y. and Nikonov, A.A. (1990) A revised lichenometric method and its application to dating past earthquakes. *Arctic and Alpine Research*, **22**, 375–88.

Smith, A.G. (1981) The Neolithic. In *The Environment in British Prehistory* (edited by I.G. Simmons and M.J. Tooley), Duckworth, London, 125–209.

Smith, A.G. (1984) Newferry and the Boreal–Atlantic tradition. *New Phytologist*, **98**, 35–55.

Smith, A.G. and Cloutman, E.W. (1988) Reconstruction of Holocene vegetation history in three dimensions at Waun-Fignen-Felen, an upland site in South Wales. *Philosophical Transactions of the Royal Society, London*, **B322**, 159–219.

Smith, A.G., Whittle, A., Cloutman, E.W. and Morgan, R.A. (1989) Mesolithic and Neolithic activity and environmental impact on the southeast Fen-edge in Cambridgeshire. *Proceedings of the Prehistoric Society*, **55**, 207–49.

Smith, A.T. (1995) Environmental factors affecting global atmospheric methane concentrations. *Progress in Physical Geography*, **19**, 322–35.

Smith, B.D. (1995a) *The Emergence of Agriculture*. Scientific American Library, New York.

Smith, B.D. (1995b) Seed plant domestication in eastern North America. In *Last Hunters – First Farmers* (edited by T.D. Price and A.B. Gebauer), School of American Research Press, Santa Fe, 193–213.

Smith, B.D. (1997a) Reconsidering the Ocampo caves and the era of incipient cultivation in Mesoamerica. *Latin American Antiquity*, **8**, 342–83.

Smith, B.D. (1997b) The initial domestication of *Cucurbita pepo* in the Americas 10,000 years ago. *Science*, **276**, 932–4.

Smith, C.J. (1980) *Ecology of the English Chalk*. Academic Press, London.

Smith, D.E. (1997) Sea-level change in Scotland during the Devensian and Holocene. In *Reflections on the Ice Age in Scotland* (edited by J.E. Gordon), Scottish Association of Geography Teachers and Scottish Natural Heritage, Glasgow, 136–51.

Smith, D.E. (2002) The Storegga disaster. *Current Archaeology*, **179**, 468–71.

Smith, D.E., Cullingford, R.A. and Firth, C.R. (2000) Patterns of isostatic land uplift during the Holocene: evidence from mainland Scotland. *The Holocene*, **10**, 489–501.

Smith, D.E., Cullingford, R.A. and Haggart, B.A. (1985) A major coastal flood during the Holocene in eastern Scotland. *Eiszeitalter und Gegenwart*, **35**, 109–18.

Smith, D.G. (1994) Glacial Lake McConnell: palaeogeography, age duration and associated river deltas, Mackenzie River basin, western Canada. *Quaternary Science Reviews*, **13**, 829–44.

Smith, D.J. (1993) Solifluction and climate in the Holocene: a North American perspective. In *Solifluction and Climatic Variations in the Holocene* (edited by B. Frenzel), Fischer Verlag, Stuttgart, 123–41.

Smith, E.A., Vonder Haar, T.H., Hickey, J.H. and Maschhoff, R. (1983) The nature of the short-period fluctuations in solar irradiance received by the earth. *Climatic Change*, **5**, 211–35.

Smith, K. (2001) *Environmental Hazards: Assessing risk and reducing disaster*. Routledge, London.

Snowball, I. and Thompson, R. (1992) A mineral magnetic study of Holocene sediment yields and depositional patterns in the Llyn Geirionydd catchment, North Wales. *The Holocene*, **2**, 238–48.

Soffer, O. (1990) The Russian Plain at the Last Glacial Maximum. In *The World at 18,000 BP* (edited by O. Soffer and C. Gamble), Unwin Hyman, London, 228–54.

Soffer, O. and Gamble, C. (eds) (1990) *The World at 18,000 BP*. Unwin Hyman, London.

Solem, T. (1986) Age, origin and development of blanket mires in Sør-Trøndelag, central Norway. *Boreas*, **15**, 101–15.

Solem, T. (1989) Blanket mire formation at Haramsøy, Møre og Romsdal, western Norway. *Boreas*, **18**, 221–35.

Sonnett, C.P. and Finney, S.A. (1990) The spectrum of radiocarbon. *Philosophical Transactions of the Royal Society, London*, **A330**, 413–26.

Sonninen, E. and Jungner, H. (1995) Stable carbon isotopes in tree-rings of a Scots pine from northern Finland. *Paläoklimaforschung*, **15**, 121–8.

Speight, M.C.D. (1991) *Saproxylic Invertebrates and their Conservation*. Council of Europe, Strasbourg.

Spindler, K. (1993) *The Man in the Ice*. Weidenfeld & Nicolson, London.

Spurk, M., Friedrich, M., Hofmann, J., Remmele, S., Frenzel, B., Leuschner, H.H. and Kromer, B. (1998) Revisions and extension of the Hohenheim oak and pine chronologies: new evidence about the timing of the Younger Dryas/Preboreal transition. *Radiocarbon*, **40**, 1107–16.

Stafford, T.W., Semken, H.A., Graham, R.W., Flippel, W.F., Makova, A., Smirnov, N.G. and Southon, J. (1999) First accelerator mass spectrometry ^{14}C dates documenting contemporaneity of nonanalog species in late Pleistocene mammal communities. *Geology*, **27** (10), 903–6.

Starkel, L. (1985) Lateglacial and Postglacial history of river valleys in Europe as a reflection of climatic changes. *Zeitschrift für Gletscherkunde und Glazialgeologie*, **21**, 159–64.

Starkel, L. (1991) The Vistula River valley: a case study for central Europe. In *Temperate Palaeohydrology* (edited by L. Starkel, J.B. Thornes and K.J. Gregory), John Wiley, Chichester and New York, 171–88.

Starkel, L. (2002) Change in the frequency of extreme events as the indicator of climatic change in the Holocene (in fluvial systems). *Quaternary International*, **91**, 25–32.

Starkel, L., Gregory, K.J. and Thornes, J.B. (eds) (1991) *Temperate Palaeohydrology*. John Wiley, Chichester and New York.

Stauffer B. (1989) Dating of ice by radioactive isotopes. In *The Environmental Record in Ice Sheets and Glaciers* (edited by H. Oeschger and C.C. Langway, Jnr), John Wiley & Sons, Chichester and New York, 123–39.

Steele, J., Gamble, C. and Sluckin, T. (2000) Estimating the rate of Palaeoindian expansion into South America. In *People as Agents of Environmental Change* (edited by R.A. Nicholson and T.P. O'Connor), Oxbow, Oxford, 125–32.

Steensberg, A. (1979) *Draved: An Experiment in Stone Age Agriculture: Burning, sowing and harvesting*. National Museum of Denmark, Copenhagen.

Steers, J.A. (1971) The East Coast floods, 31 January–1 February 1953. In *Applied Coastal Geomorphology* (edited by J.A. Steers), Macmillan, London, 198–224.

Stein, J.K. (ed.) (1992) *Deciphering a Shell Midden*. Academic Press, San Diego.

Stein, J.K. (1993) Scale in archaeology, geosciences, and geoarchaeology. In *Effects of Scale on Archaeological and Geoscientific Perspectives* (edited by J.K. Stein and A.R. Linse), Geological Society of America Special Paper 283, Colorado, 1–10.

Stephens, E.P. (1956) The uprooting of trees, a forest process. *Soil Science Society of America Proceedings*, **20**, 113–16.

Stephenson, F.R. (1990) Historical evidence concerning the sun: interpretation of sunspot records during the telescopic and pretelescopic eras. *Philosophical Transactions of the Royal Society*, **A339**, London, 499–512.

Sternberg, R.S. (2001) Magnetic properties and archaeomagnetism. In *Handbook of Archaeological Sciences* (edited by D.R. Brothwell and A.M. Pollard), John Wiley, Chichester and New York, 73–9.

Stevenson, A.C. and Birks, H.J.B. (1995) Heaths and moorlands: long-term ecological changes and interactions with climate and people. In *Heaths and Moorlands: Cultural landscapes* (edited by F.H. Thompson, A.J. Hester and M.B. Usher), HMSO, Edinburgh, 224–39.

Stewart, H. (1977) *Indian Fishing: Early methods on the northwest coast*. University of Washington Press, Vancouver.

Stewart I. (1990) *Does God play Dice? The Mathematics of Chaos*. Penguin, London.

Stewart, T.G. and England, J. (1983) Holocene sea-ice variations and palaeoenvironmental change, northernmost Ellesmere Island, N.W.T., Canada. *Arctic and Alpine Research*, **15**, 1–17.

Stoddard, J.L., Jeffries, D.S., Lükewille, A., Clair, T.A., Dillon, P.J., Driscoll, C.T., Forsius, M., Johannessen, M., Kahl, J.S., Kellogg, J.H., Kemp, A., Mannio, J., Monteith, D.T., Murdoch, P.S., Patrick, S., Rebsdorf, A., Skjelkvåle, B.L., Stainton, M.P., Traaen, T., van Dam, H., Webster, K.E., Wieting, J. and Wilander, A. (1999) Regional trends in aquatic recovery from acidification in North America and Europe. *Nature*, **401**, 575–8.

Stoddart, D.R. (1986) *On Geography*. Blackwell, Oxford.

Stoermer, E.F. and Smol, J. (1999) *The Diatoms: Applications for the environmental and earth sciences*. Cambridge University Press, London.

Stoker, M.S. and Holmes, R. (1991) Submarine end-moraines as indicators of Pleistocene ice limits off NW Britain. *Journal of the Geological Society of London*, **148**, 431–4.

Stolarski, R.S. (1988) The Antarctic ozone hole. *Scientific American*, **258**, 20–6.

Stone, P.G. and Planel, P.G. (eds) (1999) *The Constructed Past: Experimental archaeology, education and the public*. Routledge, London.

Stott, P.A. and Kettleborough, J.A. (2002) Origins and estimates of uncertainty in predictions of twenty-first century temperature rise. *Nature*, **416**, 723–6.

Stötter, J., Wastl, M., Caseldine, C. and Häberle, T. (1999) Holocene palaeoclimatic reconstruction in northern Iceland: approaches and results. *Quaternary Science Reviews*, **18**, 457–75.

Straus, L.G. (1996) The archaeology of the Pleistocene–Holocene transition in southwest Europe. In *Humans at the End of the Ice Age: The archaeology of the Pleistocene–Holocene transition* (edited by L.G. Strauss, B.V. Eriksen, J.M. Erlandson and D.R. Yesner), Plenum Press, New York and London, 83–100.

Straus, L.G., Eriksen, B.V., Erlandson, J.M. and Yesner, D.R. (eds) (1996) *Humans at the End of the Ice Age: The archaeology of the Pleistocene–Holocene transition*. Plenum Press, New York and London.

Street, M. (1991) Bedburg-Königshoven: A Pre-Boreal Mesolithic site in the Lower Rhineland, Germany. In *The Lateglacial in North-west Europe* (edited by R.N.E. Barton, A.J. Roberts and D.A. Roe), Council for British Archaeology Research Report 77, London, 256–70.

Street, M. and Terberger, T. (1999) The last Pleniglacial and the human settlement of Central Europe: new information from the Rhineland site of Wiesbaden-Igstadt. *Antiquity*, **73**, 259–72.

Street-Perrott, F.A., Mitchell, J.F.B., Marchand, D.S. and Brunner, J.S. (1990) Milankovitch and albedo forcing of the tropical monsoons: a comparison of geological evidence and numerical simulations for 9000 y BP. *Transactions of the Royal Society of Edinburgh: Earth Sciences*, **81**, 407–27.

Strickertsson, K. and Murray, A.S. (1999) Optically-stimulated luminescence dates for Late Pleistocene and Holocene sediments from Nørre Lyngby, northern Jutland, Denmark. *Quaternary Science Reviews*, **18**, 169–78.

Stringer, C.B. (1995) The evolution and distribution of later Pleistocene human populations. In *Palaeoclimate and Evolution* (edited by E.S. Vrba, G.H. Denton, T.C. Partridge and L.H. Burkle), Yale University Press, Yale, 524–31.

Stuart, A.J. (1979) Pleistocene occurrences of the European pond tortoise (*Emys orbicularis* L.) in Britain. *Boreas*, **8**, 359–71.

Stuart, A.J. (1982) *Pleistocene Vertebrates in the British Isles*. Longman, London.

Stuart, A.J. (1999) Late Pleistocene megafaunal extinctions. In *Extinctions in Near Time: Causes, contexts and consequences* (edited by R.D.E. MacPhee), Kluwer Academic Press, New York, 257–70.

Stuart, A.J. and van Wijngaarden-Bakker, L.H. (1985) Quaternary vertebrates. In *The Quaternary History of Ireland* (edited by K.J. Edwards and W.P. Warren), Academic Press, London, 221–49.

Stuart, F.M. (2001) In situ cosmogenic isotopes: principles and potential for archaeologists. In *Handbook of Archaeological Sciences* (edited by D.R. Brothwell and A.M. Pollard), John Wiley, Chichester and New York, 93–100.

Stuiver, M. and Brazunias, T. (1993) Sun, ocean, climate and atmospheric $^{14}CO_2$: an evaluation of causal and spectral relationships. *The Holocene*, **3**, 289–305.

Stuiver, M. and Grootes, P.M. (2000) GISP2 oxygen isotope ratios. *Quaternary Research*, **53**, 277–84.

Stuiver, M. and Quay, P.D. (1980) Changes in atmospheric carbon-14 attributed to a variable sun. *Science*, **207**, 11–19.

Stuiver, M., Grootes, P.M. and Brazunias, T.F. (1995) The GISP2 $\delta^{18}O$ climate record of the past 16,500 years and the role of the sun, ocean and volcanoes. *Quaternary Research*, **44**, 341–54.

Stuiver, M., Reimer, P.J., Bard, E., Beck, J.W., Burr, G.S., Hughen, K.A., Kromer, B., McCormac, G., van der Plicht, J. and Spurk, M. (1998) INTCAL98 radiocarbon age calibration, 24–0 cal BP. *Radiocarbon*, **40**, 1041–83.

Sugden, D.E. and Hulton, N. (1994) Ice volumes and climate change. In *The Changing Global Environment* (edited by N. Roberts), Blackwell, Oxford, 150–72.

Sullivan, T.J., Turner, R.S., Charles, D.F., Cummings, B.F., Smol, J.P., Schofield, C.L., Driscoll, C.T., Cosby, B.J., Birks, H.J.B., Utala, A.J., Kingston, J.C., Dixit, S.S., Bernert, J.A., Ryan, P.F. and Marmorek, D.R. (1992) Use of historical assessment for evaluation of process-based model projections of future environmental change: lake acidification in the Adirondack Mountains. *Environmental Pollution*, **77**, 253–62.

Sundquist, E.T. (1993) The global carbon dioxide budget. *Science*, **259**, 934–41.

Sutcliffe, A.J. (1970) A section of an imaginary bone cave. *Studies in Speleology*, **2**, 79–80.

Sutherland, D.G. and Gordon, J.E. (1993) The Quaternary in Scotland, In *Quaternary of Scotland*. Geological Conservation Review Series, 6, Chapman & Hall, London, 11–47.

Svenning, J.-C. (2002) A review of natural vegetation openness in north-western Europe. *Biological Conservation*, **104**, 133–48.

Svensmark, H. and Friis-Christensen, E. (1997) Variation of cosmic ray flux and global cloud coverage – a missing link in solar–climate relationships. *Journal of Atmospheric and Solar-Terrestrial Physics*, **59**, 1225–32.

Sykora, K.V. (1990) History of the impact of man on the distribution of plant species. In *Biological Invasions in Europe and the Mediterranean Basin* (edited by F. di Castri, A.J. Hansen and M. Debussche), Kluwer, Dordrecht, 37–50.

Szeicz, J.M. and MacDonald, G.M. (1996) A 930-year ring-width chronology from moisture-sensitive white spruce (*Picea glauca* Moench) in northwestern Canada. *The Holocene*, **6**, 346–52.

Tarasick, D.W., Wardle, D.I., Kerr, J.B., Bellefleur, J.J. and Davies, J. (1995) Tropospheric ozone trends over Canada: 1980–1993. *Geophysical Research Letters*, **22**, 409–12.

Taylor, D.M., Pedley, H.M., Davies, P. and Wright, M.W. (1998) Pollen and mollusc records for environmental change in central Spain during the mid- and late Holocene. *The Holocene*, **8**, 605–12.

Taylor, K.C., Lamorey, G.W., Doyle, G.A., Alley, R.B., Grootes, P.M., Mayewski, P.A., White, J.W.C. and Barlow, L.K. (1993) The 'flickering switch' of late Pleistocene climate change. *Nature*, **361**, 432–4.

Taylor, K.C., Mayewski, P.A., Alley, R.B., Brook, E.J., Gow, A.J., Grootes, P.M., Meese, D.A., Saltzman, E.S., Severinghaus, J.P., Twickler, M.S., White, J.W.C., Whitlow, S. and Zielinski, G.A. (1997) The Holocene–Younger Dryas transition recorded at Summit, Greenland. *Science*, **278**, 825–7.

Taylor, R.E. (2001) Radiocarbon dating. In *Handbook of Archaeological Science* (edited by D.R. Brothwell and A.M. Pollard), John Wiley, Chichester and New York, 23–34.

Taylor, R.E and Aitken, M.J. (eds) (1997) *Chronometric Dating in Archaeology*. Plenum, New York.

Teller, J.T. (2001) Formation of large beaches in an area of rapid differential isostatic rebound: the three-outlet control of Lake Agassiz. *Quaternary Science Reviews*, **20**, 1649–59.

Teller, J.T. and Kehaw, A.E. (eds) (1994) Lateglacial history of large proglacial lakes and meltwater runoff along the Laurentide ice sheet. *Quaternary Science Reviews*, **13**, 795–981.

Teller, J.T., Leverington, D.W. and Mann, J.D. (2002) Freshwater outbursts to the oceans from glacial Lake Agassiz and their role in climate change during the last deglaciation. *Quaternary Science Reviews*, **22**, 879–87.

Templeton, A.R. (2002) Out of Africa again and again. *Nature*, **416**, 45–51.

Tett, S.F.B., Stott, P.A., Allen, M.R., Ingram, W.J. and Mitchell, J.F.B. (1999) Causes of twentieth-century temperature change near the Earth's surface. *Nature*, **399**, 569–72.

Theodorsson, P. (1998) Norse settlement of Iceland – close to AD 700. *Norwegian Archaeological Review*, **31**, 29–38.

Thomas, J. (1988) Neolithic explanations revisited: the Mesolithic–Neolithic transition in Britain and south Scandinavia. *Proceedings of the Prehistoric Society*, **54**, 59–66.

Thomas, J. (1999) *Understanding the Neolithic*. Routledge, London.

Thompson, A. and Jones, A. (1986) Rates and causes of proglacial river terrace formation on southeast Iceland: an application of lichenometric dating techniques. *Boreas*, **15**, 231–46.

Thompson, F.H., Hester, A.J. and Usher, M.B. (1995) *Heaths and Moorland: Cultural landscapes*. HMSO, Edinburgh.

Thompson, L. (2000) Ice core evidence for climate change in the Tropics: implications for our future. *Quaternary Science Reviews*, **19**, 19–36.

Thompson, R. (1986) Palaeomagnetic dating. In *Handbook of Holocene Palaeoecology and Palaeohydrology* (edited by B.E. Berglund), John Wiley, Chichester and New York, 313–27.

Thompson, R. and Maher, B.A. (1995) Age models, sediment fluxes and palaeoclimatic reconstructions for the Chinese loess and palaeosol sequences. *Geophysical Journal International*, **123**, 611–22.

Thompson, R. and Oldfield, F. (1986) *Environmental Magnetism*. Allen & Unwin, London.

Thompson, R.S., Whitlock, C., Bartlein, P.J., Harrison, S.P. and Spaulding, W.G. (1993) Climatic changes in the Western United States since 18,000 B.P. In *Global Climates since the Last Glacial Maximum* (edited by H.E. Wright, Jnr, J.E. Kutzbach, T. Webb III, W.F. Ruddiman, F.A. Street-Perrott and P.J. Bartlein), University of Minnesota Press, Minneapolis, 514–35.

Thorarinsson, S. (1971) Damage caused by tephra fall in some big Icelandic eruptions and its relation to the thickness of the tephra layers. In *Acta of the First International Scientific Congress on the Volcano of Thera*, September 1969 (edited by A. Kaloyeropoyloy), 213–36.

Thorarinsson, S. (1979) On the damage caused by volcanic eruptions with special reference to tephra and gases. In *Volcanic Activity and Human Ecology* (edited by P.D. Sheets and D.K. Grayson), Academic Press, New York, 125–59.

Thorarinsson, S. (1981) Greetings from Iceland. Ash-fall and volcanic aerosols in Scandinavia. *Geografiska Annaler*, **63A**, 109–18.

Thorbahn, P.F. and Cox, D.C. (1988) The effect of estuary formation on prehistoric settlement in southern Rhode Island. In *Holocene Human Ecology in Northeastern North America* (edited by G.P. Nicholas), Plenum, New York, 167–82.

Tilley, C. (1994) *A Phenomenology of Landscape*. Berg, Oxford.

Tipping, R. (1996) Microscopic charcoal records, inferred human activity and climate change in the Mesolithic of northernmost Scotland. In *The Early Prehistory of Scotland* (edited by T. Pollard and A. Morrison), Edinburgh University Press, Edinburgh, 39–61.

Tipping, R. (1998) Cereal cultivation on the Anglo-Scottish border during the 'Little Ice Age'. In *Life on the Edge: Human settlement and marginality* (edited by C.M. Mills and G. Coles), Oxbow Books, Oxford, 1–12.

Tipping, R. (2002) Climatic variability and 'marginal' settlement in upland British landscapes: a re-evaluation. *Landscapes*, **3**, 10–28.

Tipping, R., Buchanan, J., Davies, A. and Tisdall, E. (1999) Woodland biodiversity, palaeohuman ecology and some implications for conservation management. *Journal of Biogeography*, **26**, 33–44.

Tolonen, K. (1986) Rhizopod analysis. In *Handbook of Holocene Palaeocology and Palaeohydrology* (edited by B.E. Berglund), John Wiley, Chichester and New York, 645–66.

Tooley, M.J. (1978) Sea-level changes in Northern England during the Flandrian Stage. In *The Climatic Scene* (edited by M.J. Tooley and G.M. Sheail), Allen & Unwin, London, 206–34.

Tooley, M.J. (1985) *Sea-level Changes in Northern England during the Flandrian Stage*. Clarendon Press, Oxford.

Torrence, R., Pavlides, C., Jackson, P. and Webb, J. (2000) Volcanic disasters and cultural discontinuities in Holocene time, in West New Britain, Papua New Guinea. In *The Archaeology of Geological Catastrophes* (edited by W.J. McGuire, D.R. Griffiths and P.L. Hancock), The Geological Society Special Publication No. 171, London, 225–44.

Toy, T.J., Foster, G.R. and Renard, K.G. (2002) *Soil Erosion: Processes, prediction, measurement, and control*. John Wiley & Sons, Inc., New York.

Troels-Smith, J. (1960) Ivy, mistletoe and elm. Climatic indicators – fodder plants. A contribution to the interpretation of the pollen zone border VII–VIII. *Danmarks Geologiske Undersøgelse ll.*, Raekke 4, **4**, 1–32.

Trotter, M.M. and McCullock, B. (1984) Moas, men and middens. In *Quaternary Extinctions* (edited by P.S. Martin and R.G. Klein), University of Arizona Press, London, 708–27.

Turner, C. (1970) The Middle Pleistocene deposits at Marks Tey, Essex. *Philosophical Transactions of the Royal Society of London B*, **257**, 373–440.

Turner, C. and West, R.G. (1968) The subdivision and zonation of interglacial periods. *Eiszeitalter und Gegenwart*, **19**, 93–101.

Turner, N. (1997) Traditional ecological knowledge. In *The Rainforests of Home: Profile of a North American bioregion* (edited by P.K. Schoonmaker, B. von Hagen and E.C. Wolf), Island Press, Washington, DC, 275–98.

Turner, N. (1999) 'Time to burn'. In *Indians, Fire and the Land* (edited by R. Boyd), Oregon State University Press, Corvallis, Oregon, 185–218.

Turner, R.C. and Scaife, R.G. (eds) (1995) *Bog Bodies: New discoveries and new perspectives*. British Museum Press, London.

Turney, C.S.M. (1998) Extraction of rhyolitic component of Vedde microtephra from minerogenic lake sediments. *Journal of Paleolimnology*, **19**, 199–206.

Turney, C.S.M., Bird, M.I., Fifileld, L.K., Roberts, R.G., Smith, M., Dortch, C.E., Grün, R., Lawson, E., Ayliffe, L.K., Miller, G.H., Dortch, J. and Cresswell, R.G. (2001) Early human occupation at Devil's Lair, southwestern Australia 50,000 years ago. *Quaternary Research*, **55**, 3–13.

Tzedakis, P.C. and Bennett, K.D. (1995) Interglacial vegetation succession: a view from southern Europe. *Quaternary Science Reviews*, **14**, 967–82.

Tzedakis, P.C., Andrieu, V., Beaulieu, J.L. de., Crowhurst, S., Follieri, M., Hooghiemstra, H. Magri, D., Reille, M., Sadori, L., Shackleton, N.J. and Wijmstra, T. (1997) Comparison of terrestrial and marine records of changing climate of the last 500,000 years. *Earth and Planetary Science Letters*, **150**, 171–6.

UNEP/WMO (1998) *Scientific Assessment of Ozone Depletion, 1998*. World Meteorological Organisation, Geneva.

Valen, V., Mangerud, J., Larsen, E. and Hufthammer, A.K. (1996) Sedimentology and stratigraphy in the cave Hamnsundhelleren, western Norway. *Journal of Quaternary Science*, **11**, 185–202.

Valentine, K.W.G. and Dalrymple, J.B. (1975) The identification, lateral variation and chronology of two buried palaeocatenas at Woodland Spa and West Runton, England. *Quaternary Research*, **5**, 551–91.

Valladas, H., Cachier, H., Maurice, P., Bernaldo de Quiros, F., Clottes, J., Valdes, V.C., Uzquianzo, P. and Arnold, M. (1992) Direct radiocarbon dates of prehistoric paintings at the Altamira. *Nature*, **357**, 68–70.

Valladas, H., Tisnérat-Laborde, N., Cachier, H., Arnold, M., Berenaldo de Quirós, F., Cabrera-Valdés, V., Clottes, J., Courtin, J., Fortea-Pérez, J.J., Gonzáles-Sainz, C. and Moure-Romanillo, A. (2001) Radiocarbon dates for Paleolithic cave paintings. *Radiocarbon*, **43** (2B), 977–86.

van Andel, T.H. and Runnels, C.N. (1995) The earliest farmers in Europe. *Antiquity*, **69**, 481–500.

Vandenberghe, J. (1993) Changing fluvial processes under changing periglacial conditions. *Zeitschrift für Geomorphologie, Supplement Band*, **88**, 17–18.

Vandenberghe, J. (1995) Timescales, climate and river development. *Quaternary Science Reviews*, **14**, 631–8.

Vandenberghe, J. (2001) A typology of Pleistocene cold-based rivers. *Quaternary International*, **79**, 111–21.

Vandenberghe, J. and Pissart, A. (1993) Permafrost changes in Europe during the Last Glacial. *Permafrost and Periglacial Processes*, **4**, 121–35.

Vandenberghe, J., Huizer, B.S., Mücher, H. and Laan, W. (1998a) Short climatic oscillations in a western European loess sequence (Kesselt, Belgium). *Journal of Quaternary Science*, **13**, 471–86.

Vandenberghe, J., Kasse, C., Bohnke, S.J.P. and Kozarski, S. (1994) Climate-related river activity at the Weichselian–Holocene transition: a comparative study of the Warta and Maas rivers. *Terra Nova*, **6**, 476–85.

Vandenberghe, J., Kasse, K. and Coope, G.R. (eds) (1998b) Palaeoclimate of the last interglacial–glacial cycle in western and central Europe. *Journal of Quaternary Science*, **13**, 361–497.

van den Dool, H.M., Krijnen, H.J. and Schuurmans, C.J.E. (1978) Average winter temperatures at De Bilt (The Netherlands), 1634–1977. *Climatic Change*, **1**, 319–30.

van der Leeuw, S. (1994) Social and environmental change. *Cambridge Environmental Journal*, **4**, 130–9.

van der Luen, J.C., Tang, X. and Tevini, M. (1995) Environmental effects of ozone depletion: 1994 assessment. *Ambio*, **24**, 138–42.

van der Merwe, N.J. (1992) Light stable isotopes and the reconstruction of prehistoric diets. In *New Developments in Archaeological Science* (edited by A.M. Pollard), Oxford University Press, *Proceedings of the British Academy*, 77, Oxford, 247–64.

van der Plicht, J. (2002) Calibration of the ^{14}C time scale; towards the complete dating range. *Netherlands Journal of Geoscience*, **81**, 85–96.

van der Sanden, W. (1996) *Through Nature to Eternity: The bog bodies of northwest Europe*. Batavian Lion International, Amsterdam.

van Geel, B. (1999) Letter to the Editor: comment on 'a large California flood and correlative global climatic events 400 years ago' (Schimmelmann *et al.* 1998). *Quaternary Research*, **51**, 108–10.

van Geel, B. and Middeldorp, A.A. (1988) Vegetational history of Carbuty Bog (Co. Kildare, Ireland) during the last 850 years and a test of the temperature indicator value of $^{2}H/^{1}H$ measurements of peat samples in relation to historical sources and meteorological data. *New Phytologist*, **109**, 377–92.

van Geel, B. and Renssen, H. (1998) Abrupt climatic change around 2,650 BP in north-west Europe: evidence for climatic teleconnections and a tentative explanation. In *Water, Environment and Society in Times of Climatic Change* (edited by A.S. Issar and N. Brown), Kluwer, Dordrecht, 21–41.

van Geel, B., Buurman, J. and Waterbolk, H.T. (1996) Archaeological and palaeoecological indications of an abrupt climate change in The Netherlands, and evidence for climatological teleconnections around 2650 BP. *Journal of Quaternary Science*, **11**, 451–60.

van Geel, B., Raspopov, O.M., van der Plicht, J. and Renssen, H. (1998) Solar forcing of abrupt climatic change around 850 calendar years BC. In *Natural Catastrophes during Bronze Age Civilisations* (edited by B.J. Peiser, T. Palmer and M.G. Bailey), BAR International Series, **728**, 162–8.

van Geel, B., van der Plicht, J. and Renssen, H. (2003) Major ^{14}C excursions during the late glacial and early Holocene: changes in ocean ventilation or solar forcing of climate change? *Quaternary International*, **105**, 71–6.

van Gijn, A.L. and Waterbolk, H.T. (1984) The colonisation of the salt marshes of Friesland and Groningen: the possibility of a transhumant prelude. *Palaeohistoria*, **26**, 101–22.

van Huissteden, J. and Vandenberghe, J. (1988) Changing fluvial style of periglacial lowland rivers during the Weichselian Pleniglacial in the eastern Netherlands. *Zeitschift für Geomorphologie, Supplement Band*, **71**, 131–46.

van Vliet-Lonoë, B. (1990) The genesis and age of the argillic horizon in Weichselian loess of northwestern Europe. *Quaternary International*, **5**, 49–56.

Vartanyan, S.L., Garutt, V.E. and Sher, A.V. (1993) Holocene dwarf mammoths from Wrangel Island in the Siberian Arctic. *Nature*, **362**, 337–40.

Vasari, Y., Glückert, G., Hicks, S., Hyvärinen, H., Simola, H. and Vuorela, I. (1996) Finland. In *Palaeoecological Events during the Last 15,000 Years* (edited by B.E. Berglund, H.J.B. Birks, M. Ralska-Jasiewiczowa and H.E. Wright), John Wiley, Chichester and New York, 281–352.

Vaughan, D.G. and Spouge, J.R. (2001) Risk estimation of collapse of the West Antarctic ice sheet. *Climate Change* (2002), **52**, 65–91.

Vayda, A.P. and McCay, B.J. (1978) New directions in ecology and ecological anthropology. In *Human Behaviour and Adaptation* (edited by V. Reynolds and N. Blurton Jones), Taylor & Francis, London, 33–51.

Veillette, J.J., Dyke, A.S. and Roy, M. (1999) Ice-flow evolution of the Labrador sector of the Laurentide Ice Sheet: a review, with new evidence from northern Quebec. *Quaternary Science Reviews*, **18**, 993–1020.

Vera, F.W.M. (1997) Metaphors for the wilderness: oak, hazel, cattle and horse. Ph.D. thesis Agricultural University, Wageningen, 440 pp.

Vera, F.W.M. (2000) *Grazing Ecology and Forest History*. CABI Publishing, Wallingford, Oxon.

Vermeulen, F. and de Dapper, M. (eds) (2000) *Geoarchaeology of the Landscapes of Classical Antiquity*. Stichting Babasch, Leiden.

Vignaud, P., Duringer, P., Mackaye, H.T., Likius, A., Blondel, C., Boisserie, J.-R., de Bonis, L., Eisenmann, V., Etienne, M.-E., Geraads, D., Guy, F., Lehmann, T., Lihoreau, F., Lopez-Martinez, N., Mourer-Chauviré, C., Otero, O., Rage, J.-C., Schuster, M., Viriot, L., Zazzo, A. and Brunet, M. (2002) Geology and palaeontology of the Upper Miocene Toros-Menalla hominid locality, Chad. *Nature*, **418**, 152–5.

Viles, H.A. (1989) The greenhouse effect, sea-level rise and coastal geomorphology. *Progress in Physical Geography*, **13**, 452–61.

Vita-Finzi, C. (1969) *The Mediterranean Valleys*. Cambridge University Press, London.

von Droste, B., Plachter, H. and Rossler, M. (1995) *Cultural Landscapes of Universal Value*. Fischer, Stuttgart.

von Grafenstein, U, Erlenkeuser, H., Brauer, A., Jouzel, J. and Johnsen, S.J. (1999) A mid-European decadal isotope-climate record from 15,500 to 5000 years BP. *Science*, **284**, 654–7.

Vrba, E.S. (1995) On the connection between palaeoclimate and evolution. In *Palaeoclimate and Evolution* (edited by E.S. Vrba, G.H. Denton, T.C. Partridge and L.H. Burkle), Yale University Press, Yale, 24–48.

Vuorela, I. (1986) Palynological and historical evidence of slash and burn cultivation in south Finland. In *Anthropogenic Indicators in Pollen Diagrams* (edited by K.-E. Behre), A.A. Balkema, Rotterdam, 53–64.

Vuorela, I. and Hicks, S. (1996) Human impact on the natural landscape in Finland: a review of the pollen evidence. In *Landscapes and Life* (edited by A.M. Robertsson, S. Hicks, A. Åkerlund, J. Risberg and T. Hackens), PACT 50 11.13, Belgium, 245–58.

Wagner, F., Bohncke, S.J.P., Dilcher, D.L., Kürschner, W.M., van Geel, B. and Visscher, H. (1999) Century-scale shifts in early Holocene atmospheric CO_2 concentrations. *Science*, **284**, 1971–3.

Wagner, G.A. (1998) *Age Determination of Young Rocks and Artifacts: Physical and chemical clocks in Quaternary geology and archaeology*. Springer-Verlag, Heidelberg and Berlin.

Wagner, G.E. (1996) Feast or famine? Seasonal diet at a Fort Ancient community. In *Case Studies in Environmental Archaeology* (edited by E.J. Reitz, L.A. Newsom and S.J. Scudder), Plenum Press, London, 255–71.

Wagstaff, M. (1992) Agricultural terraces: the Vasilikos Valley, Cyprus. In *Past and Present Soil Erosion* (edited by M.G. Bell and J. Boardman), Oxbow Monograph 22, Oxford, 155–62.

Wainwright, G. (2000) Time please. *Antiquity*, **74** (286), 909–43.

Walden, J., Oldfield, F. and Smith, J. (1999) *Environmental Magnetism: A practical guide*. Technical Guide No. 6, Quaternary Research Association, London.

Walden J., Smith, J.P. and Dackombe, R.V. (1992) Mineral magnetic analyses as a means of lithostratigraphic correlation and provenance indication of glacial diamicts: intra- and inter-unit variation. *Journal of Quaternary Science*, **7**, 257–70.

Waldren, W.H. (1982) *Balearic Prehistoric Ecology and Culture*. British Archaeological Reports IS 149, parts 1–3, Oxford.

Walker, D. and Singh, G. (1993) Earliest palynological records of human impact on the world's vegetation. In *Climate Change and Human Impact on the Landscape* (edited by F.M. Chambers), Chapman & Hall, London, 101–8.

Walker, I.R., Smol, J.P., Engstron, D.R. and Birks, H.J.B. (1991) An assessment of Chironomidae as quantitative indicators of past climate change. *Canadian Journal of Fisheries and Aquatic Sciences*, **48**, 975–87.

Walker, M.J.C. (1984) Pollen analysis and Quaternary research in Scotland. *Quaternary Science Reviews*, **3**, 369–404.

Walker, M.J.C. (1995) Climatic changes in Europe during the last glacial–interglacial transition. *Quaternary International*, **28**, 63–76.

Walker, M.J.C. and Lowe, J.J. (1990) Reconstructing the environmental history of the last glacial–interglacial transition: evidence from the Isle of Skye, Inner Hebrides, Scotland. *Quaternary Science Reviews*, **9**, 15–49.

Walker, M.J.C., Björck, S., Lowe, J.J., Cwynar, L.C., Johnsen, S., Knudsen, K-L., Wohlfarth, B. and INTIMATE Group (1999) Isotopic 'events' in the GRIP ice core: a stratotype for the Late Pleistocene. *Quaternary Science Reviews*, **18**, 1143–51.

Walker, M.J.C., Bryant, C., Coope, G.R., Harkness, D.D., Lowe, J.J. and Scott, E.M. (2001) Towards a radiocarbon chronology for the Lateglacial: sample selection strategies. *Radiocarbon*, **43**, 1007–20.

Walker, M.J.C., Coope, G.R., Sheldrick, C., Turney, C.S.M., Lowe, J.J., Blockley, S.P.E. and Harkness,

D.D. (2003) Devensian Lateglacial environmental changes in Britain: a multi-proxy record from Llanilid, South Wales, UK. *Quaternary Science Reviews*, **22**, 475–520.

Walker, M.J.C., Griffiths, H.I., Ringwood, V. and Evans, J.G. (1993) An early-Holocene pollen, mollusc and ostracod sequence from lake marl at Llangorse Lake, South Wales, UK. *The Holocene*, **3**, 138–49.

Waller, M. and Hamilton, S. (2000) Vegetation history of the English chalklands: a mid-Holocene pollen sequence from the Caburn, East Sussex. *Journal of Quaternary Science*, **15**, 253–72.

Walters, J.C. (1994) Ice-wedge casts and relict polygonal patterned ground in north-east Iowa, USA. *Permafrost and Periglacial Processes*, **5**, 269–82.

Walther, G.R., Post, E., Convey, P., Menzel, A., Parmesan, C., Beebee, T.J.C., Fromentin, J.M., Hoegh-Goldberg, O. and Bairlein, F. (2002) Ecological responses to recent climate change. *Nature*, **416**, 389–95.

Wanner, H., Pfister, C., Bràzdil, R., Frich, P., Fruydendahl, K., Jonsson, T., Kington, J., Lamb, H.H., Rosenorn, S. and Wishman, E. (1995) Wintertime European circulation patterns during the late Maunder Minimum cooling period (1675–1704). *Theoretical and Applied Climatology*, **51**, 167–75.

Warner, B.G. (1990a) Plant macrofossils. In *Methods in Quaternary Ecology* (edited by B.E. Warner), Geoscience Canada, Reprint Series 5, 53–64.

Warner, B.G. (1990b) Testate amoebae (Protozoa). In *Methods in Quaternary Ecology* (edited by B.E. Warner), Geoscience Canada, Reprint Series **5**, 65–74.

Warner, B.G. and Charman, D.J. (1994) Holocene changes on a peatland in northwestern Ontario, Canada interpreted from testate amoebae (Protozoa) analysis. *Boreas*, **23**, 259–69.

Warren, A. (2002) Erosion by wind, global hot spots. In *Encyclopedia of Soil Science* (edited by R. Lal), Marcel Dekker, New York, 508–11.

Warrick, R.A., Le Provost, C., Meier, M.F., Oerlemans, J., and Woodworth, P.L. (1996) Changes in sea level. In *Climate Change 1995: The science of climate change* (edited by J.T. Houghton, L.G. Meira Filho, B.A. Callander, N. Harris, A. Kattenberg and K. Maskell), Cambridge University Press, London, 359–405.

Washburn, A.L. (1979) *Geocryology*. Edward Arnold, London.

Wasyilikowa, K. (1986) Analysis of fossil seeds and fruits. In *Handbook of Holocene Palaeoecology and Palaeohydrology* (edited by B.E. Berglund), John Wiley & Sons, Chichester and New York, 571–90.

Waters, M.R. (1996) *Principles of Geoarchaeology: A North American perspective*. University of Arizona Press, Tucson.

Watson, R.T., Zinyowera, M.C., Moss, R.H. and Dokken, D.J. (eds) (1996) *Climate Change 1995. Impacts, Adaptations and Mitigations of Climate Change: Scientific-technical analyses*. Cambridge University Press, Cambridge.

Wattez, J., Courty, M.A. and MacPhail, R.I. (1989) Burnt organomineral deposits related to animal and human activities in prehistoric caves. In *Soil Micromorphology* (edited by L. Douglas), Elsevier, Amsterdam, 431–9.

Watts, M. (1983) On the poverty of theory: natural hazards research in context. In *Interpretations of Calamity* (edited by K. Hewitt), Allen & Unwin, London, 231–62.

Watts, W.A., Hansen, B.C.S. and Grimm, E.C. (1992) Camel Lake: a 40,000-yr record of vegetational and forest history from northwestern Florida. *Ecology*, **73**, 1056–66.

Wayne, W.J. (1991) Ice wedge casts of Wisconsinan age in eastern Nebraska. *Permafrost and Periglacial Processes*, **2**, 211–23.

Webb, R. E. (1998) Problems with radiometric 'time': dating the initial human colonisation of Sahul. *Radiocarbon*, **40**, 749–58.

Webb, T., Bartlein, P.J., Harrison, S.P. and Anderson, K.H. (1993) Vegetation, lake levels, and climate in eastern North America for the past 18,000 years. In *Global Climates since the Last Glacial Maximum* (edited by H.E. Wright, Jr, J.E. Kutzbach, T. Webb III, W.F. Ruddiman, F.A. Street-Perrott and P.J. Bartlein), University of Minnesota Press, Minneapolis, 415–67.

Webb, T. III, Cushing, E.J. and Wright, H.E. Jr (1984) Holocene changes in the vegetation of the Midwest. In *Late Quaternary Environments of the United States. 2. The Holocene* (edited by H.E. Wright, Jnr), Longman, London, 142–65.

Welinder, S. (1990) Mesolithic forest clearance in Scandinavia. In *The Mesolithic in Europe* (edited by C. Bonsall), John Donald, Edinburgh, 362–6.

West, R.G. (1970) Pleistocene history of the British flora. In *Studies in the Vegetational History of the British Isles* (edited by D. Walker and R.G. West), Cambridge University Press, London, 1–11.

West, R.G. (2000) *Plant Life of the Quaternary Cold Stages: Evidence from the British Isles*. Cambridge University Press, Cambridge.

White, J.W.C., Barlow, L.K., Fisher, D., Grootes, P., Jouzel, J., Johnsen, S. and Mayewski, P.A. (1997) The climate signal in stable isotopes of snow from Summit, Greenland: results of comparisons with modern climate observations. *Journal of Geophysical Research*, **102**, 26425–40.

White, R. (1999) Indian fires in the northern Rockies. In *Indians, Fire and the Land* (edited by R. Boyd), Oregon State University Press, Corvallis, Oregon, 36–49.

Whitehouse, N.J. (2000) Forest fires and insects: palaeoentomological research from a subfossil burnt forest. *Palaeogeography, Palaeoclimatology, Palaeoecology*, **164**, 231–46.

Whitehouse, N.J. (2004) Mire ontogeny, environmental and climatic change inferred from fossil beetle successions from Hatfield Moors, eastern England. *The Holocene*, **14**, 79–93.

Whittington, G. (1994) *Bruckenthalia spiculifolia* (Salisb.) Reichenb. (Ericaceae) in the Late Quaternary of western Europe. *Quaternary Science Reviews*, **13**, 761–8.

Whittington, G. and Edwards, K.J. (1999) Landscape scale soil pollen analysis. In *Holocene Environments of Prehistoric Britain* (edited by K.J. Edwards and J.P. Sadler), *Quaternary Proceedings*, **7**, 595–604.

Whittington, G. and Hall, A.M. (2002) The Tolsta Interstadial, Scotland: correlations with D-O cycles GI-8 to GI-5? *Quaternary Science Reviews*, **21**, 901–15.

Whittle, A. (1996) *Europe in the Neolithic*. Cambridge University Press, Cambridge.

Whittle, A. (2003) *The Archaeology of People: Dimensions of Neolithic life*. Routledge, London.

Wigley, T.M.L. and Kelly, P.M. (1990) Holocene climatic changes, ^{14}C wiggles and variation in solar irradiance. *Philosophical Transactions of the Royal Society, London*, **A330**, 547–60.

Wilkinson, T.J. (2003) *Archaeological Landscapes of the Near East*. University of Arizona Press, Tucson.

Willems, W.J.H. (1998) Archaeology and heritage management in Europe: trends and developments. *European Journal of Archaeology*, **1**, 293–311.

Willemse, N.W. and Törnqvist, T.E. (1999) Holocene century-scale temperature variability from west Greenland lake records. *Geology*, **27**, 580–4.

Willerslev, E., Hansen, A.J., Binladen, J., Brand, T.B., Gilbert, M.T.P., Shapiro, B., Bunce, M., Wiuf, C., Gilichinksy, D.A. and Cooper, A. (2003) Diverse plant and animal genetic records from Holocene and Pleistocene sediments. *Science*, **300**, 791–5.

Williams, D.F., Thunnell, R.C., Tappa, E., Rio, D. and Raffi, I. (1988) Chronology of the Pleistocene oxygen isotope record: 0–1.88 m.y. BP. *Palaeogeography, Palaeoclimatology, Palaeoecology*, 64, 221–40.

Williams, M. (1990) Clearing of the forests. In *The Making of the American Landscape* (edited by M.P. Conzen), Unwin Hyman, Boston, 146–68.

Williams, M. (2003) *Deforesting the Earth*. University of Chicago Press, Chicago.

Willis, K.J. and Bennett, K.D. (1994) The Neolithic transition – fact or fiction? Palaeoecological evidence from the Balkans. *The Holocene*, **4**, 326–30.

Wills, W.H. (1995) Archaic foraging and the beginning of food production in the American southwest. In *Last Hunters – First Farmers* (edited by T.D. Price and A.B. Gebauer), School of American Research Press, Santa Fe, 215–43.

Wilson, C. (1985) The Little Ice Age on eastern Hudson/James Bay: the summer weather and climate at Great Whale, Fort George and Eastmain, 1814–1821, as derived from Hudson's Bay Company records. *Syllogeus*, **55**, 147–90.

Wilson, R.J.S. and Luckmann, B.H. (2003) Dendroclimatic reconstruction of maximum summer temperatures from upper treeline sites in interior British Columbia, Canada. *The Holocene*, **13**, 851–61.

Wiltshire, P.E.J. and Moore, P.D. (1983) Palaeovegetation and palaeohydrology in upland Britain. In *Background to Palaeohydrology* (edited by K.J. Gregory), John Wiley, Chichester and New York, 433–52.

Wobst, H.W. (1990) Afterword: minitime and megaspace in the Palaeolithic at 18K and otherwise. In *The World at 18,000* BP (edited by O. Soffer and C. Gamble), Unwin Hyman, London, 331–43.

Wolf, T.C.W. and Thiede, J. (1991) History of terrigenous sedimentation during the last 10 my in the North Atlantic (ODP-Legs 104, 105, and DSDP Leg-81). *Marine Geology*, **101**, 83–102.

Wolff, E. and EPICA Dome C 2001–2 science and drilling teams (2002) Extending the ice core record beyond half a million years. *EOS, Transactions of the American Geophysical Union*, **83**, 509–12.

Wolff, E. and Peel, D. (1985) The record of global pollution in polar snow and ice. *Nature*, **313**, 535–40.

Woodland, W.A., Charman, D.J. and Sims, P.C. (1998) Quantitative estimates of water tables and soil moisture in Holocene peatlands from testate amoebae. *The Holocene*, **8**, 261–73.

Woodman, P., McCarthy, M. and Monaghan, N. (1997) The Irish Quaternary Forum Project. *Quaternary Science Review*, **16** (1), 129–60.

Woodward, J.C. (1995) Patterns of erosion and suspended sediment yield in Mediterranean river basins. In *Sediment and Water Quality in River Catchments* (edited by I.D.L. Foster, A.M. Foster, A.M. Gurnell and B.W. Webb), John Wiley & Sons, Chichester and New York, 365–89.

Woodward, J.C. and Goldberg, P. (2001) The sedimentary records in Mediterranean rockshelters and caves: archives of environmental change. *Geoarchaeology*, **16**, 327–54.

Woodworth, P.L., Tsimplis, N.M., Flather, R.A. and Shennan, I. (1999) A review of the trends observed in British Isles mean sea level data measured by tide gauges. *Geophysical Journal International*, **136**, 651–70.

Worster, D. (1990) The ecology of order and chaos. *Environmental History Review*, **14**, 1–18.

Wright, H.E. and Clark, J.S. (1994) Charcoal analysis of varved lake sediments. In *Laminated Sediments* (edited by S. Hicks, U. Miller and M. Saarnisto), PACT 41, Belgium, 125–30.

Wright, H.E. Jnr, Kutzbach, J.E., Webb, T. III, Ruddiman, W.F., Street-Perrott, F.A. and Bartlein, P.J. (eds) (1993) *Global Climates since the Last Glacial Maximum*. University of Minnesota Press, Minneapolis.

Wymer, J.J. (1999) *The Lower Palaeolithic Occupation of Britain*. Wessex Archaeology, Salisbury.

Yalden, D.W. (2001) Mammals as climate indicators. In *Handbook of Archaeological Sciences* (edited by D.R. Brothwell and A.M. Pollard), John Wiley & Sons, Chichester and New York, 147–54.

Yates, D.T. (1999) Bronze Age field systems in the Thames Valley. *Oxford Journal of Archaeology*, **18**, 157–70.

Yen, D.E. (1989) The domestication of environment. In *Foraging and Farming* (edited by D.R. Harris and G.C. Hillman), Unwin Hyman, London, 55–75.

Yiou, P., Fuhrer, K., Meeker, L.D., Jouzel, J., Johnsen, S. and Mayewski, P. (1997) Paleoclimate variability inferred from the spectral analysis of Greenland and Antarctic ice cores. *Journal of Geophysical Research*, **102**, 26441–54.

Yokoyama, Y., Lambeck, K., de Dekker, P., Johnston, P. and Fifield, L.K. (2000) Timing of the Last Glacial Maximum from observed sea-level data minima. *Nature*, **406**, 713–16.

Zagwijn, W.H. (1984) The formation of the Younger Dunes on the west coast of the Netherlands (AD 1000–1600). In *Geological Changes in the Western Netherlands during the Period 1000–1300 AD* (edited by H.J.A. Berendensen and W.H. Zagwijn), *Geologie en Mijnbouw*, **3**, 259–68.

Zagwijn, W.H. (1994) Reconstruction of climate change during the Holocene in western and central Europe based on pollen records of indicator species. *Vegetation History and Archaeobotany*, **3**, 65–88.

Zagwijn, W.H. (1996) The Cromerian Complex Stage of The Netherlands and correlation with other areas in Europe. In *The Early Middle Pleistocene in Europe* (edited by C. Turner), Balkema, Rotterdam, 145–72.

Zangger, E. (1992) Neolithic to present soil erosion in Greece. In *Past and Present Soil Erosion* (edited by M.G. Bell and J. Boardman), Oxbow Monograph 22, Oxford, 133–48.

Zazula, G.D., Froese, D.G., Schweger, C.E., Mathewes, R.W., Beaudoin, A.B., Telkall, A.M., Harington, C.R. and Westgate, J.A. (2003) Ice-age steppe vegetation in east Beringia. *Nature*, **423**, 603.

Zielinski, G.A. (2000) Use of paleo-records in determining variability within the volcanism–climate system. *Quaternary Science Reviews*, **19**, 417–38.

Zielinski, G.A., Mayewski, P.A., Meeker, L.D., Grönvold, K., Germani, M.S., Whitlow, S., Twickler, M.S. and Taylor, K. (1997) Volcanic aerosol records and tephrochronology of the Summit, Greenland, ice cores. *Journal of Geophysical Research*, **102**, 26625–40.

Zohary, D. (1989) Domestication of the Southwest Asian Neolithic crop assemblage of cereals, pulses and flax: the evidence from the living plants. In *Foraging and Farming* (edited by D.R. Harris and G.C. Hillman), Unwin Hyman, London, 358–73.

Zohary, D. (1996) The mode of domestication of the founder crops of Southwest Asian agriculture. In *The Origins and Spread of Agriculture and Pastoralism in Eurasia* (edited by D.R. Harris), UCL Press, London, 142–58.

Zohary, D. and Hopf, M. (1994) *Domestication of Plants in the Old World*. Oxford University Press, Oxford.

Zvelebil, M. (1986) Mesolithic prelude and Neolithic revolution. In *Hunters in Transition* (edited by M. Zvelebil), Cambridge University Press, Cambridge, 5–16.

Zvelebil, M. and Rowley-Conwy, P. (1986) Foragers and farmers in Atlantic Europe. In *Hunters in Transition* (edited by M. Zvelebil), Cambridge University Press, London, 67–93.

Zwiers, F.W. (2002) The 20-year forecast. *Nature*, **416**, 690–1.

Index

Illustrations are shown in italics.